OCCUPATIONAL HEALTH

and

MANTALENT DEVELOPMENT

OCCUPATIONAL HEALTH

and

MANTALENT DEVELOPMENT

by

ROBERT COLLIER PAGE
M.D., F.A.C.P., M.R.S.H.

Diplomate of the American Board of Internal Medicine; Diplomate of the American Board of Preventive Medicine; Associate Clinical Professor of Industrial Medicine, New York University Post-Graduate Medical School; Fellow of the New York Academy of Medicine; President of the Industrial Medical Association 1953-1954; Chairman of the Board of Trustees of the Occupational Health Institute 1953-1954; General Medical Director of Jersey Standard 1944-1954; Medical Management Consultant, Vice-President of the Colonial Research Institute; Chairman and Co-ordinator of the Medical Advisory Board of the Institute of Gerontology; Chairman of the Board, Mantalent Limited.

1968 Printing
WARREN H. GREEN, INC.
St. Louis, Missouri

© Physicians' Record Company 1963

Copyright 1963

in the U.S. and Canada

by

PHYSICIANS' RECORD COMPANY

All rights reserved. No part of this book may be reproduced in any form without permission in writing from the publisher, except by a reviewer who wishes to quote brief passages in connection with a review written for inclusion in a magazine or newspaper.

Copyright protection under the Articles of the

Copyright Convention of the

Pan-American Republics and the United States

Library of Congress Catalog Card
Number: 62-9730

PRINTED IN THE UNITED STATES OF AMERICA BY
PHYSICIANS' RECORD COMPANY, BERWYN, ILLINOIS

THE WORKER

DEDICATED TO THE WORKER

— the employed man or woman — whose health and leisure time, because he makes his way and earns his pay, must be constructively dealt with. He is entitled to a health program that will:

1. *Prevent all preventable health problems;*

2. *Take care of incipient health problems at a time when they can be corrected at a reasonable cost;*

3. *Keep him vertical and productive for the maximum period;*

4. *Treat him and habilitate him in the event of an unforeseeable major medical catastrophe under a sound indemnity plan; and*

5. *Educate him in the construction as well as maintenance of health.*

FOREWORD

I

THIS TEXT WILL SERVE as a guide for cost-conscious management.

Competition for mantalent is largely a matter of one company being "a better place to work" than another. This is one basic reason why up-to-date management is paying increasing attention to the problem of dealing with and maintaining human beings.

To the cost analyst it is rapidly becoming apparent that there is another more urgent reason for a constructive mantalent development program.

Mantalent is a costly investment. Even at the lowest echelon, it takes time and money to train a man to the point where he is worth as much to his company as his company is investing in him. Every time a man leaves owing to ill health or disgruntlement, he represents a total loss in dollars and cents. Every time a man leaves, he creates a vacancy which someone must be trained at considerable cost to fill.

Management deals with employees and executives at each stage of their working career. Each stage has been studied with a high-powered magnifying lens. The well-briefed personnel executive has at his fingertips a tremendous amount of data to help him appreciate and understand any given individual at any given stage of his career in business.

One vital bit of information about each person without which all of the data becomes meaningless has to do with the individual's health. Without health, one cannot be expected to work at top efficiency.

It is health that determines whether the individual will have the mental desire and the physical adjustment to perform to the best of his ability and to arrive serenely at the final stage of retirement with the happy knowledge that he has paid his way and earned his pay.

Throughout this text the author outlines a most practical approach to the problem of health maintenance and mantalent development which is both personal and managerial.

JAMES H. RAND
Industrialist

FOREWORD
II

DR. PAGE'S EXHAUSTIVE STUDY of occupational health pinpoints the failures of management and labor generally, with respect to their responsibilities for employee occupational health. The widespread incidence of management grants of contributions, based upon payroll, to union-management trust funds to provide health and welfare fringe benefits to employees covered by collective bargaining agreements between management and labor, has not carried with it a sufficient sense of responsibility on the part of management to follow through and insure that the concerned employees receive a maximum return in terms of occupational health for the dollars contributed by the employer in their behalf.

The primary concern of management appears to have been generally limited by typical collective bargaining considerations of keeping the wage and related operating costs down. Once agreement has been reached as to how much management would contribute, the mere presence of management-labor trust fund plans has not secured for the concerned employees effective usage of the health and welfare trust funds; nor has organized labor's interest and participation in trust fund planning and administration secured adequate quality medical coverage.

The health and welfare provisions seem to have followed the traditional line of providing "horizontal" medical care, where remedy rather than prevention and early detection is the major objective. Experience has proved that even such limited objectives have been subject to a continuous cost rise without any substantial return (as Dr. Page clearly points out) in terms of elimination or decrease of worker absenteeism or other loss of productivity due to illness, disability and absence of the work force.

It is high time that management recognize its full responsibilities and primary interests in achieving maximum health for its work force and join with labor in securing qualified medical consultative aid on the subject of occupational therapy which will be aimed at prevention of illness and retention of health as a positive approach to the objectives sought in health and welfare plans.

Dr. Page's strong and fully documented argument favoring adoption of "vertical" health carries with it a persuasive ring of truth. It may well be that the adequate understanding of the obvious requires belaboring on every level and in all directions.

HARRY H. RAINS
Attorney at Law
Industrial Relations Counsellor

FOREWORD

III

WORK GIVES MEANING AND DIGNITY to man's life. When man likes his work and performs it in pleasant surroundings, with no risk to his health, work is then a source of great happiness for him.

Because of the necessity to feed, clothe and shelter himself, man was compelled to work as early as prehistoric times, and the evolution of his work guided the evolution of society from the times of primitive hunting and farming cultures until the advent of manufacturing and trade many centuries later. When man began to work in groups or teams, and his working tools became more and more complex and the conditions of his work more crowded and confined, the risks in most of his occupations increased. The original occupational risks of the prehistoric hunter, whose bones have revealed to us that he suffered fractures and traumas, were also the same risks involved in later human occupations — hunting, farming, manufacturing — and in the surroundings in which he practiced these occupations — at first, jungles, swamps, the outdoors; and, later, mines, factories, workshops.

The industrial revolution, technologically the most decisive step in the history of work, and the introduction of the steam engine unleashed innumerable new menaces to the health of the worker. And the very first flutters of that industrial revolution coincided with the publication of the first works on occupational risks and diseases. In the fifteenth century, Ulrich Ellenbog published the first pamphlet about the noxious effects of the poisonous fumes of metals on goldsmiths; one century later, Paracelsus wrote the first monograph about the diseases of miners; and in 1700 Ramazzini wrote the first textbook — a magnificent one — on occupational diseases.

Modern society has witnessed a great increase in occupational diseases, both in those that threaten the worker "from without," such as inhalation of dust and fumes, infections, and traumatic accidents caused by machinery, and in those that attack him "from within," such as pathologic stress and psychosomatic and circulatory diseases. To industry the sick worker represents a disaster, for, besides the fact that the disabled man hampers the progress of society by his not being able to occupy his legitimate position in that society, a poor health economy unavoidably leads to a poor business economy. This has stimulated the advance of industrial medicine, especially in its preventive aspects, as applied to the multiple facets of work in the modern world. A new science has been developed to protect not only the health of the working man, but also the economy and prosperity of the society that enjoys the fruits of his toil.

And now Dr. Robert Collier Page, an eminent international authority, who has worked so widely and so well in this field, offers in this text, monumental in horizon, length, and scope, a new and original philosophical vision (particularly his concepts of "vertical" health and "horizontal" sickness) of what can be done today and what could be done tomorrow, not only to cure the sick worker, but also to protect his health. The approach of this text is multidimensional, both in the sense that it integrates what the physician, the sanitary engineer, the public health officer, the social worker, the psychologist, and the sociologist can do, from their different points of view, in favor of occupational health, and in the sense that it studies the problem in all its complex magnitude. For this reason alone, this book richly deserves a place in the history of industrial medicine.

Dr. Robert Collier Page's efforts, a fine example of dogged industriousness and keen and humanitarian scientific vision, mark a memorable hour in the history of the struggle to protect and improve the health of the working man, not only for his own sake, but for that of the society to which he belongs.

FÉLIX MARTÍ-IBÁÑEZ, M.D.
Editor-in-Chief of the medical newsmagazines, MD, MD *of* Canada, and MD en Español

FOREWORD
IV

MY FORMAL ENTRY in the field of occupational health was during the post-depression years of the 1930's. The late James A. Britton, a broad-visioned clinician interested in chest diseases, occupational health, social welfare and above all the *individual* patient, gathered in a university atmosphere, three different yet complementing personalities in whom he instilled a lasting interest in the present specialty of *occupational health*.

Bob Page and I survived and remained physicians in industry. For us this was an easy and interesting maturation. We had only to grow with the social reform and retain the patient orientation of the traditionally trained clinician. This was still the era of the *total* patient concept and the *art* of medicine.

Now the science of medicine with its specialty concept relegates the art of medicine to a secondary importance. I submit that the middle ground is desirable; we are still evaluating and treating a *patient,* not a disease. The social setting has changed, the doctor servant is now the doctor scientist, the patient is now the interposed middle man between the scientist and the third party payer. The patient no longer accepts health maintenance as an individual privilege and responsibility but *demands* it as a right and condition of modern society. The proper attitude is somewhere between in any democratic culture.

About 90 per cent of physicians in the United States derive some income from industrial sources, yet less than 5 per cent devote their full-time professional services to the medical needs of private enterprise. Most occupational medicine is practiced by part-time physicians. These need direction and education to comprehend the full implication of their professional responsibilities. This should be instituted in the undergraduate medical education, but apparently there is not room in the already overfilled curriculum.

Bob and I learned by experience (on-the-job training if you wish). Perhaps you, too, can if you will but peruse the pages of this text. It is a recount of: the growth and developments in occupational health as a *medical specialty;* the life and professional career of one of the controversial characters of the developmental era of our industriosocial reform; the call to rebirth of the holoistic concept of man, his illnesses and his health; the effect of the socioeconomic developments and stresses upon our culture; the development of the team approach and understanding to the ever growing *health* needs of those employed.

I can argue with Bob concerning many of the presentations, but I can never doubt the *sincerity* of his approach nor the *validity* of his general conclusions. I too can ask, are we yet ready? Are all of us thus dedicated? You, likewise, will ask questions, but in so doing, this text will have served its purpose (you will gain en route information) and added insight to *vertical health as opposed to horizontal sickness.*

EUGENE L. WALSH, Ph.D., M.D., F.A.C.P.
Associate Professor of Medicine
Northwestern University Medical School
Chicago, Illinois

PREFACE

OCCUPATIONAL HEALTH AND MANTALENT DEVELOPMENT, a text for the student of humanology and a reference work for members of the professions, representatives of management and labor whose interests encompass selection, job placement, health and welfare policies, mantalent development, and retirement practices, has been planned to point up the ever growing significance of human ecology; man plus health = job; man minus health = no job.

It is a book for all members of the humanological team. It will be of interest to practitioners of long standing and fill a long-existing gap in the libraries of universities, medical schools, and hospitals.

The author is conscious that the approach is in many respects unusual, even unorthodox, but remembering the words of Professor Matthew J. Stewart that new views would receive sympathetic consideration, the author was encouraged to put his own views forward as a contribution to the closer understanding of *total health* as applied to the needs of the working man and woman.

Because of the ever-increasing scope of occupational health and the extent of interaction between man and his working environment, the author has endeavored to give some weight to historical considerations, sometimes possibly at the cost of discursiveness and repetition, but he is strongly of the opinion that the method followed in this text is amply justified by the character and significance of the material presented.

The technique of intermingling strictly medical problems and matters pertinent to selection, job placement, and mantalent development is the result of the author's privileged experiences in the world of business. It is the author's signal hope that the two disciplines heretofore separate and often opposed will become united in the true interests of the *total* man.

The writing of this text could not have been accomplished without the help generously given from many quarters. In the first place, the author would like to pay tribute to his professional associates and to that steadily increasing number of doctors, nurses, and paramedical personnel who are confining their professional interests to the total health of the employed and to the members of management and labor who, with integrity, understanding, and perseverance, take an active and sincere interest in matters pertinent to the total health needs of our working population.

The author wishes to make special mention of his publisher, the late Mr. J. W. Voller, Sr., and his editorial staff, whose cooperation leaves nothing more to be desired; to Doctor Howard Craig, Director of the New York Academy of Medicine, and the various members of his library staff who have been untiring in their cooperative efforts in ferreting out from the archives reference material for this work. For pin-point editorial assistance, the author is indebted to Robert Green; to James H. Rand who reviewed the manuscript from the standpoint of private enterprise; to Harry Rains who has considered the work from the viewpoint of labor relations; to Doctor Félix Martí-Ibáñez who has recorded his thoughts in terms of medical history; and to Doctor Eugene Walsh who points up its significance as applied to the medical needs of industry.

R. C. P.

TABLE OF CONTENTS

Chapter   *Page*

Introduction

Occupational Health Is Everybody's Business	2
Occupational Health Is Only Part of the Whole	2
The Professional Consultant	2
The Members of the Team	3
Reference Materials	4
The Captain of the Team	4
The Seven Stages of Progression	5

1. The Whole Man

The Basic Functions of Competitive Enterprise	7
Humanology in Business	8
Self-Interest and Humanology	9
Man as a Total Being	10
The Socid	10
The four sides of man's nature	11
Five Case Studies	12
The man of management	12
The worker	15
The secretary	17
The union man	19
The scientist	21
All Are Individuals	24
The Ideal Condition of Man Is One of Wholeness	25
Man's Gregarious Habits	26
The Importance of Morale	27
Man's Potential — And His Performance	27
The Influence of Environment on Morale	29
The Importance of Early Detection	29
The Alternative	30
What the Public Hears	30
The True Story	30
The Concern of All	31

Chapter	Page
Episodic vs. Constructive Medicine	31
Inadequacy of the Curative Aspects of Health	31
The Hospital Horizontal-Care System	32
The Trustee and His Professional Advisors	32
Horizontal-Care Insurance Plans	33
Constructive Health Maintenance	33

2. *Evolution of Occupational Medicine*

Expediency Creates a Need	36
Progress Was Slow and Painful	37
The Emerging Role of Management	38
Notable Contributions by Foresighted Medical Practitioners	38
The Birth of the National Safety Council	39
The Student Becomes Informed	40
World War I and Manpower	41
A Positive Program in the Offing	42
Industrial Hygiene	43
Occupational Health Abroad	43
Collective Bargaining and Health	49
Clause XXII-medical and hospital attention	49
The Coal-Mining Industry	51
Consumer Interest	51
The Automotive Industry	52
Employer payment for health security protection	52
Basic program of benefits	53
Joint administration	53
The Rubber Industry	53
The Problem of Private Enterprise in Medical Affairs	54
Solution Depends on the Medical Consultant	54
Stockholder Interest	55
Health Insurance	56
What Actually Happened	56
The Older Worker	57
The Doctor's Fee	57
No Coverage for Preventive or Diagnostic Services	58
Catastrophic Illness	58
Labor Is Heard	59
The Role of the Medical Consultant	60
The Chinese Custom	61

TABLE OF CONTENTS XXIII

Chapter *Page*
    Group Clinic Idea 61
    There Must Be No Coercion 62
    Health Control a Production Control 63
    Services at Dispensary Level 65
    Types and Scopes of Present-Day Service Units 65
    The Ideal Facility 67
    Marine Medical Service 68
    Health Facilities for Small Businesses 69
        The health-conservation center 69
        Associated health-conservation units 69
        Individual health-conservation contracts 69
    Employee-Sponsored Plans 73
        The Stanacola plan 73
        The New York Hotel Trades Council and Hotel
            Association Insurance Fund 74
        The Toledo Willys unit plan 75
        Medical Program of the Endicott Johnson Corporation 75
    Occupational Medicine Has Come of Age 76

3. *Organizations Interested in the Health Needs of the Employed Person*

    International Committee on Occupational Health Services
        of The World Medical Association 83
    Council on Industrial Health of the American Medical
        Association 85
    The World Health Organization 86
    International Labor Organization 87
    The International Commission on Industrial Medicine 88
    American Association of Railway Surgeons 89
    Conference Board of Physicians in Industrial Practice 90
    The American Academy of Occupational Medicine 90
    Additional Allied Organizations 91
    Industrial Hygiene 91
    Air Hygiene Foundation of America 92
    The Industrial Hygiene Association of the American
        Public Health Association 93
    The Industrial Hygiene Association 93
    The American Conference of Government Industrial
        Hygienists 94

Chapter	Page
The American Association of Industrial Nurses	94
The Industrial Medical Association	97
Occupational Health Institute	99

4. Management Sets the Policy for Occupational Health

Functions of a Line or Operating Officer	106
Functions of a Staff Officer	106
Danger of Conflict	106
What Is a Policy?	110
The Virtues of Flexibility and Consistency	110
Formulating the Occupational Health Policy	111
Occupational Health Program Stimulates Human Relations	112
History of Management Policy: Expediency	113
Nonoccupational Illness	116
The Short-Term Absentee	116
The Rising Costs of Legitimate Sick Time	118
Private insurance companies	118
Cooperatives	119
Service organizations	119
Health — The Core of All Commitments	121
Gap Between Real and Unreal	124
Essentials of a Medical Policy	127
Preventive medicine	127
Educative medicine	127
Constructive medicine (diagnosis and early detection)	127
Curative medicine	128
Habilitative medicine	129

5. Corporate Giving for Health Causes

Business and Health Causes	131
The Committee Wall	133
Guiding Principles of Action	133
How to Consider Health Causes	135
Research	135
Professional and technical education	135
Public education in matters of health	135
Community facilities	136
Community Facilities and the Corporation Pocketbook	137

TABLE OF CONTENTS XXV

Chapter *Page*

Statistical Facts Have a Sobering Effect 137
The Policy Makers 138
The Result of Ineffectual Planning 139
The Hill-Burton Construction Act 140
Facilities for the Horizontal Patient 141
Trustees Are Men of Business 142
The Humanological Principles of Hospital Design 143
The Architect's Role 144
Other Members of the Team 146
It Is a Management Function 146
Each New Tried and True Technological Device Must
    Be Added 147
There Will Be Changes 149
Flexible, Movable Interiors 150
Hospital Management Methods 150
What the Consumer Wants Is Health 151
The Hospital for Tomorrow's Needs 152
Housing the Chronically Ill and Aged 156
Education as a Health Cause 156
Attention Drawn to one Specific Disease Entity 157
Problem Drinking as a Specific Health Cause 159
Finding the Problem Drinker — A Basic Medicolegal
    Problem 159
Stake of Community and Law-Enforcement Officials 160
Unknown Facts, Which Should, Must, and Can Be
    Determined 161
A Fundamental Approach — Preventive in Nature 163
Human Relations 164
Purely Philanthropic Interest 165
The Family Service Association 165
Unhappiness 166
Medical Education 166
Fundamental Research as a Health Cause 167
Genetics 168
The Best Laboratory Is Within Industry 169
Competitive Enterprise Holds the Future of American
    Medicine in Its Hands 169

Chapter *Page*

6. *Labor Dictates Medical Policy*

The Organized Employee of Today	174
The Required Health Program	176
A Major Medical Catastrophe	176
In Search of a Solution	177
Assembly Line Doctors	177
Need for Constructive Philosophy	178
Historical Development of Collective Bargaining	179
More Today Than Yesterday; More Tomorrow Than Today	179
John L. Lewis and Sidney Hillman	181
Today's Labor Leader Must Be an Educated, Intelligent, Imaginative Administrator	182
The Three Echelons of the Typical American Union	182
The Federation of Unions	183
The Mechanics of Union Operation	183
Both Sides Should Bargain Strongly	185
Fringe Benefits	186
Disability Compensation	186
A "Four-Way Stretch"	188
Actual Monetary Factor	188
Fringe Benefit Goals	189
Health Benefits	189
The Five Categories of Health Plans	190
Standards of Medical Service	190
Inherent Problems	191
Requirements of a Complete Program	191
Competent Advice Essential for Quality Controls	192
Essential Basic Information Required	192
The Three Different Meanings of Administration	194
Legal Advice Essential	195
The Early Health Insurance Programs	195
Health Insurance to Pay Doctors' Bills	197
Major Medical Expense	198
Constructive Medical Care Means Maintenance Along the Way	199
Sweeping the Dust Under the Carpet	199
Walter Reuther	200
Pertinent Data for Effective Evaluation	203
Total Health for the Employed Individual	204
Policy Required	205

Chapter	Page

7. The Professional Members of the Humanological Team

The Shaping of Humanological Policy	209
Point of No Return	209
Interpretation	211
Members of the Humanological Team	212
Special Qualifications Essential	212
Principles of occupational medicine	213
Industrial physiology	213
Industrial hygiene	213
Occupational pathology and toxicology	213
Special medical problems	213
Accidents at work	214
Occupational psychology	214
Preventive medicine	214
Industrial technology	214
Medicolegal problems, social security	214
Organization and administration	214
Statistical methods	214
The Doctor	214
Logic	215
Ethics	216
Epistemology	218
Sociology	220
Economics	221
Human Ecology	222
The Human Being	224
The Five Potential Types of Patients	225
Constructive Medicine	226
"Physician, Know Thyself"	227
The Humanologist as a Specialist	229
The Doctor for the Needs of Industry	230
The Registered Nurse	231
Qualifications and background	232
Her place in the scheme of things	233
Her position in the organizational plan	233
Liaison with other members of the professional arm of the humanological team	233
Liaison with resources in the community at large	234
Fundamental Principles of Occupational Nursing	234

Chapter	Page
Code for Professional Nurses	236
The Occupational Psychologist	237
A Student of Behavior	238
The Role of Psychologist in Industry	239
Mental Hygiene	239
A Positive Human Engineering Program	239
The Eight Principal Facets	239
Selection and Placement	240
Routine pre-employment test battery	240
Caterpillar psycho-graph profile	240
Special test batteries	241
Induction	241
Interviewing and Counseling	241
Training of Interviewers	242
Lectures on personality and applied psychology	242
Supervised interviewing	243
"Headache clinics"	243
Reading assignments	244
Coordinating all data on applicant	244
Individual conferences with the personnel consultant	244
Training of Supervisors	244
Mental Hygiene Education	245
Articles in employees' magazine	245
Bibliotherapy	245
Visuotherapy	245
Social Service	246
Research	246
Broad Goals of Mental Hygiene	246
The Industrial Hygiene Engineer and His Co-workers	247
The Qualifications of an Industrial Hygiene Engineer	250
Education and experience	250
Knowledge, skills, and abilities	250
Industrial Hygiene in Action	251
A Classical Example of Fundamental Research	251
History and Highlights of the Team Approach	253
Specific Industrial Hygiene Recommendations	255
Restrict the number of employees exposed	255
Make medical examinations every three to six months	255
Select the employees	255

TABLE OF CONTENTS   XXIX

Chapter   *Page*
    Remove employee from exposure if lesion develops................256
    Treat lesions promptly..256
    Wear protective clothing..256
    Do not touch scrotum or face with oil-soaked hands..............256
    Remove oil from the skin immediately............................256
    Take a shower bath daily before leaving the plant..............256
    Put on clean work clothes daily..................................257
    Provide special facilities for cleaning work clothes..............257
    Reduce exposure points in plant to a minimum..................257
    Identify equipment containing these oils........................257
    Inform employees of potential hazard..........................257
    Supervise precautionary measures in each plant..................257
    Make a periodic survey of plants handling these oils............258
    Eliminate possibility of inhalation of these oils..................258
    Provide the medical and industrial hygiene personnel
        and equipment necessary to carry out these
        recommendations..258
    Investigate blends of these oils..................................258
    Make reports..259
  The Broader Aspects of Industrial Hygiene Peculiar to the
    Petroleum Worker..259
  Summary of Study..262
  Hygiene Problems of Other Industries..............................263
    Hygienic standards..264

8. ***A Positive Program of Occupational Health***
  A Company Safety Program..270
  Physiological Properties of the Work Environment..................271
  The Three Levels of an Occupational Health Program............272
  The Dispensary Level..272
  The Registered Nurse Is the Key Person..........................272
  Standard Orders for Nurses in Industry............................273
  Airs, Waters and Places..274
  Public Health..275
  Public Health and Preventive Medicine............................276
  Epidemiology..277
  Immunization..278
  Requirements for Overseas..279
  Immunization Against Various Specific Diseases..................280
    Smallpox..280

Chapter	Page
Typhoid fever	282
Typhus fever	283
Cholera	283
Yellow fever	283
Plague	284
Poliomyelitis	284
Tetanus	285
Lesser Diseases	285
The common cold	285
Influenza	287
Judgment Requisite	289
Malaria	292
Yaws	296
Mental Health	297
Absenteeism	298
Social Adjustment of the Absentee	299
"The Half Man"	299
The Role of the Supervisor	301
The Nurse and the Supervisor	302
Space Medicine	305
Counseling	307
Constructive Versus Educative Medicine	308
Personal Hygiene — Obesity	309
Educative Medicine at the Dispensary Level	309
Moonlighting	310
Dermatoses	313
Medical Services at Diagnostic Level	314
The Placement Examination	314
Environmental History	315
Able to perform any duties	316
Able to perform limited duties	316
Able to perform any duties when defects have been corrected	316
Able to perform only limited duties when defects are corrected, if correction is possible	316
Logic as to Place of Performance	317
Placing the Medically Handicapped	317
The Periodic Health Inventory	319
Iatrogenic Diseases	320
Positive Health Versus Negative Health	321

TABLE OF CONTENTS   XXXI

Chapter   *Page*
    Scope of the Periodic Health Inventory 321
        Basic data 321
        Summary of interval record 321
        Environmental history 322
        Physical examination 322
        Consultations 322
        Impression and discussion 322
        Disposition 322
    Cancer — Significance of Early Detection 323
    Glaucoma 325
    Diverticulitis and Mucous Colitis 326
    Tuberculosis (Pulmonary) 327
    Bronchitis and Bronchiectasis 328
    Arthritis 329
    Deafness 329
    Back Pain 330
    Proctologic Disease 331
    The Concept of Disease 331
    The Benefits to Be Derived 332
    Individualization Essential 335
    Educative Medicine at Diagnostic Level 336
    Constructive Medicine Entirely Lacking 337
    Constructive Medicine at Its Best 337
    Community Relationship 339
    Consumer Effects 340
    Role of Humanologist 340
    Costs of a Constructive Health Maintenance Program 341
        Variability of needs and costs 341
        Some estimates of total costs 342
        Background philosophy for the foregoing remarks 343
        Basic purpose of a constructive health
            maintenance program 346

9. **The Seven Cycles of Life**

    The Fourth Dimension 352
    Evaluation of Man in the Fourth Dimension 353
    History of the Individual 353

XXXII   TABLE OF CENTENTS

Chapter   *Page*
   Period of Parental Dominance..354
      Conditioned reflexes sensitized to given situations..........355
      Duality..355
   Period of Growing Independence...357
   Period of Adult Experimentation..359
   The First Ten Years of Employment......................................362
   Period of Acquisition...362
   Period of Peak Attainment...364
   Period of Substitution and Compromise................................367
   Golden Period of Reminiscence and Leisure........................371
   Types of Employees...374

10. Recruitment

   Why Continuous Recruitment?...379
   Objects of Recruitment..380
   Job Classification and Definition...381
   Limitations of Job Classifications...382
   Sources of Manpower...382
      Employees within the organization itself...........................383
      Persons nominated by present employees.........................383
      Persons filing personnel applications..................................385
      Immigrant workers...387
      Rural-urban migration...388
      Women..391
      Older workers...395
      Schools and colleges...396
      Labor organizations..397
      Employment services...397
      Miscellaneous sources, such as lodges, churches,
         clubs, prison associations, and the like........................399
      Advertising...399
   The Positive Approach...400

11. Selection and Placement

   No Mechanical Short Cut Will Do the Job............................402
   A Trained Humanologist Is Needed..403
      The selectee's wants...404
      True evaluation of the selectee..404
   Occupational Medicine and the Placement Interview.........406

TABLE OF CONTENTS   XXXIII

Chapter   *Page*
   Bridging the Gap 406
   The Significance of the First Ten Years of Employment 408
   Handling the Unsuitable 408
   Filling Out Forms 409
   Psychological Tests 410
   Significance of Character, Personality, Aptitudes,
      and Limitations 412
   Underplacement and Overplacement 414
   Misassignment Produces Medical Symptoms 417
   The Significance of Stress 418
   Physiological Limit and Emotional End Point 419
   Fatigue 420
      On-the-job factors 421
      Off-the-job factors 421
   Poor Placement as a Cause of Fatigue 422
   Situational State 426
   Symptoms of Situational State 426
   Case Histories 429
   Approach to Therapy 432
   Nonoccupational Disability 434
   Importance of Placement Examinations 435
   Complexity of Nonoccupational Disability 435
   Summary 436

12. *Training*

   This Age of Automation 440
   Art of Training a Science 441
   Every Employed Person Requires Training Now 442
   Significance of the Permissible Margin of Error 442
   Mantalent Versus Manpower 443
   Minimum Standards of Employability 443
   All Individuals with Few Exceptions Can Be Trained 444
   What Was Learned in World War II 444
   Training Research 445
   Technique for Living 445
   Health Safeguards 446
   Training Methods and Health Education 447
   Pride a Prime Motivating Force 449

Chapter	Page
Motivation and Training	450
Training the Long Service Employee	451
Resentment	452
How to Deal with Training Resistance	453
A Training Program	454
Training Needs	454
Constant Aspects of a Given Job	455
Precedent and Individual Preference	456
Actual Training Areas	456
Training on the Job	457
Habits and Age	457
Morale Factor	458
Purposes of Training	458

13. Motivation and Promotion

Every Stage of Employment Has Its Motivation Aspects	460
Fundamentals of All Human Motivation	461
Physiological and Social Motives	461
Nine Basic Human Needs	462
Food and physical welfare needs	462
Personal development needs	463
Desire for achievement	463
Desire for activity, variety, and novelty	463
Need for release from emotional stress	464
Need for security of status	464
Need for worthy group membership	465
Need for sense of personal worth	465
Need for a sense of participation	466
Influencing Factors and Basic Motives	466
Money Is an Important Motive	467
Title a Motive	467
Fear as a Motive	467
Gratitude and Happiness as Motives	468
Goals Become More Complex	469
Each Employee Must Have a Goal	469
Envy or Pride as a Motive	470
Promotion and the Unpromotable	471
Team Spirit and False Security	472
Communications	472

TABLE OF CONTENTS   XXXV

Chapter   *Page*
    Mechanical Efficiency Does Not Guarantee Human
        Efficiency..473
    Essential Factors for the Development of Team Spirit............473
    Functional Disorders...475
    Functional Disorders Must Be Detected Early.........................476
    The Distress Itself and Its Cause..478
    Functional Disorders May Affect any Part of the Body.........480
    Migraine Headache..480
    Headaches May Be Nonfunctional..481
    Headaches with a Physical Cause...482
    Treatment of Tension Headache..483
    Obesity as a Functional Disorder...483
    What Experience Teaches...485
    Problem Drinking and Personality Disorders..........................486
    Peptic Ulcer..486
    Technique of Honest Self-Analysis...489
        Where do I stand health-wise?..489
    Explanation a Counseling Function..490
    Communication..491
    A Layman's Confirmation..491

14. *Supervision (Part 1)*
    Definition of a Supervisor..493
    Meaning of "Supervisor"..493
    Supervisor Must Be a Vanguard..495
    Health Requirements of the Supervisor Himself....................496
    Meaning Of Emotional Maturity..499
    Six Basic Principles of Maturity..500
        Self-acceptance..500
        Acceptance of others..501
        Sense of humor...502
        Appreciation of simple pleasures..503
        Enjoyment of the present..504
        Appreciation of work...505
    A Check List of Emotional Growth..507
    The Specific Value of Such Tests..508
    Cult of the Positive..509
    State of Becoming..509
    Translating Policies and Objectives in Terms of
        Individual People..510

Chapter	Page
Responsibilities to Top Management Versus Subordinates	514
A Static Impression of the Individual	515
The Use of Gentle Reins	516
Cracking the Whip	516
Basic Principles of Good Discipline	517
Human relations	517
Discipline that prevents	517
Discipline that controls	517
Discipline that punishes	518
Considerations in deciding on penalty	518
Did action result in attaining desired objective?	518
Supervisor's Responsibilities	519

15. Supervision (Part 2)

True Sickness Absenteeism	520
Five Basic Motives for Attendance	521
Five General Motives for Nonattendance	521
Questions to Be Answered	523
Major Illness and the Family Doctor	524
The Job Factor	524
Competing Interests Outside the Plant	525
Problem Drinking	526
Definition of the Problem Drinker	526
Problem Drinking Is Still Essentially a Mystery	526
Statistics Are of Questionable Value	527
Four Categories of Problem Drinking	528
Pathological intoxication	528
Environmental habituation	528
Dipsomania	528
A neuropsychiatric problem associated with an independent symptom of alcohol addiction	529
Three Types of Problem Drinkers	529
The chronic excessive drinker	529
The compulsive drinker	529
The chronic alcohol addict	529
Problem Acknowledged by Top Management	530
The Result of Cover-Up	532
The Problem Drinking Detection Unit	532
The Periodic Health Audit	534

TABLE OF CONTENTS   XXXVII

Chapter   *Page*
   The Supervisor Is Affected by Nonoccupational Illness 534
   Nuisance Diseases 534
   Absences Attributable to Degenerative Disease 535
   What a Careful Analysis Will Show 535
   Respiratory Diseases 536
   Gastrointestinal Disturbances 536
   Degenerative Disease and the Periodic Health Audit 537
   What a Thorough Health Audit Includes 539
   The Supervisor's Responsibility 540
   The Supervisor's Role in Preventive Psychiatry 542
   Normality Is a Delicate Concept 542
   The Interpretation of Atypical Behavior 543
   The Supervisor's Function 545
   The Quality of Supervision 545
   A Constructive Approach 545
   The Key to Supervision 547

16. *Retirement*

   Retirement a Form of Severance 550
   To Replace Inadequate Concepts with Dynamic Ones 551
   Established Usages and Precedents 551
   For the Good of the Service 552
   Total and Permanent Disability 553
   Total and Permanent Incapacity 554
   A Team Approach Will Bring Realistic Answers 556
   The Four Subcategories 556
   The Twilight Zone 556
   The Absent-Prone Cause Precedents to Be Set 557
   A Decision Requires Specific Answers 557
   Standards of Efficiency Can Be Graded Realistically 558
   The Desired Goals 558
   Expediency-Based Thinking 559
   Points of Similarity 561
   Contributing Factors 561
   Normal Retirement 562
   Retirement Planning Can Be Constructive 562
   What Retirement Means 563
   Seven Sins Against Older People 564

Chapter	Page
The Humanologist, as Well as the Potential Retiree, Must Understand	565
The Least Important Part of Retirement	566
Aging in Harness	566
Arbitrary Figures for Bookkeeping Only	567
Chronological Age	567
Physiological Age	567
The Development of Reliable Standards — The Gateway to Flexible Retirement	569
The Theorists' Point of View	570
The 40-Year Old Is a Potential Candidate for a Health Setback	571
Circulatory Problems and Adjustment	572
The Circulatory System No Great Mystery	572
The Humanologist — A Counselor	573
How Heart Myths Get Into Currency	574
Facts for the Layman to Understand	576
The Mechanical Types of Heart Disorder	577
Medical Terminology Complicates the Picture	578
No Single Symptom Adequately States the Case	580
Recommendations Must Be Stated Clearly	582
Constructive Retirement Planning	583
The Little Insidious Mishaps	584
Health Planning and Retirement Planning Go Hand in Hand	586
Accent on Mantalent Development	589

17. A Constructive Mantalent Development Program

	Page
Specific Concerns of Humanology	593
1 Recruitment	594
2 Selection and Placement	595
3 Training	595
4 Motivation and Promotion	596
5 Supervision	599
6 Direction	600
7 Relinquishment of Power	602
Retirement Planning	602
Requirements for Inaugurating a Constructive Humanological Program	604

TABLE OF CONTENTS   XXXIX

Chapter   *Page*
    General Functions and Qualifications of Principal
        Team Members..604
            Management representative......................................604
            Professional advisor...605
            Economic realist..605
    The Team Approach..605
    What a Humanology Team Can Accomplish..........................606
    Methods..609
    Evaluation Center..610
        Purpose...610
        Requirements..610
        Comment..610
        Advantages...611
        Space and Equipment...611

18. *Reports*

    Guiding Principles..612
    Glossary...613
        Lost-time cases..613
        Frequency rate...613
        Noneffective rate...614
        Days of disability..614
        Days of disability per case...614
        Days of disability per person..614
    Files..614
    Filing Guide...616
    Diagnosis and Codes...617
    Standardization for Statistical Purposes..............................624
    Statistical Analyses...625
    Monthly Report of Medical Activities..................................629
    Suggested Topics to Be Discussed...629
        Administrative medicine..629
        Preventive medicine..630
        Constructive medicine..630
        Educative medicine...630
        Curative medicine..630
        Mantalent development...630
    Preparation and Distribution..631
    Confidential Medical Information...631

Chapter	Page
Summary	631
The Placement Health Inventory	633
The Periodic Health Inventory	636
Special Examinations for Employees	638
Maintenance of Adequate Nutrition	639
Reassignment of Duties for Medical Reasons	639
Premature Medical Retirement (Because of Illness)	639
Periodic Health Inventory Summary	640
Estimating Expenses and Initial Capital Requirements	641
Annual Report	641
Suggested Topics to Be Discussed	643
Administrative medicine	643
Preventive medicine	644
Constructive medicine	644
Educative medicine	644
Curative medicine	644
Mantalent development	644
Preparation and Distribution	648

INDEXES

Author's Index	651
Subject Index	657

LIST OF ILLUSTRATIONS

Facing Page

The Worker .... Dedication

GRAPHS

Graph   *Page*

I   The Physical Growth and Ages of Man 28
II   The Psychic Growth and Ages of Man 28
III Obstetrical Deliveries — Hospital Operated by an American Corporation Operating Abroad 46
IV Medical Expense 47
V   Medical Expense 48
VI South American Affiliates — Total Medical Expense 48
VII Amebic Dysentery 114
VIII Dollars for Health 132
IX Hospital Costs and Total Population 196
X   Effect of DDT Program on Malaria 294
XI Situational States I 427
XII Situational States II 427
XIII Situational States III 428
XIV Special Executive Examinations 538

TEXT ILLUSTRATIONS

Figure   *Page*

1 Man Is a Tetrahedron 11
2 Industrial Mobile Unit, *outside of factory* 70
3 Industrial Mobile Unit, *close-up view* 71
4 Interior Plan of Mobile Unit 72
5 Certificate of the Industrial Medical Association 98
6 Horizontal in Design — Streamlined for Efficiency 152
7 Vertical in Design — Streamlined for Efficiency 153
8 Both Horizontal and Vertical in Design 154

LIST OF ILLUSTRATIONS

Figure   *Page*

9. Approximate Number of Patients Handled Daily by Each Unit....155
10. International Certificates of Vaccination.................281
11. Example of an Open Health Letter.......................312
12. The Able Executive Who Surrounds Himself With Men of Similar Potential Abilities.....................375
13. The Executive Who Gets the Job Done for a While but Fails to Delegate Responsibility, to the Consternation of all as the Years Roll By.....................376
14. Everyone Apparently Working at Full Speed but Accomplishment in the Reverse.....................377
15. Segment of Stomach and Intestine.......................487
16. The Circulatory System...............................575
17. It's Your Choice.....................................600
18. Employee Register Card...............................625
19. Monthly Statistical Summary..........................632
20. Placement Examination *(Front)*......................634
21. Placement Examination *(Back)*.......................635
22. Periodic Health Inventory *(Front)*..................637
23. Periodic Health Inventory *(Back)*...................638
24. Confidential Report..................................640
25. Suggested Work Sheet.................................642
26. Guide for Evaluating Mantalent Potential *(Front)*...645
27. Guide for Evaluating Mantalent Potential *(Back)*....646
28. Qualities Associated with Mantalent Performance *(Front)*...647
29. Qualities Associated with Mantalent Performance *(Back)*....648

TABLES

Table   *Page*

I. Results of Survey.....................................260
II. A Health Education Calendar..........................311

TEXT

I will in the first place readily acknowledge that one of the worst Maladies, which a Man in any Trade, or Way of Living, ever can fall into is for a Man to be sick of his Trade. If a man has a Disaffection to the Business that he has been brought up to and must live upon, 'tis what will expose him to many and grievous Temptations and hold him in a sort of perpetual Imprisonment. Man, beg of God a heart reconciled unto thy Business, and if He has bestow'd such a Heart upon thee, as to take Delight in thy daily Labour, be very thankful for such a Mercy!

>Cotton Mather in "Angel of Bethesda," 1724, published by Beall and Shryock, 1954.

Introduction

THIS IS A TEXTBOOK on the subject of *occupational health*. The term calls for definition, but there is probably no precise definition that can cover all conceivable uses of the term. The stockholder, represented by the management of competitive enterprise, in the last analysis pays the bill for occupational health.[1,2] As a result, he may insist upon one definition. Organized labor, which represents the not easily discounted wishes of the majority of those on the receiving end of occupational health policies, may insist on an entirely different definition. The professional practitioners—medical consultants, nurses, industrial hygiene engineers, and various specialists in the field of occupational health—may advocate a third definition. Moreover, there may be disagreement among each of the above-mentioned groups.

One reason for this elusiveness of meaning is that occupational health has rapidly changed in scope.[3] At the turn of the century the chief concern of the practitioner of occupational medicine was with the specific "diseases of occupation." Today it is generally agreed that the ideal of occupational health is to keep well people healthy, happy, and productive. The only disagreements occur over how this is to be accomplished and who is to be responsible for what.

[1] SAMUEL L. H. BURK, "Management, Medicine and the Future." An address to the Occupational Health Institute, Philadelphia, 1956.

[2] BENSON FORD, "The American Road to Health." An address to the Blue Cross Plan of Wisconsin, Milwaukee, 1953.

[3] ROBERT COLLIER PAGE, M.D., "Where Do We Stand Now with Regard to Socialized Medicine and Private Enterprise in Providing Adequate Health for all Our People Throughout the United States?" An address to the Economics Club, Detroit, May 25, 1953.

Occupational Health Is Everybody's Business

Perhaps one of the hindrances to perfect concord is the lingering tendency to regard occupational health as a narrow specialty. More perfect agreement may come about when it is brought to light that occupational health is everybody's business. Everyone is affected by it, directly or indirectly. Everyone has a specific role to play and a specific responsibility with regard to occupational health.

Everyone either works for a living or is wholly or partially dependent upon someone who does. Even the independently wealthy person is dependent for his enjoyment of life upon the goods and services provided by someone who works. The quality of work performed by one person can affect the comfort or happiness of a great many people. This quality can act upon and be acted upon by other aspects of the total environment of the individual worker himself.

Occupational Health Is Only Part of the Whole

This text will endeavor to discuss occupational health in terms of the total man or woman who works for a living. It will show that the *medical* aspect of occupational health is only a part of the whole, and it will set logical boundaries to the functions and responsibilities of the strictly medical arm of the occupational health team. It will show how this team—more aptly called the *humanology* team—came into being over the years. It will point to the present-day functions and objectives of its members from management, labor, and the professions and discuss its various future possibilities.

The Professional Consultant

The term "medical director" will be studiously avoided. The doctor in industry does not direct; he is not a representative of management, nor is he an employee paid to

INTRODUCTION   3

render complete medical care to ill workers. He is a consultant. Management and labor come to him for his expert advice in matters having to do with keeping well workers healthy and productive. In giving his advice, he draws on his knowledge and appreciation of the medical resources of the community: the specialists and general practitioners trained to give certain types of definitive medical care. One of his first functions in going to a community is to familiarize himself with these resources and incorporate them into the humanology team, or, if they are not available, move heaven and earth to get the cooperation of his company and the community to do what is necessary to *make* them available. But it is important to remember that, except in emergencies, where quick action is necessary, he remains a consultant, uncommitted either to management or labor and concerned principally with the construction and maintenance of health of the vertical individual at a job which he likes, needs, and can perform capably. Therapeutic resources within the community can and will take over when the individual ceases to be vertical or is otherwise incompetent to perform his job. There are approximately 37 different kinds of specialists in the medical field. The consultant in occupational health must understand the functions of each of these, in addition to the functions of the lay members of the humanology team—the personnel director, for example, or the shop steward—but he must be very careful not to usurp any of these functions.

The Members of the Team

The present work will clarify and particularize broad concepts with repeated reference to individual case studies which will demonstrate the roles played by the several members of the team: (1) the man at policy-making level, representative of management or labor; (2) the medical consultant; (3) the clinician with his profound interest in therapeutics; (4) the registered nurse; (5) the psychologist; (6) the industrial hygiene engineer and other paramedical people who have specialized parts to play in a humanological program of health construction and maintenance designed to enable employees to live better, more happily, more productively, and perhaps longer.

Reference Materials

There are numerous reference works on *specific* health problems peculiar to occupational environments. Hunter's *The Diseases of Occupations,* Cluver's *Social Medicine,* Johnstone's *Occupational Medicine and Industrial Hygiene,* and Stieglitz's *A Future for Preventive Medicine* are monumental examples. This book will not attempt to repeat in detail the voluminous data that can be readily looked up when needed in any of these authoritative works. Subjects nowhere else included will be dealt with in some detail. Specific reference to proper authorities will be made when the occasion arises. The author is fully cognizant of the tremendous amount of reading that must be completed by the student preparing to fill the health needs of tomorrow. For this reason, the author has assembled pertinent facts which, in his judgment, will be of value to everyone whose chosen role has to do with the environmental health needs of the employed person.

The Captain of the Team

Obviously, no one person can possibly master all of the disciplines involved in achieving the goal of occupational health. No single person can be a past master of logic, ethics, epistemology, sociology, human ecology, psychology, medicine, economics, and sound management practice combined. That is why a team approach is necessary. Each member of the team can utilize his particular skills under the over-

INTRODUCTION

all direction of one person—ideally, one who has passed his peak of attainment in the business world, who has maintained his health, and who is now prepared to enter his life stage of substitution and compromise and to relinquish the power and material wealth he has worked so hard to accumulate in the past. Such an individual can be counted on to understand the importance of human relations, and may now wish to devote his remaining active years to humanology in its broadest sense.

Under an organizational setup of this sort, occupational health for all can be a reality at a price the individual consumer can afford to pay, and yet comprise health services of the highest quality.

The Seven Stages of Progression

The various facets of this compelling challenge will be dealt with in subsequent chapters in terms of the seven stages of progression of the employed person: (1) recruitment; (2) selection and placement; (3) training; (4) motivation and promotion; (5) supervisory responsibility; (6) executive responsibility; and (7) relinquishment of responsibility and power prior to retirement.

It is the fervent hope of the author that more and more individuals will focus their interests on the health needs of the employed person. A healthy person is a happy person. A healthy, happy individual whose job is in keeping with his likes and capabilities is an efficient, productive person and a good citizen of his country.

If those trained in the various aspects of humanology can develop an appreciation of accepted and acknowledged principles of action plus the ability to put these principles into practice, the picture of the happy, healthy, productive employed person can become a universal one. Your role in making this possible is outlined in the following pages.

REFERENCES

ALLEN, RAYMOND B., M.D. "Medicine — A Lifelong Study." *Journal of the American Medical Association,* CLXX (No. 18, August 29, 1959).

CLUVER, EUSTACE HENRY, M.D., D.P.H. *Social Medicine.* Johannesburg (South Africa): Central News Agency, 1951.

DIMOND, E. GRAY, M.D. "The Family Physician The General Practitioner and the Internist." *Journal of the American Medical Association,* CLXXI (No. 11, November 14, 1959).

HUNTER, DONALD, M.D. *The Diseases of Operations.* Boston: Little, Brown and Company, 1955.

JOHNSTONE, RUTHERFORD THOMPSON, M.D. *Occupational Medicine and Industrial Hygiene.* St. Louis: C. V. Mosby Company, 1948.

KESTNBAUM, MEYER. "The Modern Corporation and the Nation's Health." *Industrial Medicine and Surgery,* XXIII (No. 5, May, 1954), pp. 221-223.

MURRAY, DWIGHT M., M.D. "Who's Responsible for Health?" *New York State Journal of Medicine,* LVII (No. 9, May 1, 1957), pp. 1673-1676.

PAGE, ROBERT COLLIER, M.D. "Doctors and the Federal Medical System." *GP* (The American Academy of General Practice), IV (December, 1951), pp. 107-108.

STIEGLITZ, EDWARD JULIUS, M.D. *A Future for Preventive Medicine.* New York: The Commonwealth Fund, 1945.

1
The Whole Man

MEDICAL SCIENCE, for the most part, was founded upon the basic physical sciences. As a result of unquestioned success by this approach, many have come to think that the borders of materialism correspond to and fix the limits of medical science. This is in error. Of late, scientists have begun to admit the reality of "unreal" things. As a result of a better understanding of the psyche, there is positive evidence of progress toward ridding the world in general, and the employed person in particular, of emotional unrest and unhappiness. Codes of morals, canons of taste, systems of social modes, standards of value, practical and efficient principles of economy—all are now within the purview of occupational medicine. As a result, it is now possible soundly to augment the strictly material approach to man's total health. The material contributions of science in general, and of medical science as presently applied to the total health needs of the employed person and his family, have compounded the complexity of living because the confines of science have been set at the borders of materialism. Physical science has been such a profound benefactor that scientists in general, and employers in particular, have given little thought to psychic matters.

The Basic Functions of Competitive Enterprise

The cornerstone of our economic system is the producing organization—the industry—whose primary function has been generally considered to be the production and distribution of goods and services. This is the purely economic function. Competitive enterprise, following the lead of Henry Ford after the turn of the century, has acquired—not always willingly—a second function: that of employing and providing the wherewithal for the consumers who will avail themselves of the goods and services produced and distributed. This is the sociological function. Its importance is growing, owing to a number of factors: demands of organized labor, government intervention, and the enlightened self-interest on the part of industry itself.

Humanology in Business

Enlightened self-interest on the part of competitive enterprise has paved the way for the enlightened study of man. Nudged from time to time by the other factors, it has been responsible for the phenomenal development of what might be called *humanology* in business. Where formerly the management team was made up of production men and distribution men, backed by specialized staff experts in the bookkeeping, advertising, and the legal aspects of production and distribution, there is today an important segment of management which concerns itself almost entirely with *human beings*—with placing them, training them, motivating them, understanding them.

The members of this team may be called *humanologists*. They include individuals in charge of personnel and human relations, medical consultants, the registered nurse, the industrial hygiene engineer, the toxicologist, the psychologist, and the safety engineer. In many companies, individuals whose primary function is to produce—foremen, for example, or department supervisors—may be called upon to be part-time humanologists. This is important since such individuals are necessarily closer to the human beings in question on a day-to-day basis than the individual in charge of personnel, the part-time medical consultant, the retained industrial hygienist, the plant nurse, or the safety engineer who may have other duties to perform.

Self-Interest and Humanology

With the astonishing development of technological devices which began in 1900, it is understandable why many business minds looked forward to the day when the fallible human element, with all its unpredictable quirks, might be wholly replaced by the infallible, utterly predictable machine. Man would become relatively unimportant except at the rarefied level of the planner, the administrator, and the entrepreneur.

What actually has happened is in keeping with anticipated advances in our knowledge of man. As machines improved, the people operating them became more alert to their individual roles and rights. This will continue to be the case. Automation can never supersede the human element.

A spokesman for the Industrial Hygiene Foundation of the Mellon Institute is on record to this fact in the following words:

A company's employees have always been its most valuable assets, and automation is not going to change this. In fact, more responsibility will be vested in employees than heretofore. There will be a larger number of technical jobs, including maintenance and repair of the complex machinery and monitoring of automatic equipment. This will call for greater skills and more training and education. Management personnel will be called upon to make major decisions. Errors in judgment would be costly.

In addition, each employee will represent a much larger capital investment than heretofore. In 1954, the capital investment per worker in the highly automatized chemical industry was twice that of industry as a whole, or $26,000; and in some plants it is considerably more. The present investment per employee in the electric power-generation industry, which is almost completely automatized, is in excess of $106,000.

And not only does automation increase the value of the human operator, it introduces peculiar human problems as well. Automation is not just a further advancement of mechanization. In mechanization man and machine function together, one as an extension of the other, in a sense. In automation, man supervises the machine but is not involved in its actual operation.

But much more than the method of production is involved. Automation signifies a whole new way of life, with bio-social values as well as technological ones. It will change the living and working habits of people, alter educational patterns in our schools, and raise our standard of living to new heights.

Man as a Total Being

Man—his wants, his expectations, his actual needs—these in the long run will actually determine the exact scope of humanology. Edgar H. Norris offers a thought-provoking method of viewing man as a total being:

If we think of man diagrammatically in terms of plane geometry, he is *ideally* an equilateral triangle. One side represents his body; another, his mind; and the third, his spirit.

However, man is not a plane; he is a solid with three dimensions. Geometrically, therefore, man is an equilateral pyramid—a pyramid resting *not* upon a base, but balanced precariously on *one* of its four apices. Man is a poised pyramid revolving on that apex where the sides representing body, mind and spirit meet. Equilibrium, we may suppose, is maintained by a sort of internal gyroscope whose action is most efficient when body, mind and spirit are proportionately developed and harmoniously related.

THE SOCID

But there is a fourth side on the pyramid! The fourth face is man's social side; it is that face of the pyramid which is uppermost when all is well with man. Let us call it the *socid*.

Thus, man is a tetrahedron. And since geometrically man is an equilateral tetrahedron with body, mind, spirit *and* with a social side, too, human health and well-being do not depend upon religion, philosophy, intelligence, emotion, reason, logic or sociology—not upon any of these *alone* any more than upon the materialism of physical science alone.

THE WHOLE MAN 11

THE FOUR SIDES OF MAN'S NATURE

The body . . . is what we have come to believe man to be . . . . Thus far, medical science has invested most of its labor on the body, and much surely has been learned. One fact is pre-eminent: Though there are huge differences between life and death, the body is the same whether it be dead or living . . . .

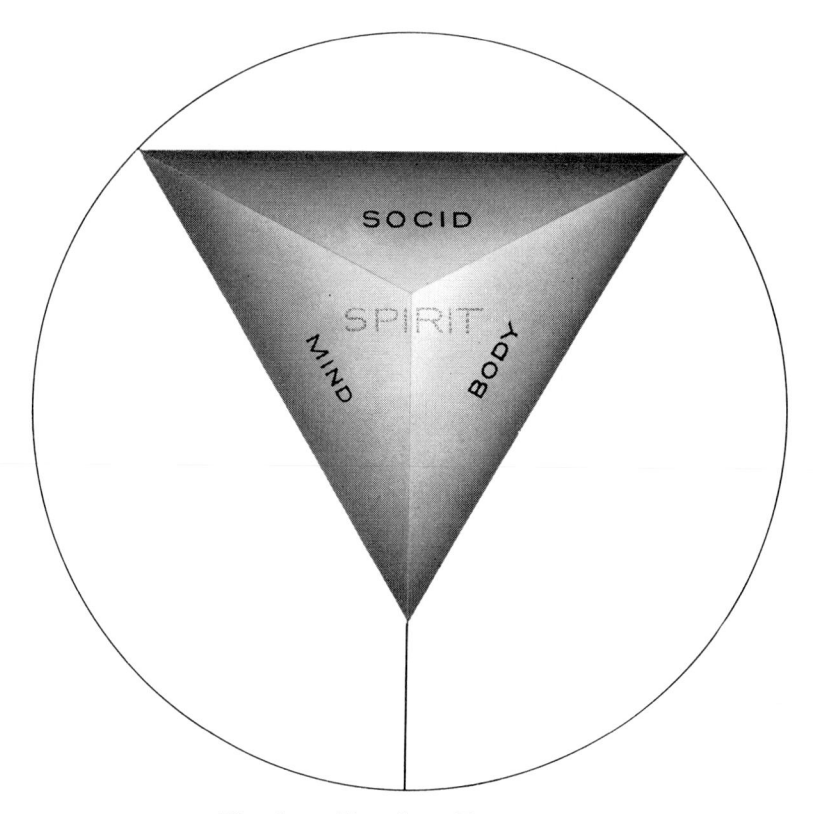

Fig. 1 — MAN IS A TETRAHEDRON

The body is *where we live;* the mind is *what we are.* The spirit of man is the *why* of all man's doing; the body is the mechanic; the mind, the interpreter; and the spirit—the breath of life—is the animator of living. The socid is *how* man lives with his fellows. It is the game of life in which bodies, minds and spirits play; a game whose gains, losses and outcome are determined by the degree of health of the player's bodies, minds and spirits . . . .

Just as there is a pathology of the body, there is a pathology of the mind, of the spirit, and of the socid. The four sides of man's pyramid are so joined together that debility of one may be offset or countered by the strength of another; or conversely, damaging distortions and disease of one side may horridly disfigure the others . . . . Research that would determine the nature of man, *the whole man,* must concern itself equally with the body, the mind, the spirit, and the socid . . . .

Five Case Studies

Analysis of case studies of five more or less typical representatives of human beings in differing occupational environments will make the goals of humanology more vivid. The individuals under analysis are George B., aged 55, first vice-president of a large manufacturing concern; Pedro Y., 40, an employee of a large American mining concern at one of its tropical overseas locations; Anne T., 38, a skilled stenographer; Walter McK., 45, a shop steward and punch-press operator; and Dr. Wharton L., 50, a director of specialized chemical research for a pharmaceutical concern.

THE MAN OF MANAGEMENT

Mr. George B. has outgrown the arrogant youthful assumptions that all obstacles may be overcome by a simple combination of aggressiveness, concentration, and perseverance, that the basic motive for human endeavor is money plus "an occasional kick in the tail," that the only significant reasons for nonachievement are (1) plain laziness, (2) malingering, and (3) stupidity.

Since his fiftieth birthday he has, on at least one occasion, been induced to tackle an assignment that was patently out of his line and has in consequence been obliged to admit defeat. On at least one other occasion, he has been impelled to abandon, for health reasons, a project that would have ordinarily fallen within his personal limitations. If this can happen to him, he reasons, it can happen to anyone. Human motives, capabilities, and limitations are more complex than he had been trained to realize.

Though he is far from being disabled, his health is not good. He is about 20 pounds overweight, is subject to attacks of indigestion and

THE WHOLE MAN 13

constipation, and has atherosclerosis, which has lately revealed itself in moderately severe attacks of angina pectoris.

His dietary habits are irregular. His breakfast usually consists of various proprietary pills for his digestion plus a cup of black coffee. The Martini he says he has before lunch is more often two than one. After he has had these, he looks at his watch, is amazed at how late it is, gulps down his entree without tasting it, and rushes back to the office. Before supper he consumes, on the average, 4 ounces of whiskey. Supper itself is likely to be enormous and highly caloric. He is particularly partial to mashed potatoes and gravy. At some time in the course of a restless night, he gets up and raids the refrigerator. He smokes about two and a half packs of cigarettes daily.

He was enthusiastic about his work until the age of 48, at which time he was elevated to his present position. Now he is still engrossed in his job, but he is not so enthusiastic about it. On the tables of organization, he is second in command. Actually, as the president is aging and slated for retirement shortly, Mr. B. often finds himself bearing the full responsibility for final decisions. In spite of this, he is fully aware that when the president retires, another younger man is slated to step into his shoes. He is not particularly resentful about this, but he feels that the president-to-be ought to be called upon to share some of the burden of responsibility which he is currently expected to bear alone.

He finds himself on the dealing as well as the receiving end of humanological endeavor in his company. In his executive capacity, he makes the final decisions on company policy having to do with motivating human beings, placing them properly, dealing with their problems, developing them, and maintaining them at peak capacity. In the long run, the policies he makes will eventually have an effect on himself and his own particular adjustment to his job. The directors of the company, viewing him through the spectacles of these policies—through his own spectacles, so to speak—see him as rather a knotty problem himself. There is nobody at the moment fully qualified to assume his responsibilities, yet, clearly, they are beginning to tell on him. It is necessary to find some way to adjust his job to him in such a way that the company itself is not imperiled. At the same time, it is necessary to establish a procedure by which he may hand the reins to someone else, bit by bit; so gradually, indeed, that when the successor finally has the reins entirely in his grasp, no one can tell just how or when the transition actually took place. Solutions to this problem—and Mr. B. is fully aware of it—will be based on painstakingly formulated company policy which will apply not only to Mr. B., but to all human participants in the corporate endeavor who have passed their peak of performance and are ready to be gently decelerated and tactfully replaced.

Thinking about humanologistic policy, Mr. B. has become greatly impressed with the importance of health—or, rather, illness—in relation to work performance. Optimum health has no place in Mr. B.'s thinking; he has never heard the term. Absence of health is what concerns him, particularly when it requires treatment, recuperation, or rehabilitation away from the job. Never having been in serious financial straits, he is not personally concerned with sickness as far as costs are concerned and does not carry health insurance. He understands the importance of the cost factor to most of the company's employees and has been instrumental in committing the company to a comparatively liberal medical benefits plan, but outside of this he isn't much exercised over the costs of hospitalization or medical care.

What does exercise him is the thought of the time lost in the course of illness. He was away from the office over a week following his initial attack of angina at age 52. Though medication had modified the severity of subsequent attacks, he had lost a total of seven working days in the intervening years and was seriously alarmed at the prospect of a future attack—or a coronary occlusion—that might hospitalize him for a month or more. At present, he simply did not have a month to spare; the corporation could not afford a two-week absence on his part. This was a specter. It haunted him more than a casual acquaintance could have realized. Moreover, being human, he tended to think of other people in terms of himself; he was certain that every individual in the plant was haunted by the specter of time lost. His idea of the ideal and perhaps never to be evolved company policy concerning health was twofold:

1. Every man should be immediately replaceable in case of sudden illness.
2. It should be made possible for everyone stricken by illness to get back to the plant at the earliest possible time, even if the individual in question had to come in a wheel chair and was only capable of working half an hour a day. (He did not mean this as a slave-driving gesture; he sincerely believed the average recuperating individual would have higher morale if he could come back to the plant as soon as physically possible, even if only to "look things over and say 'hi' to the gang.")

In short, Mr. B. is intelligent; his goals are well chosen, but no one has alerted him to the proper way to achieve these goals. His management of his own life, particularly off the job, indicates that he has not been adequately instructed as to *construction* of health and *prevention* of illness. He thinks in terms of treatment, of hospitals, of plans for rehabilitation. He stands in need of wise advice and guidance—of the sort that a truly constructive company medical consultant could give him. Being intelligent, he would respond, put his own house in order, and institute constructive policies that would pay dividends in his company's mantalent development program.

THE WORKER[1]

Pedro Y., unlike Mr. B., is wholly on the receiving end of humanology—well-meaning but misguided humanology at that. He is the product of industry's search into remote areas for raw materials.

In the beginning, it was always the same, whether it was the gold mines of Africa, the deserts of Arabia, the rubber fields of Indonesia, the fruit lands of Central America, or the oil and steel regions of South America. In his remote corner of the world, Pedro looked up one day from his overloaded burro and saw—as his counterparts of other far-off places had seen or would yet see in their time—a group of odd-looking, highly charged strangers. These new arrivals were carrying weird things of all shapes and sizes—tools, surveying instruments, maps—things Pedro had never seen before, and he shook his head and wondered what it was all about.

Later that day, when he reached his village, he found that the word had already spread: a huge mining concern was coming to the region to tap the resources that had lain undisturbed under the soil for eons. It was reported that the new industry would need the services of hundreds, perhaps thousands, of people from the area. They would be trained to run giant new machines and would be paid well—fabulously well in comparison with former standards in the area.

It was plain that changes were going to be made—some of them drastic and a little frightening. Thousands of workers had to be recruited in this wild, thinly populated area. This meant that people had to be rounded up over a wide area and brought many miles to work. Pedro's little village was 20 miles from the site of the projected plant. Commutation was no solution. The terrain was mountainous, with

---

[1] ROBERT COLLIER PAGE, M.D., "Medical Problems of the Oil Industry in the Tropics." An address to the Ninth International Congress of Industrial Medicine in London, England, September 13-17, 1948.

tangled, almost impenetrable valleys. It was a good two-day trip by burro. Therefore, the new plant had to become a town in itself, providing food, shelter, and clothing for its employees.

In other words, though the lure of good wages was irresistible, it was almost counterbalanced in the minds of Pedro and his compatriots by the fear of being wrenched suddenly into an entirely new kind of world, into an alien and unknown pattern of living.

Ground was leveled, barracks went up, and more and more workers moved in with their families. Water was piped in. A commissary was stocked with groceries and other necessaries. Pedro was living better than he ever had.

Before long it developed that the plant had to offer more than food, shelter, and clothing. Tropical diseases were prevalent. Not only were an alarming number of man-hours lost owing to these diseases, many employees were *never* in good health. They were lethargic, moody, forever ready to drop off to sleep or to stare at nothing in sullen silence.

It was therefore necessary to establish facilities for immunization. This meant importation of medical personnel as well as supplies, and as long as this was necessary, the company decided it might as well go the whole hog and build, man, and equip a full-scale *hospital*. This would serve the additional purpose of protecting the company from unwarranted accident claims.

The company considered itself thoroughly enlightened and benevolent in this matter. Why it was neither enlightened nor particularly benevolent will be discussed in a later chapter.

In any event, it worked beautifully at first. For the first time in his life Pedro knew what it was to feel truly *well*. Now, at 40, he says he feels younger than he did at 17.

The hospital has proved to have another function, that of recruiting agent. Pedro and his family, for example, are walking examples of the benefits of good health when they go to visit friends in their old village. So are the thousand-odd other workers and their families. Non-plant people who see them are attracted to the plant; hence the company has little trouble replacing workers who quit or are discharged.

To Pedro, medical care is a miracle wrought by the company. In his mind, it is the company's sole reason for existence. If, at some later date, some budget-conscious board of directors should decide to trim the number of services at Pedro's disposal, he will be puzzled and unhappy. Perhaps he will even quit.

It doesn't occur to him—and never will—that he might have some responsibility for maintaining his own health or paying for the services of a physician. The idea that it might be possible for him to perform the miracle of guarding his own health would strike him as insane, and, as for paying doctor's bills, a doctor was a high company official, magnificently salaried, who would surely be above taking money from one so lowly as Pedro.

THE SECRETARY

Miss Anne T., the 38-year-old stenographer, is like Pedro in looking on medicine as a miracle, but the miracle she has in mind takes place in the research laboratory, not in the doctor's office. The doctor is merely a middleman whose sole function is to push these miracles into one by means of pills or hypodermic needles. He is rather an exasperating middleman; he tends to be miserly with the miracles at his disposal, oddly insisting sometimes that the illness one knows one has is either psychosomatic or nonexistent and not responsive to medication.

Unlike Pedro, however, Miss T. does not expect to get something for nothing. It would not occur to her to seek free medical care; she is covered as fully as possible by Blue Cross and Blue Shield; she also carries a major medical expense policy written by a large private insurance company. She is no advocate of socialized medicine.

Miss T. is a graduate of a major women's college. After college she attended secretarial school. She began working for her present company 12 years ago, when it was just a shoe-string operation. Her first boss was the founder and, at that time, the only executive of the company. As the years went by and the company expanded—a little too rapidly for comfort—she became indispensable to her boss, for whom she had deep respect. Twice during these periods she had proposals of marriage, but at neither time did she feel she could desert her boss, who was undergoing crisis after crisis in the attempt to build up a staff of junior executives to whom he could delegate the mushrooming burden of responsibilities. Most of this time, it was not unusual for her to put in 12 to 14 hours a day. The boss appreciated her loyalty and paid her well, but she had no life to speak of, outside the office.

Two years ago her boss died. It was a sudden death following a cerebral accident. For a few hectic months, it looked as though his company might have to follow him since none of his junior executives were properly prepared to step into his shoes. Production was high; sales were high; but no one had a grasp of the operation as a whole. Some errors in judgment, such as buying certain raw materials when the price was unfavorable and underestimating the production cost of a new line of products, nearly caused the firm to be swamped by its leading competitor. However, a combination of luck, a timely bank loan, and the acquisition of some high-grade talent from rival firms enabled the ship to right itself.

Miss T. is now something of a fifth wheel. As private secretary to the late boss, she had antagonized most of the junior executives; she had felt it her duty to protect him from what seemed excessive demands and impositions on their part. Moreover, she had felt that their "incompetence" was wrecking his health. She still blames them for his death, and she had secretly hoped that the ensuing period of chaos would end in corporate disaster.

Each of the remaining top executives has his own girl Friday. An effort is made to keep Miss T. busy taking dictation when someone's private secretary is otherwise occupied, typing reports, retyping botched or annotated material, even filing. She does what she is asked because she has to, but she is not happy. She knows that she has been accused of "inefficiency" behind her back. She is not afraid of being fired, but she feels that it is a matter of months before she will be obliged to accept a pay cut. She has done a small amount of pavement-pounding, hoping to locate a job in happier surroundings, but she has had no luck. What weighs against her most, in addition to her age, is the fact that while she insists justly that she is potentially as efficient as ever in terms of work, she has an *inefficient attitude*, which is evident at a glance. She is somewhat slovenly in her appearance; her voice and the expression of her face are both flat and resigned. She reflects the shabbiness of her life and her future prospects.

Organically, nothing can be found wrong with her. Her habits are exemplary. She neither smokes nor drinks; she is almost neurotically fussy about proper diet and proper sleep. Nevertheless, she is a great trial to private practitioners for miles around; she goes from one to another until she can find one who, in an unguarded moment, will blurt out that there *might* be something organically wrong with her that can be cured by some shot, potion, or surgical operation. Her ambition is to find a doctor who will say that she has tuberculosis. She herself knows she has it, but all doctors are quacks, out to get your money without doing any work. She had a brief moment of happiness a year ago when a doctor suggested the possibility of pleurisy and gave her a prescription for it, but the prescription didn't work; so she went on to the next doctor, who insulted her by suggesting that

THE WHOLE MAN   19

her real illness might be unhappiness. Miss T.'s real name is legion. Her future is bleak.

THE UNION MAN

Walter McK., the 45-year-old union shop steward and punch-press operator, is unlike all the examples cited previously in that he is not only a skillful man who enjoys his work and his life in general, but also in that he knows what he wants and needs in health services. The only thing he doesn't know in this respect is exactly how these needed services can be practically organized; hence he is not yet in a position to express himself in terms of union demands. He and his colleagues are doing some heavy thinking and wrangling on the subject, and it is only a matter of a few years, or perhaps months, before concrete demands will be formulated—and, in the long run, acceded to.

This man, who was raised on a farm and whose farmer father is still alive, hale, and hearty at the age of 87, has an appreciation of what *optimum* health can mean and how much it can be influenced by environment. Before he was born, his father had run a blacksmith shop in a farm-market town with a population of roughly 20,000. The shop had made money, but his father had hated the work; he had inherited it from *his* father. The profits from the shop went into drink as fast as they came in, and McK., Sr., had been a common spectacle in the gutters of Gin Row along the railroad tracks or in the town police court. By some miracle, he had met a girl who was willing to marry him, provided he gave up boozing, sold his shop, and went into farming, where his real roots had always been.

McK.'s father had been a positive example of one aspect of health construction. His mother had been a negative example of one aspect of health maintenance. At age 50, she had died of cancer. There had been innumerable warning signals, but she had kept putting off going to the doctor until it was too late to do anything positive.

With these examples in mind, McK. has been something of a hair shirt to his foreman, who, nonetheless, respects him for his skill and common sense and is willing upon occasion to overlook it if McK. seems to be overstepping his authority as shop steward.

McK. is constantly taking him aside to say something like this:

"You can't put that kid on that job. That's slave stuff; he's too good for it. Let him mess around with some tools, and he'll pay for himself in no time."

Or

"You better pull old Johnson off that machine and make him go to the dispensary. Tell him you won't let him back in here without a medical okay. If we don't get him to a doctor while he can still walk, we're going to have to carry him."

Or

"Why don't you lay down the law to Hank there. What the hell good is he to anybody with a head as big as a watermelon. Tell him, the next time he comes in here with a hangover you'll send him home to sleep it off—without pay. Maybe that will straighten him out; if you wait till he starts having his lunch across the road at Joe's Gin Mill, it will be too late for anything to straighten him out."

McK.'s company offers an up-to-date dispensary, a liberal group plan covering certain off-the-job medical and hospital expenses, a fair pension plan, and good wages. Moreover, the company contributes heavily to the local facilities for medical care. McK. is not a chronic "sorehead." He appreciates what is being done, but he senses that it is not enough. He feels that there should be more attention paid to the health of the vertical man. He feels that something should be done to scout out incipient problems implicit in the environment or the individual physiology or personality before they become full-blown health problems. He does not feel, as Pedro takes for granted, that his company ought to go into the hospital business, beyond the maintaining of a plant dispensary for emergency on-the-job care. Neither does he believe that the full burden of off-the-job medical care, whether diagnostic or therapeutic, ought to be borne by the company. He is a proud man and pays his own bills.

He does believe, however, that his company has a vested interest in the prevention of illness on the part of its employees, that it could profitably bear the burden of health education, and that it is the local agent to take the leadership in its community in organizing to obtain facilities for the construction of health in the *vertical* man. As for health insurance, he would like something that would cover prevention and diagnosis (including routine periodic medical examination), as well as after-the-fact reimbursement for treatment and hospitalization.

McK.'s thinking is typical of responsible labor-union thinking today. The Pedro type of thinking still exists in some quarters of the United States, where demands are being made for full-scale medical treatment to be provided wholly by business, but individuals like McK. know that such an arrangement would only injure the goose that lays the golden eggs without providing the employees with the kind of service they most need.

When McK.'s ideas are fully clarified as to details, there is no doubt that they will be translated into demands, which will undoubtedly be met, in the long haul, by either industry or government.

THE SCIENTIST

Dr. Wharton L., the 50-year-old director of chemical research, is a man who, as long as he can maintain what is left of his health, need not worry about holding his job and drawing a better-than-average salary without being overconscientious about his job performance.

About 15 years ago he had to make a momentous choice. When he received his Ph.D., he was regarded as a brilliant "comer" by his professional colleagues and mentors. It was taken for granted that he would have a noteworthy, if not necessarily lucrative, career as a scientist. His talents seemed to indicate a bent for pure rather than applied science, and several major universities offered him posts which would provide him with the wherewithal to further his career.

His present company, a pharmaceutical concern, outbid all the others as far as salary was concerned. He was assured that he would not have to submerge his individuality and that he would be given a reasonably free hand once he had learned the corporate ropes in experimentation research. Of course, there would be limits. He would have to "fit into" the company, and his activities must be at least somewhat in keeping with the company's goal, which was the profitable manufacture and sale of pharmaceutical supplies. In other words, however basic and fundamental his research efforts might be and however far afield these might logically lead him, he must always keep his mind's eye on the pharmaceutical angle and the corner drug store.

This seemed fair enough at first. He was working under a brilliant chemical researcher whom he admired and who was independent as a hog on ice. Occasionally some lay vice-president in the administrative end of the business would criticize his superior for losing sight of the immediate profit angle in an important series of experiments or would order him to follow a procedure that any scientist could tell was unsound and would not have the result anticipated by the layman of management. At such times, Dr. L.'s superior would set his jaw stubbornly and demand that the matter be taken to the president—or even to the board of directors. He knew that the president, who was a firm believer in fundamental research for long-range results, would always back him up.

But when the president died of undetected cancer 15 years ago, the meddling vice-president stepped into his shoes, Dr. L.'s superior was retired (or fired, with part pay), and Dr. L. was called into the new president's office.

This individual laid it on the line:

"Listen," he said, "don't get me wrong. I'm not against science. I think it's wonderful. But this isn't an endowed university. We can't afford to support research for its own sake, much as we might like to. This is a business, and it has to run at a profit or not run at all. Every year we put hundreds of thousands of dollars into your department, and in order to justify itself your department has to turn out even more hundreds of thousands of dollars' worth of formulas for salable pharmaceutical supplies. If you have any time left over after that, you may spend it in any form of research that seems proper to you, but up until then you play ball with the sales end and the production end of this business. That goes even if they ask you to do things you think are scientifically silly. Our end of the business may not know much about science, but we know what will turn over and make money."

Dr. L. found that he really couldn't quarrel with this point of view. From the point of view of his employer it seemed sound, if somewhat short-sighted. He realized that he could do one of three things: (1) resign and seek a position at less pay at a university; (2) follow the footsteps of his former superior, carry the fight for his "scientific integrity" all the way to the board of directors, knowing that he would lose in the end; or (3) play ball, becoming a sort of "short-order cook" for sales, and trade in the esteem of his professional colleagues for an administrative title and an easy, well-paid job.

He made choice number three. It would have been the right choice for a great many average or better-than-average men in his chosen field. Whether it was the proper choice for one of his outstanding talents is debatable. Now that it is too late, he is tormented by doubt, complicated by a sense of having wasted his life.

His company has treated him well. Two years ago, it became known that he was suffering from a degenerative circulatory disease which would inevitably shorten his life. The course of the disease could be decelerated by means of proper treatment plus cooperation on his part, but could not be completely arrested, let alone reversed. When he was told of this, his first reaction was to think painfully of the little time left to him to accomplish all the things he had dreamed of accomplishing in his university days.

He began drinking heavily after work and talking about his company's responsibility for "wasting his life." Eventually his "wasted life" and his disease became one and the same thing in his mind, and he began broadcasting his conviction that his illness was brought about directly by his company.

He became difficult to work with owing to his continual state of bitterness. He insulted his junior colleagues, accused them of toadying and of scientific incompetence. Long-suffering management gave him a year's sabbatical to straighten himself out. This, plus some strong language on the part of his family doctor, has enabled him at least to curb his tendency to be "difficult," but he is not altogether tractable. He has a tendency to mutter to himself when upset, and he has not been known to smile or show gratitude for favors received. He is egocentric, and, though married and the father of two boys, he does not seem to be warmly liked by anyone. It is not likely that he will last out another five years in harness.

What Dr. L. thinks he wants out of life is health and time for pure scientific studies. It is possible he is deluding himself. Until the time of his health setback, he had made no attempt to devote himself to any scientific activities outside the nontaxing ones involved in his job, though he had occasionally expressed dissatisfaction with his way of life and his isolation from kindred spirits. There is no particular reason to believe that, given health and time *now*, he would conduct himself in any different manner. It is probable that his wants for the present and future are beside the point. He has "had" it. As for his needs, common sense and insight are no doubt the most important, but he resists these, and in all probability will continue to do so. The best that can be hoped is that his family and co-workers will find ways of bringing him as much peace of mind as possible—a thankless task —in the few years that are left to him.

As for the past, some of his former professional colleagues say that since he made his bed, he ought to lie in it, with no further ado on anyone's part. This is true as far as it goes, but a great many highly talented men—some wholly unsuited for the corporate life—are being induced every year to make a similar bed. There are probably a number of people capable of steering these individuals properly at the crucial time, but somehow they refrain from doing so. Dr. L. says that when he first contemplated joining his present company, some of his scientific friends tried to dissuade him, but his reply was: "Aw, you guys are just jealous. You wish you could make as much money in 20 years as I'll be making in one."

Perhaps if someone from business—or, better yet, someone like his former superior who had once made the same choice—had talked to him like a Dutch uncle, his story would have had a happier ending. Someone whose knowledge of the problem was truly impressive might have said to him:

"Sit down, Wharton, and try to analyze yourself. What do you really want out of life? To achieve something in the field of science, right? Well, this job you have been offered is an honorable one, and it will give you money, but it will literally *finish* you as a scientist. As an esteemed member of your profession you will go backward, not forward, because there is no place for your peculiar brand of talent in our corporate organization. Think carefully, and decide whether the additional money is worth the price you have to pay."

In this era of expanding technology, American business must draw more and more from science; hence, people like Dr. Wharton L. are more and more in demand. The competitive bidding for them grows hotter and hotter. The humanologist needs to keep cool in all this heat and determine whether and where these brilliant individualists can be fit into the corporate scheme with the least damage to themselves and the organization.

All Are Individuals

All of the above cases are those of actual persons who have, for one reason or another, come to the attention of some medical person associated with an industrial humanological team. All are individuals reacting to occupational environment. To try to interpret the health problems of any one of them as though that occupational environment did not exist would be fruitless. It would be just as fruitless to try to interpret their problems, their wants, and their needs from the standpoint of occupation alone, as though they had no existence whatever except during working hours. It is important that the man of management, Mr. B., has a neurotic wife who drinks too much and runs up extravagant bills. It is important that Pedro's first son is feeble-minded. It is important that Miss T., the stenographer, was molded by a perfectionist mother. It is important that McK., the shop steward, is the star member of his neighborhood barbershop quartet. It is important that Dr. L.'s roommate in college has never bothered to answer his letters or Christmas cards.

In other words, it is not enough merely to see the individual as a whole. One should see him also in terms of a whole environment.

Ideally, a whole man is master of all facets of his environment, or he can obtain reward enough from one or more facets to compensate for scant desserts from some other facet. The ideal man, for example, may have to put up with a disagreeable job just to make a living, but he may be able to make up for this by leading a particularly stimulating life away from work. A person with an inevitably difficult home life may find wholeness in a rich occupational existence.

Real life tends to fall short of the ideal. With most individuals, if one facet of life is unsatisfactory, the other facets not only will not compensate for it, but may in time be poisoned by it. The cowering Bob Cratchit in the accounting department may imbibe on the way home and take it out on his wife and children. The boss who never gets the last word at home may insist upon it at the office to the utter confusion of everyone concerned.

The Ideal Condition of Man Is One of Wholeness

Ill health results when social and physical environment fall short of adequacy. There are two causes for the pathological disorganization of the human individual: (1) genetic; (2) environmental.

Genetic causes account for inherent defects in physiological makeup. Such defects are relative to the individual's willingness or unwillingness to adjust to them, to come to terms with them. As T. M. Frank says: "We might call health that state of well-being in which disease and disability do not intrude to alter one's plans or mode of living. . . . Often the pathologist says it is not a wonder that the patient died, but amazing that he lived long enough to accumulate all these physical handicaps. And the friends testify that this individual went his way in perfect health, serenely unconscious of his deteriorating body, radiating cheerfulness. Who will say this one did not have health?"

Many conditions formerly considered to be hereditary in origin are now known to be largely or entirely the result of some sort of social inadequacy, e.g., insufficiency of necessary food constituents.

Certain truly hereditary conditions (transmitted as dominant or recessive characteristics, as envisioned in Mendel's Law) may be exacerbated by a faulty social environment. Schizophrenia, for example, would not cause overt ill health in an environment which was physically and psychologically satisfactory.

It is important for the humanologist to understand the significance of genetic flaws in relation to environment. A goal of humanology is to manipulate the environmental element to such an extent that the genetic element is not exaggerated beyond its true significance. It is possible to prevent or ameliorate potential problems due to genetic flaws.

Environmental pathological disorganization accounts for the vast bulk of ill health. Such disorganization is considered to be the direct result of preventable social inadequacy.

"If all the preventable diseases were prevented," says E. H. Cluver, "then only a few hospital beds would be necessary for the 'genetic' cases which include atrophy due to senility . . . . The large proportion of the buildings in our communities which are at present set aside as hospitals are costly evidence of social failure. They accommodate very largely those members who have become disorganized by forces that could be eliminated in a society which implemented the preventive knowledge at its disposal."

This is the charge of the humanology team.

Man's Gregarious Habits

Man's ill health is closely linked up with his gregarious habit, his urge to congregate with his fellows in tightly packed communities. As a consequence, housing conditions are reflected in the health of civilized man.

Eating habits, which frequently have a social base, can exert a profound influence on man's health or lack of health. In this country, the common tendency, backed by social usage and tradition, is to ingest more calories than are needed for normal energy requirements. Owing to automation, most jobs today are either sedentary or semisedentary; yet the average diet is geared to the needs of the farmer or ditch-digger. A detailed report on a periodic examination of 8,000 workers showed that obesity (20 or more pounds overweight) was the chief variation from acceptable health standards. Undernourishment is an even more serious problem in some areas, notably in tropical zones. Death is not necessarily the immediate result of malnutrition, but life is savorless and easy work almost unbearably onerous to the ill-fed human being.

The Importance of Morale

Morale—the will to live productively and successfully—is a vitally important factor not only in health, but in the ability to perform. Everyone is capable of doing more than he is doing, and of doing it better, provided he chooses his goals and activities wisely and does not attempt to overstep his inborn limitations. Unfortunately, the individual who realizes his full physical and mental potentials is the exception rather than the rule.

Man's Potential—and His Performance

J. W. Still believes that not more than 5 per cent of the population follows the upper curves which indicate both physical and psychic potentials. The rest become physically flabby before they are out of their 30's and make no effort to furnish their minds with anything more substantial than the contents of one or two daily newspapers, three or four slick periodicals, and a light paper-back novel. It has been

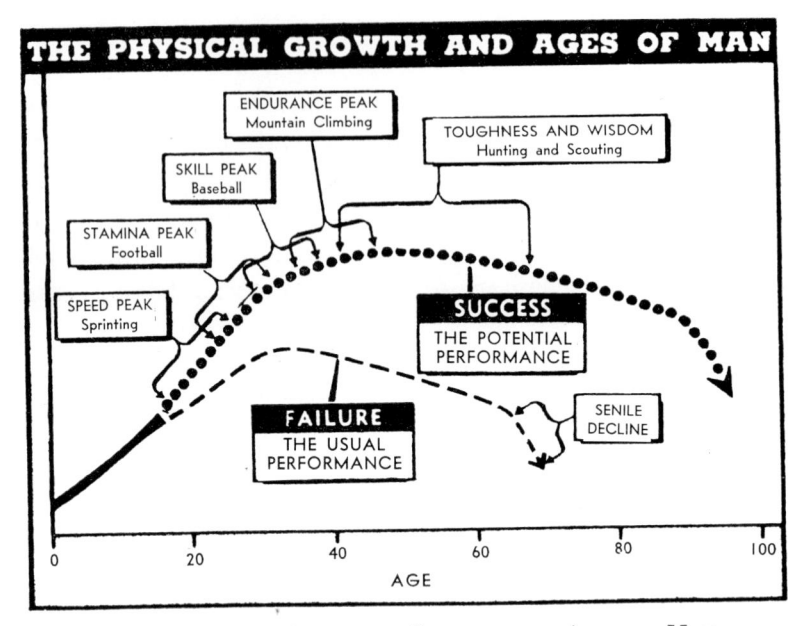

28   OCCUPATIONAL HEALTH AND MANTALENT DEVELOPMENT

Graph I — The Physical Growth and Ages of Man

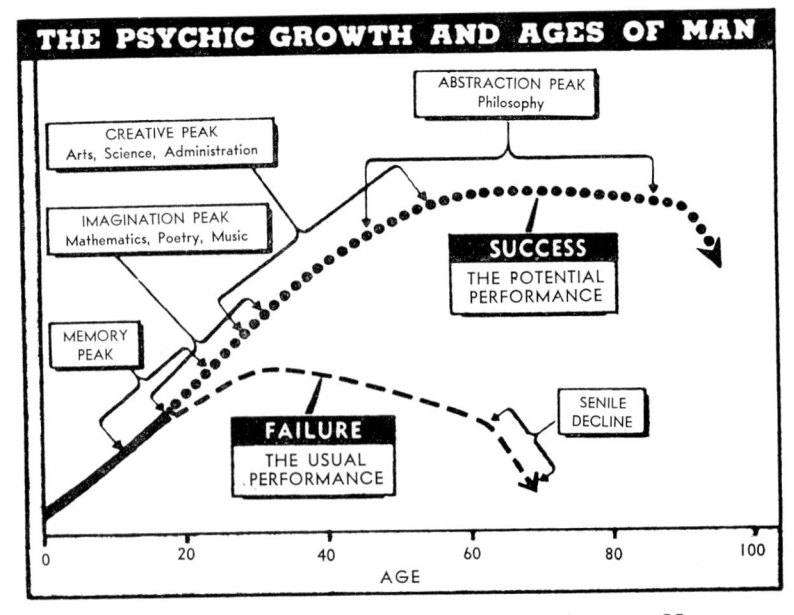

Graph II — The Psychic Growth and Ages of Man

said that the beginning of senility is marked by the death of curiosity. People who lose a burning desire to learn in their 30's become old before their time. It has not been proved that morale actually increases longevity, but there is reason to believe that utter lack of it can decrease longevity. Robert Louis Stevenson noted this fact in a tour of the South Sea islands in the latter half of the nineteenth century. New ways of life, imposed by white men, had taken away the will to live of many natives, and entire populations of islands were dying out from no determinable cause except sheer melancholia.

The Influence of Environment on Morale

Future professional members of the humanology team will investigate and measure the influence of environment on morale and devise ways to motivate men and women to realize their true potential. They will find it necessary to become imbued with the spirit of the beloved family doctor. The medical members of the team will be practitioners of constructive medicine. They will be trained to consider man as a whole human being; they will understand man's physiology and psychology, and they will know that changes in man have a way of affecting his whole environment, and vice versa. The continuing vitality of professional education will depend upon the ability of educators to understand and transmit these concepts.

The Importance of Early Detection

The primary purpose of each member of the humanology team must be to anticipate and prevent human problems, particularly those pertaining to health, by detecting potential illnesses and maladjustments early enough to do something positive, constructive, and curative, at a cost in keeping with the individual's ability to pay, and at the same time to assist all individuals in maintaining a normal curve of health, adjustment, and productivity.

The Alternative

The alternative is to wait until damage has been done, until health, adjustment, and productivity are marred, perhaps beyond repair, until amelioration of the problem (to say nothing of cure) is likely to be beyond the financial resources of the individual concerned. Such an approach can only complicate existing humanological problems, not alleviate them; ultimately it will invite federal aid and subsidy as the burden becomes too great for free enterprise.

A most learned educator, Dr. James Conant of Harvard, summed up the problem and illuminated the potential social repercussions when he said: "To train enough professionals to keep all the people well by curing disease would involve the expenditure of so large a fraction of the income of the nation as to make the cure less attractive to the taxpayers than the disease."

What the Public Hears

The miracles of modern medicine are highly publicized. The press is full of accounts of such occurrences as operations on living hearts, transplantations of vital organs, skillful manipulation of the delicate material of the brain. New wonder drugs find space on the front page with clocklike regularity. Banks of human replacement parts are reported with awe by feature writers for slick mass-circulation magazines.

The True Story

These dramatic events are indeed highlights in the progress of modern science. But the really big story that doesn't make headlines is the story of the changing concept of medicine which is behind the dramatic highlights. Gone forever is the stage in medicine when a complete health center was carried in the little black bag of the general practitioner. The concept of humanology now takes in the whole man in his environment, not merely his medically measurable aches and pains. The five case histories cited earlier in the chapter prove that disease is not limited to an anatomical part. It affects the entire organism. A disease which may be located in a given tissue actually affects not only the function of the entire organism, but also may set in motion a subtle progression of reverberations—psychologic, sociologic, and economic.

The Concern of All

Accordingly, though the responsibility of maintaining good health is the individual's personal responsibility (a fact which cannot be overemphasized), the health of the whole man concerns all who are interested in human welfare.

Episodic vs. Constructive Medicine

This understanding has enormous consequences. The average physician, for example, still practices episodic medicine. For lack of facilities, he is unable to relate the isolated disease to the problems of the whole body, and for lack of time he is unable to evaluate the patient's entire personality, which includes his physical, mental, and social state. The result of this has been an unfortunate professional tendency to overemphasize the importance of curative care in contradistinction to preventive and *constructive* care. But, this concept is changing. Constructive medicine is coming into its own.

Inadequacy of the Curative Aspects of Health

In the meanwhile, the curative concept of health has had ample opportunity to create its own peculiar Frankenstein monster—the hospital horizontal-care system. Laymen, oversold on cure, have helped make hospital care the focal point of all medical service. The hospital is regarded as the epitome of medical service, though, in many instances, it is actually the most costly and poorly organized of all branches of the health service.

As a result, the average American seeks his answer to health need in available hospital and other insurance plans and perhaps waits for government or his community or labor union to provide the broader program which he senses should be available.

Unfortunately, the insurance and benefit plans in which both management and labor tend to put their faith will never completely fill the bill. They do not hit at the heart of the matter. Illness alone costs workers 400 to 500 million working days a year. About 95 per cent of these absences are off-the-job illnesses that are mostly preventable in nature.

The Hospital Horizontal-Care System

The hospital system designed to combat this state of affairs is costly and extremely difficult to manage. About 69 per cent of all the hospitals for acutely ill patients are of less than 100-bed capacity, which, authorities agree, make it impossible for them to be operated economically. Expansion by additions to existing units ordinarily does little more than compound the obsolescence. A survey of any metropolitan area will discover many hospitals in structures obsolete and inadequate to meet the needs of present-day medicine.

The Trustee and His Professional Advisors

Some 50,000 laymen in the United States, most of them businessmen, act as trustees for voluntary nonprofit hospitals. These legally and morally responsible individuals tend to rely entirely on the professional people to run the hospital.

Professional people are rarely trained or skilled in economics or practical management efficiency.

Horizontal-Care Insurance Plans

In well-managed hospitals, horizontal-care insurance plans still tend to pile burden upon burden on the shoulders of the consumer of medical services. Such plans, with their emphasis on cure, are in reality not "health plans" at all. They represent sickness-cost insurance to prepay the cost of definitive medical care. They are not constructive in that they offer nothing which would tend to keep well people well and, hence, out of hospital beds. In fact, they tend to put more and more people *in* horizontal position for elective medical and surgical care. Simple diagnostic or therapeutic procedures which could be managed easily in the doctor's office are carried out in the hospital if the patient happens to be covered by hospitalization. As a result, horizontal bed-care costs will continue to climb and climb. The average man, who finds that it costs him more and more to keep "covered," naturally wonders if, maybe, there might not be something after all in this so-called "socialized" medicine.

Constructive Health Maintenance[2]

The value of a positive health program becomes even more evident in the light of the foregoing facts. It is basically good business to keep all employed persons healthy and therefore happy and productive. This is what is really needed in place of the mushrooming corrective and curative plans which have in the past been *sold* to labor and management. In the future, *trained* professionals will be asked to sit in at executive sessions, and proper weight will be given to their

---

[2] Robert Collier Page, M.D., "Industry, Potential Contribution to the Improvement of World Health." An address at the National Conference on World Health Organization in Washington, D.C., April 7, 1953.

expert opinions. As a result, management and labor both will begin to learn how to get increasingly better results for less and less money—results which will be in the long-range interest of the employed person and his family, and of the economy as a whole.

REFERENCES

ASHE, WILLIAM F., M.D. "Education for Industrial (Occupational) Medical Practice." *Journal of Occupational Medicine*, II (No. 7, July, 1960), p. 305.

"Automation: Feedback to a Better Economy." *Chemical and Engineering News*, XXXIII (October 31, 1955), pp. 4648-4650.

BATESON, WILLIAM, PH.D. *Mendel's Principles of Heredity.* Cambridge (England): Cambridge University Press, 1909.

BULGER, T. J., M.D., and FLEMING, A. J., M.D. "Cancer Detection Activities in Industry." (11 year follow up report) *Journal of Occupational Medicine*, (April, 1959), p. 211.

CLUVER, EUSTACE HENRY, M.D., D.P.H. "The Causes of Human Disorganization." *Social Medicine.* Johannesburg (South Africa): Central News Agency, 1951, pp. 1-30.

CONANT, JAMES BRYANT. "Foreword." Included in James S. Simmons' *Public Health in the World Today.* Cambridge (Massachusetts): Harvard University Press, 1949.

DASTUR, H. P. "The Concept of Integrated Medicine." *Industrial Medicine and Surgery*, XXVI (No. 7, July, 1957), p. 336.

FELTON, JEAN SPENCER, M.D. "The Teaching of Occupational Medicine in the United States and Canada 1960." *Journal of Occupational Medicine*, II (No. 7, July, 1960), p. 315.

MCGEE, LEMUEL CLYDE, M.D. *Manual of Industrial Medicine.* 3rd edition. Philadelphia: University of Pennsylvania Press, 1956.

Medical Services in Industry — A Symposium. (KAMMER, A. G., M.D., "Viewpoint of the Profession"; MONK, R. M., "Viewpoint of Management"; MALLERY, O. T., JR., M.D., "The Industrial Physician — Past, Present and Future"; DRINKER, PHILIP, "Industrial Nurse, Past, Present and Future"; BROWN, MARY L., R.N., "The Industrial Nurse, Past, Present and Future"; PAGE, ROBERT COLLIER, M.D., "Report on a Survey of Medical Services in Industry.") Chicago: The American Foundation of Occupational Health (The Occupational Health Institute), (No. 3, September, 1952) : *Industrial Medicine and Surgery,* XXI (June, 1952), pp. 282-296.

NORRIS, EDGAR HUGHES, M.D. *Medicine Rededicated; Health by Community Enterprise.* Detroit: The American Federation of Medical Centers, 1951, Chapter XII, pp. 210-216.

PAGE, ROBERT COLLIER, M.D. "Present Day Medical Needs of Industry." *Surgical and Medical News.* Bombay, (India), (July, 1955).

——————. "The General Practitioner in an Industrial Era." *Ontario Medical Review,* XXIV (No. 6, June, 1957), p. 415.

SCHOENLEBER, ALVIN WILLIS, M.D. *Doctors in Oil; A Story of Medical Pioneers in Oil Fields Around the World.* New York: Standard Oil Company (New Jersey), 1950.

STEVENSON, ROBERT LOUIS. *In the South Seas.* New York; C. Scribner's Sons, 1896.

STILL, JOSEPH W., M.D., M.P.H. "Man's Potential and His Performance." *New York Times Magazine,* (November 24, 1957), p. 37.

WALMER, C. RICHARD, M.D. "Medical Aspects of Automation — Influence on Industrial Medical Program." *Industrial Medicine and Surgery,* XXV (May, 1956), pp. 216-220.

WILLIAMS, C. "Trends in Industrial Research." *Battelle Technical Review,* (September, 1955).

2

Evolution of Occupational Medicine

OCCUPATIONAL MEDICINE is not new. In the fifth century B.C. Hippocrates studied and treated occupational lead poisoning. Ramazzini, in the late seventeenth and early eighteenth centuries, advised his fellow physicians to ask their patients questions regarding the nature of their occupations. His investigations into the nature of occupational diseases earned him the title of "Father of Occupational Medicine."

Following in Ramazzini's footsteps, a number of medical scholars began studying the effects of occupation on health. The first American work in this field, entitled *On the Influence of Trades, Professions and Occupations in the Production of Disease,* appeared in 1837. It would never have occurred to anyone in this burgeoning era of industry that there could possibly be any place for a man of medicine in a house of business; so the bulk of information for the literature of occupational medicine had to be collected after symptoms were full blown, and often at a considerable distance from the actual working environment. The ulcers or the hypertension of the harassed front-office go-getter certainly did *not* get into the literature.

Expediency Creates a Need

In those days, a workman who became measurably ill as a direct and recognized consequence of his occupation was considered simply out of luck. Shortly after the Civil War, with the tremendous expansion of railroading indicative of a growing nation, the railroad companies began employing doctors, partly to do something about the alarming number of railroad injuries often due to wrecks, and partly to make sure the men selected to operate the trains would not collapse from some unsuspected illness a thousand miles from home base. In 1884, the Pennsylvania Steel Company hired the first plant surgeon. In the same year, in Germany, Bismarck introduced the first industrial insurance act, under which clearcut occupational illnesses or accidents became compensable. In 1895, the Vermont Marble Company, at Proctor, Vermont, employed the first industrial nurse. In 1910, the first American Congress on Industrial Diseases was held, and the State of New York adopted a workmen's compensation law. In 1911, ten more states followed suit, and New York and California made it compulsory to report occupational disease. In 1914, the United States Public Health Service established an office on industrial hygiene and sanitation, and a section covering industrial hygiene was organized in the American Public Health Association. Two years later, the American Association of Industrial Physicians and Surgeons was founded.

Progress Was Slow and Painful

By the time of World War I, practically no one would have understood the now generally accepted definition of occupational medicine: ". . . that branch of medicine which deals with the relationship of man to his occupation for the purpose of the prevention of disease and injury and the promotion of optimal health, productivity, and social adjustment." The above definition was adopted by a group of industrialists, medical educators, and physicians in industry during a three-day symposium late in 1955 at Pittsburgh, Pennsylvania, sponsored jointly by the National Fund for Medical Education and the American Medical Association's Council on Medical Education and Hospitals and Industrial Health.

Industrial medical service prior to 1920 consisted of first aid and caring for injuries resulting from trauma. The full-time company doctor was looked upon as someone who did not have the intelligence or ability to build up a private practice. All too often he was simply expected to be available when and if illness did occur; his primary function was considered to be that of a "finger-wrapper." Other local medical men had a tendency to regard the "contract" physician as an interloper undercutting their own practices; the medical profession as a whole was not entirely enthusiastic about the development of occupational medicine. On the other hand, it is not wholly accurate to say that management built occupational medicine in the teeth of the resistance of the medical profession. Old-line owner-managers of small plants tended to be slow to accept the concept that companies had responsibility for the health and welfare of the employee.

The Emerging Role of Management[1]

By and large, it is safe to say that management has yielded to some pressure, initiated some plans on its own, made some false moves, made some wise ones, listened to some good advice, ignored some good advice, acted on some bad advice, and, in general, moved ahead in a muddling but sure fashion.

Notable Contributions by Foresighted Medical Practitioners

As early as 1908, Dr. Harry Mock, a private-practicing Chicago surgeon and consultant to local industry, developed the all-important pre-employment physical examination.

---
[1] WILLIAM M. SEYMOUR, "The Problems of Industrial Health as Seen by the Industrialist." An address at Conference on the Education of Physicians for Industry, Pittsburgh, Pennsylvania, December 6, 1955.

A primary consideration in the dawning days of occupational medicine was the handling and avoidance of claims arising from compensable on-the-job accidents or occupational disease. After the turn of the century, with the development of the assembly line and the new incentive for rapid-volume production, the lethal aspects of manufacturing and mining machinery became increasingly pronounced. Management realized that something had to be done in the interests of legal invulnerability if not of simple humanity.

The Birth of the National Safety Council

The first step was to set up facilities to render emergency care when accidents happened. This was the era of the previously mentioned "finger-wrapper." But after-the-fact treatment did not prevent accidents from occurring, nor did it stem the rapid increase in incidence of industrial accidents. Backed by public opinion, more and more states enacted stringent compensation laws which made it increasingly costly for the stockholder when accidents happened, whether or not a competent "finger-wrapper" was immediately at the scene to patch up the accident victim.

The next obvious step was to put into effect measures which would prevent accidents or, at any rate, make them less likely to occur. The National Safety Council spearheaded the campaign for industrial safety, and corporation heads in leading industries became ready converts to the cause of cutting down industrial hazards. Engineers were trained in methods of renovating installations from the point of view of worker safety. Techniques were standardized, precautionary methods developed, and workers trained in the limitations of the equipment to be used.

Results of this organized attack were soon apparent. Accident curves were halted in their climb—then started to decline. The trend for the past 40 years has been consistently downward. In coal mining, for example, approximately five miners were killed in work accidents for every million tons of coal mined in 1913. In 1950, the average was down to slightly over one major injury for each million tons.

In railroading, the lowering of accident rates was even more pronounced. In 1913, one worker in every five hundred was injured on the job. In 1950, the ratio was down to one in three thousand.

Similarly, occupational diseases were being curbed by industry-wide action under the pressure of governmental action and law. Phosphorus necrosis, for example, had been a hazard of the match industry since the first match factories in Germany and Austria were built around 1833. It took the Hughes-Esch Act of 1912, which imposed a prohibitive tax on poisonous phosphorous matches manufactured in the United States, to banish the dreaded "phossy jaw" as a hazard to American workmen in the industry.

The Student Becomes Informed

Along with general medical progress, industrial problems began to receive serious attention from a growing number of physicians of real stature. Osler's book *Principles and Practice of Medicine,* published late in the nineteenth century, included chapters on lead poisoning and other industrial hazards. Through the early years of the twentieth century, Kober gave great impetus to study and activity in the field of industrial hygiene. Hayhurst added immeasurably to the literature on industrial poisons. The studies of other physicians—Hamilton on lead, Crowder on ventilation, for example—and the positive contributions of government medical men, such as Schereschewsky of the United States Public Health Service and Patterson of the Pennsylvania Department of Industry and Labor, all helped bring about a general offensive against industrial hazards and diseases. The interest in occupational health began to grow.

World War I and Manpower

This offensive against industrial hazards and diseases was accelerated with the coming of World War I. This war was the first all-out war calling for all the manpower resources of the nation. It was no longer possible to rely on a standing army and navy plus a handful of gunsmiths and ball-and-powder manufacturers. Practically every man not in uniform was needed in some industrial operation essential to the war effort. Thus the health of the individual American on the job became a critical factor almost overnight.

There was an upsurge of social and scientific interest in industrial medicine, hygiene, and safety. The war ended, but the interest in conservation of health and promotion of productivity continued, with the result that the incidence of occupational disease declined as dramatically as did the incidence of industrial accidents.

This decline, plus postwar relaxation of the production rush, made for moments of idleness on the part of the company doctor who had been hired full time to concern himself wholly with occupational illness. Companies began to dispense with full-time doctors in the 1920's, not knowing how to obtain practical use from them during the hours they were not treating specific occupational illnesses and accidents. The part-time company doctor became the rule during this period. He operated at certain specified hours in company facilities, often on a fee-for-service basis, augmenting his income sometimes by working for a number of companies, sometimes by setting up a private practice.

Many of the companies who were getting rid of their full-time doctors found themselves with elaborate medical facilities on their hands. These were obviously not being utilized to capacity. In the case of the railroads, as with certain Texas oil operations, it was found to be expedient to let the industrial medical plant be merged with, or absorbed by, facilities in the community at large.

The battle to date had been essentially a negative one, waged *against* certain specific industrial accident and illness problems, but not necessarily *for* the broad goals of positive health and productivity.

In other words, Johnny the sandhog, by the 1930's, was considerably less likely than his 1890 prototype to be crushed in a cave-in or die of silicosis, but he could still be saddled with a number of preventable off-the-job health problems which could have a markedly deleterious effect on his work performance; yet nobody much cared about this, outside of Johnny himself, except as a disciplinary problem when poor job performance became too marked.

A Positive Program in the Offing

Progress—positive progress—was in the offing. Here and there an astute manager or wise company doctor or nurse noted and called attention to the desirability of preventing *off*-the-job health problems or reassessing problems (such as ulcers) formerly considered to be totally unrelated to the job. Gradually there was dawning a realization of a clear-cut distinction between the related fields of *industrial hygiene* and *occupational medicine*.

The medical profession played a vital role in the discovery, classification, and evaluation of the industrial health hazards in metals, solvents, radiation, and toxic organic materials; and measures for preventing or dealing with crippling epidemics of contagious illness are also largely owing to the medical profession. Moreover, the profession continues to play a major role in the diagnosis and determination of the occupational genesis of specific health problems, as well as in treatment when they occur. However, once the proper preventive measures against these hazards have been established, the matter of putting them into execution is an engineering problem, not a medical one. Thus, as the "finger-wrapper" was replaced by the plant nurse, so was the medical specialist in specifically occupational health hazards replaced by the industrial hygiene engineer.

The role of the industrial hygiene engineer as part of the humanology team will be discussed in detail in Chapter 7. Suffice it to say here that, while he must of necessity work in close cooperation with the medical profession, he is not of it. Quite properly, he usually has an educational background of engineering or chemistry and holds a Ph.D. or B.S. rather than an M.D. degree.

Industrial Hygiene

The development of industrial hygiene engineering was eventually to pave the way for the modern concept of occupational health which frees the medical consultant from the negative preoccupation with multifarious specific health hazards and permits him to plan and advise positively for the health and productivity of the total person. However, it had a different effect at first in that it resulted in a temporary "technological unemployment" for many doctors in industry.

Occupational Health Abroad

While this was happening in the United States and in other industrialized countries, something altogether different was happening in remote areas, such as the tropics, where industry had to go to satisfy its growing hunger for raw materials: oil, coal, tin, rubber, iron, copper, and many others. In the case history of Pedro (Chapter 1), it was seen how considerations of communicable diseases among groups of workers herded together in company barracks, plus concern over claims arising from industrial accidents, were catalyzing agents in the initial establishment of on-the-scene medical facilities.

There were other considerations, too. The engineers, geologists, surveyors, chemists, drillers, and managers who came to these remote areas were accustomed to a quality of food, lodging, and medical facilities that was not available there unless supplied by the company. If the operation was a big one that was not doomed to be abandoned after hasty exhaustion of a limited supply of natural resources, these civilized outsiders planned to bring their wives and families to the scene. This was understood and approved by the company, but it entailed the establishment of modern homes, hospitals, theaters, schools, and water-purification systems.

Obviously, if the wives and families of these men could be induced to stay at the scene of operations, the husbands would be able to concentrate more effectively on the work at hand without pining and worrying about home. At first these up-to-date improvements were mainly for the foreigner, although medical facilities and personnel brought in from outside were available to Pedro and his compatriots insofar as contagious disease and occupational disability were concerned.

Before long, however, the imported doctor, his nurses, and his elaborate company-financed facilities were not doing much in the way of emergency treatment. The care of industrial accidents had decreased in certain instances to a vanishing point. Parallel with this, the contagious and environmental diseases of the area (malaria, typhoid, amebiasis, etc.) were being rapidly wiped out.

In other words, if doctors were not literally "sitting on their hands," they certainly were not keeping busy fulfilling their primary functions of controlling disease and offering emergency medical care. Back home, their counterparts were either going into private practice or dividing up their practice among a number of companies. Here, of course, that was out of the question. There was only one company and no field for a paying private practice.

Besides, management liked the idea of having complete medical facilities available. It created a sense of security in that it removed at least one ground for complaint. Hence a medical program was formulated, vaguely, in an unplanned way; often the principles of the program were not even recorded on paper. Usually the medical policy, such as it was, reflected certain ideas of some of the people in management. One official, perhaps, looked upon complete medical care of the individual as a gracious gesture which the company could well afford. Another official, perhaps higher in rank, might decide to go him one better and advocate complete care not only for the employee, but for his family as well. In any event, there is no denying that the company's intentions were honorable. Perhaps it might have been possible for the companies to have encouraged and subsidized community facilities and the indoctrination of local medical and paramedical personnel to operate them. As at home, it was the exception for the management representative in charge to seek competent professional advice and guidance at recognized and accepted university teaching level in the area. Had this been done, medical facilities built in an unplanned way might well have become at least partially self-supporting in time as payrolls grew and local inhabitants became accustomed to paying for their own medical care. In one area, this method was actually inaugurated; then management, on acquiring the operation, put in a "cradle-to-grave" policy that had been their established practice elsewhere.

In general, however, such a constructive solution probably would not have been put into effect even if someone had thought of it at the outset. For one thing, the imported experts and managers and their families wanted the kind of medical care, food, housing, and entertainment they were accustomed to at home, and while the husbands might have to learn the local language in order to communicate with the local employees, the wives often wanted no part of such an arrangement. This was one reason why it was expedient at the beginning for the company to import all of its services from home. Another reason was that the finding and training

46   OCCUPATIONAL HEALTH AND MANTALENT DEVELOPMENT

of local medical personnel, plus the motivating of the national population to utilize (voluntarily) health services, was obviously a slow, complicated, and delicate business, however well it might pay off in the long run. It was simpler to cut the Gordian knot, so to speak, and to bring in the whole works from outside.

The accompanying graphs show what happened in one industry—the oil industry.[2]

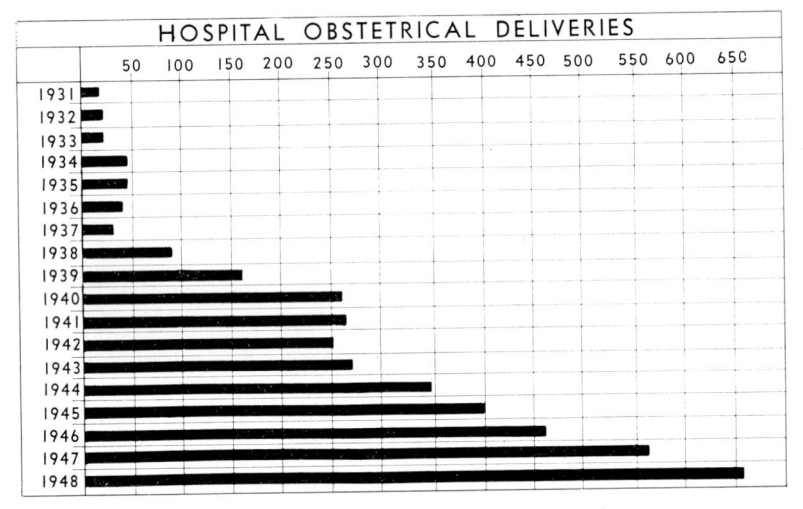

Graph III — OBSTETRICAL DELIVERIES—HOSPITAL OPERATED BY AN AMERICAN CORPORATION OPERATING ABROAD

By 1949, the Standard Oil Company (New Jersey) found itself medically responsible in tropical America for 36,751 people in Peru, 18,783 in Colombia, 73,705 in Venezuela, and 21,298 in Aruba. Of these, the number of actual employees was as follows: Peru, 4,905; Colombia, 7,138; Venezuela, 17,598; and Aruba, 8,105.

Wherever there were dispensaries, patients were appearing in ever-increasing numbers. The company hospitals were bulg-

---

[2] ROBERT COLLIER PAGE, "Review of Medical Problems of Medical Care in the United States." An address presented at Seminar on Industrial Medicine sponsored by Harvard School of Public Health, April 3 and 4, 1953.

EVOLUTION OF OCCUPATIONAL MEDICINE 47

ing with invalids, and the majority of these were not employees but their dependents—wives, children, aunts, uncles, grandparents, concubines. The beds formerly filled with ma-

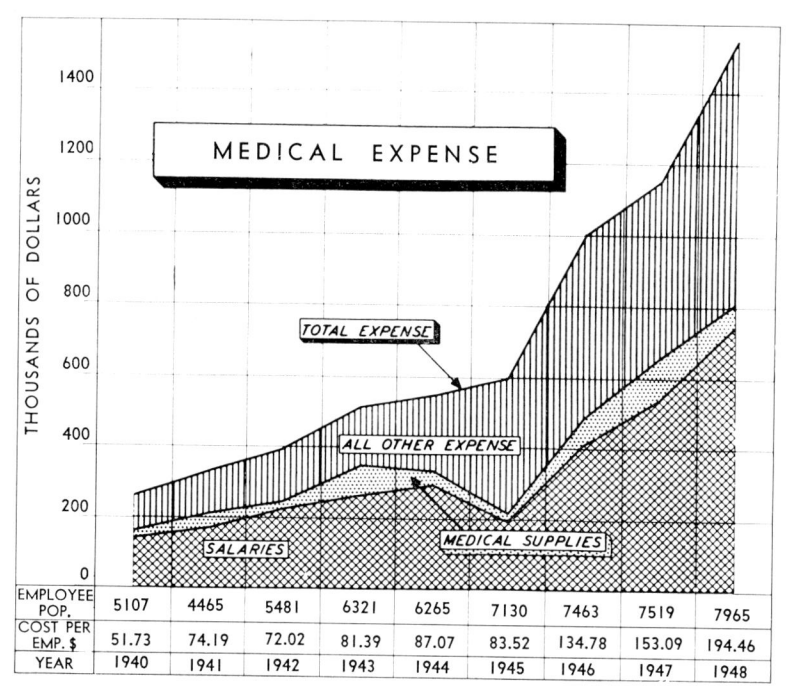

	1940	1941	1942	1943	1944	1945	1946	1947	1948
EMPLOYEE POP.	5107	4465	5481	6321	6265	7130	7463	7519	7965
COST PER EMP. $	51.73	74.19	72.02	81.39	87.07	83.52	134.78	153.09	194.46

Graph IV — MEDICAL EXPENSE

laria victims were being kept warm by nationals in childbirth, women who, in many cases, had borne several children previously at home but who were attracted to the hospital by the prospect of clean sheets and good food.

This sort of thing, once started, was almost impossible to stop. As was seen in the case of Pedro in the previous chapter, these heretofore unknown medical services came to be taken for granted. They were not accepted as a gracious gesture, but as a requisite working condition, such as, for example, proper lighting and time out for lunch.

MEDICAL EXPENSE

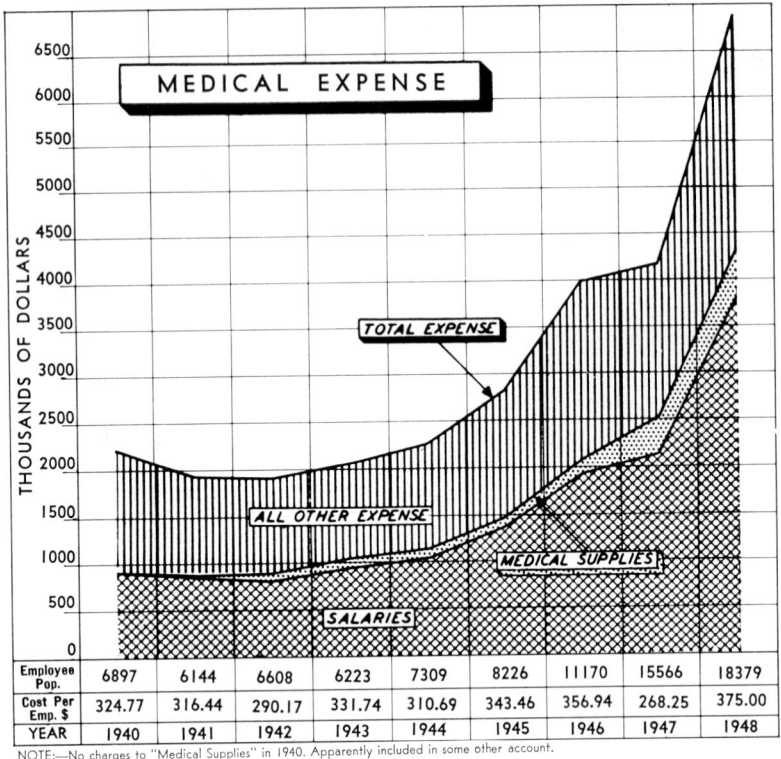

	1940	1941	1942	1943	1944	1945	1946	1947	1948
Employee Pop.	6897	6144	6608	6223	7309	8226	11170	15566	18379
Cost Per Emp. $	324.77	316.44	290.17	331.74	310.69	343.46	356.94	268.25	375.00
YEAR	1940	1941	1942	1943	1944	1945	1946	1947	1948

NOTE:—No charges to "Medical Supplies" in 1940. Apparently included in some other account.

Graph V — MEDICAL EXPENSE

SOUTH AMERICAN AFFILIATES
TOTAL MEDICAL EXPENSE

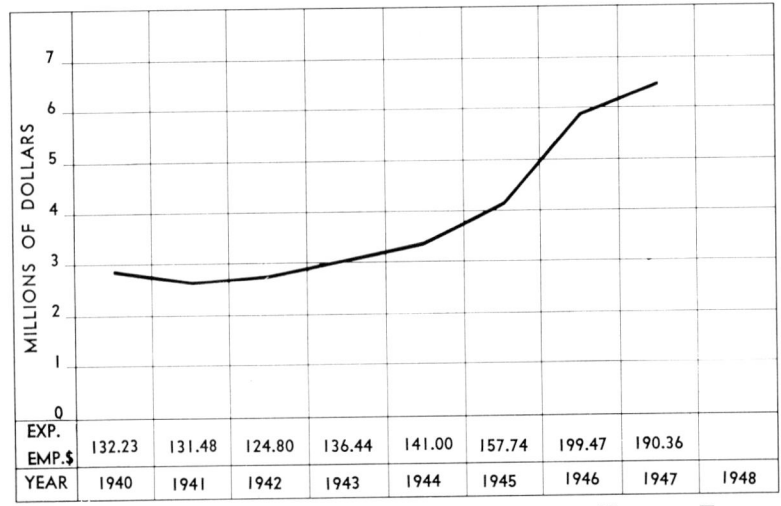

	1940	1941	1942	1943	1944	1945	1946	1947	1948
EXP. EMP.$	132.23	131.48	124.80	136.44	141.00	157.74	199.47	190.36	
YEAR	1940	1941	1942	1943	1944	1945	1946	1947	1948

Graph VI — SOUTH AMERICAN AFFILIATES — TOTAL MEDICAL EXPENSE

Collective Bargaining and Health

In fact, medical service became a keystone of collective bargaining. The powerful and well-established labor organizations in the United States would no doubt boggle at the following demands put forth in August, 1952, by a Peruvian labor group and considered by competent medical authorities not in the least astonishing or outrageous:

CLAUSE XXII—MEDICAL AND HOSPITAL ATTENTION

a) The Company will provide, free of charge, hospital service, medical attention, and medicines to: the employee, his wife or concubine, his legitimate, illegitimate, or adopted children, his sons and daughters-in-law, his parents, his grandparents, his tutors, brothers, and uncles in first degree.

c) The Company will keep at its service the national physicians who, during the time they have carried out their professional activities, have shown efficiency, dignity, and respect towards the employee and his family.

e) The Company physicians will provide daily medical attention to the employees and their families in Negritos, Lagunitos, Verdun Alto, and other sections. These physicians will have specialized personnel at their service, the necessary care being taken so that patients are attended by persons of the same sex.

f) The Company will provide round-trip transportation to the employees and their families when they have to go to the hospital or the polyclinic. This transportation will be provided either at the request of the physicians or because an employee or a member of his family is feeling sick.

g) With reference to the natural or professional diseases that afflict the employees, and the incapacity arising as a consequence of industrial work, the Company physicians will be obliged to report to the syndicate within a reasonable time.

i) The Company will intensify the campaign against tuberculosis and social diseases. To this effect, the Company will use the new and recently discovered drugs against tuberculosis.

j) When the syndicates denounce irregularities in the hospital or medical services, either in hospitals, dispensaries, or pharmacies, the Company will immediately take the necessary steps to stop these irregularities, and should it be established that the cause of these irregularities was due to a member of the professional staff or a member of the auxiliary personnel, they will dismiss him immediately.

k) When, for any given circumstances, the Company physicians are unable to obtain the recovery of an employee's health, or of his family members, the patients will be able to secure the services of a private physician, and in these cases the Company will be responsible for the payment of the private physician's fees as well as for the medicines which he should prescribe and any other expenses involved in the treatment. The employee and his family members will have a right to these facilities after a reasonable period of time.

r) The Company physicians cannot force an employee or a member of his family to submit himself against his will to any medical examinations of an "intimate" character.

Pedro and his compatriots can scarcely be blamed for demanding as a natural right that which the company has led them to take for granted. Moreover, his demands cannot be ignored. If the company tries to backtrack from the position of medical responsibility into which it has gotten itself, some ambitious politician will maneuver himself into power by offering medical services at a government level. The company, through taxes, will still foot the lion's share of the bill; it may, in fact, having no control over government fiscal policy, find itself bearing an even greater burden than it does while it is still in a position to keep an eye on medical care expenses.

In the United States, there is now a concerted move on the part of labor toward complete company-supplied medical services. In certain areas of the economy, for example, where medical care in the past has been of notably poor quality, there is a growing pressure on industry for the provision of "something better."

Many industries—notably the railroads—were enlightened enough in the past to work cooperatively with communities in making medical services available. In some cases, these industries began by financing these services entirely, then proceeded to shift the load gradually on to the volunteer agencies and/or governmental agencies in the communities. In other cases, industries began and continued on a 50-50 or

75-25 or "we'll match what you put up" basis. These industries are now in a relatively happy position and are not plagued with demands that they provide cradle-to-grave medical service for all employees and their families, though they are expected and expect, in the interest of employee health and productivity and good community relations, to share the financial load on some kind of basis.

The Coal-Mining Industry

The industries that are reaping various degrees of trouble today in the United States are those which operated in hinterland areas where poor quality or no medical services existed and which made little effort to improve conditions. Coal-mining towns, for example, were often in remote, desolate areas which did not attract doctors. Most of these towns could not support hospitals or even simple clinics without some kind of aid, either from industry or government. Ill health and ill will were being fostered in these almost doctorless towns. Some of the mine owners were alert to this and made lone efforts to correct conditions in scattered localities. The industry as a whole, however, did not pull together. As a result, the workers themselves took the initiative.

In the late 1950's, 40 cents per ton of mined coal sold went into the Health and Welfare Fund of the United Mine Workers' Union. There is no difference between medical services available to the worker and those available to his family; it is *total* medical and surgical care paid for by the Health and Welfare Fund.

Consumer Interest

The obvious trouble with arrangements of this sort is that the extra costs must be passed on to the consumer. These costs tend to increase in geometric progression as new developments in medical care make old procedures and equipment obsolete. The danger is that the laborer who derives his income from mining, manufacturing, or processing a product that is in competition with another product—as coal is with oil, for example—may find that he has priced his product out of competition. The consumer reaches a point where he can absorb no further cost increases.

However, it is difficult to be too critical of the harassed worker in this case. Curative medical care can come to appear to be the most important thing in the world if one has seen a child die of a ruptured appendix, or watched a neighbor's wife go through an extraordinarily difficult childbirth, or nursed one's own wife through days of a mysterious raging fever—all without the comfort of a doctor or nurse. Understandably, the demands of such people have been for cure or the money to pay for cure.

Even in large cities, where complex medical facilities and highly trained medical personnel are readily available at all times, the demand for company- or government-subsidized medical care is beginning to come into being where the costs of such care are beyond the ability of the average organized laborer to pay. If a disabling illness is not considered compensable, he is no better off than the isolated coal miner. Medical care may be within a stone's throw, but he cannot afford it and must learn to make the most of being a burden on his family and the community at large.

The Automotive Industry

Typical of such demands, increasing in volume from year to year, are those published not long ago in a newspaper distributed to the members of the United Auto Workers, C.I.O. The key demand, in terms of union emphasis, is labeled "Health Security." It is as follows:

A. EMPLOYER PAYMENT FOR HEALTH SECURITY PROTECTION

Non-contributory financing of Health Security benefits for active and retired members and their families.

B. BASIC PROGRAM OF BENEFITS
  1. Hospital Services. Comprehensive in-patient hospital services paid in full. Semi-private room and board; all other in-hospital services and emergency out-patient care.
  2. Surgical-Medical Care. A program to provide full payment for surgical and in-hospital medical services. Provisions shall be made to give workers the option to enroll in comprehensive medical care plans by paying the excess in cost, if any, on a contributory basis.
  3. Improved Disability Protection. Improved temporary disability protection to enable sick or injured workers to maintain an acceptable standard of living for themselves and their families during periods of disability.
  4. Extended Coverage. Extended protection during periods of layoff covered by guaranteed wage payments and for periods of disability and other work interruptions caused by factors beyond the worker's control.

C. JOINT ADMINISTRATION
  An equal voice in the selection of carriers and in all phases of administration of the health security program, including review of contested claims.

The same theme is echoed by *Ford Facts,* a weekly paper issued by Local 600 of the United Auto Workers, C.I.O., the largest union local in the world. The president of the local is quoted as saying, at the conclusion of a statement regarding forthcoming demands: "We must have full coverage—medical, surgical, hospitalization, life and sickness-benefits insurance—paid for by the Company."

The Rubber Industry

In Akron, Ohio, center of the rubber industry, organized labor has been even more aggressive than the automobile workers on the medical insurance front. The so-called Akron Plan consists of company-paid hospital and surgical insurance for 100,000 rubber and aircraft workers, together with their families. It is the first such plan, fully financed by employers, to cover all the top companies in one industry.

In Summit County, in which the city of Akron is the major component, nearly 300,000 persons are covered by the program. This is approximately 75 per cent of the county's population. Goodyear, Firestone, Goodrich, General Tire, and smaller rubber makers in 1950 paid $12,000,000 to finance the medical plan. By 1970, as subtle inflation and normal population growth manifest themselves, the actual annual cost will be in just proportion.

A typical contract today covers hospital room and board for a maximum of 120 days; $3 a day for a physician's visit in the hospital; payment of all medical extras, such as bandages, drugs, and medicines; and surgical benefits scaling up to $250 per operation.

The Problem of Private Enterprise in Medical Affairs

All of this leads to a consideration of just how deeply an industry can go into the medical business and still act in the best interests of its employees as well as its stockholders. It is generally agreed that business has an investment in the health and productivity of its employees. What is of interest at this point is what steps have been taken to realize dividends on this investment; what steps have been forced by labor; what steps have gone in the wrong direction; what steps in the right direction; and what kind of course can and should be charted for the future.

Solution Depends on the Medical Consultant

It is of the utmost importance for the medical consultant in industry to find answers to these questions. He has perhaps been consulted too little in the past; before new mistakes are built on the foundations of old ones, he will be the one whose voice will have to be heard. In the long run, he is the one best qualified to advise both management and labor as to the goals that will be best for both.

The complete company medical setup, as seen in the case of Pedro, is too much for any industry to handle. It is one thing for an industry to foot the bills for community medical care; it is something else again when industry itself attempts to provide that care.

The nineteenth century alternative, in which industry took no responsibility whatever for the safety and health of those employed, is obviously out of the question in this age of stringent compensation laws. Nor does industry wholly discharge its responsibility merely by taking all the necessary steps to eliminate legally compensable accidents and occupational diseases. Stiff competition—not only for markets but for manpower—means narrowing profit margins. This is simple economics. It is simply the result of an effort to keep prices as low as, or lower than, those of one's competitor to attract the consumer and to keep wages as high as, or higher than, those of one's competitor to attract productive people. Narrowing profit margins necessitate a bigger volume of business and a minimum of waste. Top managers and supervisors must be alert and competent; salesmen must be aggressive and eager; cost accountants must be meticulous and shrewd; clerical workers must be accurate and quick; production workers must be lively, enthusiastic, and on the job. All of this calls for more than a simple absence of occupational illness; it calls for positive health. Absenteeism figures begin to cause immeasurable damage as competition grows more and more intense.

Stockholder Interest

Clearly, then, in the interest of its stockholders, a business will do well to take *some* concrete action regarding the total health of its people. Moreover, stockholders or no, those of its people who happen to be organized within the ranks of labor expect and demand that *some* concrete action be taken.

In general, the action that has been taken (or demanded), over and above the installation of industrial hygiene standards for the prevention of occupational accidents and diseases, has had to do with insurance plans of various types, paid in part or in full by the employer and covering costs of certain aspects of medical care.

Health Insurance

The early health-insurance programs were designed to cover (or partially cover) hospital costs, which were, for most people, the most expensive part of medical care. It was hoped that this coverage would make medical care available to anyone who could pay the then negligible premiums. Group plans covering all the employees of participating industrial concerns extended this coverage to the majority of American employed people, and it looked as though we were on the threshold of a new era of health for the rank and file.

What Actually Happened

The result of this was that hospitals soon found themselves crowded with people who had never before been able to afford hospital care. These institutions had to expand or burst, and in the process of expanding they had to keep up with the rapid advances of medical technology. Moreover, they had to compete for manpower at a time when this commodity was in short supply. As a result, hospital costs zoomed upward faster than insurance benefits could keep up. It became necessary for the underwriters either to raise premium rates or reduce benefits. In many cases they did both.

At the same time that the participants in these plans were becoming sharply aware of the fact that hospitalization was becoming increasingly expensive in spite of their insurance, they were also finding out that hospitalization was only a small part of the medical care picture.

The Older Worker[3]

Older people, in particular, were coming to the realization that the kind of medical care most likely to count was the kind designed to keep them *out* of the hospital. An aging circulatory system, for example, may not call for hospitalization until it is too late for any kind of effective medical care; in this case what is more to the point is early diagnosis and detection so that complications may be prevented while the individual is still vertical and on the job.

The Doctor's Fee

Even among younger people it became apparent that coverage of hospital costs alone was not the answer to the medical care problem. The majority of these people could go for years without requiring hospitalization. Their main problem was paying doctors' bills for the sort of problems that were likely to be treated not in the hospital, but at home or in the doctor's office. Sometimes a doctor, realizing that a patient might be financially embarrassed, would send him to the nearest hospital for simple diagnostic services that could be carried out just as easily in the office. This would enable the patient to take advantage of his hospitalization coverage, but it also tended to complicate the overcrowded condition of the hospital.

Inevitably there grew a demand for a type of health insurance that would pay doctors' bills as well as hospital expenses. Such coverage was bound to be complicated; there were innumerable possibilities for abuse; hence the benefits had to be severely limited. For example, it was extremely difficult, if not impossible, to prevent unscrupulous doctors from offering a minimum of medical service for the maximum fee possible under the terms of the coverage. There

---
[3] G. WARFIELD HOBBS, "Brighter Prospects for Senior Citizens." An address at the Connecticut State Conference on Aging, Hartford, Connecticut, May 20, 1960.

was nothing to prevent the insured from seeking incompetent help in order to get the most for his allowance, in terms of quantity, not quality, medical care.

No Coverage for Preventive or Diagnostic Services

Moreover, since plans of this sort provided only *funds* for medical services, with no control over the kind of medical services obtained, it was virtually impossible to provide coverage for preventive or diagnostic services. Under a number of conceivable circumstances it might be altogether impossible to prove whether a payment for diagnostic or preventive services was valid, legitimate, and reimbursable. Hence, coverage was limited to after-the-fact treatment, which is often too late to be fully effective from the point of view of positive health.

Catastrophic Illness

Premium payments on insurance for major medical expense, involving what might be called catastrophe, such as the need for heart surgery, installation of major prosthetic appliances, excessive periods of hospital rehabilitation, etc., are necessarily too high for the average pocketbook. For this reason, the majority of people have been willing to take a chance on the big disasters while they have protected themselves against the small ones. As has been pointed out, many of the small ones have tended to be almost too small to demand any medical attention whatever, but since they were covered by insurance, they were assiduously seen to. The man with a minor sprain of the thumb availed himself of elaborate medical care, including a night at the hospital for x-ray examination, since his insurance was paying the bills. Unwittingly, he was doing his bit to compound the confusion in the general medical care picture; at the same time, he was helping boost health-insurance rates and reduce available medical benefits.

Then, one day, this hypothetical individual had a serious health setback. He was, perhaps, in his middle 40's and was suffering from a previously undetected condition which had to do with the aging process. If this condition had been detected in its incipiency, it might have been corrected or placed under control. At this late stage, however, the condition had progressed to the point where it was largely irreversible. It was necessary for the patient to undergo a long and costly treatment which would keep him away from gainful occupation for several months. Moreover, even after discharge from the hospital he would have to face life as a chronic invalid and make radical adjustments in his work assignment.

Now he realized that, in spite of his coverage (even if he had been willing and able to supplement his regular sickness and accident insurance with major-medical-expense insurance), he had not really assured himself of any kind of effective medical care.

Labor Is Heard

The United Auto Workers of America have sponsored a widespread community-health plan in Detroit which appears to go right to the heart of the matter in that it endeavors to furnish the policyholders with prepaid quality *medical service* (including diagnostic and preventive services) rather than simple indemnity for *medical costs*. The distinction is important. The emphasis is now no longer on the financial aspect, where it used to be, but on the service aspect, where it belongs.

The details of the United Auto Workers' plan and the philosophy that brought it into being will be discussed further in a subsequent chapter devoted to the role of organized labor in occupational health. The plan has been mentioned superficially here to illuminate an important trend.

Evidently, to those who feel that the positive health and productivity of those employed is a desirable goal, the ready availability of quality medical care (rather than mere monetary indemnity) is becoming a matter of prime concern. The interest of the United Auto Workers is a clear indication of labor's interest. A number of corporations, such as Kaiser, on the West Coast, have established medical service schemes which give a clear indication of management's growing interest.

Experience has proved that quality medical service has to include diagnostic and preventive services. Not only are prevention and early detection of incipient disease the cheapest guarantee of good health, but in many cases they comprise the *only* guarantee of good health.

The Role of the Medical Consultant

The role of the medical consultant of the humanology team is to see to it that the policies ultimately agreed upon by management and labor will furnish the best kind of medical care without in any way committing industry to be the actual purveyor of medicine, as in the tropical area cited above. There is as yet no pat formula that will cover every conceivable industrial situation and type of community in the United States. In any event, industry must work with private facilities outside industry, perhaps on a partial subsidy basis, and limit in-plant facilities to the rendering of emergency care.

This means that one of the first obligations of the medical consultant, as has been suggested elsewhere, is to scrutinize the available resources of the community. The medical consultant must determine the following facts:

1. Are the community hospitals adequate to render quality care at a price the employed individual can afford to pay (or within the limitations of his insurance coverage)?

2. Are there adequate diagnostic facilities for the early detection of incipient health problems in the vertical employee?
3. Does the employed person have ready access to high-quality doctors: general practitioners, specialists, consultants, and others, when necessary? Are arrangements available by which he can afford to pay for the services he requires?

Having secured the answers to the above questions, the medical consultant is in a position to advise industry as to the course it can and should take to assure high standards of health and productivity without involving the stockholder in the corporate practice of medicine.

The Chinese Custom

There is a classic legend of unknown source and doubtful reliability that the Chinese, in some remote golden age, had a custom of paying their doctor a flat fee every year, for which emolument he undertook to keep his patients well. If he failed in this, he had to forfeit his annual fee.

Group Clinic Idea

Variations of this arrangement have been proposed and tried out in certain parts of the United States. For the sake of economy, these arrangements usually are based on the group-clinic idea. The insured person pays a certain amount, depending upon the size of his family and, in some cases, upon the scale of services he wants. This fee entitles him to a prearranged schedule of regular diagnostic examinations, plus routine treatment for illnesses (certain illnesses—alcoholism, for example—may be excluded). It further entitles him to hospital care and the services of specialists in the group when necessary.

These plans have not been uniformly successful. Where they have been most likely to succeed has been in heavily populated areas, where medical resources abound. The problem of quality control is an ever-present one; the industrial medical consultant can judge whether health services of this nature in any given area are entitled to support and participation on the part of his industrial clients.

There Must Be No Coercion

No medical consultant is in a position to go out on a limb for any specific setup of medical services, sight unseen. If quality medical services do not exist in a given community at a price the employed person can afford to pay, the consultant may determine where and how such services may be obtained and whether such services should be subsidized wholly or partly or not at all by the client company or companies.

If quality services are abundant, he is in no position to say: "The employees of Company X should avail themselves of the services of the Clinic Y, even though other available services are equally good." This is a matter that is properly left to the discretion of the individual employee.

This may seem at first sight to limit industry to the point that, however real its interest in employee health and productivity, there is nothing really positive that it can do about it. The actual truth is that, in spite of a determination to stay out of the hospital business and a rigid hands-off attitude regarding community medical service, men of management can do more about the health of the employed in a positive way than any other health-oriented group, be it government or the medical profession itself. Having seen to it that community health facilities and services are of high quality, and having done what it can to make available the best possible indemnity plan to reimburse employees in case of serious illness, the actual policy-maker can add the one element that is missing in most health plans—the vital element. He can provide early detection and prevention of incipient disease by underwriting the costs of routine medical evaluation for all employees. The individual employee should be allowed to choose his own examiner, though the company has a right to pose some of the questions to be asked in a thoroughgoing examination. If such subsidized checkups become a part of standard policy, it may be expected that a large part (perhaps the majority) of the employee population will have a clear idea of where to go for examination. Here is where the consultant will have an opportunity to exercise a certain amount of quality control, but, for the sake of community relations, he will bend over backward to avoid being dogmatic in his recommendations. If any of those employees who do not ask for advice should happen to seek out services that are patently unsatisfactory, the consultant is in a position to find this out and offer corrective advice. In no circumstances will he offer coercion. For example, it is conceivable that an employee, for reasons of religious bias or ignorance, will seek out the diagnostic services of an unqualified person (perhaps a naturopath or faith healer). The consultant, or plant nurse, can say to the individual tactfully: "No doubt your 'doctor' is fine for certain purposes, but only a qualified person can give you all the helpful information you need in an examination like this. Since you have nothing to lose, why don't you go to Dr. A or Dr. B for an *additional* checkup?"

If the individual employee continues to be refractory, which is highly unlikely in this age of mass communication and widespread literacy, the company must respect his bias and leave him to his own devices. In any event, he represents an infinitesimal fraction of the population.

Health Control a Production Control

There is no reason for the policy-makers to balk at the cost of periodic medical evaluation. If the stockholders are not willing to pay to keep their employees well, they are eventually going to have to pay for them when they are sick. The handwriting is on the wall.

Moreover, by keeping itself posted on employee health, management acquires an invaluable tool for production control. If there were no routine maintenance or depreciation funds on machinery, it would be impossible for a business to predict production rates as far as a month ahead, and the company that cannot accurately predict its rate and cost of production at least a year ahead will inevitably lose out to the company that can and does.

The sudden and unplanned stoppage or the insidious decrease in productivity in a human being can create as many over-all production headaches as the same occurrences in a piece of machinery. But this is a negative reason for industry-sponsored routine health examinations. There is a positive one, too. Year-to-year observation of the health changes in an individual who is being assessed simultaneously as to his job performance can shed bright light on hitherto mysterious factors of man's relation to his environment. Many things can be learned about the individual that could not be learned by other methods. His attitudes, abilities, potentialities, and rate of improvement or deterioration can be determined by this method better than by any other, since a *complete* picture of the total man is always on file. The doctor alone sees only a partial picture, as does the supervisor or personnel manager alone.

Further, there is no other satisfactory method of gauging physiological, in contradistinction to chronological, aging. The facts pertaining to health are correlated continuously with the facts relating to job performance. This correlation is the gauge that is needed. Forthcoming studies in human ecology will shed light in this direction. This will be discussed further in a subsequent chapter.

Services at Dispensary Level

While the ideal is for private enterprise to be entirely free from the corporate practice of medicine, certain hard realities will probably always make it essential for any work unit, from medium to large size, to offer some medical services on its own premises. While the function of a dispensary in the past was primarily to offer first aid in the event of accidents, other types of service of equal value are now given. The man suffering from a violent migraine headache as well as the female with severe menstrual cramps can be adequately serviced at dispensary level and in many instances continue to work throughout the day.

The medical consultant will need to know just where to draw the line in the planning, equipping, and manning of such medical facilities.

Types and Scopes of Present-Day Service Units

Three major types of services are to be considered, namely:
1. Complete self-contained units for large plants (more than 1,000 employees).
2. Shared or limited facilities for small plants.
3. Medical facilities sponsored and financed principally by the employees themselves.

While the practice of occupational medicine is, by currently accepted definition, limited to consultation, education, and treatment of emergency trauma, there has been a tendency for large and small plants alike to employ specialists in diagnosis and even in various branches of therapy. This is also true of employee groups establishing their own medical facilities.

A national survey discloses current inconsistency of approach on the part of an organization with seven board-certified men on its permanent payroll who, while running a very active dispensary, do little more than perform routine annual physical examinations for approximately 2,800 white-collar workers, while another concern employing 15,000 workers (both blue- and white-collar) has a single full-time doctor who renders emergency care and routine pre-employment examinations with the assistance of first-aid men.

This is indicative of a short-range view of general medical policy, which explains the following course of events:

In the course of periodic examination, it develops that employee A evidences a need for further study by a specialist in gastroenterology, while employee B is clearly in need of psychiatric evaluation. Company X agrees to pay the bills of both these specialists, who are chosen freely by the employees themselves, upon authoritative advice.

When the gastroenterologist and the psychiatrist submit their bills, which are in keeping with standard fees and by no means exorbitant, a company financial expert takes a look and remarks:

"This is costing us money. Why can't our own medical department take care of this sort of thing? Think of the money we'd save."

So a specialist is added to the staff of the medical department. For a while he is kept busy examining concerned employees. Substantial savings are evident—*on paper*. Before long, however, the medical specialist, primarily interested and trained in therapeutics, becomes bored—there are fewer problems which offer him a challenge. From a recompense point of view he is a satisfied employee; *however,* from a quality-performance point of view his actual worth may well be open to question.

With such a precedent established, another company may well follow suit and add also a physician qualified in a subspecialty, e.g., cardiology. It may well follow that the chairman of the union local at Company Z takes a look at Company Y and reports to the trustees of the Union Welfare Fund: "Looks like we're missing a bet down at the union diagnostic clinic. Over at Company Y they've added a heart specialist to their staff."

The Ideal Facility

The ideal facility for a large plant is the centrally located emergency vertical-care area staffed by a full-time registered nurse, with as many assistants as are justified by the average daily patient load. Some companies are large enough, and the patient load great enough, to maintain a full-time doctor whose sole function is to treat on-the-job trauma and to direct the medical affairs of the emergency health unit. Where there are fewer than 1,000 employees, an incidence of on-the-job trauma high enough to keep a full-time doctor occupied can be looked upon as evidence of some shortcoming in the industrial hygiene, accident-prevention, or company-placement program.

Typical equipment for a well-planned large-plant emergency clinic includes a clinical laboratory and an x-ray unit for placement, chest, and skeletal surveys and for post-traumatic diagnostic studies. If the operation is large enough to justify a full-time doctor, the clinic will also have:
1. Several temporary cots.
2. All-purpose industrial chairs for hand and arm dressings.
3. Universal hand splints.
4. A physiotherapy department with a trained therapist.
5. Resuscitation apparatus.

Some companies operating over extensive areas have found it expedient to have pick-up stations where patients can be collected by bus and brought in for first aid.

Definitive treatment is rigidly restricted in scope. The clinic is equipped to carry out the following functions:
1. Treat lacerations, with local anesthesia and débridement when necessary.

2. Apply splints.
3. Wash puncture wounds and administer tetanus antitoxin or booster shots, if indicated.
4. Treat minor burns (in which 10 per cent of the body surface—and no vital area—is involved).
5. Amputate fingers and repair severed dorsal tendons.
6. Treat simple fractures and dislocations.
7. Furnish rehabilitative services, such as massage, planned exercises, heat treatments, and the like.
8. Provide first aid for eye injuries (complicated injuries to be referred to an outside ophthalmologist).

The majority of plants are far too small to be able to afford facilities such as these. In the United States, the occupational health needs of small businesses are taken care of by union health centers, or by the group effort of several businesses, or through individual contracts for part-time medical services. In France, a law was enacted in 1946 making preventive occupational medical service mandatory for all plants regardless of size, with the result that many businesses either cooperated in joint polyclinics or engaged the part-time services of a peripatetic physician. In Italy, the government maintains health centers; workers there may also be treated at university work clinics.

Marine Medical Service

In the United States, the solution to the problem has largely been left to free enterprise, although government has stepped into marine medical service. Under admiralty law, a shipowner is obliged to provide medical service to any seaman who has become sick or injured while in service of the ship. If the seaman is an outpatient, he is entitled to subsistence. This covers any injury, on or off the ship, provided it is not the result of "willful misbehavior or vice." Marine hospitals operated by the United States Public Health Service are free to seamen; the shipowner is required to pay a nominal sum for each day any of his seamen are hospitalized. In some ports not large enough to justify a marine hospital, the United States Public Health Service maintains marine wards in general hospitals. The Service further publishes a manual entitled *The Ship's Medicine Chest and First Aid at Sea*, which is a guide for shipboard medical treatment and offers a standard list of drugs and medical supplies.

The character of the medical service at sea is set by the shipowner and is governed by the size of the vessel, crew complement, number of passengers, and length of voyage. The average tanker, for example, carries a Purser/Pharmacist's Mate. Larger ships may have a doctor.

Health Facilities for Small Businesses

There are three principal types of small-business health facilities:

1. *The health-conservation center.*—This consists of a centralized dispensary or clinic used by a number of participating companies. It affords facilities for physical examinations, and may further offer mobile units to carry equipment and staff from company to company.

2. *Associated health-conservation units.*—These consist of units in each company served by one doctor, with no central clinic and no mobile equipment. Under this plan each company has a dispensary with a full- or part-time nurse. The doctor visits each on a fixed itinerary known to all participants so that he may be called from any place in case of an emergency.

3. *Individual health-conservation contracts.*—These come into play when individual small companies find it inexpedient to establish health-maintenance services in cooperation with other companies. Under such a contract a company may avail itself of the part-time services of a doctor and may either equip its own first-aid room with a full- or part-time nurse or utilize the doctor's office if it is handy to the plant.

Fig. 2 — Industrial Mobile Unit, *outside of a factory*

EVOLUTION OF OCCUPATIONAL MEDICINE 71

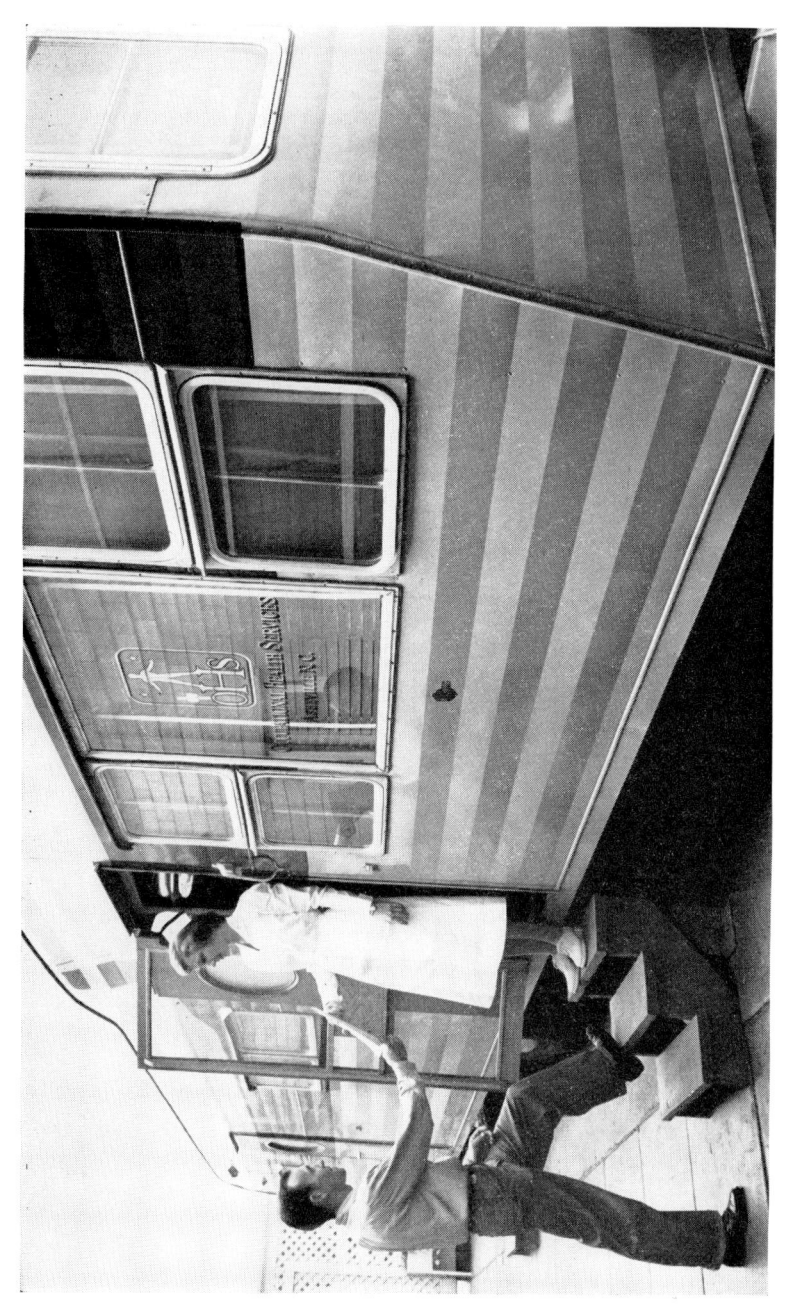

Fig. 3 — Mobile Unit, *close-up view*

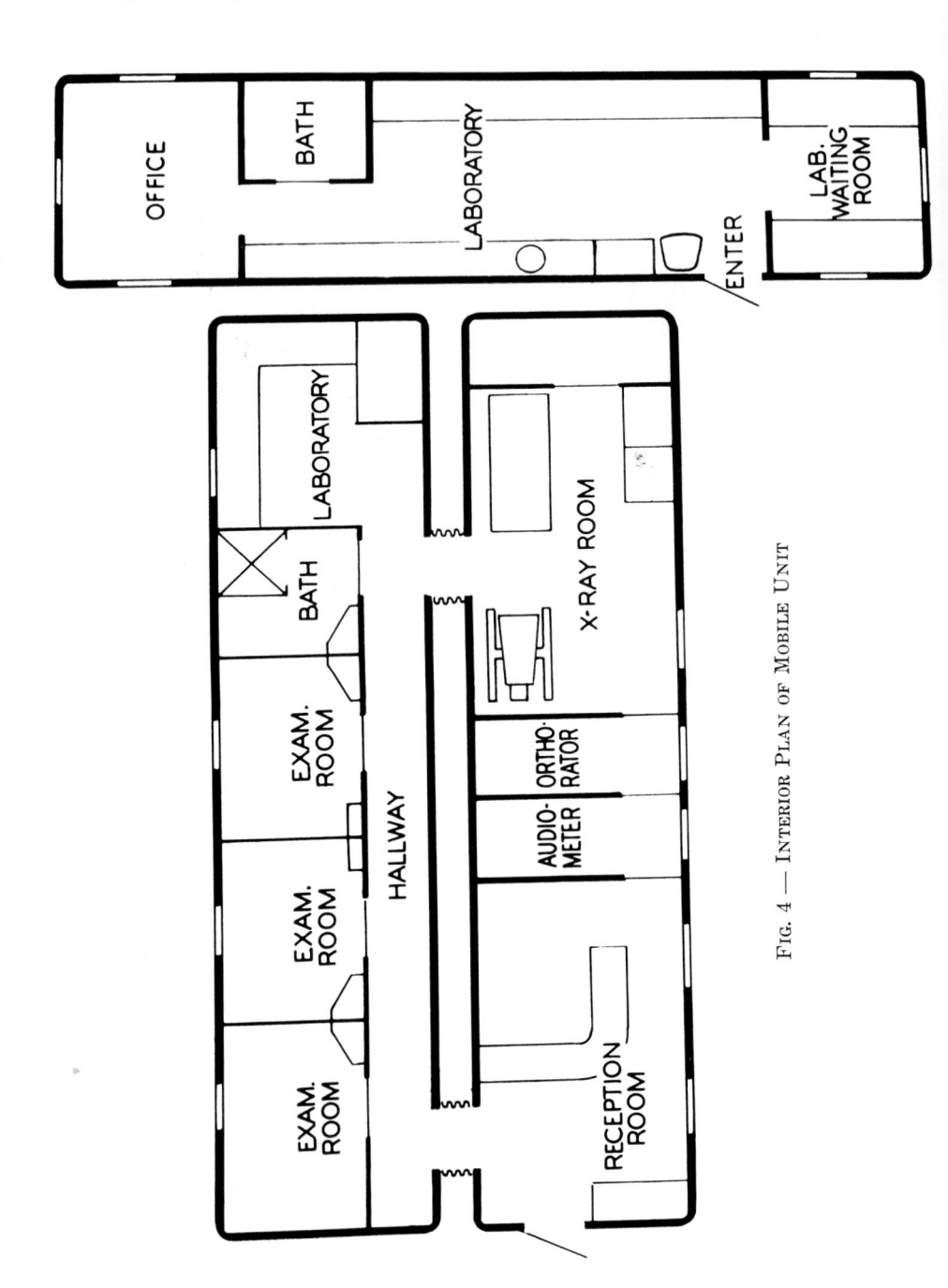

FIG. 4 — INTERIOR PLAN OF MOBILE UNIT

In certain areas, groups of small businesses have joined to avail themselves of the full diagnostic and, in some cases, therapeutic services of specific group clinics. The principles of one such group have been expressed as follows:

1. To provide continuous personal scheduled medical service for industry, including care and supervision of employees with occupational disease and trauma; to furnish preplacement, periodic and retirement medical examinations, industrial hygiene services, and consultation regarding safety, sanitation, compensation law, communicable disease and rehabilitation.
2. To make possible health conservation through counseling on personal health problems plus evaluation of same and referral to the appropriate outside medical agency for treatment.
3. To control absenteeism.
4. To keep and study confidential medical records.
5. To present and interpret significant statistics to management.
6. To supervise nursing service and to purchase and maintain medical supplies.

Employee-Sponsored Plans

Four examples out of many illustrating the growing trend toward employee-sponsored facilities are presented in the following pages:

1. *The Stanacola Plan.*—This consists of a clinic owned and financed by the employees of Stanacola Corporation in Louisiana. It has been financed largely out of payroll deductions, augmented from time to time by special company contributions. Organized in 1924, with 2,200 employees and their families participating, the group built the clinic in 1931. Currently the staff consists of several full-time doctors, including surgeons, pediatricians, ear, nose, and throat and obstetrics and gynecology specialists, as well as a part-time roentgenologist and general practitioners. The paramedical staff includes graduate as well as nongraduate nurses and laboratory and x-ray technicians. Facilities comprise clinical and x-ray laboratories and a pharmacy.

2. *The New York Hotel Trades Council and Hotel Association Insurance Fund.*—This organization began as a group-insurance plan for employees and was launched by the New York City Hotel industry in 1945. As a result of collective bargaining, a health center was established in 1950 by management and labor. Here emergency, diagnostic, and therapeutic medical care is given to 35,000 hotel workers located in the five boroughs of New York. In addition, the employees covered by indemnity insurance to which they are entitled by collective bargaining agreements receive hospitalization and horizontal care from a professional staff appointed and subsequently approved by a board of professional advisors. In 1960 the total medical staff of 147 consisted of the following:

General surgery 24	Dermatology and syphilology 8
Obstetrics and gynecology 8	Clinic physicians 14
Ophthalmology 9	Gastroenterology 2
Orthopedics 11	Alternate clinic physicians 6
Otorhinolaryngology 6	Neurology and psychiatry 4
Peripheral vascular 4	Emergency clinic physicians 2
Proctology 1	Emergency house-call physicians 10
Thoracic surgery 2	
Urology 12	Pulmonary and tropical disease 2
Radiology, pathology, and physical medicine 6	Gastroenterology 2
Internists and physicians in charge 10	Allergy 4

These doctors work on a full, part-time, hourly, or fee-for-service basis. There is a full-time medical director in charge. There is also a full-time lay administrator with equal status.

This open-end health service unit is sponsored by the union, paid for by hotel owners, and governed and administered by a board made up of an equal number of representatives of hotel owners and unions. The health center is serviced by a carefully selected professional staff.

EVOLUTION OF OCCUPATIONAL MEDICINE

Historically, the health center arose out of the fact that it was impossible for employees of hotels to receive competent medical care, owing in part to their low salary scale and in part to the fact that their hours of work in many instances prevented them from attending any of the free clinics at city, hospital, or medical school level.

3. *The Toledo Willys Unit Plan.*—This plan was organized by the Willys Unit, Local 12, United Auto Workers, C.I.O., in Toledo, Ohio. Representatives of this organization first sought the advice of the Toledo and Lucas County Academy of Medicine in shaping policy for a diagnostic unit which the local had voted to establish out of funds received in a retroactive wage settlement. In 1955, the clinic was completely organized, with a radiologist, two general medical associates, a biochemist, a junior medical associate, and a junior pediatrician on the staff. Other members of the team included a business administrator, a head nurse, two blood-bank attendants, a medical technologist, an x-ray technologist, and a maintenance and clerical staff. Local members and their families paid 20 per cent of fees which the Council of the Academy of Medicine of the area has established as standard for the services performed. The remaining 80 per cent came out of the welfare fund of the local, to which the company was a contributor. Services are purely diagnostic and preventive. Outside consultants and therapeutic practitioners are obtained through the cooperation of the Academy.

4. *Medical Program of the Endicott Johnson Corporation.*— One of the most comprehensive medical care plans in the United States, resembling the total-care package deals that are sponsored by American and European corporations operating abroad, is that of the Endicott Johnson Corporation, shoe manufacturers of Endicott, Binghamton, Johnson City, and Oswego, New York. *The plan is financed entirely by the Company and carried by the Company as an operating cost.* It is associated financially and administratively with other

fringe benefits, such as old-age pensions, widows' allowances, general relief, loan service, and nonoccupational disability insurance. Under the terms of the plan, the employees themselves, whether active or retired, and their dependent children under 19 years of age are eligible for total medical care. Roughly, half a million persons are covered.

Two clinics for the care of the "vertical" patient are maintained at different locations. The services offered are complete and unlimited. The specialties represented are general surgery, gynecology and obstetrics, chest surgery, orthopedic surgery, genitourinary surgery, neurosurgery, otolaryngology, ophthalmology, dermatology, internal medicine, pediatrics, roentgenology, pathology, dentistry, and podiatry. Drugs and medicines are dispensed from company-owned pharmacies free of charge. Hospitalization is fully covered; this includes all extra services, such as operating room, delivery room, special medicines, laboratory, and x-ray service. The plan further covers ambulance service. A dietitian acts in a consulting capacity for individuals who are on specially prescribed diets. It was estimated that the cost of this plan per worker was $142 in 1956, and one may assume that the costs have followed the nationwide upward trend in the ensuing years.

Occupational Medicine Has Come of Age

Clearly, occupational medicine has come of age, hampered though it may be from time to time by hastily conceived policies or misunderstanding on the part of the medical profession as well as that of management and labor.

The teaching of occupational medicine, with its various ramifications, is a subject which has been becoming more prominent in the public eye since 1935. The following universities: Cincinnati, New York, Harvard, Pittsburgh, Miami, Oklahoma, Ohio, and Michigan, have led the way in the instruction of occupational medicine. The programs offered are somewhat diversified. Some offer courses which entitle candidates to degrees—either Doctor of Public Health or Master of Public Health. Harvard offers a course leading to a degree of Master in Industrial Hygiene. At Michigan, field training is offered in conjunction with local industry while the student still remains in direct association with the university.

Perhaps the most important phase in the evolution of occupational medicine was accomplished with its relatively recent formal recognition in this country as a medical specialty in the society of medical sciences. Physicians qualifying as specialists in this field are now certified as such by the American Board of *Preventive* Medicine.

This surely points up the definition of occupational medicine as "that branch of medicine which deals with the relationship of man to his occupation for the purpose of the prevention of disease and injury and the promotion of optimal health, productivity, and social adjustment."

REFERENCES

ADAMS, JAMES M., M.D. "Stanacola Medical Care Plant Plan." The *Journal of the American Medical Association*, CXXVI (No. 6, October 7, 1944), pp. 333-335.

ALBERT, ROY E., M.D. "Criteria for the Diagnosis of Occupational Illness." *Industrial Medicine and Surgery*, XXIV (October, 1955), pp. 436-439.

ALLMAN, DAVID B., M.D. "Medicine's Role in Financing Health Care Costs." The *Journal of the American Medical Association*, CLXV (No. 12, November 23, 1957), pp. 1571-1573.

ANDREWS, JOHN B. "Phosphorus Poisoning in the Manufacture of Matches." *American Association for Labor Legislation*, Publication No. 10, (June 10, 1910), p. 16.

BERRY, CLYDE M. "The Mercury Hazard and Its Control in the Refinery Laboratories." The *Medical Bulletin*, the Standard Oil Company (New Jersey), XIV (March, 1954), pp. 76-80.

BRADLEY, WILLIAM R. "Industrial Hygiene Considerations in Plant Location and Design." *Chemical and Engineering News*, XXIX (March 26, 1951), pp. 1196-1200.

BRINDLE, JAMES. "A Labor View of Medical Care." *New York State Journal of Medicine,* LVII (May 1, 1957), pp. 1660-1663.

BUCHAN, RONALD F., M.D. "The New Look in Occupational Medicine." *Industrial Medicine and Surgery,* XXV (No. 6, June, 1956), pp. 285-288.

CLARKE, ROBERT J., and EWING, DAVID W. "New Approach to Employee Health Programs." The *Harvard Business Review,* XXVIII (No. 4, July, 1950), pp. 109-124.

COOK, WARREN A. "Engineering Phases of Plant Health Control." *Chemical and Engineering News,* XXIX (March 26, 1951), pp. 1517-1518.

CROWDER, THOMAS REID, M.D. *Communicable Diseases and Travel.* U.S. Public Health Bulletin, No. 129. Washington, D.C.: Government Printing Office, 1922.

FOLLMANN, JOSEPH F., JR. "Major Medical Expense Insurance — Its Development and Problems." The *Journal of the American Medical Association,* CLXV (No. 12, November 23, 1957), pp. 1578-1586.

FOULGER, JOHN H., M.D. "Criteria for the Diagnosis of Occupational Illness." (Solvents) *Industrial Medicine and Surgery,* XXIV (October, 1955), pp. 432-436.

FOX, JACK V. "What You Should Know About Health Insurance." *New York World Telegram and Sun; Saturday Feature Magazine,* (October 5, 1957).

GRAY, ROBERT D. *Frontiers of Industrial Relations.* California Institute of Technology (Industrial Relations Section) Pasadena, California, (September, 1959).

GUIDOTTI, F. P., M.D. "Direct Medical Service Plan in the Hotel Industry." The *Journal of the American Medical Association,* CLII (No. 9, June 27, 1953) pp. 788-792.

"Guiding Principles of Occupational Medicine." The *Journal of the American Medical Association,* CLV (No. 4, May 22, 1954), pp. 364-365.

HAMILTON, ALICE, M.D., and HARDY, HARRIET LOUISE, M.D. *Industrial Toxicology.* 2nd edition, New York: P. B. Hoeber, 1949.

HATCH, THEODORE F., M.S. "Training in Industrial Hygiene in the United States." *Journal of Occupational Medicine,* II (No. 7, July, 1960), p. 321.

HAYHURST, EMERY ROE, M.D. *Occupation and Disease of Middle Life.* Philadelphia: F. A. Davis Company, 1923.

HEWITT, J. G., M.D. "Industrial Medicine Goes to Sea." *Industrial Medicine and Surgery*, XXI (No. 11, November, 1952), pp. 535-537.

HORAN, JOSEPH C., M.D. "Non-Cancellable and Guaranteed Renewable Insurance — Underwriting Considerations." The *Journal of the American Medical Association*, CLXV (No. 12, November 23, 1957), pp. 1592-1601.

IRWIN, P. C. "Economic Tolerance." The *Journal of the American Medical Association*, CLXV (No. 12, November 23, 1957), pp. 1574-1577.

KEHOE, ROBERT A., M.D. "Criteria for the Diagnosis of Occupational Illness." (Metals and Other Inorganic Substances) *Industrial Medicine and Surgery*, XXIV (October, 1955), pp. 427-432.

KOBER, GEORGE MARTIN, M.D. *A Plea for the Prevention of Tuberculosis, 1894.* Pamphlet 12381, New York: New York Academy of Medicine, 1894.

LEGGE, ROBERT T., M.D. "Chronological Events in Industrial Medicine, 1900-1916." Unpublished Manuscript, Library of Industrial Medicine Association, Chicago, Illinois.

MCANALLY, WILLIAM F., M.D. "A Mutual Health Association for Industrial Workers." The *Medical Bulletin*, the Standard Oil Company (New Jersey), IV (April, 1939), pp. 1-6.

MCCAHAN, J. F., M.D. "Industry's Challenging Opportunity." *Medical Advance*, published by the National Fund for Medical Education, IV (Fall, 1956).

MCCORD, CAREY P. "Occupational Health Publications in the U.S. Prior to 1900." *Industrial Medicine and Surgery*, XXIV (August, 1955), pp. 363-368.

—————. "Lead and Lead Poisoning in Early America." *Industrial Medicine and Surgery*, XXII (No. 11, November, 1953), pp. 534-539: XXII (No. 12, December, 1953), pp. 573-577: XXIII (No. 3, March, 1954), pp. 120-125.

MCNAMARA, WILLIAM J., M.D. "The Role of the Medical Director in Major Medical Expense Insurance." The *Journal of the American Medical Association*, CLXV (No. 12, November 23, 1957), pp. 1586-1591.

MADDOCK, CHARLES S., LL.B. "The Physician and Workmen's Compensation." *Manual of Industrial Medicine*. Philadelphia: University of Pennsylvania Press, 1956.

MEYER, NORMAN C., M.D., and BENNETT, RICHARD J., M.D. "Management and Scope of Treatment in the Surgical Department of a Large

Industry." *Industrial Medicine and Surgery*, XXV (May, 1956), pp. 213-214.

MOCK, HARRY EDGER. "Industrial Medicine and Surgery: The New Specialty." The *Journal of the American Medical Association*, LXVIII (January 6, 1917), pp. 1-11.

NATIONAL SAFETY COUNCIL. *A Final Report of the Committee, Chemical and Rubber Section on Benzol, May, 1926.* New York: National Bureau of Casualty and Surety Underwriters, 1926.

OSLER, SIR WILLIAM. *The Principles and Practice of Medicine.* New York: D. Appleton and Company, 1892.

PAGE, ROBERT COLLIER, M.D. "Industry Calls in the Doctor." The *Harvard Business Review*, Vol. 31 (September, 1953), pp. 109-117.

──────────. "Report of a Survey of Medical Services in Industry." *Industrial Medicine and Surgery*, XXI (No. 6, June, 1952), pp. 293-296.

Phosphorus Necrosis. Hughes-Esch Act of 1912. American Association for Labor Legislation, pamphlet 40. New York: New York Academy of Medicine.

ROBERTSON, LOGAN T., M.D. "Industrial Health Programs." *Manufacturers' Record*, (Asheville, N.C.), (November, 1956).

SAWYER, WM. A., M.D. "The Role of Organized Labor in Occupational Health." *Industrial Medicine and Surgery*, XXVIII (No. 7, July, 1959), pp. 329-331.

SCHEPERS, G. W. H., M.D. "Theories of the Causes of Silicosis." *Industrial Medicine and Surgery*, XXIX (Part 1, No. 7; Part 2, No. 8, July, August, 1960).

SCHERESCHEWSKY, J. W., M.D. *Industrial Insurance Bulletin*, (No. 197, June 5, 1914), United States Public Health Service.

SHILLING, JEROME W., M.D. *Occupational Medicine, Yesterday, Today and Tomorrow.* Chicago: Library of the Association of Industrial Physicians and Surgeons, 1955.

SMYTH, HENRY F., JR. "The Field of Chemical Hygiene." *Chemical and Engineering News*, XXIX (March 26, 1951), pp. 1196-1197.

STERNER, JAMES H., M.D. "The Medical Aspects of the Industrial Hygiene Program." *Chemical and Engineering News*, XXIX (March 26, 1951), pp. 1399-1401.

TABERSHAW, IRVING R., M.D. "Medical Progress: Industrial Medicine." *New England Journal of Medicine*, CCXXXVII (No. 9, August 28, 1947), pp. 313-320.

TAHKA, ALEKSIS, M.D., and NORO, LEO, M.D. "Medical Services in Small Plants." The *Medical Bulletin,* The Standard Oil Company (New Jersey), XIII (No. 3, October, 1953), pp. 304-307.

THOMPSON, DORIS M. "Trailer Clinic Supplements Medical Program." *Management Record,* published by the National Industrial Conference Board, XVII (No. 4, April, 1955).

Union Demands, Syndicate 11. (Clausula 22) Talara (Peru), August, 1952.

WAMPLER, FREDERICK JACOB. *The Principles and Practice of Industrial Medicine.* Baltimore: The Williams and Wilkins Company, 1943.

WILKINS, G. F. "President's Page." *Industrial Medicine and Surgery,* XX (July, 1953), pp. 7 and 325.

WILLYS UNIT LOCAL 12-UAW-CIO. *Brochure on Diagnostic Clinic.* Toledo: Ohio, 1955.

WOODY, MCIVER, M.D. "The Present Practice of Industrial Medicine." The *Medical Bulletin,* the Standard Oil Company (New Jersey), VI (No. 3, April, 1945), pp. 169-176.

YINGLING, DORIS B., R.N. "Educational Possibilities for Industrial Nurses." *Journal of Occupational Medicine,* II (No. 7, July, 1960), p. 325.

ZAPP, JOHN A., JR., PH.D. "Criteria for the Diagnosis of Occupational Illness." *Industrial Medicine and Surgery,* XXIV (October, 1955), pp. 439-442.

3

Organizations Interested in the Health Needs of the Employed Person

A MAN-MADE COMMODITY comes into existence either because there is a wish for it, perhaps subconscious, or because there is a clear-cut need for it, which will be father to the wish as soon as the need has been recognized. However, given the unspoken wish and the unrecognized need, the commodity does not appear out of thin air. Some person or group of people must become alert to the wish or the need and have the vision to shape a commodity which will conform to both.

This was the case with occupational medicine. As has been shown, there was a wish for it, more or less unexpressed, on the part of the employed, and there was a demonstrable, though unrecognized, need for it on the part of those who stood to profit from increased productivity on the part of the employed. These were the determining factors in the history of the growth of occupational medicine; they will determine the direction of whatever future progress there may be in this field.

However, the speciality did not spring into existence simply because a group of employed individuals suddenly woke up and said: "We want it," or because a conclave of boards of directors and stockholders suddenly pounded the conference table and shouted: "We need it!" Someone had to be sensitive to the wants and needs that flourished out of sight around the corner of history; someone had to measure these wants and needs and shape a product in conformance with them; someone had to interpret these wants and needs to employers and employees alike and convince them that the commodity was worth acquiring; someone had to see to it that the commodity was indeed worth acquiring, that it continued to keep pace with wants and needs in a changing and increasingly challenging market, and that there were safeguards against those who would undercut the commodity with plausible but substandard facsimiles.

Credit for this must go to that vast and ever-growing number of organizations, professional associations, voluntary societies, study groups, governmental units, foundations, institutes, and the like, which have devoted time and talent to the health needs of those employed and have cooperated to establish minimum standards, seek out goals, and set limits for the science of occupational humanology.

International Committee on Occupational Health Services of the World Medical Association

Broadest in its scope, functionally as well as geographically, is the International Committee on Occupational Health Services of the World Medical Association. This committee, formed in 1954, endeavors to act as the professional coordinating body for the principal international groups in the occupational health field in terms of the relationships of these groups to the actual functions of physicians in industry and to national medical societies engaged in setting standards and advancing educational projects. Without such a coordinating body, committees, associations, and other units functioning in the same field often find themselves either at cross purposes or duplicating each other's efforts.

Specifically, the World Medical Association's International Committee concerns itself with eight principal aspects of occupational health:

1. Organization and administration (pertaining to policies, establishment of procedures, details of examinations, records, channels of communication, and personnel)
2. Preventive medicine and public health
3. Industrial hygiene
4. Health education
5. Occupational medical education and research
6. Medical emergencies (pertaining to emergency care for the injured and first-time treatment for the employee ill on the job, limitations on treatment of nonoccupational health problems, referral to community resources for definitive care, cooperation between occupational health services and community resources)
7. Rehabilitation
8. Workmen's compensation

With regard to these aspects, the International Committee considers its function to be as follows:

1. To keep the existing international organizations, such as the World Health Organization, the International Labor Organization, and the Permanent International Committee on Industrial Medicine, aware of the World Medical Association's interest in this field and to collect from these specific organizations information as to their particular activities in this field.
2. To contact the member associations from various nationalities to ascertain the status of occupational health services in each country and to receive recommendations as to possible courses of improvement.
3. To make reports based on the information obtained from national and international sources.
4. To formulate recommendations and principles for seeing to it that something constructive is done with reference to each of the eight above-mentioned specific areas of concern.

5. To act as a clearing house on all international aspects of occupational health services.

In a sense, this organization may be considered the international voice of the individual practitioner of occupational medicine. It is not to be confused with the World Health Organization. The latter, which will be dealt with later, is part of the United Nations and represents governments in the field of environmental health. The World Medical Association is nongovernmental.

Council on Industrial Health of the American Medical Association

On a national scale, the counterpart of the World Medical Association is the American Medical Association. In 1937, this organization created the Council on Industrial Health, which aims to coordinate all medical efforts nationally in the field.

Representation on the Council is drawn from the major segments of medical practice which have direct application —industrial medical administration, clinical medicine, and preventive medicine. Members are recognized in the field and have a background of practical experience, plus a good knowledge of theoretical, research, educational, and intradisciplinary aspects of health as applied to the needs of the worker. Annual congresses are held regularly in different cities in the United States. A major enterprise is education of the medical profession in the significance of occupational health. Reports are issued from time to time describing the basic aims of occupational health, the physician's place in the industrial picture, the techniques of administering an industrial medical program, and the establishment of constructive relations with other affiliated groups in or out of the humanology team. The Council has further published, or stimulated the publication of, many statements on specific occupational health and hygiene problems. It has worked with other councils, bureaus, and sections of the American Medical Association to upgrade standards of industrial medicine, surgery, and hygiene, to intensify interest in health education and legislation, and to prevent unnecessary duplication or conflict among organizations interested in the field.

The World Health Organization

The World Health Organization is the international body which coordinates all health activities which fall within the scope of governments. This includes the collection and publication of world-wide statistics on health, the marshaling of widely scattered forces in the war against communicable diseases and epidemics, the administration of international sanitary regulations, the standardization of disease classifications, and the preparation of an international pharmacopeia. In other words, whatever can be done at a governmental level in the field of disease prevention is a matter of concern to the World Health Organization.

An example of the importance of this organization is seen in the gradual turning of the tide in the battle against malaria—the world's greatest single health problem, with an incidence and frequency of mortality greater than that of any other disease. It has been known for many years that effective headway against the ravages of this disease would be made only with some form of international effort. The malaria-bearing mosquito is no respecter of national boundaries. Scrupulous public health measures in one nation could be canceled out overnight by a total lack of such measures in the immediately adjacent country.

The efforts of the World Health Organization have not only coordinated public sanitary measures, such as spraying, draining of swamps, establishment of quarantines, etc.; they have also elicited accurate statistics to highlight the international gravity of the problem.

The World Health Organization has become increasingly concerned with occupational health. Its European Seminar on Occupational Health, held at Milan in 1953, resulted in the shaping of a program to seek closer world-wide cooperation between public health and industrial hygiene services, to endeavor to shape international agreements on safety and other regulations, and to fix internationally accepted definitions for terms used in the literature of occupational health.

Policies of the World Health Organization are established by the World Health Assembly. This body meets annually, adopts conventions and agreements, and appoints the Director General of the World Health Organization, who, in turn, appoints his staff.

International Labor Organization

Another international body which is necessarily keenly alert to occupational health is the International Labor Organization. This group, founded in 1919, is an intergovernmental agency consisting of representatives of government, management, and organized labor in the participating nations. It entered into relationship with the United Nations as a specialized agency in 1946.

Its stated purpose is "to promote social justice in all the countries of the world." To this end it collects and disseminates information about labor and social conditions, formulates international standards, and supervises their national application. It also engages in operational activities and provides technical assistance in carrying out social and economic development programs.

The supreme body of the organization is the International Labor Conference, which constitutes a world forum for labor. National delegations include two representatives of the government of each participating nation, one representative of management, and one of organized labor. The Governing

Body is made up of 16 representatives of government, 8 of management, and 8 of labor. This body functions as the organization's executive council.

An International Labor Office is maintained in Geneva, Switzerland. This is a combined secretariat, operational headquarters, world information center, and publishing house. Operationally, the International Labor Organization provides governments with expert advice and technical assistance in matters connected with labor and social policy. For this purpose, it has established in various parts of the world centers for assistance. to governments in such matters as building up employment services, increasing productivity, developing training facilities, and administering social security programs.

An example of the interest of the International Labor Organization in occupational health is embodied in a series of resolutions adopted by the organization's Petroleum Committee at a session held in 1950 in Geneva. The first resolution was to suggest to governments of states concerned and to health and research institutions, as well as to petroleum companies operating in those countries, that they cooperate in the study, localization, preventing, and combating of prevalent diseases and in the improvement of public health conditions. The second resolution had to do with the establishment of minimum standards of medical services through joint efforts of public authorities and the petroleum companies. The third called for a study to draw the line between occupational and nonoccupational diseases in the petroleum industry, with particular reference to tropical diseases.

The International Commission on Industrial Medicine

The International Commission on Industrial Medicine is an association of specialists not representative of any officially designated organization or government. This group interests itself primarily in technical aspects of occupational health on a world-wide basis.

Originally known as the Permanent International Committee on Industrial Medicine, the group first came into being in Milan in 1906, established headquarters in Geneva, and held meetings every three years in different countries. In 1954, the body was reorganized in Naples under its present name and a new constitution. Its stated aims are as follows:

1. To work for the establishment of international and, "where possible," national Congresses on Industrial Medicine.
2. To collaborate with all international bodies having similar aims.
3. To study new developments in industrial medicine as well as in the economic and social sciences which may be of value in terms of occupational health.
4. To draw the attention of public authorities to the results of research in industrial medicine, to promote the teaching of occupational medicine, and to stimulate public interest in the solid achievements in the field.
5. To recommend to learned societies fruitful and suitable discussion topics relative to occupational medicine.

American Association of Railway Surgeons

In the United States, the pioneer among occupational health organizations is the American Association of Railway Surgeons, founded in 1888. Long before the advent of workmen's compensation, this group pledged itself to make use of only the best among consultants in the treatment of railroad injuries, which were mostly due to wrecks or similar physical emergencies.

The use of the word "surgeon" in the title is traditional. Actually, a Railway Chief Surgeon is considerably removed from the practice of surgery. His functions tend to be largely administrative and consultative.

Conference Board of Physicians in Industrial Practice

Another, more recent pioneer organization, particularly in the development of the philosophy of occupational medicine, was the Conference Board of Physicians in Industrial Practice. This organization was among the first to recognize and make public the concept of the total man at work as industry's most complicated machine and most neglected asset.

In 1916, a business research association, the National Industrial Conference Board, was established, and the Conference Board of Physicians was called upon to advise in matters pertaining to humanology. It continued to serve in this capacity until 1937, when it ceased to function as an independent organization and was absorbed by the National Industrial Conference Board. The latter organization continues to maintain an interest in occupational health. A senior research specialist is assigned full time to the subject. As part of its regular conference program covering several aspects of business, the National Industrial Conference Board schedules panel discussions on industrial health and hygiene. A course on medical and health plans has been included as part of the curriculum of its personnel courses conducted semiannually for personnel officers of member firms.

The American Academy of Occupational Medicine

The American Academy of Occupational Medicine was formed in 1946. Its members are full-time practitioners of occupational medicine, and its broad objective is to enhance full-time practice in this field by:
1. Establishment of minimum standards of quality (as to performance, duties, staff, and equipment).
2. Cooperation with acceptable agencies that may promote, either directly or indirectly, the physiological well-being of the employed population.

3. Institution and encouragement of research directed toward a better understanding of the nature of occupational disease and improvement in methods of prevention and care; dissemination of the results of these scientific investigations to all interested agencies, groups, and individuals in forms suited to their needs and capacities.
4. Taking active part in the education of medical, nursing, and technical personnel in the field of occupational medicine.

Additional Allied Organizations

Organizations dealing with the occupational problems of individuals with specific handicaps may, upon occasion, be of value to the humanology team. These include the National Rehabilitation Association, which dedicates itself to the return to gainful employment of severely handicapped persons and to obtaining legislation favorable to such persons; the National Society for Crippled Children and Adults, which has as one of its objectives the education of employers as to placement and understanding of crippled persons; and the National Society for the Prevention of Blindness, which interests itself in industrial hygiene measures to eliminate conditions that might bring about the specific handicap of blindness and also fosters programs in industry and the community at large for the early detection of eye problems.

Industrial Hygiene

In the early days of occupational medicine, there was no distinction between the industrial hygienist, the safety engineer, and the plant doctor or nurse. However, as occupational medicine became a specialty, so did industrial hygiene. Organizations specializing in this field came into being, sparked by the shocked recognition in the 1920's of the diseases caused by pneumatic drills, radium, and other accouterments of a technological age.

Prior to 1932, only a handful of persons were engaged in the specialty of industrial hygiene. One of the factors that stimulated the growth of industrial hygiene divisions throughout the nation after that date was the avalanche of silicosis claims that was brought on by the depression.

Air Hygiene Foundation of America

Harvard University's School of Public Health had offered courses in industrial hygiene since the early 1920's. In 1936, a group of industries formed the Air Hygiene Foundation of America, Inc., with headquarters at Mellon Institute, Pittsburgh, Penn. Later, as this organization broadened its scope to include other aspects of environmental hygiene, it changed its name to Industrial Hygiene Foundation. This organization is staffed with qualified industrial hygiene physicians, engineers, and chemists.

Since then, a number of institutes have been established in the field. An instructive overseas example is the Institute of Occupational Health founded in 1946 in Helsinki, Finland. In addition to a medical hospital with medical specialists, bed service, and outpatient facilities, this institute has an industrial hygiene department which not only trains specialists to go into industry, but offers industrial hygiene services to businesses which cannot afford to hire a full-time hygienist. Upon request, the Institute sends a team of specialists to make a survey of the working environment, including the air, raw materials, and technical products. The employer in question is then (after laboratory analysis of air, blood, urine, and working-material samples) given a written report with advice as to the prevention of health hazards that may be found to exist. The Institute also has a physiological and psychological department engaged primarily in research into the effects of the working environment on the total individual.

Also recently inaugurated are the Institute of Industrial Health of the University of Cincinnati (1947), the Institute of Industrial Medicine of New York University (1948), and the Institute of Industrial Health of the University of Michigan (1950). A full curriculum in industrial hygiene is offered at Harvard University and at the Universities of California, Pittsburgh, and Michigan.

The Industrial Hygiene Association of the American Public Health Association

Prior to 1939, industrial hygienists had no organization of their own comparable to the professional societies open to doctors of medicine. In the early 1930's, there was no great demand for such an organization; industrial hygienists at that time were able to "make do" with the Industrial Health Section of the National Safety Council (which organization currently directs its programs more to the safety engineer than to the industrial hygienist) and the Industrial Hygiene Association of the American Public Health Association. The latter group, which still takes an active interest in industrial hygiene from the standpoint of public health, is a nongovernmental organization supported by the dues of members. It is not to be confused with the United States Public Health Service, which is a tax-supported federal agency and which has a Division of Occupational Health that carries on research in the cause and prevention of occupational illness, administers laws pertaining to pollution, epidemic illness, and the like, provides consultative services and technical personnel for industrial health programs, and disseminates educative literature, vital statistics, etc.

The Industrial Hygiene Association

Toward the end of the 1930's, the ranks of industrial hygienists had begun to swell noticeably, and it became apparent that some sort of association was needed to act as a clearing house for new concepts and discoveries, to disseminate pertinent information, and to define and set standards for the practice of industrial hygiene.

In 1939, the Industrial Hygiene Association was formed, with 160 charter members. The organization has more than quintupled in size since then. All of its members are required to have had at least three years' experience in industrial hygiene plus a degree from a school of college grade. Until 1961, the annual meeting of the Association was held jointly with that of the Industrial Medical Association, the American Conference of Governmental Industrial Hygienists, the American Association of Industrial Nurses, and the American Association of Industrial Dentists. This joint meeting is called the Industrial Health Conference. At this meeting, the Donald E. Cummings Memorial Award is presented for outstanding contributions to the knowledge or practice of industrial hygiene.

The American Conference of Government Industrial Hygienists

The American Conference of Government Industrial Hygienists (mentioned above as a participant in the Industrial Health Conference) is an organization made up of hygienists associated with official agencies. An important function of this group is the annual publication of a list of maximum allowable concentrations of atmospheric contaminants.

The American Association of Industrial Nurses

The American Association of Industrial Nurses was organized in 1942, largely owing to an earnest desire on the part of a number of nurses in industry to clear up several prevalent management misunderstandings of the nurse's true position in a business organization. Bad precedents had been established in some quarters. Employers had taken for granted in some cases that a registered nurse was automatically prepared, given the proper facilities and equipment, to satisfy a given company's occupational health needs without further indoctrination or supervision. There seems to have been a widespread lack of understanding of the limitations of the trained nurse: she might be asked to perform medical services falling outside the field in which she was licensed to practice; she might even be asked to assume responsibility for a plant's broad medical program.

A group of 16 top-line officers of companies were persuaded to act on a Management Advisory Council to help the nurses' association get its message across to industry at large. This council drafted "A Statement of Principles to Govern Management's Relationship with the Industrial Nurse," setting forth the following principles:

1. Company health services should recognize the ethical and legal limitations of nursing services.
2. The medical program of a plant should be under the eye of a physician (full-time, part-time, or consultative); the nurse should act under his direction and be responsible to him for professional activities having to do with medicine.
3. The nurse should not be called upon to render service or make decisions outside the scope of her training.
4. It should be clearly determined and agreed upon to whom the nurse should report in administrative matters.
5. Where a given nurse has administrative responsibilities concerning other nurses or employees in her department, the extent of these responsibilities should be clearly defined.
6. The nurse and the company medical consultant should participate in the formulation of management's written medical policies.

7. Openings for industrial nursing positions should be filled with an eye to the training, experience, and personal qualifications required for this specialized type of nursing.
8. Rate of pay should not be based only on going rates for community nursing services; a company nurse's salary should be in keeping with salaries paid for positions of comparable responsibility in the company, and provision should be made for pay adjustments as the nurse gains experience and becomes more valuable.
9. The company nurse should be given full information as to company policies regarding so-called fringe benefits (overtime, insurance plans, vacation pay, and the like).
10. The nurse should be encouraged to take part in the activities of her professional societies, workshops, institutes, university classes; the same policies that govern other professional and management employees should apply to the nurse with regard to (*a*) participation in organizations, meetings, etc.; (*b*) payment of dues; (*c*) time off with pay to attend meetings; and (*d*) payment of expenses incurred.
11. The function of the nurse should be explained to everyone with whom she might have to deal.
12. The nurse's administrative superior should initiate periodic conferences to determine how the medical program is functioning.
13. Management should back up the nurse in her impartial decisions and insure that her impartiality and objectivity be maintained by shielding her from labor-management controversy.

Organizations like this are bringing about a bloodless revolution in humanology. Specialists, getting together in groups, not only have an opportunity to learn from each other; they have a medium which enables even the humblest to be heard. As a result, the various members of the humanology team are arriving at a new level of influence in policy-making at management level. The doctor, nurse, industrial hygienist, toxicologist, labor, and management, having achieved an understanding of themselves as well as of each other through their own associations, societies, institutes, and foundations, are becoming more and more a functioning team.

The Industrial Medical Association

The above remarks may serve as introduction to a discussion of the major organization for American practitioners of occupational medicine: the Industrial Medical Association. This body was formed in 1915 and had its first annual meeting in Detroit, Mich., home of the infant automobile industry, in 1916. A total of 125 industrial physicians attended. Its title at that time was the American Association of Industrial Physicians and Surgeons.

The accent in 1916 was on the surgeons, not the physicians. At that time, the doctor in industry was rather a lonely man, professionally. Management looked upon him with suspicion; his colleagues in private practice looked upon him as an interloper taking advantage of an unfair opportunity to steal their potential paying patients by giving them treatment at the company premises. It didn't take much persuasion to interest him in joining an organization of his fellows.

In addition to sponsoring an annual conference on occupational health and affording its members an annual subscription to the *Journal of Industrial Medicine and Surgery,* its official organ until December, 1958, when its own journal, *The Journal of Occupational Medicine,* edited by the late Professor Kammer first came into being, it annually presents two awards: (1) the Knudsen Award, established in 1939 by William S. Knudsen, then president of General Motors Corporation, and is given to a member of the medical profession who has distinguished himself; (2) the Award for

INDUSTRIAL·MEDICAL·ASSOCIATION
HAS · APPROVED · THE · MEDICAL · SERVICE · OF
The Medical Department
of the ....
WHICH·HAS·COMPLIED·WITH·THE·STANDARD·FOR·MEDICAL·SERVICE IN · INDUSTRY · OF · THE · INDUSTRIAL · MEDICAL · ASSOCIATION AND·THE·AMERICAN·FOUNDATION·OF·OCCUPATIONAL·HEALTH

I
THIS industrial establishment has an organized medical department or service with competent medical staff, including consultants. It has policy, procedures and facilities for adequate emergency dispensary and hospital needs and personnel to assure efficient care of the ill and injured.

II
MEMBERSHIP on the medical staff is restricted to doctors and/or nurses who are (a) graduates of acceptable schools of learning with degrees of Doctor of Medicine and Registered Nurse respectively, in good standing and licensed to practice in their respective states or provinces, (b) competent in the field of medicine and nursing as applied to industry, (c) worthy in character and in matters of professional ethics.

III
FACILITIES are available for the performance of competent preplacement and voluntary periodic medical examinations in keeping with existing needs of the industrial organization.

IV
A system of accurate and complete records is maintained. Such records include, particularly, a report of injury or illness, description of physical findings, treatment, estimated period of disability, end results, as well as other information pertinent to the case or required by statute for Workmen's Compensation claims or other purposes. All medical records are regarded as confidential material, filed under medical supervision, and maintained only in the medical department.

V
EMPLOYEES requiring special care are referred to competent consultants.

VI
THE medical department or service has general supervision over the sanitation and industrial hygiene of the plant which has to do with the health and medical welfare of all employees.

This certificate is hereby granted
August 28, 1952

E. A. Irwin, M.D.
President, Industrial Medical Association

R. Dagemo
Trustee, American Foundation of Occupational Health

Fig. 5 — CERTIFICATE OF THE INDUSTRIAL MEDICAL ASSOCIATION

Health Achievement in Industry, which is given to management of firms which have inaugurated or improved comprehensive medical departments.

Occupational Health Institute

The Industrial Medical Association's most important work, however, is carried out through the Occupational Health Institute, its tax-free educational affiliate. This unique organization was first incorporated in 1945 as the American Foundation of Occupational Health, with the avowed goal "to improve the teaching of industrial medicine and occupational health in medical schools and schools of public health, and to conduct studies and surveys of facilities and educational programs in medical fields as applied to industry, as well as to obtain funds for the sponsorship of medical residencies and post-graduate training in industrial medicine."

Since 1926, the American College of Surgeons had undertaken the task of evaluating and certifying plant medical services through its Committee on Industrial Medicine and Traumatic Surgery. In 1951, this work was turned over to the American Foundation of Occupational Health.

In 1954, the Foundation was reorganized and expanded under its present title of Occupational Health Institute. It has enlarged its scope along with the size of its board of trustees, which has been increased to include as many members of management as there are representatives of medicine, nursing, and industrial hygiene. Organized labor has not yet been brought into the picture, but in view of the interest in occupational health expressed vocally by leading labor leaders, it is the consensus that it is only a matter of time before they will be asked for their advice and guidance.

The functions of the Occupational Health Institute include:
1. Evaluation and certification of medical services in industry.
2. Studies and surveys of medical services.

3. Sponsorship of medical residencies and postgraduate training in industrial medicine.

The Institute has a number of clear-cut prerequisites for certification of the medical program of a given company. These are:

1. A stated medical policy.
2. Preplacement medical examinations.
3. Periodic physical examinations of employees exposed to industrial hazards.
4. Available facilities for voluntary periodic examinations of all employees if desired.
5. A competent consulting staff.
6. Attention to be paid to sanitation, safety precaution, industrial hygiene.
7. A chief medical officer to report to member of management familiar with policies.
8. A well-equipped dispensary.
9. Properly licensed and trained medical and nursing personnel.

Where standards are met, the Occupational Health Institute awards the company its Certificate of Health Maintenance.

In addition to this, it carries on a program of education aimed at maintaining high employee health standards, provides a clearing house for information, helps companies in dealing with health problems in specific industries and localities, and holds national and regional conferences and meetings for exchange of views and advice.

Its Regional Consultants are outstanding practitioners of occupational medicine who give their time for this work. There are some 35 of these, placed in strategic regions in the United States and Canada. In addition to the Board of Trustees, the body has a Board of Industrial Advisors, made up of top-line officers in business. This is to bridge a gap that has been felt to exist between management, which establishes industrial medical policies, and organized occupational medicine, which assesses these policies.

The Institute has been a leading proponent of the philosophy of prevention or early detection of incipient disease. It defines a well-rounded company medical program as one in which the emphasis is not on treatment of specific health, nor even entirely on health maintenance, but on *construction* of health; not stasis, but continuous and measurable improvement in employee health, morale, and productivity.

The promotion of this philosophy by the Institute, its parent, and allied organizations interested in the employed person has resulted in the following trends in occupational medicine today:

1. More and more prevention as the focal point of the medical program
2. More precise detection of existing and potential exposures in the work environment
3. Expanded dispensary, laboratory, diagnostic, and industrial hygiene facilities
4. The increased pursuit of health education
5. Worker counseling
6. Broadened insurance provisions
7. Contemplation of the individual worker in his entire being; deeper appreciation of human dignity
8. Wider use of, and more dependence upon, consultants
9. Increased concern with the total work environment apart from the individual work point
10. Rehabilitation as an essential of the entire medical program
11. More physicians precisely skilled in occupational medicine
12. An increasingly larger role for the medical consultant in policy planning
13. Closer cooperation among all members of the humanology team: the policy-maker, doctor, nurse, industrial

hygienist, and psychologist

14. A more significant relationship between company and community through the programs of the humanology team

For further information regarding the various and allied organizations referred to in this chapter, the reader may wish to contact:

American Academy of Compensation Medicine, Inc., 221 West 57th Street, New York 19, New York

American Academy of Occupational Medicine, 1600 Walnut Street, Philadelphia 3, Pennsylvania, Attention: Lemsen Blaney, M.D.

American Association of Industrial Nurses, 654 Madison Avenue, New York 21, New York

American Association of Railway Surgeons, 5800 Stony Island Avenue, Chicago 37, Illinois

American Conference of Government Industrial Hygienists, Division of Special Health Services, United States Public Health Service, Washington, D.C.

American Industrial Hygiene Association, 1501 Alcoa Building, Pittsburgh, Pennsylvania, Attention: Lester V. Crawley

Central States Society of Industrial Medicine and Surgery, Room 1300, 28 East Jackson Boulevard, Chicago 4, Illinois

Commission Internationale Permanente pour la Médecine du Travail, 3 Chemin de l'Escalade, Geneva, Switzerland

Council on Industrial Health of the *American Medical Association*, 535 North Dearborn Street, Chicago 10, Illinois

Industrial Hygiene Foundation, (*Mellon Institute,* Pittsburgh), 4400 Fifth Avenue, Pittsburgh 13, Pennsylvania

Industrial Hygiene Section of the *American Public Health Association,* 1790 Broadway, New York 19, New York

Industrial Medical Association, 28 East Jackson Boulevard, Chicago, Illinois

Institute of Industrial Health, University of Cincinnati, Cincinnati, Ohio

Institute of Industrial Health, University of Michigan, Ann Arbor, Michigan

Institute of Industrial Medicine, New York University, Bellevue Medical Center, New York

Institute of Occupational Health, Haartmaninkatu 1, Helsinki, Finland

International Committee on Occupational Health Services of the *World Medical Association,* 345 East 46th Street, New York 17, New York

International Labor Organization, Geneva, Switzerland; Washington Branch: 917 Fifteenth Street, N.W., Washington 5, D.C.

National Industrial Conference Board, 460 Park Avenue, New York 22, New York

National Rehabilitation Association, 1025 Vermont Avenue, N.W., Washington, D.C.

National Society for Crippled Children and Adults, 2023 West Ogden Avenue, Chicago 12, Illinois

National Society for the Prevention of Blindness, 1790 Broadway, New York 19, New York

Occupational Health Institute, 28 East Jackson Boulevard, Chicago, Illinois

World Health Organization, Pan American Sanitary Bureau, (Regional Office for the Americas of the WHO), 1501 New Hampshire Avenue, N.W., Washington, D.C.

REFERENCES

BEATTY, C. FRANCIS. "Management and Medicine." *Proceedings of First Industrial Tropical Health Conference, Harvard School of Public Health, Boston, December 8-10, 1950.* New York: Robert W. Kelly, Publishing Corporation, 1951, pp. 118-125.

BAUER, LOUIS H., M.D. "Secretary General's Page." *World Medical Journal,* II (No. 6, November, 1955), p. 376.

NORO, LEO, M.D. *The Occupational Medical Foundation and the Institute of Occupational Health Annual Report, 1957.* Helsinki, 1958.

PAGE, ROBERT COLLIER, M.D. "World Medicine and Industry." *Industrial Medicine and Surgery,* XX (September, 1951), pp. 411-416.

RIGNEY, T. G., M.D., and ECKARDT, ROBERT, M.D. "Research in Occupational Medicine in the United States." *Journal of Occupational Medicine,* II (No. 7, July, 1960), p. 327.

SELLECK, H. B., and WHITTAKER, A. H., M.D. *Occupational Health in America — The Evolution of the Industrial Medical Association.* Wayne State University Press, 1960.

4

Management Sets the Policy for Occupational Health

THE VERY ESSENCE of occupational health is management policy. The most brilliantly conceived medical program is doomed to some degree of failure if it is not solidly rooted in economically and efficiently devised policy, laid down and directed by top-echelon line officers.

It might be pointed out here that the logical pace setters in this respect are companies which are comparatively large financially and comparatively small in terms of manpower. They have the time, money, and talent to conceive and implement a constructive, positive health program in the best interests of all concerned. The oil industry, for example, has found it relatively painless to offer its employees some rather elaborate fringe benefits in the way of medical care. Such benefits have become a matter of policy. As a result, employees of other companies look enviously and put pressure on their own employers to institute similar measures. This pressure can be nearly irresistible, and the policy of one company becomes policy for the industry at large. The same policies may be quite painful for such as the automobile and steel industries, which are huge in manpower. This explains why heavily manned businesses, as well as small businesses capital-wise, tend to be slow in instituting policy regarding the extension of fringe benefits for their employees.

Functions of a Line or Operating Officer

The terms "line officer" and "policy" call for definition. A line or operating officer of an organization is one who has full responsibility for the success or failure of his unit. He receives orders from those above him (even at his pinnacle of advancement, the line officer of the average company is subservient to the wishes of the stockholders, represented by the board of directors), and he is responsible for carrying them out by giving instructions to, and getting the cooperation of, his own subordinates. In other words, line officers are charged with the actual functioning of a company.

The professional member of the humanology team is not a line officer, even if he is given the more or less meaningless title of "Medical Director" (see Introduction). If he is on the company payroll with officer status, he is more properly considered a staff officer.

Functions of a Staff Officer

A staff officer is removed from the line of action. His function is to counsel and advise those on the line. His responsibilities do not have to do directly with accomplishment of company aims, but with getting the facts upon which decisions may be based. He must carry on continuous research in his special field in order to supply technically proper advice. He must audit line operations in terms of his specialty in order to keep his judgments up to date and to determine the current strengths and weaknesses of that aspect of operational policy which falls within his particular field.

Danger of Conflict

Problems arise when a staff officer, either on his own initiative, on orders from above, or in accordance with precedent, undertakes personally to translate his recommendations into action.

For example: A Midwest manufacturing firm, with a working force of 2,500, hired an experienced medical consultant to the full-time post of "Medical Director," with the title of Vice-President. This individual, Dr. B., was something of a crusader and felt that the company offered him a challenge worth accepting. Its medical setup seemed to him, at first appraisal, to be the most chaotic he had ever encountered.

His predecessor had looked upon occupational health as simply a more secure version of general practice and had built up a rather elaborate little therapeutic center on the company premises. His staff had consisted of a junior physician, a head nurse with two assistants, and a secretary. When he had sent a memo to the president, who was under his personal care for a cardiovascular condition, recommending that a certified internist be added to the staff, the cost-accounting department went into a tailspin.

This resulted in some harsh language at the next board meeting. A spokesman for the majority of shareholders produced figures gathered painstakingly and insisted that the medical department had proved itself to be a "white elephant." Its cost, he said, was burdensome, and its value, from a stockholder's point of view, dubious to say the least. The incidence of absenteeism, he said, had not gone down but had increased in direct proportion to the increase in cost of company medical facilities. Moreover, he added, statistics showed that over 90 per cent of the employee traffic at the dispensary was properly handled by the nursing staff alone, since it was the result either of minor accidents, headaches, colds, or digestive upsets requiring first aid or standard medicaments such as aspirin. The remaining 10 per cent, he said, consisted of (1) legitimate occupational emergencies requiring prompt surgical action or diagnosis and referral to capable community medical facilities (2 per cent); (2) compensable (some borderline) cases of chronic illness theoretically occupationally caused (1 per cent); (3) nonoccupational illness (mostly chronic) which could have been treated by private physicians (7 per cent).

It was pointed out that an undue proportion of the time of the medical staff was taken up with the treatment of specific health problems of top-line officers who were well able to afford more comprehensive care outside.

Dr. B. was called in to straighten matters out. His predecessor had resigned in a huff, taking his junior assistant along with him.

The day Dr. B. took over, the president said to him:

"Now, Doctor, you're in charge. You've got a good reputation, and you can be trusted to set up the kind of medical department that will keep the stockholders off my neck without stirring up the union men or the industrial relations people. You're a vice-president. What you say goes down there in your department. Just don't bother me about it."

Dr. B. thought at first that this was a green light to institute constructive policies on his own and carry them out himself without top-level interference. First he made a survey of the medical resources of the community. He discovered that there were excellent diagnostic and therapeutic facilities convenient to the plant and not exorbitant in cost. He blueprinted a plan whereby the company would provide medical examinations (preplacement as well as periodic) for all personnel, making use of community facilities and paying all fees involved.

Next, after careful study of various group hospital-insurance plans, he settled on one which would cover costs of unpreventable major catastrophic illness, premiums to be paid partly by the company, partly out of voluntary paycheck deductions. Ordinary hospital insurance, under this plan, was deemed to be the responsibility of the individual employee.

The medical department, now consisting of himself and a nurse-secretary, was to confine itself to emergency first-aid treatment, to consultation and advice when sought, to continuing study of the work environment and its effect on the total health of the total individual, and to acting as liaison with the medical profession in the outside community.

So far, so good. Here was at least the germ of a constructive policy. But Dr. B. discovered shortly that the president's promise: "What you say goes down there in your department," meant literally just that and nothing else. Outside his department he had no authority whatsoever that couldn't be countermanded instantly by any line officer, or even by a foreman or shop steward. He was a "director" *of one person:* his nurse.

He had no authority to agree to pay the bills for routine diagnostic examinations unless he could somehow borrow that authority from the real management of the company for as long as they might be willing to lend it to him. He could not commit the company to any kind of insurance plan. He could not wander at will in the shop, examining the working environment; he could not even take individuals aside to ask them pertinent questions without acting under the clearly seen aegis of someone extremely high in command.

Since the president had said he "didn't want to be bothered" and had made it obvious since then that that was what he had meant, Dr. B. was obliged to yield to pressure from various sides and rebuild a medical department somewhat along the lines of the "therapeutic center" of his predecessor, taking care, however, not to make too much of a good thing out of even this for fear of awakening the stockholders. *He was a frustrated individual.*

From the foregoing, it is clear that while a staff officer or consultant can and should sit in on the framing of policy and even submit his considered opinion as to what policy ought to be and why, the final framing of policy must be by the top-line officers who possess the authority to see that it is implemented and given a respectable and impregnable place in the budget of the company.

What Is a Policy?

A policy has been defined as "a statement of intention that commits management to a *general* course of action to accomplish a specific purpose." For example, when a firm announces that "each employee shall receive a periodic medical examination, frequency to be determined by medical counsel in each individual case," it is committing itself to a course of action with a definite objective in mind.

The carrying out of a policy involves attention to specific actions, which are currently interpreted as being in keeping with the terms of the policy. However, as times and situations change, so do interpretations. The specific detail that seems today to point to the desired goal may be anathema tomorrow. Hence, when policy is formulated, it must necessarily be in broad terms.

The Virtues of Flexibility and Consistency

A policy should be flexible. It is not a categorical and arbitrary dogma designed to keep a confused flock moving in a straight line; it is a guide for management representatives all along the line who are expected to use some judgment in carrying it out.

It is of the utmost importance, too, that a policy covering any particular aspect of company functioning should bear a consistent relationship to other company policies.

For example, Company X, which has a number of branch offices, has always made a point of giving each branch manager considerable latitude in running his branch. Its public relations department has let it be known in business circles that the "Company X way of doing things" is to delegate full authority as well as responsibility; the reputation gained as a result of this broad policy has attracted top-grade branch managers who have been suffering from overcentralization in other companies.

However, a recent survey has shown that while this system has brought in excellent results, many of the branch offices are snowed under with bookwork. These offices are not large enough to justify the installation of elaborate accounting and computing machines in each one, but if the bookkeeping of each office is centralized, the home office can easily afford to install the necessary machinery in one place.

Hence, out of good intentions, a scheme is devised for centralized bookkeeping, and a policy is announced. Sales slips, expense vouchers, purchase orders, and payroll records from each branch shall be sent to the home office weekly to be properly recorded and processed.

A policy of this sort might be met with cheers in another sort of company, but at Company X it creates confusion and some resentment. Branch managers say: "Here's the company that was famous for its policy of decentralization; now it's trying to check up on every move we make. It doesn't trust us any more. Apparently this decentralization idea turned out to be a bust, and they're slowly trying to pull all the authority back into the home office."

Not only the branch managers, but the line officers in the home office as well, are somewhat at a loss as to how to execute the new policy. It is not in line with the "Company X way of doing things"; hence, it is difficult to translate the policy into specific actions based on specific precedents.

Formulating the Occupational Health Policy

The medical consultant on the humanological team will do well to bear the above principles in mind when he is asked to counsel management in the formulation of policies relative to employee health. His recommendations will, of course, be based upon a study of facts, and among the facts to be studied are the company policies already in existence covering the other company activities. Policy must be tailored to the environment in which it must operate.

The consultant will also in most cases have to accept the fact that, no matter how sound his recommendations as to health policy are, he will have to *sell* them—first to top management, then to functioning officers and supervisors all the way down the line of command. A policy conceived in benevolence is not benevolent in fact if those on the receiving end do not understand it or know how to obtain benefits from it, or if those administering it do not do so wholeheartedly. The medical consultant will probably learn that the first thing he must contend with is not philosophical resistance so much as sheer inertia. The philosophical resistance that stood in the way of the early proponents of occupational medicine has long since vanished.

Occupational Health Program Stimulates Human Relations

As soon as a given management made its first, perhaps reluctant, move in the direction of an occupational health program, it found that it had tapped a hitherto unknown and surprisingly copious reservoir in the field of human relations. More people than one ordinarily realizes, have or are, worried about health problems; anyone who throws out any kind of a straw for them to clutch in the way of help and advice cannot help being pleased and astonished at the reaction. An editor of a mass-circulation magazine once remarked that every time his publication offered a popularized article on some health problem, extra help had to be taken to handle the deluge of letters from people grateful for a little help and hungry for more information concerning their health problems.

Within at least the last 10 years most of the major industrial leaders have given public recognition to the value of occupational health from a human relations point of view. Many have gone even further and expressed appreciation of occupational health in terms of productivity; the human relations aspect of occupational health was not even dreamed of in the early stages of its development.

History of Management Policy: Expediency

The history of management policy regarding health has amounted to a series of mere expediency measures. These have been, by and large, *post factum* measures to solve existing problems, rarely forward-looking measures to forestall problems that were bound to arise, as shown in the following brief review of facts set forth in a preceding chapter regarding the growth of medical policy from one expediency measure to the next:

1. Something had to be done about industrial accidents. These were particularly dramatic in the railroad industry. Newspaper headlines helped stir up public pressure on government to take steps. Compensation laws came into being.

2. The crippling effects on industry of infectious disease epidemics in certain areas made it imperative that management achieve a sanitary environment. Government action entered the picture in this respect, too.

3. Technological development brought workers into contact with new chemicals and dust-producing machinery for grinding, cutting, reaming, etc. At the same time, the expansion of industry starting with the turn of the century brought a growing proportion of the population into factories where these health hazards existed. Inevitably, the public limelight lit on the diseases and deaths from these causes and forced companies to avail themselves of the services of the industrial hygienist and the toxicologist.

AMEBIC DYSENTERY

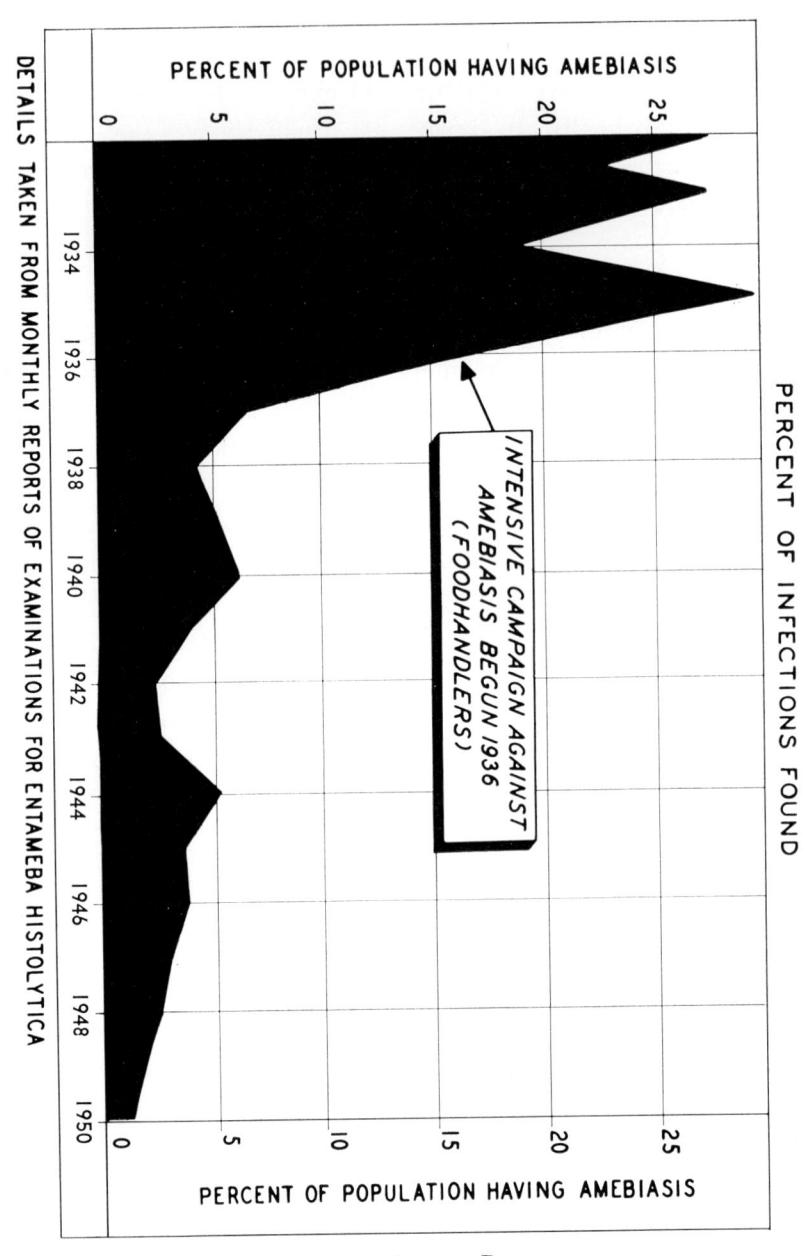

Graph VII — AMEBIC DYSENTERY

4. As compensation laws became tighter, management found it expedient to have doctors on hand, part-time or full-time, to patch up the victims of industrial accidents. Once the momentum started in this direction, it showed a tendency to accelerate under union pressure, until, in many cases, the first-aid room with full-time nurse and part-time doctor became a miniature hospital with a staff of doctors and nurses.

5. In tropical countries, where on-the-spot medical facilities were inadequate, many large research-hungry corporations found themselves jockeyed (by themselves) into a position where they were responsible for free medical care, first for the employee and finally for his entire family, including concubines, on a cradle-to-grave basis.

6. Most labor unions, prior to World War I, had incorporated into their constitutions provisions to set up welfare funds providing life-insurance coverage for members. Some of these funds also indemnified members in cases of permanent disability. In the beginning, they were maintained out of the dues of members. The necessarily rising costs of premiums in the 1920's resulted in the abandonment of most of these funds. In the meanwhile, however, a demand had been created to which industry was sensitive. The insurance companies moved in with group-insurance plans under which a whole block of employees in a given company might be signed up. The insurance salesman would sell the plan to management; the latter, in turn, would enroll as many employees as wished to join. The selling point was that this insurance was cheap. The salesman didn't have to go from door to door, racking up more turndowns than sales. He could make a big deal covering hundreds of people all in one sale. A disadvantage in some cases is that certain group policies are not convertible to individual policies in the event of retirement or termination of employment. Many modern group life-insurance programs are largely or wholly company-financed, in which case this objection is eliminated.

Nonoccupational Illness

Having moved so far in this direction, it was inevitable that management would be induced, on grounds of enlightened self-interest, to take a benevolent attitude toward nonoccupational illness. Before the 1920's, management did not feel it was necessarily obliged to pay an hourly worker for time lost owing to nonoccupational illness. However, because absenteeism due to off-the-job illness was and is many times more frequent than absenteeism owing to occupational illness, and because in chronic cases (chronic cardiovascular disease is one of the leading causes of absenteeism) sick time without pay could work a genuine hardship on an employee, the next item on the human relations agenda was clearly the establishment of clear-cut policies regarding pay for sick time. This was usually worked out by a sliding scale, based on length of service with the company. After a certain period of illness, the sick employee would go from full pay to half pay; at the end of another set period sick pay would be terminated. Short-term absences (caused perhaps by the common cold, or hangovers, or menstrual difficulties, or perhaps the funeral of a favorite aunt, staged conveniently at the local baseball park) were usually ruled out of the sick-pay picture, except, perhaps, where clear medical evidence substantiated the urgent health problem. One way to handle this was to withhold sick pay for the first week of absence. If a worker was sick for more than seven days, he was put back on the payroll, and his pay was made retroactive to the first day of his absence.

The Short-Term Absentee

It was difficult to separate the short-term stayaways with legitimate health problems from the chronic malingerers, hooky players, and ill-advised hypochondriacs. It still is. An alert supervisor can spot those whose absences fall into a sort of pattern, and if the cause of absence is *not* a readily diagnosable illness, he can, by tactful questioning, gain at least an inkling of the truth in time to see that corrective measures—disciplinary, medical, or otherwise—are taken. At the time of the seven-day-minimum-absence qualification for sick pay, the problem was in some measure self-liquidating. The average would-be hooky player tended to look upon a day's pay as more rewarding than all but a few of the ball games scheduled for the current season.

Management, however, having made other steps in the direction of benevolent human relations, began to have qualms about the short-term stayaways who might actually have suffered the claimed sieges of influenza, sore throat, or migraine headache for which they felt pay was being unfairly withheld from them. One solution might be a doctor's certificate of illness; but, then, it was obvious that anyone truly desirous of seeing a ball game or sleeping off Sunday night's binge could obtain one of these without too much effort. It would not pay to turn the company doctor or nurse into a policeman in these cases; an important relationship would be destroyed. (This fact has been proved in instances where doctors and nurses have been employed to check up on the short-term absentee.)

The solution, dictated by expediency, was to abolish the seven-day-minimum-absence provision and to offer sick pay from the first day of absence without asking too many embarrassing questions. As a result, the problem of short-term absences (a supervisory, not a medical problem) has grown over the years, and management has developed some migraine headaches of its own.

But there is no backtracking. Once a benefit of this nature has been given, the employed individual accepts it as a matter of course, and he will not stand to have the benefit taken away without causing some measure of disruption and general dissatisfaction.

The Rising Costs of Legitimate Sick Time

While management was moving toward more and more liberality in reimbursing employees for sick time, the costs of legitimate sick time, particularly when it involved hospitalization, were rising to the point where they canceled out this liberality. During this period health insurance schemes of various sorts were growing up from not too promising beginnings. Before 1930, health insurance (underwritten by private insurance companies, union funds, cooperatives, or employers) was confined to cash benefits for loss of wages, with a limited amount of *surgical* benefits. As the costs of hospitalization, surgery, and medical treatment began to soar in the late 1930's (owing partly to a rising cost of living in general, and partly to an expanding technology that made old equipment and old methods obsolete overnight in many cases), there arose a demand for increased hospital, surgical, and medical benefits.

Three main types of underwriting bodies were developed in response to the demand:

1. *Private insurance companies.*—Policies issued by these provide for payment of cash benefits in amounts and under conditions specified in a master contract. Privately underwritten policies covering hospital and surgical expenses are the most widespread, but medical care plans are coming into their own. Unfortunately, it is difficult to set up quality controls for these latter, and their scope is, of necessity, limited to definitive after-the-fact treatment. The most constructive type of coverage is for major medical expenses in the event of serious and costly health problems. Had management promoted this approach first, much effort and literally millions of dollars, in the author's opinion, would have been saved. This coverage takes up where ordinary medical and surgical coverage leaves off. Premiums for this coverage are, of course, generally high.

2. *Cooperatives.*—This type of plan maintains a group-practice where members and subscribers receive direct service in the form of periodic examinations, diagnosis, and medical treatment. The main objective is to keep members well and out of hospitals. The clinic may be owned by a group of doctors, by the members, by a labor union, or by one or more employers. The main resistance to this sort of plan may come from members of the medical profession in the community, who may look upon the clinic as monopolistic. Professional ill-will in a given community can undermine many plans of this sort.

3. *Service organizations.*—The outstanding examples are Blue Cross and Blue Shield, both nonprofit organizations. The former provides for free choice of hospital and physician. Benefits are stated on the contract in terms of hospital service rather than dollars. The latter organization is sponsored by organized medicine. There are two types of payment agreements: (*a*) the service benefit principle, whereby the physician agrees to accept an established schedule of professional fees from the plan as full payment for services rendered; (*b*) the indemnity contract, according to which the plan pays stated cash benefits and the physician is free to charge his customary fee.

Plans of these types seemed to be tailor-made for the human relations problems of management. The most feasible idea from the point of view of expediency and absence of complications, was that of simple after-the-fact indemnity to cover costs of illness. By footing all or part of the bill for such coverage, management felt that it had eliminated a major cause of employee worry and potential disgruntlement. The fact that mere indemnification, in terms of money, did not keep anybody from getting sick, and the further fact that removal of the financial miseries of sickness did not make sickness any less miserable in itself or less destructive to morale and productivity were not taken into consideration.

More to the point, of course, was the cooperative type of plan based on the philosophy of keeping its members well. One problem here was that whenever an employer or group of employers acted as prime mover in the establishment of the group clinic at the core of any such type of plan, there was unavoidable bad feeling on the part of the nonclinic members of the medical profession locally. Charges were likely to be made that management was "going into the medical business," and individual workers tended to feel that there was something subtly coercive about the whole affair, limiting them in their choice of doctors. For these reasons, management found it easier to stay away from such plans in general and concentrate on financial indemnity deals.

Through the above steps, management has acquired a number of responsibilities it cannot shed. Management has committed itself as follows:

1. To create a working environment where a minimum of accidents is likely to happen, and to offer emergency treatment and compensation when accidents do occur.
2. To create a sanitary working environment with adequate safeguards against toxic or otherwise injurious materials with which workers come into contact, and to compensate for illness brought about through injurious elements in the working environment.
3. To provide in many cases curative treatment for employees in special cases to furnish complete medical care for all employees and their families.
4. To pay employees in full, up to a stated limit, for all time lost due (actually or allegedly) to illness (unless untrue allegations are proved), and to underwrite, in whole or in part, insurance plans providing indemnity for death, disability, hospitalization, surgery, and medical care in the event of nonoccupational illness.

Obviously, with these commitments management has a tremendous financial stake in the health and well-being of those employed, above and beyond the elementary consideration of keeping them fully efficient and productive.

Health — The Core of All Commitments

At the core of all these commitments is health. They all come under total health policy, though in the main they came into being one at a time in response to specific, isolated pressures that arose as a result of a lack of any policy at all. Had management thought at the outset in terms of comprehensive health policy; had management aimed at a flexible policy which would not only maintain the health of its employees at a maximum possible level but actually *construct better health* in those of its employees considered already healthy, most of these after-the-fact and add-a-patch commitments would probably never have come to the collective bargaining table.

In the last 10 years, top men in industry have awakened to this and have spoken increasingly in terms of total policy. The following remarks are illustrative:

> Industry came to realize many years ago that an effective medical service was not limited to the handling of injuries; that its most exciting possibilities lay in steps that would prevent disease while maintaining and improving employee health. — M. J. Rathbone, President, Jersey Standard.

> It is estimated that the training of a skilled mechanic costs his employer from $1,000 to $5,000 and, naturally, the employer is interested in a low turnover. It is likewise estimated that, in the larger industries, it costs at least $200,000 to prepare an executive for his responsibilities... It costs about $200,000 to make a competent jet pilot and, unfortunately, the turnover is very heavy. So our interest in protecting the lives and health of our associates in business has not only a humanitarian justification, but, also, a very practical and financial one. — Reuben B. Robertson, Sr[1]., Chairman of the Board, Champion Paper & Fibre Company.

---

[1] REUBEN B. ROBERTSON, SR., "Opening Remarks," a report of the Third Annual Lake Logan (North Carolina) Conference, May, 1957.

Certainly today everyone agrees that the best guarantee of a healthy work force is a healthy population. Progressive American business is understanding better the close relationship between industrial medicine and community health. The environment in which a worker lives is as important as the environment in which he works. He is in the plant eight hours a day, but he is "on the town" 16 hours. So — some of the problems to which industrial medicine might well address itself . . . are (1) how better to extend the benefits of plant programs into the homes of the workers, and (2) how to work with other groups for the improvement of community health. — Meyer Kestenbaum, President, Hart Schaffner and Marx.

Corporations are today social institutions exerting a wide influence beyond the plant gates in the areas where they operate — and the community attitude, in turn, affects the corporation. On this premise, management surely has a responsibility toward the entire community. Our corporate well-being is directly related to the quality of the schools in our plant community, to the efficiency of police and fire protection, to the good operation of the health department, and to the medical and hospital care available. — Robert E. Wilson,[2] Chairman of the Board, Standard Oil Company, Indiana.

The fundamental requirements which are likely to be reflected in most well-rounded industrial medical programs . . . are:

1. The pre-placement examination . . . to determine . . . qualifications . . .
2. The periodic examination . . . to insure
    a) that no health impairment arising out of employment goes undetected and uncorrected . . .
    b) that any (non-occupational) health problem . . . is drawn to the employee's attention . . . that he may take corrective steps through his personal physician . . .
    c) that those who have health conditions incompatible with present employment are transferred to work compatible . . . or released.
3. Consultation and close coordination with those responsible for employment, placement and transfer of workers.
4. First-aid care of occupational injuries . . . adequate follow-up.
5. Diagnosis and medical care, frequently through outside specialists, of occupational illnesses.

---

[2] ROBERT E. WILSON, "Industry's Stake in Medical Education," an address at the conference meeting sponsored by the National Fund for Medical Education at Northwestern University Medical School, Chicago, February 8, 1955.

6. Consultation and close coordination with those responsible for safety to insure that information developed is used to remove, or reduce and control, causes of occupational illnesses and injuries.
7. Consultation . . . with supervisory organization to insure that unfavorable working conditions receive appropriate attention.
8. Plant hygiene service, covering . . . sanitation and toxicology.
9. Plant inspection service [to determine] adverse working conditions.
10. Cordial and constructive relations with medical profession generally.
11. A definite plan for dealing with the medically restricted employee.
12. Development of a record procedure. Medical department records may be drab, colorless and useless, or alive and vital, depending upon the point of view with which they are approached. — I. Dent Jenkins,[3] General Personnel Manager, Harrison Radiator Division, General Motors Corporation.

In a modern large corporation, which thinks in terms of career employment for most of its employees today, is it more profitable to deal with such health problems early — when there is time to help — or to let nature take its course? Obviously, there are advantages in dealing with health problems early — for the individual and for the employer. Preventive medicine can raise the health level of a whole organization. It can prevent many of the long absences which will come in later years to employees who neglected health problems at their beginning. It may add years to the lives of employees whose value may be great — and in whom the company has invested large sums of money. — Frank W. Abrams,[4] former Chairman of the Board, Jersey Standard.

The physician must know the demands of the job to which a particular employee is assigned. When he is assigned to a job whose demands are within [his] capacity, the first step of a good health maintenance program has been accomplished.... Periodic examination of established employee is the next step. Knowledge of how each worker is reacting to the type of environment in which he has been placed is basic in protecting his good health.... The responsibility of the industrial

---

[3] I. DENT JENKINS, in an address to the Industrial Medical Group at the University of Michigan, Ann Arbor, January 15, 1952.

[4] FRANK W. ABRAMS, an address at an industrial medical lunch in conjunction with the convocation of New York University's Bellevue Medical Center commemorating the Diamond Jubilee of University Hospital, March 4, 1957.

physician is one of fact-finding.... Prompt referral to the private practitioner, whose responsibility is that of treatment, frequently brings under control a condition which might otherwise have gone undetected.... The industrial physician also makes an important contribution to health education.... Community health [is] a team effort.... One of the challenging opportunities in the field of industrial medicine lies in research. Few doctors are offered the opportunity, enjoyed by the industrial physician, to observe a large and relatively stable group over an extended period of time.... Furthermore, the employee can be seen not only when he is ill but, also, when he is at work on his regular job. — H. W. Anderson,[5] Vice-President in Charge of Personnel, General Motors Corporation.

The past generation of managers spent 90 per cent of their time and energy on machines. In the future, those companies which succeed will devote 90 per cent of their time to people. — Cass Hough,[6] former Vice-President of the Daisy Manufacturing Company.

All the above quotations are worth studying. They represent constructive thought. They are only a few of the understanding remarks being made today by lay industrial leaders regarding occupational health. If one were to add them all together, one would have a comprehensive statement of what humanologically sound occupational health policy ought to be.

Gap Between Real and Unreal

Unfortunately, however, at the same time that top-line officers of corporations are making pronouncements relative to ideal industrial medical programs, organized labor, often in the same corporations, is girding itself to go after the identical "ideal" programs via the collective bargaining table. Plainly, there is a gap between real and ideal.

Policy is not a statement of what ought to be. Policy is something laid on the line by someone in authority as something that *will* be.

---

[5] H. W. Anderson, "Medicine and Manpower," an address at a symposium on Medical Education and Industry, Mellon Institute, Pittsburgh, December 8, 1955.

[6] Cass Hough, as quoted by G. W. Hobbs, at the 1958 spring meeting of the National Committee of the Aging, Washington, D.C., 1958.

However, the fact that leaders such as those quoted above have gone on record as believing in sound health policy indicates that they are thinking seriously along these lines and are subject to pressure, persuasion, or inner compulsion which may eventually lead them to implement their ideas with concrete blueprints for action. When they do, industry throughout the country will inevitably follow suit.

It must be remembered that a policy, like a constitution, commits authority to certain *philosophies* of action — not to temporary details of action. A statement of policy should be as true a hundred years from today as it is today. A society founded before the invention of the telephone or the automobile may have stated in its constitution and bylaws: "The secretary shall notify all members when it is decided to change the date of a meeting." Such a provision will make sense in the year 4000 A.D. Suppose, however, the provision had been worded: "The secretary shall ride on his horse to the homes of all members to notify them that the date of the meeting has been changed." This is not as outlandish as it sounds. Any world traveler is likely to have his snapshot album filled with charming pictures of ritual absurdities that are little more than hollow pomp and circumstance based on some antique statement of policy that was much too specific in regard to details.

Policy says: "We shall accomplish this end." It does not say: "We shall accomplish this with these specific means." Means change; ends, if intelligently conceived, do not.

The ends to be considered in the framing of an occupational health policy must be based upon the following accepted truths:

1. Employee health is of primary concern to management.
2. A company has a responsibility to help maintain employee health.

The general approaches to these ends, which may be looked upon as true at any conceivable time under any conceivable

circumstances, are:
1. The application of constructive medical procedures (stressing prevention and early detection).
2. The curative, symptomatic approach.

The characteristics of each approach are set forth in the following table:

Preventive, Constructive Medicine	*Curative, Symptomatic Medicine*
This is industry's responsibility	This, according to the American way of life, is the employee's responsibility. However, industry should show the way and enlighten the employee as to the difference between good and bad medical care
Accepted sanitation	
Institution of proved industrial hygiene methods; heat, light, ventilation, safety protection	
Placement and periodic medical audits	
Noonday lunchrooms with balanced meals	Establishment of liberal sickness benefits
Health education program to teach the employee to live within his physical and mental capabilities	Establishment of mutual health associations, local, Blue Cross, etc., in which industry and employee share alike
Retirement and annuity benefits	

Comprehensive medical policy will envision a combination of the two, with emphasis on the former. When the emphasis is placed upon constructive medicine, the cost is minimal, absenteeism is minimal, there is greater work efficiency, the over-all morale is higher, the general health is good, and there are relatively more people living to retirement age *with health.*

If, on the other hand, emphasis is placed on curative and symptomatic medicine, the cost is maximal and cannot be controlled. Absenteeism is likewise maximal, work efficiency is lowered, and morale is bad. There are relatively more neurasthenics and malingerers on the payroll, and fewer individuals continue on to retirement age.

Further, when the approach is constructive, the cost of maintenance of a creditable medical department is controlled.

If curative medicine is the primary intent, sickness benefits may soar to unlimited heights. There is little or no control over the caliber of therapeutic care received or over the quantity and quality of the therapeutic agents used.

Essentials of a Medical Policy

A medical policy need only be a simple statement in outline form. In defining the various branches of medicine it should endeavor to set ideal limits to management's involvement with each. First things should be put first.

1. *Preventive medicine.* — Medical policy should begin with a consideration of management's proper commitments with regards to preventive medicine, which is far more basic than therapy. Management's responsibility in this respect is clear-cut in the working environment. As for the home environment, management in this country will avoid, if possible, undertaking the actual medical or sanitary work involved, but will see to it that community resources are up to standard and are extended throughout the community.

2. *Educative medicine.* — The backbone of prevention is what might be termed educative medicine. Management's responsibility in this regard is threefold. First, it will provide medical guidance to the employee by whatever means are the most effective and, at the same time, feasible. Second, it will seek competent counsel in order to contribute wisely to the voluntary associations in the field of public education, without being unduly pressured to donate disproportionate amounts to those whose "educative" aspect is largely fundraising propaganda or is comparatively insignificant (see Chapter 5; "Corporate Giving for Health Causes"). Third, it will make an effort to assist in subsidizing the training of medical and paramedical personnel where such training will build the medical resources of the plant and/or community.

3. *Constructive medicine (diagnosis and early detection).* — Next after prevention, which includes education as well

as vaccination, inoculation, sanitation, nutrition, and other basic facets, occupational health policy must come to terms with diagnosis and early detection. The importance of this field of medicine need not be labored at this particular point; it has been emphasized in preceding chapters, and will be re-emphasized and thoroughly explained in appropriate subsequent chapters. Suffice it to say here that management policy, in answer to the costly problems imposed by its current commitments to curative or straight indemnity programs, will inevitably call for thorough periodic medical evaluation of all employees at company expense, preferably using community resources. In this respect, management will further commit itself to take the leadership in organizing and developing the kind of diagnostic resources that will best do the job of maintaining and constructing the health of the employed persons of the community.

4. *Curative medicine.* — Despite its prominence in the public mind, curative medicine is fourth in importance — not first — in the formulation of occupational health policy. Wherever possible, industries will stay clear out of the practice of curative medicine. The ideal policy will provide for first-aid facilities and prompt referral to the appropriate medical agency outside the plant. Circumstances alter cases, and exceptions may have to be made in this respect. Curative services may have to be temporarily provided for in remote camps and in instances where complete curative facilities have already been provided and are taken for granted, as in the tropics. Although private enterprise will have difficulty abandoning these, it can be done. In preparing blueprints for the future, however, the well-advised man of management will protect himself against further commitments in this regard. Existing industrial facilities will be converted as expeditiously as possible into community-managed facilities, even if the company involved still finds itself obliged to provide most of the financial support. In areas where proper

outside facilities do not exist, industry will find it cheaper in the long run to encourage and even finance such *outside* facilities than to establish them on its own premises.

5. *Habilitative medicine.* — Finally, medical policy will concern itself with habilitative medicine. Habilitation has too often been a hit-or-miss proposition. A worker returns to his company after a serious health setback which may have disabled him more severely than he knows or his supervisor is able to guess. He may be encouraged to go directly back to his old job and try it out for awhile, and his failure to perform may undermine his self-confidence so completely that he becomes permanently unfit for that activity. Or his employers may have an exaggerated idea of the seriousness of his setback. (Many mild cardiovascular conditions, for example, tend to be overrated by employers in the absence of competent medical counsel.) In this case he is assigned to some humdrum — and, perhaps, humiliating — post, far beneath his capacities, with resultant ill-will, misunderstanding, and often permanent deterioration of performance. Management's commitments in this respect will be under firmly based but flexible policy to seek medical advice regarding the placement of all recuperating personnel. Professional counsel will advise not only as to immediate placement, but will recommend workable rehabilitative measures and offer a prognosis for future patterns of performance. If it is determined that rehabilitative measures will be not only effective but economically feasible, management will take responsibility for seeing that they are brought into play, whether or not the original health setback was classed as occupational. In all cases, management will inform the employee fully as to medical recommendations and his future in the company.

If medical policy is built upon these principles, the future of occupational medicine will not consist of a series of expediency measures. The true goals of employee health, happiness, and productivity, already endorsed by management, will be attained because they will never be lost sight of in off-the-cuff scrabbles for petty solutions to petty, fleeting pressures. The challenge to the professional members of tomorrow's humanology team will be to hammer away at these self-evident truths until top-line officers of business throughout the country have committed industry as a whole to policies that will promote optimum occupational health now and a thousand years from now.

REFERENCES

DIETZ, DAVID, LL.D. "Speaking for Newspapers." *Condensed Report of Program 35 of the Annual Meeting of the National Health Council, March 23-25, 1954.* New York: National Health Council, 1954, pp. 19-20.

KESTENBAUM, MEYER. "The Modern Corporation and the Nation's Health." *Industrial Medicine and Surgery,* XXIII (No. 5, May, 1954), pp. 221-223.

KIRKPATRICK, A. L. "Health Insurance Programs in Industry." *Industrial Hygiene Digest,* XVIII (No. 1, November, 1954).

MOSKIN, J. ROBERT. "Speaking of Magazines." *Condensed Report of Program 35 of the Annual Meeting of the National Health Council, March 23-25, 1954.* New York: National Health Council, 1954, pp. 16-19.

PAGE, ROBERT COLLIER, M.D. "The Road Ahead." *Industrial and Tropical Health I,* by the Harvard School of Public Health, 1951, pp. 109-117, Chapter 3.

──────────. "Medicine in Industry." *Industrial Medicine and Surgery,* XXIII (No. 5, May, 1954), pp. 217-220.

PIGORS, PAUL JOHN WILLIAM, and MYERS, CHARLES ANDREW. *Personnel Administration, A Point of View and a Method.* 3rd edition. New York: McGraw-Hill Book Company, 1956, pp. 25 and 49.

RATHBONE, M. J. "Our Debt to Medical Science." The *Medical Bulletin,* the Standard Oil Company (New Jersey), XVI (No. 1, March, 1956), p. 1.

YODER, DALE, PH.D. *Personnel Management and Industrial Relations.* 3rd edition. New York: Prentice-Hall, pp. 18, 136-137.

5

Corporate Giving for Health Causes

THE PROBLEM OF CORPORATE giving belongs properly under the heading of *management policy*. This aspect of policy, however, has grown to such fantastic proportions in recent years and has acquired so many ramifications that it is necessary to take it up in a separate chapter. It is still growing. National goals are rapidly approaching the billion-dollar mark.[1]

A rather facetious example of the reaction of business to the expanding problem is seen in a letter written to the American Medical Association by a "long-suffering" corporate "Santa Claus." This individual requested the name of an impressive-sounding, *preferably incurable* disease which he could use as a basis for starting his own voluntary association: The National Association for the Eradication of (?). He concluded:[2]

> The ailments you suggest must present almost innumerable obstacles to the research scientists, since I do not want them to arrive at a successful conclusion in a mere matter of a year or two. Such an unhappy contretemps would only necessitate a fresh start, all of which would be demoralizing to staff and contributors.

Business and Health Causes

There may be more truth than poetry in this tongue-in-cheek appraisal of the situation. Innumerable voluntary asso-

---

[1] *Time*, of July 21, 1958, p. 34.

[2] H. JACKSON, Fourth National Conference on Solicitations, Detroit, Michigan, April 3, 4, 1954.

ciations[3] with "health missions" manage to find their way into the corporate pocket each year by focusing attention

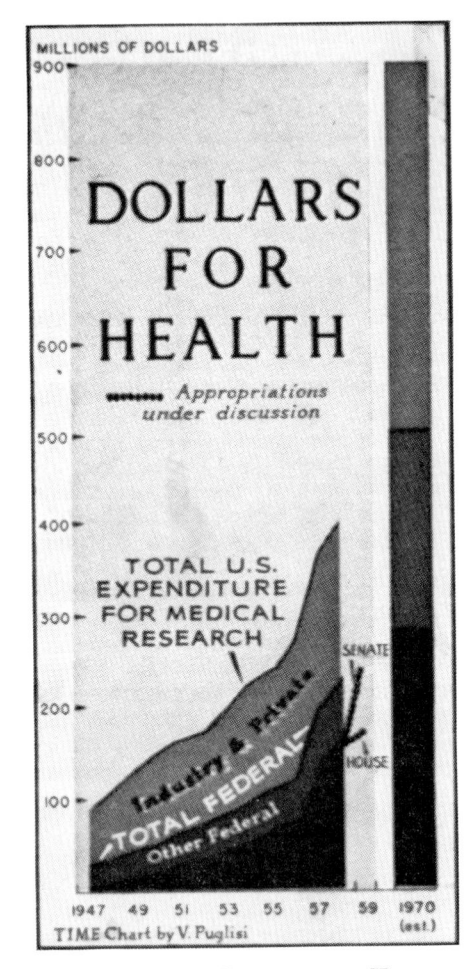

Graph VIII — DOLLARS FOR HEALTH

upon specific ailments and enlisting the aid of fund-raising experts and top men in the fields of advertising and pub-

---

[3] A. D. KELLY, M.D., "The Relationship of the Medical Professions to the Voluntary Health Organizations." An address at the Thirty-Seventh Annual Meeting of the Health League of Canada, February 13, 1958.

licity. Most of these agencies concentrate their appeal on a search for a cure (perhaps, as in the case of the sarcastic businessman, for the preferably incurable disease) or means of rehabilitating types of people that can be fitted to the requirements of tear-jerking publicity. It is rare that these appeals concentrate on prevention, which is hard to dramatize.

The modern high-echelon corporate officer can consider himself lucky if he receives as few as three requests a week for contributions to help find a cure for some kind of ailment. It has reached the point where business leaders are forced to barricade themselves behind a "committee wall."

The Committee Wall

Protective committees have a certain limited usefulness, but all too frequently the decisions they make are based upon the sales ability of the representative of the voluntary society seeking funds, or on the pressure brought to bear, or on the uninformed opinion of either the committee members or their wives. It is the exception for them to seek advice from the humanological experts best qualified to judge which health causes will produce the best result in terms of actual health. Humanology, however, is gaining ground, and it is to be expected that more and more "charity-fatigued" boards of directors will be induced to seek constructive advice in their mounting perplexity. The student of humanology will do well to be prepared with positive answers.

Guiding Principles of Action

Ideally, the corporate dollars given by business to health causes should return dividends in the form of healthy, happy personnel. To this end, guiding principles[4] need to be de-

---

[4] ROGER M. BLOUGH, a newspaper release May 17, 1955 for the United States Steel Foundation Incorporated, New York.

veloped, based upon expert knowledge of health problems and needs. The following principles will form the basis of any constructive appraisal of a health cause in search of funds:

1. Management has a practical interest in health. Human enterprise functions best in a healthy environment. All monies given should pay off, either directly or indirectly, in employee health, reduced absenteeism, increase of work efficiency and output, and improvement of community relations.
2. Management has a real interest in making it possible for employees and their families to receive *quality* medical care at a price they can afford to pay. Unless private enterprise begins thinking in these terms (and, as it has been pointed out elsewhere, the recipients of management's paychecks *are* thinking in these terms), the pressure of public dissatisfaction will inevitably bring about government subsidy and control of medical care for the rank and file. This will mean *quantity* medical care without respect for *quality,* and the cost to the taxpayer will be out of all proportion to value received. The corporation officers cannot hide behind a committee wall when the organization in search of funds happens to be the Bureau of Internal Revenue.
3. In addition to its concern for those employed and their families, management has a practical interest in the health and soundness of the community at large. A business is a part of the community in which it operates. It has customers and clients in the community. Its potential recruits are in the community. Its annuitants are in the community.
4. Wealth and power entail philanthropic obligations. It is management's responsibility to carefully evaluate the distribution of the monies spent fulfilling this important responsibility.

How to Consider Health Causes[5]

Management is approached by fund solicitors for health causes from so many different directions that a certain amount of confusion is inevitable when it comes to sorting, classifying, and evaluating the various appeals for funds. This confusion can be dispelled if one considers all possible health causes under one of the four following headings:

1. *Research.* — As new scientific facts become known, methods must be devised to put them into application. This is extremely important. It is, in fact, the basic health cause.

All aspects of medical care depend upon research. If only half the amount that is contributed annually to pay the costs of *cure* were contributed instead to fundamental research, the nation's sickness bill, which means so much in terms of industrial production, would be radically lowered.

Research is sometimes a by-product of causes whose primary mission is to raise funds for *treatment*: the Infantile Paralysis Fund, for example. Sometimes this treatment aspect gets all the drama. But in the end it is the research aspect that pays off.

2. *Professional and technical education.* — High-quality medical care for the greatest number of people depends upon the large-scale availability of thoroughly trained doctors, scientists, technicians, and paramedical personnel. Proper training means a considerable investment of time and money on the part of the individual receiving the training, and it entails a much vaster investment on somebody's part for the establishment of training facilities manned by skilled faculties. A prime example of a cause seeking funds for training of professionals is the National Fund for Medical Education.

3. *Public education in matters of health.* — This is sometimes a by-product of causes having other principal goals.

---
[5] HOMER W. TURNER, in a personal communication with the author.

For example: the Cancer Society and the Heart Association offer a certain amount of education, though their principal goals are to raise money for studies of treatment methods. The educational aspect is perhaps more valuable than the search for cure, because an educated citizenry has some of the tools it needs for the *prevention* of illness.

4. *Community facilities.* — These include two types:

    a) *For Diagnosis and Detection of Incipient Disease.* Such facilities become the cornerstone of a soundly devised program of constructive medicine which embraces preventive medicine. Diagnosis and detection are, of course, largely dependent upon research. Facilities under this heading will cooperate fully with groups engaged in research endeavors; they may even be themselves engaged in specific research projects, being ideally situated to study the etiology and the human ecological aspects of a number of given diseases, and note the progress of disease from its earliest, barely diagnosable, stages to its final full-blown chronic form. It hardly need be repeated that such facilities offer *the* solution to the cancer and mental disease problems, insofar as the majority of cases are concerned.

    b) *For Treatment, Care of the Chronically Ill, and Habilitative Services.* In terms of dollars, causes for establishment of such facilities are the best supported. They are by no means *first* in order of importance; nevertheless, there are practical reasons why corporations find themselves almost forced to support them heavily, since the dollar issue spells expediency.

These four types of health cause will be taken up one by one, in the order of their *apparent* importance in terms of dollars contributed.

Community Facilities and the Corporation Pocketbook

Any corporation with a sizeable force of employees must eventually face the fact that it has a stake in the local provisions for medical care at all levels. Employees and their families need and demand medical care, and if adequate facilities do not exist in the community, the company will be "pressured" to go into the hospital business. This spells disaster, as evidenced by the experience of corporations both at home and abroad which have done so, willingly or unwillingly. This alternative has been discussed in previous chapters.

The other alternative is for humanologically minded men of business to join with others and help to establish adequate community facilities which are independent of business.

Statistical Facts Have a Sobering Effect

Some statistical facts regarding medical care facilities may have a sobering effect on those who may be asked for advice regarding allocation of company funds toward establishment of such facilities.

The hospital horizontal-care aspect of medicine is between the fifth and tenth largest business enterprise today. In a typical five-year period, it may be expected that approximately $4,000,000,000 will be spent to furnish something over 500,000 beds, which is roughly 20 per cent of the total need.

Another figure for the humanologist to keep in mind is that, for every dollar spent in terms of original capital outlay, it takes 50¢ a year to operate and maintain the facility. In other words, a hospital which costs $3,000,000 to build spends another $3,000,000 in two years of operation.

The following statistics shed light on management's stake in medical-care facilities:

Illness alone costs workers 400,000,000 to 500,000,000 working days a year. These figures express a loss of approximately

2,000,000 men from the labor force each day. Of these unscheduled absences, one-quarter are for personal reasons and the other three-quarters for illness. Accidents and illnesses which are strictly occupational in nature account for only 5 to 6 per cent of the total absences owing to illness in general.

Consequently, both management and labor see their stake in health as a large one. Indeed, many workers receive benefits through health and welfare plans which either management or labor have established. But medical care in this category varies greatly; it is generally incomplete. In many cases, benefits are for emergency care only.

The Policy Makers

The alarming frequency with which purely corrective, curative medical plans have mushroomed indicates that the policy makers are either ill-advised or not advised at all.

In a way, it is not difficult to understand why the average American has come to see medicine in terms of episodic care. He has been led to believe that medicine has made such miraculous advances that problems can be coped with as they arise, through the agency of wonder drugs or fantastic feats of surgery.

As a result, *ex post facto* hospital care has become the focal point of all medical service. Sometimes, the man in the street senses that available hospital and insurance plans do not really solve his medical problems, but not knowing what the true solution might be, he continues to demand the only thing he is familiar with, coverage for hospital care and the treatment of full-blown illness. The urgency of his demand may be measured by the number of holders of some kind of health insurance policy — currently counted in the *hundreds of millions!* [6]

---

[6] A. M. WILSON, "Group Disability Insurance." An address at the symposium on Industrial Medicine, Harvard School of Public Health, April 3-4, 1953.

As explained elsewhere, plans of this sort are usually not health plans at all. They represent sickness-cost insurance to prepay the cost of definitive medical care. They cushion the burden of heavy expense.

By supporting demands for such plans, management has overlooked the fact that hospital and surgical insurance plans only help to compound the headaches created by lack of planning in providing medical-care facilities.

Management has responded to demands in this field by contributing increasing amounts toward the building, equipping, and manning of hospitals. It has done this in an episodic way, with no clear long-range policy except to deal with demands as they happened to come along. It has not carried out the necessary fundamental research that ought to be undertaken to determine just what kind of health-cause contributions will bring about the best results for management, for those employed and their families, and for the community at large.

The Result of Ineffectual Planning

Hospitals are bursting at their seams, and this is the reason private enterprise is being dunned daily, either in terms of capital outlay or in terms of operation and maintenance costs, for hospitals or other facilities which supposedly had highly successful fund-raising drives only two or three years back.

The contributing corporation is perplexed. The hospital to which it must contribute today was, according to the books, put on its financial feet for years to come in its last campaign. It can't possibly need funds again, so soon. And yet a look at the current books will show that it does, indeed, need funds, even more badly than it did in its last campaign.

To get down to cases: one corporation reported receiving as many as 350 requests for support of medical-care facilities within a period of a few months. One request for $25,000 was received from a hospital where only 15 private patients, who were in anyway connected with the corporation in question, had been admitted during the current year. In one case, two surgeons were running a closed hospital and were setting the stage for a 150-bed extension.

Serious-minded businessmen are beginning to wonder: what is the meaning of this? What can be done to hold the status quo? What can be done from a long-range point of view to offset this trend?

The Hill-Burton Construction Act

At least a partial answer is offered by a particular action of the federal government. Since 1946, under the Hill-Burton Construction Act, which provides federal aid where health units are needed, more than 2,000 units have been financially aided. The original effort got off in a hit-or-miss fashion, as a result of a serious bed-shortage problem following World War II, plus the rapid, unanticipated growth of hospital, medical, and surgical insurance.

By 1954, the scope of the Act had been broadened to provide categorical grants to an additional four types of medical-care facilities, including care for the *vertical* patient to relieve the cost load on horizontal-care facilities.

Under the act, any location with 10,000 population is eligible for financial aid. The following four types of facilities are eligible for federal funds;

1. Diagnostic or ambulatory treatment facilities
2. Chronic disease institutions
3. Rehabilitation centers
4. Nursing homes

To support the construction of type (1) units is a natural for private enterprise because of its stake in keeping the vertical employee vertical. Yet, to date, this activity has gotten off to a slow start.

The federal grant is for construction only and represents one-third of the construction cost. The remaining two-thirds of construction costs must be obtained from state, county, or municipal funds, or from contributions on the part of private individuals or organizations, or from a combination of any or all of these sources. Maintenance and operating costs will be the problem of the trustees of the facility in question.

Diagnostic facilities for the vertical man are the kind of facilities which industry needs most and which will pay the greatest dividends in health and effectiveness of those employed. Such facilities might also house emergency units for the sick and injured.

The more effort industry puts into promoting and financing the construction, equipping, and staffing of such units, the less need there will be for industry to deplete its resources in a vain effort to satisfy the insatiable financial demands of the kind of facilities that are geared primarily to the goal of curing and rehabilitating horizontal patients.

Facilities for the Horizontal Patient

It can easily be determined just what any given community needs in the way of facilities for the *horizontal* patient. Industry should see to it that such a determination is made before support is lent to any fund-raising campaign. Moreover, industry has a right to ask that experts examine the present status of any health unit in need of funds. Such an examination will enable the donor to determine from a straight business point of view whether it is wise to expand the facility, tear it down and start anew, or leave it as it is.

Facts pertinent to decisions of this nature should be incorporated into any request for funds, and business has a right to ask for them.

There is no set formula for giving. If a community does not have sufficient diagnostic and ambulatory treatment centers, perhaps the best idea would be for industry to give only token support to horizontal-care units until the basic need has been dealt with. Unless each donation stands up on its individual merits, it will backfire. Private enterprise must tread carefully lest its benevolent efforts compound the obsolescence of existing facilities and pave the way for unprecedented future demands.

Trustees Are Men of Business

Some 40,000 men of business act as trustees for the voluntary and teaching hospitals as they exist at present. These individuals are in a position to see that these facilities are operated on a businesslike basis so that the maximum in health services is produced for a minimum outlay of funds.

A good example of businesslike hospital management was seen a few years back in Boonton, New Jersey, when a group of businessmen actually assumed the leadership they were entitled to assume as trustees. These men directed the building of a hospital along the principles that would be used in building a factory. Cost-consciousness all along the line made possible a structure about $5,000 per bed cheaper than the average. Everything was on one floor, eliminating the cost of elevators, to name just one money-saving item.

This represents progress in cutting *construction costs*. Businessmen, who serve on hospital boards, can apply their knowledge as well in cutting operating costs. Strong emphasis on diagnostic units will make this possible, in that such units will diminish the demands that are made on the more costly functions of local hospitals. The more interest that the businessman takes in this respect, the less his purse and that of his stockholders will be depleted. In other words, a businessman serving on a board of trustees of a hospital has a clear obligation *not* to be docile. As the representative of the employed persons for whom the facility has been designed, he is in an excellent position to wield a positive influence.

A number of pilot undertakings involving a three-way study of the situation by a team of architects, trustees, and professional experts should precede any sizeable commitment of industry for the support of community facilities. Improvisation can be costly and sometimes ruinous.

First, it should be determined that the proposed facility (1) be available to employees; (2) render services at a reasonable price; (3) be duly accredited with a staff that is recognized as competent; (4) be in keeping with the community and human relations program of the contributing company or companies.

The Humanological Principles of Hospital Design

The donors will do well to bear in mind certain principles or, perhaps, more properly "ideals," of hospital design which have yet to gain widespread acceptance, but which will inevitably underlie the hospital of tomorrow. These are essentially humanological principles.

A hospital cannot be a highly profitable business, if it is to serve its humanological purpose. Neither can it be a luxury for those with the highest income or most elaborate hospital-insurance plans. Least of all can it be a plush boarding home for the ambulatory older person or the slightly ill hypochondriac. It must be dedicated to the proposition that medical care must be for the consumer *who* needs it, *when* he needs it. Moreover, its services must be offered at a price this consumer can afford to pay.

In bringing this philosophy to its fullest realization, changes will have to be made, — changes in keeping with the advance of medical science and technology. Some of the changes that will be required in the near future may be almost revolutionary in nature.

Too many health units now have become, through no fault of their own, luxury establishments for the treatment of people who are not actually ill. It takes considerable patience on the part of the dedicated hospital employee not to show displeasure when someone in need of a hospital bed is kept on the waiting list because the next available bed is now being occupied by someone in the upper-income bracket who is paying a premium price for a purely elective plastic operation.

The citizens of tomorrow's communities may be considerably less patient. They will know what they want and need in their community health facilities, and they will not scruple to stand up and be counted. Their demands will be heard and carefully heeded to, and, in the end, they will get what they want, even if some innocent bystanders are hurt in the process. That is the way free enterprise works. If a sufficient number of people want something badly enough and make their demands known loudly enough, someone who is in a position to do something about it will sooner or later come to realize that it is good business, good politics, or both to satisfy these wants.

For this reason, the hospital that is established without thorough consideration for the changing and growing demands and needs of tomorrow's consumer, whether rich or poor, may find itself obsolete before the first bond issue has been paid off. No one group can undertake the kind of forward-looking planning that is needed. It takes teamwork.[7]

The Architect's Role

The role of the architect is a case in point. He is often the forgotten man of hospital planning. Well-intentioned civic leaders and medical people put their heads together, and, after much arguing back and forth, they agree on a "dream hospital" for the community. Then they say to the architect:

---

[7] ROBERT COLLIER PAGE, M.D., "Expanded Horizons for Health Service." An address at the Eighteenth Annual Meeting of the Massachusetts Hospital Association, Boston, January 26, 1954.

"Here, make this dream a reality for such and such a strictly limited amount of money."

He is obliged to tell them sadly that either more money must be raised, or, if this is out of the question, a few dreams will have to go by the board. If he knew more about the needs and resources of the community, perhaps he could work out a suitable alternative. But, no, the money has been raised. The bond issue has been floated, and interest is accumulating. No time can be taken for further deliberations. Ground must be broken on the new project as soon as possible.

If the architect had been in on the planning from the very beginning, even at the tentative stage when it was not yet decided whether the project was even worth while talking about, things would have been different. He would have bridged the gap between dreams and reality. Such a bridge is needed at the outset of any building project.

But the architect knows that he alone is not master of all the hard realities of a hospital project. He knows, for example, that he is no demographer and, hence, not equipped to predict population trends in the community. He is not a professional in the field of labor relations and, probably, cannot evaluate the past, present, and future demands of local union people in health matters; nor can he speak with authority regarding the concessions that will or will not be obtained from management or, alternatively, from government. He is not a hospital administrator, and cannot possibly be permitted the last word on the logistics of obtaining, storing, and efficiently utilizing medical supplies, sheets, blankets, food, hardware, etc., nor can he say with any assurance just how many nurses are needed to man a hospital with a given function and a given number of beds. He is not a doctor and should not be expected to know what sizes and shapes of rooms and technological installations will help the dispenser of medical service perform his function in the best interest of the consumer. Last, but not least, he is not a hospital cost accountant, and even if he were able to predict what services the customer might conceivably demand and how much he might be expected to pay, he would find it more than difficult to determine how to keep the anticipated cost of quality operation from leaping astronomically ahead of anticipated revenues.

If it is true, as every architect realizes, that professional knowledge of construction and design techniques is not enough to build a hospital, then it is equally true that the professional skills and backgrounds of the other members of the hospital team cannot do the job singly.

Other Members of the Team

Neither the nurse, with all her experience in the delicate art of bedside care; nor the administrator, with his knowledge of logistics and operational problems; nor the doctor, with his understanding of the medical needs of the sick human being; nor the up-to-date layman, with his unequivocal wants and needs as to medical care, is equipped to perform adequately the simple primary function of erecting a roof over the head of the acutely ill person.

It Is a Management Function

Management, backed by humanological advice, is best equipped to weld all of these disparate elements into a practical functioning team.

The basic goal is to provide the ill consumer with the very highest quality medical care at a price he can afford to pay, not only tomorrow, but the day, the year, and the decade *after* tomorrow. This means that each member of the planning team must begin with the realization that what is quality medical care today will not necessarily be high-quality medical care 10 years from now, when, in all probability, the hospital will still be paying interest and principal on money borrowed today just to lay the foundation.

In order to approach the problem of planning from all possible angles, the team should include not only architects, doctors, nurses, hospital supervisors, and humanologically advised business leaders, but also well-informed representatives of organized labor, cost accountants, and practical economists, who can express educated opinions as to what the consumer will need, what he will be able to pay, what monetary resources will be available not only at the outset of the project, but in subsequent years, when deficits make an appearance. Valuable team members may be drawn also from community welfare, fraternal, and service organizations which have had past experience in successful as well as unsuccessful civic promotions.

Each New Tried and True Technological Device Must Be Added

One point on which there need be no argument is that the hospital of tomorrow will have to be prepared to incorporate every new tried and true technological device which will improve the lot of the patient, add to the efficiency of the hospital organization, and cut down on personnel, maintenance, repair, or equipment costs of the operation. Somehow it should be communicated to all concerned that the capital cost in hospital financing is only a drop in the bucket compared with maintenance and operating costs after the health unit is built. This is true even in the best planned hospitals. In a poorly planned hospital, it can spell disaster of a sort to a community with the best of intentions.

Predictions of hospital costs based on present-day standards can be dangerous. One can never be sure what new therapeutic discoveries or labor-saving inventions will come tomorrow to make today's ideal floor plans, organizational tables, or expensively purchased technological equipment as obsolete as the appurtenances of a medieval barber-surgeon's pavilion.

The hospital builder can never stand back to look at his completed masterpiece, like the artist who has put his final touch to a landscape, and say: "Ah, there! At last, finished!"

Almost every year technology comes up with some previously undreamed of device that cries to be installed in any hospital that calls itself up to date. An example of recent progress is the television nurses' check-up system, which 10 years ago would have seemed like something out of a science-fiction yarn. This apparatus consists of a television screen at the nurses' station, which can show her, at will, what is going on in any of the bedrooms connected to the system. The advantages of such a system are self-evident. In a few seconds, the nurse on duty can, without leaving her desk, tell which patients are resting comfortably and which are in urgent need of aid or attention. By making a room-to-room check on foot, it might take her up to an hour to obtain the same information; the patient checked last might well be the one now in most need of immediate attention; while she is losing time answering the unnecessary question of a child accidentally awakened in the course of a door-to-door check, another patient may be leaving the hospital through a window.

It is difficult to see how any hospital could be planned in the future without such a device. It will necessitate considerable costly remodeling of yesterday's hospitals. Other possibilities for the future stagger the imagination. It is not inconceivable, for example, that someone will produce electronic, self-balancing stretchers which glide easily from the ambulance to the emergency ward; it takes only a little imagination to envision such devices as long-distance, remote-control diagnostic equipment enabling doctors to make use of specific hospital facilities in their own offices; or elaborate new surgical machinery for the accomplishment of operations now considered too delicate to be undertaken (researchers have reported success in experimental brain surgery using high-frequency sound waves!); or nuclear power plants that will heat, light, and turn the many wheels that make a modern health unit hum with activity. Some inventor, with a real love of humanity, may even invent a bedpan that is not a medieval torture rack, though such a possibility appears to be too good to be true.

There Will Be Changes

It would be futile to attempt to outline in detail the various possible new inventions which will revolutionize hospital operation. The important thing is to take for granted that there *will* be changes, important changes, and that they will cost money. Therefore, it is in the interest of the planning team to create a plant that will combine the best features of modern medical care with the maximum of flexibility to keep pace with the necessary changes imposed by the demands of tomorrow.

The television monitoring device referred to above calls, of course, for extensive installation of new wiring. This may involve ripping out sections of wall and making a number of rooms uninhabitable while the installation is going on. It can be predicted that many of the most important technological advances in medical care will be electronic in nature and give rise to the same sort of problems.

Other advances will have to do with plumbing or heating fixtures and may, again, involve annoying alteration programs which will put valuable bed space temporarily out of use. Still other advances may entail changing entire floor plans; straightening corridors; enlarging rooms; adding rooms; converting single rooms into double rooms or vice versa; changing the entire functions of rooms; relocating kitchens, offices, operating rooms, bedrooms; temporarily removing walls to facilitate the entry of massive equipment; adding, enlarging, or subtracting doors — these are only a few of the possibilities.

Then, in addition to the alterations made necessary by technological progress, there is the routine patchwork that must be done in increasing amounts as the structure succumbs to the anticipated ravages of time. Paint becomes dirty and flakes off the wall, plaster cracks, bricks become loose, doors become warped, pipes become rusty, the insulation on the electric wiring disintegrates and dangerous short circuits develop. Trustees of many of our older hospitals are discovering that their plants are little more than "patches on patches on patches," and before too long a point is bound to be reached where it is more expedient to tear down and start from scratch than to try to add new patches.

Flexible, Movable Interiors

This is undoubtedly the reason that many hospital planners are turning to the principle of movable interiors, which, when made of superior, durable material that is easily kept clean and in condition, is a challenging answer to the problem of creeping obsolescence. Certainly, when rewiring, reinstallation of plumbing or heating fixtures, or alteration of floor plan become necessary, it is easier, faster, and less expensive to deal with a movable wall than to rip into the plaster of the old-fashioned solid wall.

It has been estimated that an average hospital, to keep abreast of the times, will have to relocate at least 5 per cent of its interior walls each year; to plan on less than this is to invite early obsolescence.

Hospital Management Methods

Such organizations as the National Society for the Advancement of Management and Industry's Advisory Board for Hospitals have been working together to study hospital management methods and to assist hospitals in streamlining their operations in accordance with sound business practice.

This is desirable not only from the standpoint of keeping in step with technological progress, but also from the standpoint of combating the problems caused by abnormal labor turnover, rising labor costs, misassignment of hospital workers, intramural strife, and the like. The people who pay for hospitals, from the patient who avails himself of medical service to the philanthropist who helps make up the annual deficit, have a right to demand that hospitals be run every bit as economically as successful corporations; provided, of course, that the "economy" is not the false sort that is affected only by tampering with the quality of the services.

Business sense can be used not only in determining how the hospital shall be built and operated, but also in determining just what the hospital *is*, to its community, and what kind of service it shall offer.

What the Consumer Wants Is Health

The product the consumer wants is health. What this means is that the distributors of medical care can and should do what any solvent businessman does—think in terms of supply and demand and make every effort to put out a product that the consumer wants at a price he is willing to pay.

The consumer wants to regain health when he is ill, but, more important, he wants to maintain it when he is well. The hospital of tomorrow will be dedicated to offering him the brand of health he wants and needs.

Because health is better maintained than regained; because an ounce of prevention is worth a ton of cure, it is inevitable that the largest section of the health unit for tomorrow's needs will be the section devoted to diagnosis and early detection of incipient illness, all of which has to do with the vertical patient. In the old-fashioned health unit this section is often tucked away where one can hardly find it, and the plant as a whole is organized as though every patient in it was the victim of an acute emergency in need of *horizontal* care.

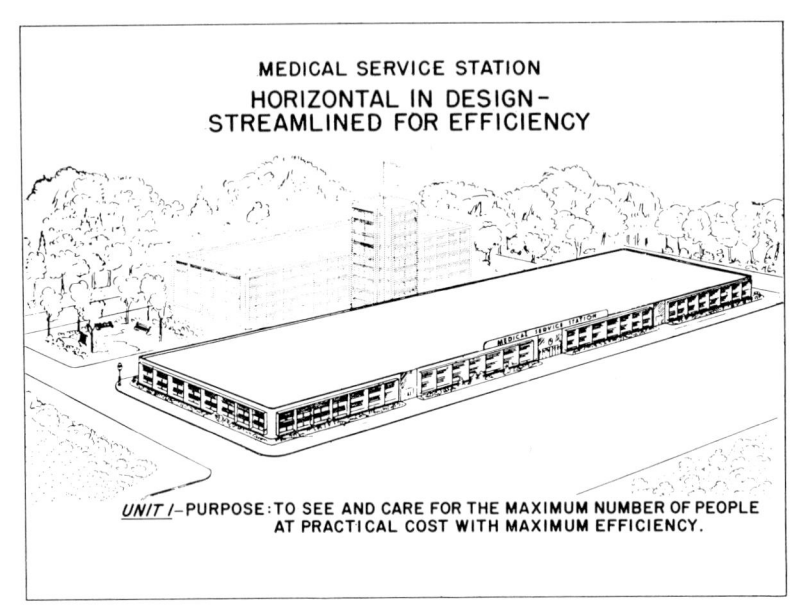

Fig. 6 — HORIZONTAL IN DESIGN—STREAMLINED FOR EFFICIENCY

The Hospital for Tomorrow's Needs

In tomorrow's health unit, this section will be entirely for the individual who is still vertical, on the job, and is not in need of a bed or definitive nursing care. It will be designed to care for a maximum number of vertical patients efficiently at controlled costs. It will be staffed largely by general doctors whose concept of medical service embraces the total needs of the given patient and his entire family. The latest in modern diagnostic and therapeutic facilities will be available at all times. Comparative ease of change will be the keynote of its original design.

The smallest unit in tomorrow's hospital will be that which is the largest and costliest in the average hospital of today and yesterday. This is the unit for the seriously ill, horizontal bed-patient.

Here, the professional staff will consist of graduate bedside nurses. A minimum number of full-time medical special-

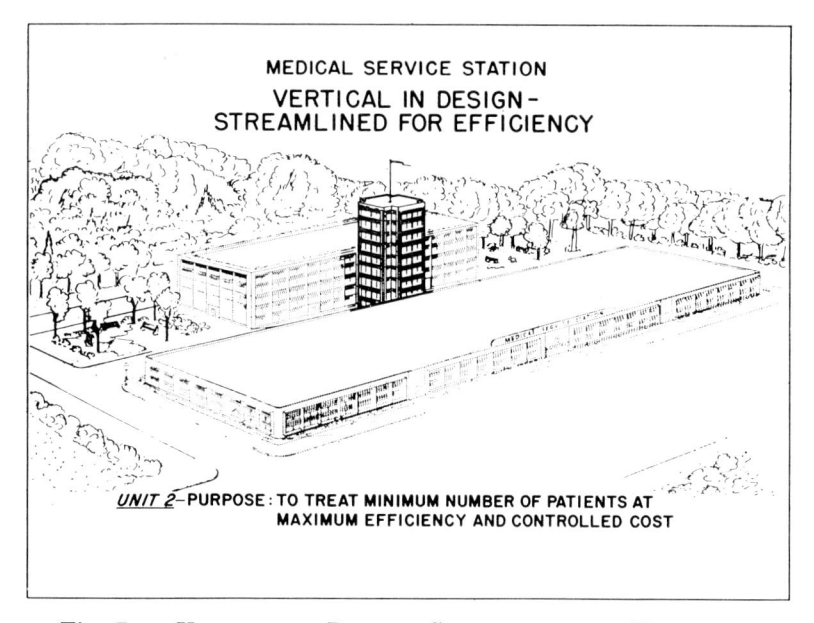

Fig. 7 — VERTICAL IN DESIGN—STREAMLINED FOR EFFICIENCY

ists needs to be on hand. Since this section is only for the patient who is acutely ill, and not for the semi-horizontal patient or the recuperating patient, it can efficiently be limited to a handful of essential facilities, thereby cutting operating costs to the bone. The unit will include the following:
1. Maximum number of private rooms geared to the actual health needs, not luxury wants, of the patient.
2. Graduate and student nurses only in attendance. When the patient is ambulatory, he is immediately transferred to another unit which does not offer complete bed care.
3. For therapeutic purposes, a minimum number of visitors will be allowed, but no flowers, the patient in

this area being too ill to appreciate such things. All of this represents a calculable potential saving in actual operating costs.

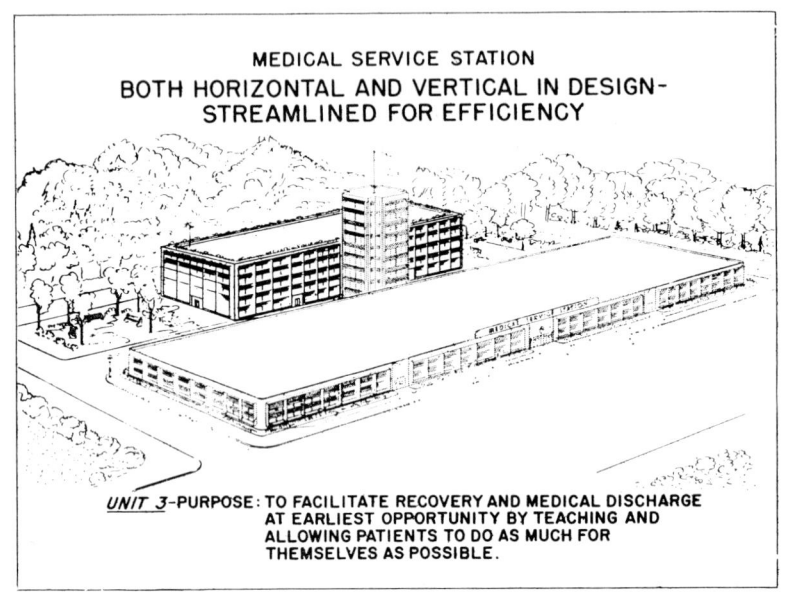

Fig. 8 — BOTH HORIZONTAL AND VERTICAL IN DESIGN

The third area of tomorrow's hospital will be for the patient who can walk around but is not yet ready to go home or back to the job. This includes recuperating patients as well as persons in need of rehabilitation or chronically ill patients, some of whom have to move back and forth sporadically between this unit and the unit for the acutely ill.

This unit will not require the services of many doctors or bedside nurses. It can be largely staffed by paramedical personnel who can round out the needs of the patients therein, with self-service being the prevailing system. A minimum number of graduate nurses would be required; practical nurses and allied nonprofessional help who would encourage the patients to do as much as possible for themselves.

By ready access to unit 1, all diagnostic or educational services would be available. Volume demand will require that this area will be considerably larger than the acute

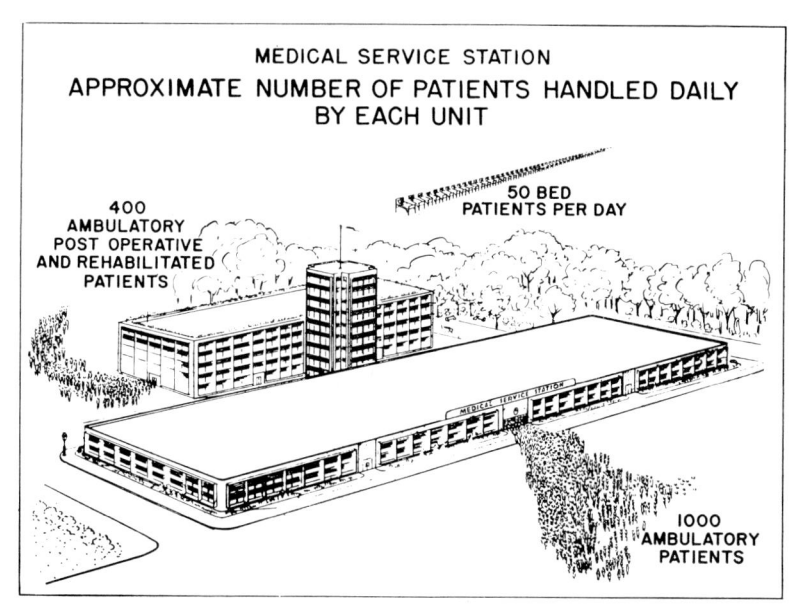

Fig. 9 — Approximate Number of Patients Handled Daily by Each Unit

illness unit, since ambulatory care is the most extensive phase of medical care.

Under most existing organizational plans, ambulatory patients are getting a number of costly bedside services, from nursing care to tray service, that they do not need. When it becomes fully appreciated that semi-horizontal patients make up the majority of persons occupying hospital beds at any given time, it will be obvious that a large part of the operating expense of hospital-bed care areas today represents an unnecessary outlay, in terms of personnel, space, and equipment. This outlay cannot and will never be economically justified.

A health facility of this sort, if constructed with a maximum of flexibility so that it can be redesigned to keep pace with the march of technology and medicine, will put top-quality medical care within reach of everyone at a reasonable price. It will not only put an end to the rising cost of medical care, but it may actually reverse the trend! Costs will diminish not only as a result of more efficient utilization of space and manpower, but also because a widespread program of diagnosis and early detection will keep horizontal bed-care needs at an absolute minimum.

Housing the Chronically Ill and Aged

The humanologist in business needs also to be alert to the situation regarding housing for the chronically ill and aged. Many long-term patients in general hospitals are people who need some medical care occasionally, but who actually do not need to be hospitalized. Yet, these people continue to hold down $25-a-day beds that are urgently needed for acutely ill, short-term patients. The solution to this problem lies in nursing or convalescent homes, which can provide a limited amount of nursing care at a much lower cost than can be found in any hospital. Patients on state or county relief can be taken care of in such homes at great relief to the taxpayer, and usually without any need to dun industry to make up deficiencies in tax revenues, provided the homes are available. In many cases, it would be to the advantage of business to find out what can be done toward making more self-supporting nursing homes available.

Education as a Health Cause

Another problem for the corporate contributor has to do with health causes which come under the heading of education.[8]

---

[8] HARRY A. BULLIS, "Business and Higher Education," an address given at the Midwest Conference on Industry and Higher Education, Chicago, December 16, 1954.

Too often the important aspect of public education, which might prove useful in raising community health standards, is swallowed up by other goals. In other words, education is a by-product. What the "cause" is doing is seeking funds for treatment. Or the "education" turns out, upon investigation, to be mainly propaganda to create notoriety or to stimulate the giving mood in the public at large.

Attention Drawn to One Specific Disease Entity

Quite often this educational material has the effect of focusing aroused public attention on one specific disease or health problem, to the point where health maintenance in general is overlooked entirely. The subject of cancer is one example of a specific disease entity. There are a great many causes of this nature and the number continues to grow.

Circulars having to do with demands for such causes fall pretty much in the same pattern. So do their methods of approach. They have a tendency to pull at the heart strings. Here is where the "target" of these appeals, the corporation, is greatly in need of the advice of an expert humanological consultant, in order to avoid becoming the unwitting, though well-intentioned, sponsor of some organizations which will in the long run tend to defeat his very purpose of giving. Here the humanologist has to differentiate between pure philanthropy and the use of corporate funds with the view of getting some return for money spent.

To return to the subject of cancer briefly, it would seem that contributors would have a right to know:
1. What percentage of the money raised is being spent on the diagnosis and early detection of cancer at a time when something positive can be done about it?
2. How much of this money is going toward acquiring an understanding of the basic mechanism of carcinogenesis, which includes etiology, together with an understanding of the food and metabolism of carcinogens, and a

more up-to-date appraisal of the biological make-up of the individual who has cancer, and how this may differ from the individual who does not have cancer?

The reports emanating from the associations dealing with cancer studies are not drawn up in such a way that one can readily obtain the answers to these questions.

A recent survey showed that much duplication of effort has existed in this field. For example, surgical dressings, blood-donor service, and certain aspects of visiting-nurse service offered by associations operating in the cancer field have also been offered by the American Red Cross. The survey also showed that while certain laymen are paid for their undertakings in the cancer effort, the professions have been expected to donate their services gratis.

Heart disease and its ramifications are known to be high on the list of diseases that kill. What is needed in this respect is a qualified team which will study and interpret diseases of the cardiovascular system in terms of the aging process.

As for mental health, a high percentage of absenteeism and lowered work efficiency occurs among those who are maladjusted to their work or home environment. The field of preventive psychiatry is of tremendous interest to business; more so, in fact, than studies having to do with full-blown mental illness.

It is difficult, however, to find out just how much time is being devoted by the various organizations in the mental health field to this preventive aspect. Most effort seems to be devoted to vast and sometimes futile efforts to find cures for those who have already suffered complete psychiatric breakdowns, or to finance institutions for the housing and care of these persons. How much more practical and economical to prevent these breakdowns by detecting them in their incipiency! Think of these advantages to industry and to the employed individual himself!

Problem Drinking as a Specific Health Cause

The literature on problem drinking[9] is full of hopeful ways and means for curing the problem drinker after his problem has become full-blown and he has created disturbance either on the job or at home. Very little, if any, effort has been put into the more constructive undertaking of finding the problem drinker before the condition has become chronic and the suffering individual unemployable. The surface of this problem has barely been scratched. Vast amounts of research into this and the narcotics problem are still in order.

Both of these community problems with specific legal, medical, and moral aspects are of vital interest to each and every member of the humanological team. The various aspects are intertwined in such a way that no affected individual can be looked at *solely* from a legal, a medical, or a moral-religious point of view.

Finding the Problem Drinker — A Basic Medicolegal Problem

The type of study which could pay humanological dividends on every dollar contributed by the corporate donor would: (1) determine the facts with regard to problem drinking and narcotics addiction; (2) effect practical and constructive changes in placement and training practices; (3) outline policy for future guidance in regard to mantalent development.

Not only the industrial humanologist and the religious or educational leader, but law officers, judges and lawmakers, and educators in the given community would have a part to play in such a proposed study. The stake of community, government, and law-enforcement officials is clear-cut.

---

[9] JOHN M. MURTAGH, "The 'Skid Row' Alcoholic." An address at the Eleventh Annual Meeting of the National Committee on Alcoholism, New York, March 29, 1956.

Stake of Community and Law-Enforcement Officials

Municipal and other local officials are constantly under fire for seeming to fail to fight vigorously against "alcoholism."

The constant reappearance of the same problem drinkers in the courts, publicized as "a disgrace to our social system," requires that effective action be taken to determine the following three points:

1. The exact role that the courts must play in alleviating this constant burden of expense upon the area budget;
2. What *is* the modern approach toward straightening out the situation, from the point of view of all groups affected — social, professional, managerial, labor;
3. A course of action for the future.

People who become intoxicated become problems that the law is expected to solve whenever they impinge upon the public (1) by keeping the neighbors awake; (2) by brawling in public places; (3) by driving automobiles dangerously; and (4) by panhandling.

It would be well for every policeman on the beat and every magistrate on the bench, to know how to evaluate and deal with any "drunk" with whom he is confronted. It is presumed that he is guided by wise laws propounded by Solons fully conversant with all aspects of the alcohol problem. Nevertheless, interested individuals with even a superficial acquaintance with the problem are generally agreed on the following points:

1. The quick municipal discipline that may be called for in one case of alcoholism may be a grave mistake in another case that appears identical on the surface.
2. The law that deters one incipient intoxicated person may have the opposite effect on another.
3. A large number of Utopian laws concocted by teetotalers have proved to be worthless in checking the problem despite time and money spent in enforcing them.

Since law enforcement agencies will ordinarily be the first ones called in cases of public misbehavior stemming from the use of alcohol or narcotics, it is essential that these agencies be given data that can be used for evaluating and dealing with the cases brought to their attention.

Laws having to do with the control of alcoholic misbehavior or the prevention of public intoxication must be based on facts, not upon unrealistic surmise.

Every policeman and certainly every judge must have some sort of yardstick which will enable him to determine whether, for the greatest good of society, the intoxicated individual should be dealt with at the legal level or by some other agency. If the individual should be dealt with at the legal level, he should be dealt with constructively. If referred to an agency, he must be referred to an effective one.

Unknown Facts, Which Should, Must, and Can Be Determined

1. The limits of law enforcement procedures in cases of public intoxication.
2. The feasibility of classifying such individuals.
3. The methods of disciplining or guiding them according to the best interests of society.
4. The effectiveness of existing statutes and usages designed to prevent, curb, or correct alcoholic misbehavior.
5. The prospects for working cooperatively with nonlegal agencies in the establishment of codes and usages that will be effective in reducing the incidence of actionable intoxication either (*a*) by preventing it, or (*b*) by rehabilitating its victims (i.e. placing square pegs in square holes).

All persons who are arrested for misbehavior when intoxicated should be studied. This has never been done thoroughly and scientifically. These individuals must be classified in a practical manner, comprehensible at the level of the rookie patrolman. Their drinking backgrounds, habits, and predictable futures must be determined by qualified researchers with reference to their total environment and health status.

A police officer, armed with the data gleaned in the proposed study, should be able to ascertain, by asking a few pertinent questions, whether he is dealing with (1) an over-enthusiastic partygoer or "visiting fireman"; (2) a neuropsychiatric drinker; (3) a dipsomaniac or sporadic binge drinker; (4) an organically-sick or senile person with low alcohol tolerance.

A magistrate should know what needs to be done with the properly classified individual. He should be able to decide whether (1) to put him in jail, (2) to turn him over to a social agency, (3) to send him to a hospital, or, (4) to report him to his employer.

The study should be organized more or less in accordance with the following outline:

    A. Sources
       1. Some focal point — perhaps a large, unused home or an antiquated institutional building where all persons legally apprehended for misbehavior when intoxicated may be sent
       2. Police court records and police blotters
       3. Medical, social and industrial authorities concerned with problem drinking
    B. Intoxicated "Offenders"
       1. Individual rights
       2. Individual health status
           a) Consideration of age and mental and physical adjustment to it
           b) Classification as a problem drinker
           c) The stage he has reached as a problem drinker
       3. Individual employment status

4. His social environment, family, background, etc.
5. His source of alcohol and drinking customs
6. His tolerance to alcohol
7. His inner resources for combatting problem
C. Effects of Alcoholic Misbehavior
1. Cost to the municipality (or county or district)
2. Number of people arrested
3. Nonlegal damage as seen by selected medical, social, or industrial experts
D. Existing Efforts to Cope with Problems
1. Legal efforts
    a) Preventive, controlling, or disciplinary statutes — History
    b) Usages, precedents, varying interpretations — History
2. Medical, sociological, and industrial efforts
    a) Early detection and prevention
    b) Treatment
    c) Rehabilitation
E. Possible Changes in Approach to Problem
1. New statutes
2. New usages and interpretations
3. Programs of cooperation with medical, social, and industrial health agencies

The cost and scope of such a project would, of course, be determined by the size of the area under scrutiny and the extent of the problem there. A preliminary investigation into the extent of the problem alone should be sufficient to point up the need for the study.

A Fundamental Approach — Preventive in Nature

Such a fundamental approach would be strictly preventive in nature. The result would entail:
1. An objective determination of the incidence of the disorder in industry. This can be accomplished by

careful case studies and a central registry system within each industry.
2. The development of a standard pattern for studying individual cases. This should include complete medical data and information about the employee's home and work environment.
3. Careful managerial study of work factors which may encourage problem drinking.
4. The development of educational programs about the alcohol problem for supervisory and medical personnel, for all workers, and, ultimately, for potential workers.
5. Research in the field of improved personnel-selection techniques to reduce the number of potential or actual problem drinkers employed. Similar development of supervisory techniques, leading to the early recognition of problem drinkers and constructive help with their problems.
6. Some type of practical support for a treatment center for problem drinking to which cases can be referred. Whether this will be community or industry sponsored will depend on the particular industry, its location, etc. In either case, it should be located away from the industry.

Human Relations

Ultimately, the solution to problem drinking lies in the realm of human relations. The first responsibility belongs to those whose job it is to accept individuals for employment and to place them according to their individual working potentialities. The medical profession's role would then become a relatively minor one, restricted to the evaluation of the part the employee's health status plays in the total picture.

Until this broad concept of the nature of problem drinking is accepted, the medical profession may, by default, play a major role. As long as this unsatisfactory situation continues, the results will not be too fruitful. Just as the basic causes of the problem are not medical in character, in the writer's opinion, so the solution is not primarily a medical one. The medical profession must accept its present palliative role only to further the basic education of all other component parts of industry in what is their proper responsibility. This is the role of each member of the humanological team.

Purely Philanthropic Interest

There are many reputable health organizations, such as the National Foundation for Infantile Paralysis, for Multiple Sclerosis, for Cerebral Palsy — to name a few — in which industry, from a stockholder's point of view, would have a purely *philanthropic interest*. One organization, the National Health Council, devotes its time to coordinating the activities of these many health causes.

The Family Service Association[10]

In the realm of pure philanthropy, an organization such as the Family Service Association is worthy of the careful consideration of the humanologically minded corporate giver. This organization helps distressed, unhappy families work out their problems while they still can be worked out. Its results are indirect but tangible in terms of community relations.

Cooperation with an organization of this sort may well be more than purely monetary. The industrial humanologist can and should, when feasible, offer his judgment and experience, as well as corporate funds, to the community endeavor to detect the early "plague-spots" of unhappiness. Unhappiness has a way of reverberating throughout whole areas.

---

[10] ROBERT COLLIER PAGE, M.D., "The High Cost of Unhappy Living." An address given at the Family Service Association of America in the Columbus Hotel, Miami, December 18, 1956.

The humanologist should spark team effort in this endeavor. Supervisory personnel, trained to spot symptoms of situational distress, should be in liaison with the welfare association which is functioning in this field. Family doctors must be brought into the team, as may family lawyers, clergymen, teachers, neighborhood policemen, and the like, all of whom are in special positions to see different aspects of "trouble a brewing" in a given family organization.

Unhappiness

Unhappiness in its infancy is a simple social problem. In full bloom, it becomes a general problem of staggering proportions. To a considerable extent it becomes a health problem in terms of situational illness, and ultimately of functional disorders. Maladjustment at first responds to common sense, but if it is allowed to burgeon, it will inevitably manifest itself in medical symptoms, which are vague and difficult to diagnose in the beginning. In time, the unhappy individual may not only present tangible, diagnosable symptoms, but he may be finding refuge in alcohol or in building psychiatric fantasies. The end result may be that he will become one of the unemployables of industry, which becomes a costly burden for society to maintain.

Hence, even "pure" philanthropy can be shown to have a dollars-and-cents value, though such a value may be difficult to translate into hard-and-exact figures.

Medical Education

A most important health cause, which demands more corporate attention than it enjoys at present, is that of medical education.

The problem is a national one. The present 83 medical schools are located in 59 cities. These 59 cities cannot be expected to meet the total cost of medical training.

Of the 121 universities[11] in the United States, 70 have medical schools. The *total* budget of the 121 universities in a particular given year was in the neighborhood of $1,500,000,000. The budget of the 70 medical schools for the same year approximated $100,000,000; or almost 10 per cent of the total budgets of the total for the entire 121!

Furthermore, in the 70 universities with medical schools, between 30 and 40 per cent of the total operating budget was required to maintain just the medical aspect of their activities. The enrollment of students, however, represented only 5 to 7 per cent of the total student body.

It is obvious, then, that medical training is the most expensive kind of higher education. It is four times as expensive as general university training. This differential is explained in part by the cost of teachers' salaries. The ratio of full-time teachers to students in medical schools is 1 to 4, and in some areas of medical training 1 to 1, as compared to a ratio of 1 to 30 in other branches of college training.

Students from 43 states attend the schools in 56 cities, and the medical graduates serve communities all over the nation.

The humanological advisor should take these facts into consideration when asked to determine what amount a given business should give to medical education.

For every dollar given to the specialized agencies in the health field, $1,000 may well be earmarked for medical education and research in a branch of the health field which has something to do with the needs of the employed.

Fundamental Research as a Health Cause

As for fundamental research, management's needs are great, but, again, as in the field of medical education, they

---

[11] KEITH G. FUNSTON, "What's Business Doing about the Fate of Our Colleges?" An address at the 125th Anniversary Convocation of the University of Richmond, Richmond, October 29, 1955.

are rarely fully appreciated. Too many people are prone to think of research in terms of a deliberate search for a specific cure. They have a picture of a doctor walking into his laboratory with the idea in mind of sitting down with his retorts and test tubes and staying there until he has discovered some long-awaited cure. Occasionally, someone stumbles onto a cure in this manner, but the greatest results have come from fundamental studies which had no specific cure in mind.

The research that is most needed by the humanologist is the kind that will tell more about the nature of man; the kind that will strive to answer the following questions: How does man live? Why does he live? What is life? What is health? What is happiness?

Genetics

An important field of basic research is genetics, which may not point dramatically toward a specific cure for a specific ailment, but which is all-important in the true understanding of a given complaint. Plants, animals, and men are what they are because of genes passed on from generation to generation. All of this is obvious to the schooled humanologist, but it often seems to be foreign to the understanding of the influential corporate contributor.

Biology, being the study of all living organisms, is, of course, at the root of all medicine. A structure of philanthropy which takes into consideration support of specific research without providing for biological studies in general is like a blueprint for a tall building without a foundation.

Fundamental research must take in more than biology. Living organisms cannot be fully grasped without complementary studies in the fields of chemistry, physics, and the like. Research in mental illness, for example, must expand to embrace behavior considered abnormal to one culture though normal in another. All fields of fundamental research must be considered as interlocked; together they take in the total environment of the total man.

This kind of knowledge, once it is obtained, will teach management how to deal with its most important resource, manpower or "mantalent," in such a way as to gain the utmost in efficient and profitable operation and to add to the well-being of mankind.

The Best Laboratory Is Within Industry

Perhaps one of the best laboratories in which to study the nature of man while he is functioning is a large industrial enterprise where he can be seen in varying situations at varying ages meeting his basic challenge, that of getting a living. Private enterprise, therefore, may be asked to contribute more than money toward research; it may be asked also to *participate* actively in research.

Industry's contributions to research today will pay off tomorrow, in terms of increased understanding of man, and ultimately in terms of dollars and cents. Its contributions to horizontal-care facilities or to causes that seek only to treat those already sick will not pay off tomorrow except in increased demands for contributions. The proportion of funds allocated to these various causes may be determined with these factors foremost in the mind of management.

Competitive Enterprise Holds the Future of American Medicine in Its Hands

Corporate giving can no longer be episodic, that is, designed merely to fill momentary gaps that seem to need to be filled at a given time. Moreover, it should be remembered that industry is not in the business of philanthropy purely as a "public relations gesture"; industry holds the future of American medicine in its hands. The progress of industry is inescapably tied up with the progress of medical care;

both have to do with the progress of health and happiness for all mankind.

The fact that management is becoming aware of the need for positive policies with regard to corporate giving to health causes is increasingly evident in statements made by business leaders. The author had occasion to address the representatives of leading corporations at a management seminar on the subject of "Corporate Giving." Following the seminar a letter was sent to all participants seeking the following information: (1) How important is the problem of corporate giving to health causes in your company? (2) How is this problem dealt with in your company? (3) How is the health aspect of manpower development evaluated and related to the company contributions program as well as to the job careers of the employees concerned?

A total of 51 letters were sent. There were 21 replies, most of them of considerable length and indicative of profound interest in some aspect of corporate giving to health problems. A great deal of dissatisfaction with most current policies was indicated.

Five of the answering letters spoke emphatically of the need for bedrock research which would cut through the periphery and get down to the fundamentals of life, health, and productivity.

Many of the correspondents stated that their companies were currently engaged in supporting peripheral, specific "applied" research, but would welcome guidance in establishing more constructive policies. The reason for this current misplaced emphasis, according to one of the correspondents, is that *specific* research excites interest at the moment; it may have dramatic appeal. If a campaign undertakes to raise funds to supply children with crutches, for example, there may be several newspaper pictures of actual youngsters receiving their crutches at an annual wind-up banquet, perhaps the company president is pictured handing out one pair of crutches, and in this case, the practical contributor has a concrete idea of what he has accomplished, or *thinks* he has.

With *fundamental* research, however, according to this correspondent, one never knows exactly where one is going to come out. The ultimate object of such research is the secret of life itself, and since no one knows what this secret is prior to its discovery, no one can express it in concrete terms, such as newspaper photographs. The practical man, then, has a tendency to ask for examples. Many top men in management believe that fundamental research always pays off, and would not be unwilling to gamble on it, but too many of their colleagues are "from Missouri." They must have a picture, and, of course, no picture is, or can be, forthcoming.

The next most pressing matter listed by the correspondents was that of re-examining and overhauling corporate philanthropic policy as a whole. Several of the representatives of business expressed awareness of a need for change in policy, but admitted that, as businessmen, they had not the man hours available to devote to thorough study and determination of the best possible course to take. (Obviously, this represents a groping toward the concept of the humanological team having a working liaison with appropriate consulting experts outside the company proper.)

Three of the correspondents spoke of company programs already launched with respect to support of medical education. There still seems to be a tendency among many corporate "benefactors" to class this educational specialty under the heading of education in general. Unfortunately, contributions to colleges, even if they all happen to have medical schools, will not necessarily go toward support of medical education unless they are so earmarked. The constructive solution is to put strings on a given contribution, specifying that it is to be used for a designated purpose, i. e., needed improvement of educational facilities, or for soundly conceived scholarships.

Taken as a group, these replies indicated that corporate thinking has made strides in recent years. Business leaders are recognizing the fact that corporate giving can be more tangible than pure philanthropy. Giving can pay dividends; the problem is how and where to give to make this possible. The general awakening of interest in fundamental research is a strong indication of the shift in corporate thinking.

Management has learned that the foundation of any business is man, not machinery, and it is beginning to direct its attention away from peripheral causes which do not answer the basic questions: What is man? How can he best be *constructed* toward optimum happiness and productivity?

What is needed now is planning, advice, guidance, study, and correlation of efforts toward a desirable end. This is a challenge for every man of management with an appreciation of humanology, as well as for each member of the humanological team.

REFERENCES

AMERICAN COUNCIL ON MEDICAL EDUCATION AND HOSPITALS. "Annual Report." The *Journal of the American Medical Association,* (November 14, 1959), p. 1507.

"Business Week Report to Executives on Foundations — Why Business." *Business Week,* (June 19, 1954), pp. 166-178.

EELLS, RICHARD SEDRIC FOX. *Corporation Giving in a Free Society.* New York: Harper and Brothers, 1956.

Fund Raising Muddle, The. "Special National Report." *Newsweek,* (June 15, 1959), p. 31.

GUNDERSEN, GUNNAR, M.D. "The Doctor's Stake in Hospital Accreditation." *Trustee,* VII (No. 5, May, 1954).

"Health Agency Donations." The *Journal of the American Medical Association,* CLXX (No. 9, June 27, 1959), pp. 146-1090.

"The Hill-Burton Act 1946." *The Journal of the American Medical Association,* CLI (February 28, 1953), p. 765.

"How Much, How Soon?" *Time,* LXXII (No. 3, July 21, 1958), pp. 34-36.

LARABEE, BYRON H. "Responsibility of Industry — in Support of Research on Tropical Disease." *Industry and Tropical Health III*, published for the Industrial Council for Tropical Health by the Harvard School of Public Health, 1957, pp. 63-67.

NATIONAL FOUNDATION FOR INFANTILE PARALYSIS. *Expanded Program No. 5*. New York: National Foundation for Infantile Paralysis, 1958.

NATIONAL FUND FOR MEDICAL EDUCATION. *The Medical School — What It means to You*. New York: National Fund for Medical Education, 1954.

PAGE, ROBERT COLLIER, M.D. "Medical Research and Private Enterprise." *Industrial Medicine and Surgery*, XXIII (No. 1, January, 1954), pp. 42-43.

——————. "How Tomorrow's Hospital Can Be Built Today." *Southern Hospitals*, XXV (No. 11, November, 1957).

PAGE, ROBERT COLLIER, M.D., and HAWKINS, EDWARD R., M.D. "Finding the Problem Drinker: Case Studies in an Industrial Health Problem." *Quarterly Journal of Studies on Alcohol*, XIV (December, 1953), pp. 586-595.

"Partners in Citizenship." *Shell News*, (June, 1955).

SLOAN, ALFRED PRITCHARD, JR. "Big Business Must Help Our Colleges." *Collier's*, CXXVII (June 2, 1951), pp. 13-15.

The Why and How of Corporate Giving. New York: National Industrial Conference Board, 1956.

6

Labor Dictates Medical Policy

A PROMINENT ECONOMIST once said, somewhat facetiously, that the United States economy is a "laboristic" one. Although it was realized by the author of this statement that this is not entirely true because production of wealth in this country is still owing to a largely free, capitalistic economy, still there is no doubt that organized labor has been vastly influential not only in the distribution of this wealth in terms of purchasing power, but in establishing the humanological rules applying to the methods of wealth production.

By mid-century, the number of union members in the United States had come close to the 20,000,000 mark. This figure represented more than a third of the nonagricultural working population. Union membership continues to increase, as new areas of employment are organized, not only in numbers, as the population increases, but also in proportion to the total population.

The Organized Employee of Today

The organized employee of today is quite different from his counterpart of 50 years ago, whose wants and interests were wholly basic in the demand for food, shelter, and clothing, and whose intellect tended to be suppressed owing to a lack of education and a challenging occupation.

The modern worker is likely to be a high-school product. The amount of secondary schooling in his background has increased rapidly within recent years. For example, as recently as 1940, the average worker had not advanced past the first semester of his freshman year in high school. By 1957, the average worker was found to have gone as far as the middle of the second semester of his junior year. A Census Bureau report for the same year showed that 18 per cent of workers had at least one year of college training, while over 9 per cent were college graduates.

Being better educated, the present-day union member is necessarily a more complicated person than his nineteenth-century counterpart. It may be assumed that he has come to take elemental necessities of food, shelter, and clothing very much for granted. His wants and interests now have to do with matters the "wage slave" of an earlier day would have considered out of reach and hence vain fancies. The modern worker is interested in such matters as happiness and health. He assumes that his job will offer him the wherewithal for bare existence. Over and above this, he wants and is willing to fight for the pay and the leisure time to pursue "the good life" plus the health to enjoy it.

Owing to his superior education, the modern worker is not so naive as to believe that his employer can give him happiness tied in a neat package. The "pursuit of happiness" is a personal matter; the most he can demand in this respect is the time and recompense which will enable him to carry on his own pursuit according to his individual rights.

With regard to health, however, he has come to realize that his needs and consequent demands cannot be expressed so simply in terms of wages and hours. Health to enjoy the good life is more than a matter of sick-leave provisions or sickness indemnification plans. This sort of thing he considers as elemental as food, shelter, and clothing; he takes for granted. Even when sick-leave provisions and indemnification plans are comparatively liberal, they do not really solve the health problem in terms of the standard of living he must come to accept as a matter of course.

The Required Health Program

The modern worker feels that he is entitled to and requires a health program (1) to prevent all preventable health problems; (2) to detect incipient health problems at a time when they can be corrected at a reasonable cost; (3) to keep him vertical and productive for the maximum possible period; (4) to treat him and rehabilitate him in the event of unforeseeable major medical catastrophes; (5) to educate him in the construction as well as maintenance of optimum health.

With an eye to his budget, plus "a little bit o' luck," the North American worker of today can live better than his counterpart of any other conceivable area in the world or era in time. Costs of living have climbed to unprecedented levels, *but* wages, productivity, and purchasing power have more than kept pace. This has been owing in part to technological progress and in part to the efforts of organized labor.

A Major Medical Catastrophe

But there is a catch. If wages have more or less kept up with the rise of normal living, the costs of hospital or other *after-the-fact* medical care have not kept pace. It is not economically likely that they will ever do so. When the average worker runs out of luck healthwise, or when he or someone in his family is put out of action by a major medical catastrophe or is limited in action by a creeping chronic illness, he is in a worse position than his grandfather would have been. He is, by the same token, not as well situated as his more poorly paid counterpart in some underdeveloped part of the world. All of his years of economic gain, frugality, and budget-conscious accumulation of the accouterments of the good life are wiped out in a blow.

Thoughtful leaders[1,2,3] of organized labor have recognized this appalling fact for many years, and have pursued or discussed a variety of possible solutions.

In Search of a Solution

Until recent years, there have been two principal types of solutions sought through collective bargaining, political action, or cooperative effort. These have been: (1) Complete after-the-fact medical and hospital care to be furnished free or at a nominal cost (*a*) by the employer, or (*b*) by government, or (*c*) by the union, or (*d*) by a combination (*a* and *c*); (2) insurance plans partially subsidized out of company profits, taxes, or union welfare funds to indemnify the costs of medical treatment and/or hospitalization.

Assembly Line Doctors

A more detailed study of these "solutions" later in this chapter will show that not only have they failed to increase the quality or lower the costs of medical care, but in the long run they have accomplished exactly the opposite. They will continue to do so. The modern hospital is perpetually overcrowded and understaffed. Any plan which makes it financially easy for an individual to avail himself of hospital bed space, whether he needs it badly or not, can only compound this condition. Costs of hospital care must rise accordingly, and no matter who is footing the immediate bill, the cost must eventually come back to the user of medical services in increased premium rates, taxes, union dues,

---

[1] CLINTON S. GOLDEN, "The Problems of Health as Seen by Industrial Workers." An address before the Conference on Education of Physicians for Industry at the Mellon Institute, Pittsburgh, December 6, 1955.

[2] FREDERICK D. MOTT, M.D., "Why the Community Health Association of Detroit Was Organized." An address to the Economic Club of Detroit, February 10, 1958.

[3] NATIONAL CONFERENCE ON LABOR HEALTH SERVICES, American Labor Health Association, Washington, D.C., June 16-17, 1958.

and prices of goods and services. Doctors besieged by subsidized patients with multifarious minor health problems must, of necessity, become "assembly-line doctors" and let quality service go by the board. This has been seen in England under the National Health Service[4]; it has been seen in tropical areas where foreign corporations have, for the sake of expediency, undertaken to dispense complete medical care "package deals" to all workers and their families. The quality of the medical care is in inverse ratio to the demand for it; the cost, approximately geometric in ratio to the demand, is passed on in one way or another to the consumer, either in the form of direct levies or indirect payroll limitations or inflationary costs of other necessities.

Need for Constructive Philosophy

Educated labor leaders have become aware of this and are beginning to move over to the right track by seeking medical care with strong *quality* controls, instead of simple indemnification or mass-produced medical service. The danger here is that, even with the most rigid quality controls, medical care can turn into an over-priced commodity, or go the other way and lose in quality, unless "quality medical care" is properly defined at the outset and established according to a constructive philosophy.

---

[4] Cost of National Health Service 1955-56 — £557,800,000; Cost of National Health Service 1956-57 — £611,700,000. The "Daily Telegraph & Morning Post," June 11, 1958 (London, England), reports:

"The Ministry of Health, after nearly 10 years' experience of nationalized hospitals, is known to favor far-reaching changes in the staffing structure. The staffing arrangements should, it believes, offer a satisfactory career to all specialists and other physicians and surgeons in the service.

"All is far from well in the service today. Perhaps the chief cause of discontent is the plight of 'time expired' senior registrars, who after years of special training see no prospect of their reaching consultant status.

"This is having its effect on the recruitment of future consultants. There is already a serious lack of junior staff in outlying hospitals. Only the presence of Commonwealth and foreign doctors has prevented a breakdown of the service in some parts."

Historical Development of Collective Bargaining

Before going into a discussion of current constructive thinking in this regard and the trends of today and tomorrow in enlightened collective bargaining, it may be illuminating to study the background of this important moving force, organized labor, and the historical development of its major tool, collective bargaining. The humanologist is playing an increasingly important role in the shaping of this tool to constructive ends; it behooves him to be thoroughly familiar with all its ramifications.

Although local craft unions of specially trained workers, particularly in areas where demand exceeded supply, existed prior to the Civil War, the American "labor movement" cannot be said to have really gotten underway until the formation of the American Federation of Labor in 1881. Another contemporary labor movement, dedicated to radical political reform and overthrow of the "big interests," was carried out under the aegis of the Knights of Labor, but the revolutionary nature of this group inhibited its growth and organizational strength in the United States. After a few unsuccessful strikes, this organization went into a decline.

More Today Than Yesterday; More Tomorrow Than Today

Samuel Gompers, one of the leading organizers of the rival American Federation of Labor, was hard-headed enough to realize that the way for any labor organization to make headway at that hostile period was for it to work realistically within the system that prevailed rather than to destroy itself in a vain effort to overthrow that system. His idea was that organized labor should work for higher wages and better work conditions little by little, and not commit itself to an effort to attain ultimate goals at once. The slogan was "more today than yesterday; more tomorrow than today."

An important principle of Gompers was federally based organization by craft specialty. The central organization of each national union covering a particular craft specialty had sole jurisdiction over its members. In other words, once the national union had obtained recognized jurisdiction over any given local group of workers, the latter unit could not be drawn into the undermining jurisdiction of another, possibly management-sponsored, organization. As for government, Gompers believed that it should stay out of collective bargaining altogether, even if an incumbent administration was known to be friendly to labor. He felt that it was unwise to commit labor to any one political party.

The American Federation of Labor reached its first peak of importance in World War I, with some 5,000,000 members. It looked then as though federally based labor organizations were about to win acceptance by both government and industry as legitimate and inevitable factors in the economic structure of the United States. True, there was some rather violent antiunion sentiment among business leaders. Certain management organizations showed an aggressive spirit. But the American Federation of Labor leaders were encouraged by the rapid advances of the recent past and believed that the continued growth of unionism could not be halted or even slowed down by this sort of opposition.

The seemingly permanent prosperity of the 1920's proved to be a surprise ally to the antiunion forces. Employment was high; prices were stable. As a result, discontent was low. Potential union members could not be interested in organization; many were dabbling in the stock market on a microscopic scale and did not even identify themselves with labor.

As a result of this, plus the death of Gompers in 1924, the movement lost considerable momentum. By the time of the depression, membership had fallen off in the amount of roughly 2,000,000 workers.

John L. Lewis and Sidney Hillman

A new impetus to the union movement was provided in 1935 by John L. Lewis of the United Mine Workers and Sidney Hillman of the Amalgamated Clothing Workers Union, who joined forces to organize the mass-production industries on an industry-wide basis instead of along the national craft-specialty lines laid out by Gompers. It was felt that the traditional American Federation of Labor setup was too unwieldy in a crisis. The steel industry, for example, had included about 40 different craft-specialty units, each under the jurisdiction of a different national union organization at the time of the 1919 steel strike, and many labor leaders believed that this fact was the cause of the failure of that strike.

The committee formed by Lewis and Hillman was launched under the aegis of the American Federation of Labor, but was expelled from the federation for taking issue with its traditional principle of centralized craft unions. The committee then became the nucleus of the Congress of Industrial Organizations, with Lewis as president. By aggressive and energetic action, this new movement soon organized nearly all of the major mass-production industries, despite bitter opposition. The American Federation of Labor, profiting from this example, began to build up industry-wide organizations, though the craft unions continued to play an important part in the federation.

For a number of years, warfare existed between the American Federation of Labor and the Congress of Industrial Organizations. In 1955, they finally decided to merge. The union movement had arrived. Since that time through combined cooperation labor has shown, and is showing a concerted and constructive interest in the *true health* of the employed individual.

Today's Labor Leader Must Be an Educated, Intelligent, Imaginative Administrator

In achieving its present status, organized labor raised the standard of income and educational background of the average worker, but at the same time it created new and unprecedented complex problems for itself. To begin with, the modern worker in the United States, as pointed out earlier, takes for granted the basic necessities of life; his educated demands are for the good things that raise life above the level of mere existence: health, pleasant surroundings, leisure time, interesting and satisfying work. There is no longer a basic need for the labor leader of yesterday, the tough-minded realist who knew how to organize and win a scrap. Today's labor leader must be an educated, intelligent, imaginative administrator. The "knock-down, drag-out fight" for essentials is past. Labor is accepted; its demands are heeded; and the need is for representatives who can understand the true wants of today's worker and translate them into demands for realistically conceived programs of action that will tend to satisfy those wants.

The modern labor leader is, in fact, more of an administrator than a combat general. Union welfare funds contain hundreds of millions of dollars, and membership rosters are in the millions; it takes an expert management man to run these organizations.

The Three Echelons of the Typical American Union

The typical American union is made up of three echelons, which include (1) the local union; (2) the national (or international) union; and (3) the federation of national unions.

The individual union member pays his dues and expresses his grievances to the local in his plant or locality. For practical purposes, the local usually *is* his union. Actually, however, a portion of his dues goes to national headquarters, and his local must conform to the policies established at national level. The business agent of the local, whose union job may be a full-time one, may be on a salary provided by the national organization. Locals vary in size from about 50 members in smaller plants or communities to the more than 30,000 members of the Ford local of the United Automobile Workers in Detroit, Michigan.

The Federation of Unions

The federation of unions, the American Federation of Labor - Congress of Industrial Organizations for example, wields considerably less power than the national member unions themselves. Its primary function is to act as spokesman for the national unions, and, though it tends to increase in power over the years, it is still a loose federation. The national member unions remain jealous of what they consider to be federation encroachments upon their sovereignty and exclusive jurisdiction.

The Mechanics of Union Operation

The actual mechanics of union operation in a given company should be familiar to the humanologist. The Massachusetts Institute of Technology's Economics Professor, Paul A. Samuelson, presents the following illuminating picture of the workings of collective bargaining in his *Economics: An Introductory Analysis* (McGraw-Hill, 1958):

Suppose you work in a factory that has just been organized. An AFL union has petitioned the National Labor Relations Board (NLRB) for an election to determine the exclusive bargaining agent in this plant. You mark a secret ballot in favor of the union; it wins more votes than, let's say, an existing so-called "company union." A day is set for your union representatives to meet with representatives of management at the bargaining table.

Seated at the table will probably be a vice-president in charge of industrial relations; with him will be attorneys from a law firm that specializes in the labor field. On the union side will be the local business agent of the union and a small committee of union officers, and handling the negotiations will be an expert from union headquarters. Probably he is neither a lawyer nor a professional economist, but the economic research staff of the union has helped him to prepare an extensive brief backing up the union's demands.

The union brief will analyze the company's profit position and published accounting records in great detail; it will present data showing that the president of the company receives $150,000 a year; etc. It will even contain national figures on the changes in the cost of living and may even present local data to show that rents in that area have risen beyond the national average. The brief will also compare wages in that plant with those prevailing in other plants and other industries. It will bristle with theoretical arguments showing the necessity for higher wages to maintain workers' consumption and national full employment. The employer's brief will have carefully prepared rebuttals to each of these points. If this is an important industry, both union and management will be prepared with publicity releases to be given to the press.

Hourly wage rates will not be the only issue in bargaining. In addition, the union may be asking for a dues "checkoff" (whereby union dues are automatically deducted from the payroll of union members) The union may be bargaining for a "union shop," requiring all employees to become union members. Pension and health-insurance demands may be discussed at the bargaining table. In many industries where piece rates prevail, the structure of rates will be an important subject for negotiation; the exact work load — how many looms each man will attend, etc. — may be discussed, and the general problem of how fast technological improvements shall be adopted will enter into the final contract. The seniority rights of workers and a grievance procedure for handling cases of discharge — these and many other problems will come into the collective bargaining.

Indeed, management has become worried over the inroads that organized labor has been trying to make into what it regards as its prerogatives. Many employers claim they can no longer run their business in the way that they feel is best. They find it hard to hire whom they will, fire for just cause, determine work methods, decide on the order in which people will be laid off, etc. They feel that every new decision occasions a meeting of a new committee; and time that could be better spent on producing output must be devoted to labor relations. They claim the worker acts as if he has a right to any job he has held for some time. Such critics complain that a good many unions oppose incentive wage schemes, insist upon rigid seniority, discourage efficient work methods, and seriously limit the autonomy of management. A recent casebook on collective bargaining devotes more space to issues arising out of workers' rights in jobs than to any other single subject.

But at last the contract, covering many pages of fine print, is signed. Everything is set down in black and white, including provisions for grievances that arise during the life of the contract; often too, there will be provisions for *arbitration* over issues that arise under it, with each side agreeing in advance to accept the decision of an impartial outside arbitrator. The usual life of a contract may be a year, with provisions made for reopening negotiations for a new contract under specified conditions.[5]

Collective bargaining is a complicated business — a matter of give and take.

Both Sides Should Bargain Strongly

The humanologist must avoid becoming one-sided in his thinking when collective bargaining is in order. All concerned are best served if both sides bargain strongly. There is likely to be right on both sides. When labor is the only hard bargainer, it may obtain unrealistic concessions which will sap the financial structure of the business in question, perhaps even kill the goose that lays the golden eggs. On the other hand, if management is the only hard bargainer, the wants and needs of the employee will fail to find expression; in the long run, morale and productivity will suffer, as will the company's humanological reputation. Such a situation may bring about a drastic reorganization of labor, and explode, eventually, in the form of protracted labor-management strife.

What the humanologist will realize is that, while organized labor and entrenched management will always have points of contention and will be fundamentally hostile toward

---
[5] "By permission from *Economics*, 4th Ed., by Paul A. Samuelson. Copyright, 1958. McGraw-Hill Book Co."

each other, they can still coexist on a basis of reasonable equality of position, and in a sense be partners in the furtherance of the economic enterprise which maintains both.

Fringe Benefits

As indicated above, the history of tomorrow's collective bargaining will have to do with what are called "fringe benefits." There will continue to be adjustment of basic pay scales in keeping with the costs of living, but these may be expected to be more or less routine. The controversial guaranteed annual wage, which has been won in modified form by the United Automobile Workers and which will be sought with increasing intensity by other unions, may seem to fit poorly under the heading of fringe benefits, but actually, under existing contracts this concession is not a guaranteed wage at all, but a supplementary unemployment compensation fund. The company sets aside a certain amount of money for each man-hour of work, which will be used to supplement state unemployment compensation during the following year whenever an employee has to be laid off. There is no raise in pay involved for the employee, except insofar as he is financially benefited while unemployed. What it provides primarily is the intangible factor of emotional security, which is of necessity more important to the productivity of the lower-echelon worker, particularly to the unpromotable, than it is to the dynamic upper-echelon policy-maker who lives by taking calculated risks.

Disability Compensation

One of the first of the so-called fringe benefits was disability compensation. At common law, a hundred years or so ago, a workman who was injured had no right to recover compensation from his employer unless he could prove that the latter was negligent. This negligence had to consist of affirmative negligent action, which rarely existed owing to the fact that the employer, in most cases, was not in the vicinity of the working place or the workman himself, or in the failure of the employer to provide a safe working place or safe tools for the workman. The fact that another workman was negligent could not be used to prove negligence on the part of the employer.

Labor's first victories in this area were achieved through government action. The first act in the nature of an over-all workmen's compensation law was adopted in Germany in 1884. In 1897, England enlarged the German idea and abolished the common law insofar as most industrial accidents were concerned, establishing the entirely new concept of workmen's compensation. Liability under this new concept depended not upon who individually was at fault for the accident, but solely upon whether the accident occurred out of the employment while the worker was engaged therein. The phrase "personal injury by accident arising out of and in the course of the employment" which is found in almost all of our present acts, first appeared in the English law of 1897. Following the passage of this law, demand for enactment of similar legislation arose elsewhere in Europe and in the United States.

The essential purpose of the present-day Workmen's Compensation Act is to establish a method whereby a workman who suffers an injury arising out of and in the course of his employment can receive an indemnity based upon his wages and receive it independently of considerations of his employer's negligence or his own contributory negligence. The worker's rights at common law have to do only with the employer's and employee's fault; under the workmen's compensation laws the only test is the relationship of the injury to the job.

The growth of workmen's compensation was violently opposed. It is still considered highly controversial in some circles. Nevertheless, workmen's compensation is generally conceded to be here to stay.

A "Four-Way Stretch"

There is what might be called a "four-way stretch" in the present growth of fringe benefits. The four kinds of growth are seen (1) in new types of benefits; (2) in higher costs of existing benefits; (3) in increased duration of benefit payments; (4) in increased numbers of individuals covered.

Actual Monetary Factor

In 1929, benefits under this broad heading cost American business about one cent an hour per employee. By 1955, this cost had increased to *25 cents an hour per employee.* At the present rate of increase, it is estimated that these costs will have climbed to $1.00 per hour, and will represent 30 per cent of the total paid out in wages and salaries.

The 1955 breakdown showed that out of the 25 cents per worker per hour paid out in fringe benefits, 4.5¢ covered such legally required payments as Old-Age and Survivors Insurance, Unemployment Compensation, Workmen's Compensation, etc. The next larger item was 5.5¢ disbursed for agreed-upon payments, such as retirement pensions, various types of insurance, and the like. Another 4.5¢ was figured in terms of time and covered rest periods, paid lunch periods, etc. The largest item, 8.8¢, was for days not worked, as in paid vacations, holidays, sick leave, and the like. Bonuses, profit sharing and other extras accounted for the remaining 1.7¢.

Costs of fringe benefits will increase almost automatically, even without further bargaining, owing to policies already established, which tie payments for specific benefits to current costs rather than to any arbitrarily fixed figure. For example, as insurance costs go up, the employer will pay out more to keep up the premiums on agreed-upon coverage.

Fringe Benefit Goals

This does not mean that organized labor does not have a number of clear-cut fringe benefit goals which *will* call for collective bargaining. The principal targets of the AFL-CIO economists are:
1. *Health benefits, including insurance* (to be discussed in detail below);
2. *Pensions:* vesting rights, so that an employee will retain equity in a pension plan when he leaves a company before retirement; medical protection after retirement; continuation of pension payments to widows of pensioners;
3. *Increased leisure:* longer vacations and more holidays as well as shorter daily or weekly working hours.

Health Benefits

"Health benefits" is a broad term, covering such distinct types of benefits as life insurance, retirement pensions, and general health care. Here, it seems appropriate to concentrate on the aspect of health benefits having to do with nonoccupational injury or illness. This aspect is considered to be largely out of the domain of government, in the United States at any rate, and it is, therefore, found frequently on the agenda for the bargaining table. Some states, notably California, Rhode Island, and New Jersey, have set up systems granting benefits to workers in connection with nonoccupational illness or injury, but these are necessarily limited in scope.

The first collective bargaining agreement to provide for nonoccupational sickness and accident benefits was negotiated as early as 1926, but such an agreement was considered a rarity prior to World War II. As wages were frozen during the war, and bargaining was restricted in this direction, many labor groups found it expedient to press for nonfrozen benefits. Although there was considerable controversy as to whether health-insurance benefits were legally subject to the same ceilings that applied to wages, a number of such contracts were negotiated, and the War Labor Board took no official action against them.

There are few industries today in which health-benefit plans of some sort are not found in collective agreements.

The Five Categories of Health Plans

Health plans may be rated under the following five broad categories: (1) control, (2) type of benefits, (3) eligibility for benefits, (4) scope or extent of benefits, (5) standards of medical service.

Under *control* there are two main types of plans: (1) commercial insurance plans; (2) nonprofit plans.

Types of benefits are: (1) cash indemnity; (2) services rendered; (3) a combination of both.

Eligibility for benefits is frequently restricted as to age, occupation, income, or physical condition. Sometimes there is a length-of-service restriction. Benefits may vary according to employees' earnings. Most plans adopted as a result of collective bargaining contain a minimum of restrictions.

Scope or extent of benefits ranges from plans limited to a single type of benefit all the way to comprehensive package deals providing complete care for employees and their families. The average plan falls somewhere in between, with benefits limited according to type of illness, length of illness, predetermined waiting period, or agreed-upon cash indemnity ceiling.

Standards of Medical Service

Standards of medical service determine whether in the long run any medical benefits plan will prove to be worth the money invested in it. Indemnity plans do not attempt to deal with this problem, with the result that quality medical service may be nonexistent in the community where the covered employee lives. In such a case, his health benefits are meaningless.

Inherent Problems

After employers and unions have agreed to some sort of a health program, the problems are, in the main: (1) what kind of a program to select; (2) how to write the agreement into the formal contract; (3) how to handle the financing of the plan. The agreement may involve the establishment of a health benefit setup by one or both of the contracting parties, or it may involve subscribing to a plan already existing in the area.

Requirements of a Complete Program

A plan organized by the contracting parties may provide cash indemnity or service benefits. The latter type of program may take one of several forms. A complete program would require three features:
1. Contracting for the medical services of a panel of general practitioners and specialists for home, office, and hospital practice;
2. Control or ownership of a hospital;
3. Establishment or sponsorship of a clinic with diagnostic and therapeutic facilities.

In actual practice, a complete program such as the above may not be feasible. Parts of such a program may be combined with other arrangements having to do with indemnity payments or cooperation with other community organizations.

When the contracting parties agree to take advantage of a pre-existing indemnity or service program, they will find themselves faced with one of the five standard alternatives:

1. Commercial insurance indemnity plans providing policies fairly well standardized among companies, which may be purchased separately or combined in packages;
2. Blue Cross or similar hospitalization plans organized by hospital associations;
3. Cash indemnity or service plans providing surgical benefits, and sometimes including other (nonsurgical) medical benefits, sponsored on a nonprofit basis by local or state medical societies;
4. Group practice plans controlled by physicians, which frequently provide comprehensive services;
5. Group-practice plans controlled by consumers, i.e., subscribers, or by other arrangements, with varying degrees of administrative responsibility in the hands of nonmedical persons.

Competent Advice Essential for Quality Controls

Selection of a suitable plan with effective quality controls should not be undertaken without competent advice. A qualified medical consultant with experience in the field of occupational humanology will prevent many costly and time-consuming *post facto* reappraisals. His services should be sought at least during the bargaining stage, if organized labor has not already obtained humanological counsel in the preparation of its agenda for the bargaining table.

Essential Basic Information Required

Preliminary to planning any health program, unions and employers will find it helpful, if not essential, to correlate the following items of basic information:
1. Size of the employee group to be covered and its normal average earnings
2. Special health needs of the group
3. Composition of the group according to sex, age, and marital status

4. Geographical concentration of the group
5. Hospital and medical facilities available in the area
6. Costs prevailing in the community for hospital services, common surgical operations, and physicians' home and office visits
7. Number of physicians in the community and the possibilities for group medical practice
8. Premium costs and benefit provisions of standard commercial group insurance policies available in the region
9. The services provided and rates charged by all service plans in the locality, such as health cooperatives, hospital associations, medical societies, and associated physicians

The main issues in collective bargaining for a health-benefit plan may have to do with the amount of employer contribution to the program and the extent of financial responsibility of the participating employee. Sometimes it is expedient to sign a basic contract with an open clause covering a health-benefit program *to be* filled in later. A special management-union committee can be formed at the bargaining table with powers (1) to investigate available health programs, (2) to recommend a specific plan, (3) to work out the details of the plan selected, and (4) to suggest later modifications of the plan within the limits of the basic contract.

The principal items usually included in the basic contract are a statement of the decision to set up a plan and certain arrangements in connection with its financing and administration. There are a number of possible methods of financing health plans. The employer may handle the entire cost through contributions to a special fund or by outright purchase of insurance policies. The union may contribute to the cost, or a percentage of the cost may be met by employees through regular wage deductions.

Employer contributions may be determined in one of the following ways; (1) by per capita contributions, (2) by percentage of payroll, (3) by percentage of sales revenue, (4) by lump sum, or (5) by tax or royalty on production.

Employer contributions tend to vary between 1 and 5 per cent of the payroll. There is no customarily established amount. The question is established by the nature of the business, the relationship between labor and management, and the types of services demanded.

The Three Different Meanings of Administration

Administration has three different meanings with regard to health-benefit programs agreed upon at the bargaining table. When the word "administration" is written into policy, it should be clearly understood which type of administration, whether it be (1) administration of actual operation of the plan, (2) administration of a fund earmarked for indemnity, or (3) handling of day-to-day details (complaints, processing of claims, etc.).

Administration of actual operation is usually in the hands of the group, hospital, or clinic providing the service, or the agency underwriting the indemnity setup. It is rarely expedient for either the management of a business or staff of a unit or organized labor to attempt to become too deeply involved in the running of a medical or insurance business. The most effective system has proved to be that under which the management-labor committee sponsors and, if necessary, provides financial support for outside health agencies, which meet predetermined standards of quality, *without* attempting to absorb or otherwise take away independent status of the given agency.

Administration of *funds,* under present federal law, must be administered by boards which have equal representation from union and management and include provision for settling of deadlocks by some neutral party. A trained humanologist would be the ideal neutral party.

Day-to-day details, such as processing employees' indemnity claims, are handled through union offices, employer offices, or by the insuring company. Problems tend to arise under any arrangement in connection with procedures for filing claims, requesting services, routing payments, and informing employees of their benefit rights. Active participation by the union and/or employer in these procedures often is necessary for smoothness and efficiency. Workers may fail to get what they are entitled to if they are not informed about their rights under the program. Differences may also arise between union and employer over the interpretation of the collective bargaining contract; between union or employer and underwriting company over proper application of the terms of an indemnity policy; and between union or employer and medical service establishment over quality or standards of service.

Legal Advice Essential

Adequate legal advice is necessary in the early stages of establishing funds or in organizing group-practice plans. In some states, laws may specifically restrict the operation of certain types of plans. On the other hand, laws in some states offer specific governmental aid in establishing certain types of plans.

A general evaluation of the two main types of health-benefit plans, i.e., indemnity vs. services, would, in the light of past experience, tend to give preference to the latter, provided proper quality controls exist. In the past, emphasis has been largely on indemnity plans.

The Early Health Insurance Programs

The early health-insurance programs were designed to cover, or partially cover, hospital costs, which were, for most people, the most expensive part of medical care. It was hoped that this coverage would make medical care available to anyone who could afford to pay the then negligible premiums. Group plans covering all the employees of participating

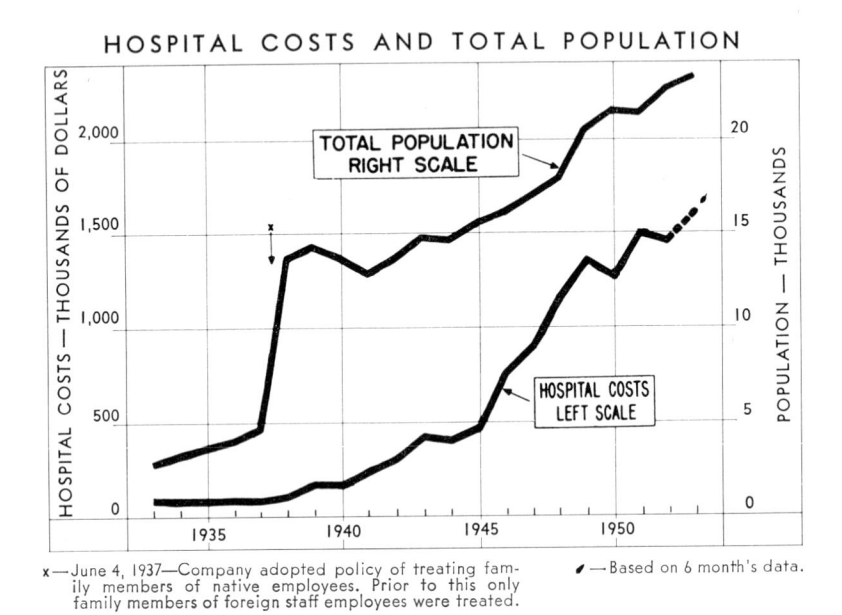

HOSPITAL COSTS AND TOTAL POPULATION

x—June 4, 1937—Company adopted policy of treating family members of native employees. Prior to this only family members of foreign staff employees were treated.

◢ —Based on 6 month's data.

Graph IX — Hospital Costs and Total Population

industrial concerns extended this coverage to the majority of American employed people, and it looked as though this country might be on the threshold of a new era of health for the rank and file.

As a result, hospitals soon found themselves crowded with people who had never before been able to afford hospital care. These institutions had to expand or burst, and in the process of expanding they had to keep up with the rapid advances of medical technology. Moreover, they had to compete for manpower at a time when this commodity was in short supply. Hence, hospital costs zoomed upward faster than insurance benefits could keep up. It became necessary for the underwriters either to raise premium rates or reduce benefits. In many cases, they did both.

At the same time that the participants in those plans were becoming sharply aware of the fact that hospitalization was becoming more and more expensive in spite of their insurance, they were also finding out that hospitalization was only a small part of the medical care picture.

Older people were finding that the kind of medical care they needed most was preventive. A chronic disease having to do with the aging process may not become serious enough for hospitalization until the time has passed for effective therapy.

Even among younger persons it became apparent that coverage of hospital costs alone was no satisfactory answer to the medical care problem. The majority of these people could go for years without requiring hospitalization. Their main problem was doctors' bills for the sort of problems that usually could be treated at home or in the doctor's office. Sometimes a doctor, realizing that a patient might be financially embarrassed, would send him to the nearest hospital for some simple diagnostic service that could have been done without undue difficulty in the doctor's office. This would help the patient immediately by enabling him to take advantage of his hospital insurance, but it also tended to complicate the already serious overcrowded condition of the hospital by encouraging admissions.

Health Insurance to Pay Doctors' Bills

There then grew a demand for a type of health insurance that would pay doctors' bills as well as hospital expenses. Such coverage was bound to have delicate complications; there were many possibilities for abuse. Hence the coverage had to be rather severely limited.

For example: it was difficult, if not impossible, to prevent unscrupulous physicians from offering a minimum of medical service for the maximum fee possible under the terms of the coverage. There was nothing to prevent the insured from seeking incompetent medical counsel in order to get the most for his allowances in terms of quantity, not quality, care.

Moreover, since plans of this sort provided only *funds* for medical services, with no control over the *kind* of medical services obtained. It was virtually impossible to provide coverage for preventive or diagnostic measures. One could scarcely *prove* that a payment for diagnostic or preventive services was valid, legitimate, and hence reimbursable. Therefore, coverage was limited to after-the-fact therapy, which is often too late to be complete, particularly where a degenerative disease is involved.

Major Medical Expense

Premium payments for major medical expense, involving what might be called catastrophes, such as the need for heart surgery, installation of major prosthetic appliances, excessive periods of hospital rehabilitation, and the like, were necessarily too high for the average wallet. For this reason, the majority of people were willing to take a chance on the big disasters while they protected themselves against the small ones. As has been pointed out, many of the small ones were almost too small to demand any kind of medical attention, but since they were covered by insurance, they were assiduously seen to. The man with a minor sprain of the thumb might avail himself of elaborate medical care, including a night at the hospital for x-ray examination, as long as his insurance company was paying the bills. Unwittingly he was doing his bit to compound the confusion in the general medical care picture; he was helping to boost health-insurance rates and reduce available medical benefits.

Then, perhaps, one day this same individual had a serious health setback. At this time he was, say, 45 years old and suffering from a previously undetected degenerative condition having to do with aging. Had this condition been detected earlier, it might have been corrected or placed under sure control. Now it was too late. The patient was obliged to undergo a long-drawn-out and costly treatment which would keep him away from gainful occupation for many months. After discharge from the hospital, he would have to face life as a chronic invalid and make some radical adjustments in his work schedule.

At this stage, he realized that, in spite of his coverage, even if he had been willing and able to supplement his regular sickness and accident insurance with major medical coverage, he had not really assured himself of any kind of effective medical care.

As a result of an accumulation of cases like the above, the demand today is only secondarily for the money with which to pay doctors and hospitals. The growing demand is for *medical care itself!*

Constructive Medical Care Means Maintenance Along the Way

It cannot be reiterated too often that constructive medical care means *maintenance along the way;* not just *repair after the fact.* A good machinist keeps his equipment in shape from day to day; he doesn't wait for a machine to break down before oiling it. The same principles hold true with regard to the maintenance of human beings.

Sweeping the Dust Under the Carpet

This new awareness has manifested itself in two principal types of endeavor: one positive, the other negative. The latter, negative "solution" is a hang-over from the days when the workingman's only effective weapon was his voting power. This "solution" involves "passing the buck" to government. As a problem-solver, this sort of action is like sweeping the dust under the carpet. For the time being it is out of sight; therefore it is presumed not to exist. It is not long, however, before the problem asserts its existence with a vengeance.

This has happened time and again in those countries where compulsory federal insurance is in force, and it is difficult to imagine it *not* happening, despite all the finely chiseled verbal sophistries of the proponents of government-in-medicine. First, the demand for medical attention increases between 50 and 100 per cent, and the large part of the demand has to do with comparatively insignificant illness. The charge of unnecessary medical attention cannot be made against the law when government, supposedly of, by, and for the people, is paying the bill. Hence it is impossible to prevent abuses. No sound insurance plan can survive under such conditions.

Irrespective of cost, there is a much more serious penalty to be reckoned with. This has to do with the *quality* of service to be rendered. Doctors' offices, under federal medicine, are forever crowded. Individual patients receive only scant attention; hospitals have long waiting lists.

Walter Reuther

The *positive* approach to the problem is typified by the medical service plan which the United Automobile Workers have been building in the Michigan area under the leadership of union President Walter Reuther.[6] The fundamentals of this plan and the underlying philosophy are set forth in the following remarks by Leonard Woodcock,[7] Vice-President of the United Automobile Workers, presented before the Labor Health Administrators, November 14, 1956, in Atlantic City, New Jersey.

[6] WALTER P. REUTHER, an address delivered by L. Woodcock before the 1957 Biddle Lecture before Annual Convention of Michigan State Medical Society.

[7] LEONARD WOODCOCK, "A Look at the Future of Direct Service Medical Care Plans." An address before the annual Meeting of the Association of Labor Health Administrators, Atlantic City, November 14, 1956.

In most areas of the country, the best available medical benefits are limited to surgical and in-hospital medical care. We have never been reconciled to and do not accept the over-emphasis on surgical and in-hospital medical care. We have learned the hard way the lesson that fee-for-service plans of this type offer an irresistible temptation to some practitioners to perform unnecessary surgery. We have watched the existence of surgical insurance accelerate the step-by-step inflation of surgical fees . . . .

Since 1947 the medical care component of the Consumer Price Index has increased at about twice the rate of the index as a whole. The cost of hospital service has increased even more and there is no end in sight. It is unthinkable that any responsible union, employer or insurer can attempt to go much further into this field of medical care without trying to pin down just what the doctor's remuneration is to be for performing a given service. . . .

Cash indemnity and reimbursement plans offered by insurance companies have been even worse. It has been extremely difficult to get insurance companies to cover retired members at reasonable cost. Benefits have been hedged in with rigid cash limits, offering economic protection to the employer and the insurance company, but very little to the worker who has to meet the total bill, whatever it is. We now see a tendency to deal with these problems by latter-day indemnity plans embodying grandiose pretensions such as major medical insurance. Workers are expected to agree to take most of the burden upon themselves and away from the insurance. It is hoped that somehow everyday medical needs will disappear into thin air. Co-insurance and deductible measures will prove illusory as cost controls contribute to further inflation of medical costs. The people need *health insurance to provide valid health services* and not added economic pressures when they are sick. . . .

Tough-minded management and practical labor unions are coming to the inescapable conclusion that something must be done about the organization of health care in order to bring the advantages of modern medicine to the American people, and that a great extension of prepayment is the only practical way of making progress in health security. . . .

Thus we come by the hard route of experience to the conclusion that direct service, group practice, comprehensive medical care plans will best serve the needs of our people. The UAW is now engaged in an effort to establish such a plan in the Michigan area.

On October 11 of this year, at the invitation of President Reuther, a group of Michigan citizens met and formed a non-profit membership association which proposes to develop a community wide program to furnish comprehensive prepaid health services of high quality, based on more efficient utilization of existing facilities and personnel, a rational organization of personal health services, and a method of financing geared to the positive promotion of health.

A new plan starting at this time has before it the successful pioneering of a great many earlier efforts. The work of the groups represented in the ALHA is of inestimable value. . . .

The accumulated experience has established beyond question some of the essential features of a sound medical care program. We anticipate that medical care under our new plan will be provided through doctors practicing in groups; wherever possible the groups will be associated with hospitals. The details of the program are yet to be worked out. . . . The association is planning to make available . . . all essential health care, including care in the doctor's office, the clinic, the hospital, and the home; including diagnostic, preventive, health maintenance and rehabilitation services, as well as treatment for illness. . . .

The fact that auto workers are not geographically separated from other concentrations of working people (was) among the factors which resulted in the UAW's decision to work with the rest of the community to develop this plan and to open enrollment to all groups in the community when feasible. We recognize that, whenever labor ventures into the broader community, it does so at a certain risk. Although the constituency of many community projects comes overwhelmingly from labor, labor's hopes and objectives frequently receive inadequate attention when it comes to setting policy. Still, community efforts are often the best way to realize labor's goals.

Details of implementation in the above exposition may be controversial, but no humanologically oriented representative of either labor or management can quarrel with the over-all philosophy expressed. The member of organized labor, like every other employed person at every echelon, wants *health services*; not sickness indemnity.

Not every union has the resources of the United Automobile Workers. The above-described setup might be considered too elaborate for some communities, in view of financial limitations not only in the available union welfare funds, but in the economic resources of the community employers and local governing bodies.

There is no need, in such a situation, for organized labor to make the old mistake of seeking as much in the way of *quantity* service as can be eked out of available resources. Neither is the alternative of federal intervention a sound one from the point of view of the best interest of the worker.

The well-advised labor leader in such an area will take advantage of expert humanological counsel. He will put *quality* medical service at the top of the agenda. This is putting first things first. The realist has neither the time nor the patience for "pie in the sky." He pursues the objective that is immediately ahead of him and takes up future objectives as he comes to them.

Pertinent Data for Effective Evaluation

In the last analysis, proper evaluation of any constructive medical plan can be made only by the medically trained consultant with experience in occupational humanology. However, there is a certain amount of data having a bearing on effective evaluation which can be compiled by the union, or union-management, committee interested in the establishment of a plan. These data should answer at least the following:

1. Is the plan operated for profit or incorporated as a nonprofit organization?
2. Is the plan designed to pay cash indemnity or to render service in return for prepayments?
3. To what types of health conditions do the provisions apply?
4. What are the type, scope, and amount, and duration of benefits or services?
5. What are the methods of organizing professional services, and what are the methods and rates of payment to the participating members of the professions?
6. What are the methods of organizing hospitalization and the methods and rates of payment to the participating hospitals?

7. What are the prepayment rates, extra charges for services, and additional obligations? Who bears the expenses and to what extent?
8. Where is administrative control vested, what is the composition of the administrative bodies, and what are their powers, duties, and functions?
9. What is the total number and the sex and age distribution of the persons enrolled at a given date?
10. What is the total number of participating professional persons (break-down by type of practice) and of beds in participating hospitals (break-down by type of service)?
11. What is the number of eligible persons, by sex and age, who have received specified benefits or services during a certain period of time?
12. What is the number of specified benefits or services received by the eligible persons during a certain period of time?
13. What is the total earned income and the other income of the plan during a certain period of time?
14. What are the total expenditures for benefits or services and for administration, and what contingency reserves have been set aside during a certain period of time?

In evaluating these data, special attention will be given by the humanologist to the kind and extent of care given in comparison to the average annual family income of those covered. The findings will show to what extent the plans encourage the prevention of disease, the significance of psychosomatic problems, and promotion of good health, early diagnosis, and quality treatment.

Total Health for the Employed Individual

Finding a satisfactory solution to the *economic* problem of illness is only one aspect of adequate health care for the country's working population. Experience has borne this out.

The most practical objective of organized labor is *total health* for the employed individual. This is within the resources of the average industry and community, if money, time, and energy are not wasted in terms of short-term peripheral solutions. Collective bargaining, when backed by sound advice, goes straight to the heart of the matter, which is health itself, and recognizes that medical care plans make up only part of the total picture. Among other important programs, which cannot be considered as isolated entities, are health education, industrial hygiene and safety, professional medical education and research, public assistance to the aged and handicapped, public health, and workmen's compensation.

This objective of total health must be understood and clearly defined before the labor leader can indulge in Utopian reveries about "package deals" covering not only the employee but his entire family, and including every aspect of medical care from face lifting and wart removal to emergency surgery. It will be found that, if the employed person is kept healthy and fully productive in a strong economy, the other, Utopian-seeming objectives will become unnecessary. The problems involved will solve themselves. Those unreached objectives that still appear urgent will be more easily attained from an established foundation of *quality* health service.

This is in keeping with Gompers' classic principle: "More today than yesterday; more tomorrow than today." It is the principle of "one step at a time," where each step upward is not so large as to throw the climber off balance or bring him to uncertain, untested ground, but large enough to improve his position significantly and careful enough to bring him to a sure, firm footing.

Policy Required

To sum up: organized labor has finally reached a position where, if it does not actually write health policy, it can say what policy it wants written and be listened to with respect.

Backed by competent humanological advice, the policy that will be hammered out in tomorrow's mills of collective bargaining will result in medical care programs that will:
1. Prevent all preventable health problems;
2. Detect incipient health problems at a time when they can be corrected at reasonable cost;
3. Keep the employed person vertical and in a job compatible with his medical abilities to perform;
4. Treat and rehabilitate the employed person in the event of major medical catastrophies;
5. Educate him in the construction as well as the maintenance of health.

REFERENCES

AMERICAN MEDICAL ASSOCIATION. *A Survey of Union Health Centers.* Chicago: American Medical Association, 1953.

BAEHR, GEORGE, M.D. "The Family Physician the Central Figure in Prepaid Group Practice." The *American Journal of Public Health,* XLIII (February, 1953), pp. 131-137.

──────────. "Professional Services under Medical Care Insurance." The *American Journal of Public Health,* XLI (February, 1951), pp. 139-146.

BRINDLE, JAMES. *Medical Care for Industrial Workers: A Labor Viewpoint.* New York: American Public Health Association, 1956.

──────────. *Views of Labor on Medical Care.* Detroit: UAW-CIO, 1955.

──────────. *Statement at Hearing of Governor's Study Commission on Prepaid Hospital Care Plans, Detroit, March 28, 1956.* Detroit: UAW-CIO, 1956.

BRINDLE, JAMES; MCGUINNESS, AIMS C., M.D.; AND HESS, ELMER, M.D. "Public Forum on Distribution of Medical Care." *New York State Journal of Medicine,* LVII (May 1, 1957), pp. 1660-1672.

BRUNN, JOHN M. "Health Programs in Collective Bargaining." *University of Illinois, Institute of Labor and Industrial Relations, Bulletin,* XLVI (No. 46, February, 1949).

DRAPER, WARREN F., M.D. *The Quest of the UMWA Welfare and Retirement Fund for the Best Medical Care Obtainable for Its Beneficiaries.* Washington, D.C.: UMWA, 1957.

DRAPER, WARREN F., M.D. *Thoughts and Experiences on Medical Care.* Washington, D.C.: UMWA, 1953.

FIDLER, ANNA. "Reports and Inquiries: Some Aspects of Collective Agreements in the Oil Industry." *International Labour Review,* (Geneva, Switzerland), LXIX (No. 4, April, 1954).

FOUNDATION ON EMPLOYEE HEALTH MEDICAL CARE and WELFARE, INCORPORATED. *Study No. 1.* New York: Foundation on Employee Health Medical Care and Welfare, Incorporated, 1957.

GOODMAN, KENNARD E., and MOORE, WILLIAM L. *Economics in Everyday Life.* New York: Ginn and Company, 1952, Chapters XXIV and XXVI.

INDUSTRIAL MEDICAL ASSOCIATION. *1954 and 1955 Reports of Committee on Workmen's Compensation and Insurance to Officers and Board of Directors of the Industrial Medical Association.* Chicago: Industrial Medical Association, 1955.

KILANEA SUGAR PLANTATION COMPANY. *Medical Plan 1954-1956.* Honolulu (Hawaii): Kilanea Sugar Plantation Company, 1956.

MADDOCK, CHARLES S., LL.B. "The Physician and Workmen's Compensation." *Manual of Industrial Medicine.* ed. Lemuel C. McGee, M.D. 3rd Ed. Philadelphia: University of Pennsylvania Press, 1956.

MAKOVER, HENRY B., M.D. "The Quality of Medical Care." The *American Journal of Public Health,* XLI (July, 1951), pp. 824-831.

PAGE, ROBERT COLLIER, M.D. "The Road to Abundant Health." The *Health Magazine,* The Health League of Canada (Toronto), XXVI (March-April, 1958), pp. 10-11.

POLLACK, JEROME. *Financing Medical Care in the Long Pull.* Detroit: UAW-CIO, 1956.

PRICE, LEO, M.D. "International Ladies Garment Workers' Union." *Industrial Medicine and Surgery,* XXII (October, 1953), pp. 489-497.

SACKMAN, MORRIS. "Welfare Collective Bargaining in Action — A Case Study." Cornell University. *New York State School of Industrial and Labor Relations Bulletin,* July 3, 1949.

SAMUELSON, PAUL A. *Economics: An Introductory Analysis.* New York: McGraw-Hill Book Company, 1958, Chapters 8 and 28.

SAWYER, WILLIAM A., M.D. "Union Health and Welfare Programs." *Group Practice,* (January, 1958).

——————. "Labor's Health Goals." *Journal of Occupational Medicine,* I (No. 9, September, 1959).

SLAVIK, FRED. *Distribution of Medical Care Costs and Benefits under Four Collectively Bargained Insurance Plans.* Bulletin 37 of New York State School of Industrial and Labor Relations, Cornell University.

TOOMEY, WILLIAM A., JR. "Labor Relations Aspects of Industrial Medicine." *Journal of Industrial Medicine and Surgery,* XXIX (No. 2, February, 1960), p. 53.

UNITED MINE WORKERS OF AMERICA'S WELFARE AND RETIREMENT FUND. *Report of Year Ending June 30, 1955.* Washington, D.C.: UMWA, 1955.

UNITED STATES DEPARTMENT OF LABOR. *Older Workers Under Collective Bargaining.* Bulletins 1 and 2, (September and October, 1956).

WEINERMAN, E. RICHARD, M.D., M.P.H. *An Appraisal of Medical Care in Group Health Centers.* San Francisco: Labor Council.

WILLYS UNIT, LOCAL 12, UAW-CIO, TOLEDO, OHIO. *Report on Diagnostic Clinic, 1955.* Toledo: Willys Unit, Local 12, UAW-CIO, 1955.

7

The Professional Members of the Humanological Team

THE PRECEDING CHAPTERS have shown how humanological policies have come into being, partly through genuine enlightenment on the part of management, partly through competition for manpower, partly through strong collective bargaining on the part of labor, partly through government action as a result of the pressure of public opinion, partly through accident. This evolutionary process has not been rapid; nor is it yet by any means complete. Goals for the future have been touched on in the foregoing chapters and will be explored further in subsequent chapters.

The Shaping of Humanological Policy

It is important to note that the professional member of the humanological team has had little to do with the shaping of humanological policy. There has been a tendency in recent years for some progressive representatives of both management and labor to seek the advice of a professional member in matters humanological, but in most cases, when this advice is sought, it has to do with policy already written and with precedents already set.

Point of No Return

The young man or woman preparing for a career as part of the professional arm of the humanological team will do well to be forewarned with a realistic understanding of this situation. From this point of no return, the members of the professional arm of the humanological team must learn to understand the present state of affairs and to interpret existing policy in terms of the best interests of the employed person. By so doing, he will fulfill the recorded wishes of both labor and management in the *one* field in which there is a mutual and personal interest, namely *health*. Experience and judgment in matters involving the multiple interests of the total man are requisite to success in this endeavor.

On the other hand, the wide-eyed Don Quixote who rides into the fray gallantly determined to bring about the new golden age of occupational health will end up a bewildered, embittered, and frustrated individual. He may as well find out early in his training that he is not *now* wanted as a policy maker and probably never will be. He is not even asked to define humanology or to determine precisely what his own job specifications shall be. These matters are worked out by management and labor, who find somewhere between the conflicting interests of maximum profits for stockholders and maximum benefits for those employed an expedient compromise which must suffice, for the time being, as humanological policy. The humanologist, who must see to the execution of this policy, is rarely asked even to sit at the bargaining table at which this compromise is being painfully sought and shaped into policy. Even more rarely is he asked for his advice as to the agenda which either bargaining party ought to bring to this table.

It is conceivable that "once in a blue moon" some bright and eager young medical consultant will be presented with a "clean slate." The president will say to him: "Doctor, we are starting from scratch. We have no policy to speak of for dealing with our employed human beings. We want you to help us frame such a policy." Such a situation would be reminiscent of the popular cartoon feature of the late H. T.

Webster: "The Thrill That Comes Once in a Lifetime." The great majority of budding humanologists are well advised to treat the possibility of such occurrences as "such stuff as dreams are made of."

Interpretation

The best the average professional member of the humanology team can do is familiarize himself thoroughly with such policies and precedents as happen to exist in the industrial situation in which he finds himself. His job is to help management and labor interpret and implement existing policies and precedents in such a way as to enable all concerned to derive maximum benefit from them.

The following simile may clarify the humanologist's position:

X and Y were having afternoon coffee at a downtown restaurant. X was in a hurry, because he had an important appointment within 10 minutes at an office building about three blocks away from the restaurant. Since it was important that he look his best at this appointment, he was freshly barbered and was wearing a brand new hat.

Just as he had finished his coffee and had risen to say goodbye to Y, a sudden cloudburst came up.

"Good heavens," he said. "I'll look like a drowned rat. Fortunately I brought a rain coat, so my suit won't suffer too much, but my hat will be a wreck. Y, you're supposed to be a pretty intelligent character. What do you suggest? Don't say a taxi, because I haven't got the money, and I know you haven't either."

Y might have made a few cutting remarks about X's inconsistency in coming out without an umbrella on a day sufficiently ominous to induce him to wear a raincoat, particularly on a day when he had taken pains to be elegantly barbered and hatted. He might have observed that it seemed to him thoughtless of X to be without taxi fare on such a crucial afternoon, when any number of unforeseen occurrences might have interfered with his keeping his appointment on schedule.

Y realized that such talk would do nothing to help X out of his present predicament. X could not possibly get hold of either umbrella or taxi money at this time without being ruinously late for his appointment. Y also realized that there was nothing he could do about inducing X to adopt more constructive policies for the future. X being the kind of person he was, he would undoubtedly repeat the performance at some later date. He had not asked Y to "make him over" but simply to offer him a solution to the situation as it existed.

Y had no umbrella himself. As it happened, the best solution handy was the daily paper, which Y gave to X to hold over his hat to ward off the raindrops. It wasn't as good as an umbrella, but it served its purpose.

Y was in the position of the professional member of the humanology team, though of course his problem was considerably simpler. The professional humanologist is given complex problems which many of his colleagues not in the industrial field would find impossible to solve within the framework of existing policies and precedents. This is why it is necessary for him to have special aptitudes, interests, and educational background.

Members of the Humanological Team

The members of the professional arm of the humanological team include (1) the doctor, (2) the registered nurse, (3) the psychologist, and (4) the industrial hygiene engineer, assisted by various experts in specific occupational hazards. This, however, does not mean that management can go out and obtain the services of *any* consulting physician, nurse, psychologist, or hygienist and rest in the confidence that it has added a functioning professional arm to its humanological team.

Special Qualifications Essential

This arm, as a unit, must have special qualifications. Each member of the arm need not have personally all of the qualifications that the group as a whole must have, but he must have at least an appreciation and fundamental understanding of all of the fields of knowledge that may have to be brought to bear in the solving of any given humanological problem.

The job specifications of this unit have been created over the years, growing from precedent to precedent, as a result of action on the part of both management and labor. Perhaps the most exhaustive list of qualifications for the professional arm of the humanological team is that drawn up in 1957 by a joint committee on occupational health of the International Labor Organization and the World Health Organization. These qualifications, as set forth in Technical Report No. 135 of the World Health Organization, include:

1. *Principles of occupational medicine:* historical background; scope and aims; general resources for occupational medicine, public health, and medical care; structure and function of industry; industrial and labor policies and relations.
2. *Industrial physiology:* muscular and mental work; energy expenditure; fatigue, monotony, rhythm of work, and rest pauses; physiological organization of work, human engineering; nutritional problems.
3. *Industrial hygiene:* environmental hygiene; hygiene of premises — temperature and humidity, ventilation, lighting, noise; air pollution and harmful types of exposure to gases, vapors, fumes, dusts, and their control; maximum allowable concentration; personal hygiene and personal protective equipment; sanitary facilities.
4. *Occupational pathology and toxicology:* general principles of industrial toxicology; diseases due to chemical agents, physical agents, dusts, biological agents; occupational cancers; occupational skin diseases; occupational allergies.
5. *Special medical problems:* specific pathology by trade or branch of industry; relationship between employment and nonoccupational disease; methods for assessment of disability; medical aspects of vocational rehabilitation; psychoneurosis related to work or injury.

6. *Accidents at work:* Causes of accidents; principles of accident prevention; first aid and treatment; rehabilitation of the injured worker.
7. *Occupational psychology:* psychological appraisal, and assessment of aptitudes; mental health and human relations.
8. *Preventive medicine:* pre-employment and periodic medical examination; health counseling; vaccinations and immunizations; the care of special working groups, including young workers, old workers, women, and handicapped workers; leisure, sports, addictions to alcohol, tobacco, etc.; education and propaganda.
9. *Industrial technology:* work organization; industrial processes; job analysis.
10. *Medicolegal problems, social security:* labor legislation; social insurance; workmen's compensation; reporting and notification; medical ethics; liaison with other physicians and with health organizations.
11. *Organization and administration:* organization of occupational health services; nursing services in industry; administrative and economic problems; medical records and reports.
12. *Statistical methods.*

It is clear that the individual with even a superficial understanding of all of the above-listed departments of knowledge must have a profound interest in the human being.

The Doctor

This must be particularly true of the captain of the professional arm of the humanological team, the *doctor*. He is no ordinary doctor. His knowledge of each of the above fields must be detailed and backed by enthusiastic appreciation. It goes without saying that he must be highly educated, that he must undergo years of expensive training and arduous apprenticeship. The extent and effectiveness of his training can be measured by formal examinations and by assessment of actual performance. What may be even more important, however, than what is gained from the training itself are the qualities and interests which the potential medical humanologist acquires before actual medical training and which serve as the foundation stones upon which all subsequent knowledge and wisdom is erected. These qualities and interests encompass the following six basic disciplines.

Logic

The first basic discipline is logic. The eminent eighteenth-century theologian, Isaac Watts, described logic in 1724 in a manner which is still valid, proving that this branch of knowledge is fundamental, basic, and timeless. It is not dependent upon the contemporary advance in scientific knowledge. His description was as follows:

Logic is the art of using Reason ("reason," including all the intellectual powers of man) well in our inquiries after truth, and the communication of it to others.

Reason is the glory of human nature, and one of the chief eminences whereby we are raised above our fellow-creatures, the brutes, in this lower world.

Reason, as to the power and principle of it, is the common gift of God to all men, though all are not favored with it by nature in an equal degree; but the acquired improvements of it, in different men, make a much greater distinction between them than nature had made. I could even venture to say, that the improvement of reason hath raised the learned and prudent in the European world, almost as much above the Hottentots, and other savages of Africa, as those savages are by nature superior to the birds, the beasts, and the fishes.

Now the design of *Logic* is to teach us the right use of our reason, or intellectual powers, and the improvement of them in ourselves and others. This is not only necessary in order to attain any competent knowledge in the sciences, or the affairs of learning, but to govern both the greater and the meaner actions of life. It is the cultivation of our reason by which we are better enabled to distinguish good from evil, as well as truth from falsehood; and both these are matters of the highest importance, whether we regard this life or the life to come.

Logic is more than just "horse sense." It is a discipline of the intellect that has been developed, step-by-step, over the centuries, through the mental trial-and-error efforts of wise men. It is not the creation of a single mind, nor can any single mind, without recourse to the fund of thought created by other minds, master all the techniques of right reasoning. Hence, the individual who respects logic will approach it with humility. Mastery of the subject will by no means lead to intellectual arrogance. The true logician knows that a line of reasoning that is seemingly flawless in its logic may nevertheless be wrong. Science has advanced through the efforts of logical people who had the humility and patience to test hypotheses even when they seemed to run counter to reason.

Ethics

Ethics is the second basic discipline. This may be considered a practical science having to do with the correct choice of a mode of behavior or specific courses of action that can be considered "good" *per se,* whether or not such mode of behavior or course of action satisfies the actual momentary wants of the individual chooser. Ethics has been defined generally as "the science which deals with those acts that proceed from the deliberative will of man, especially as they are ordered to the ultimate end of man." The purpose of this science, it has been said, is to enable the adept to *live well as a human being.*

A great deal is said about specialized codes of ethics for specific groups, professional or otherwise. One hears about medical ethics and legal ethics, for example, as though they were highly specialized codes, apart from ethics in general.

It is true that the principles of ethical behavior within individual professions have actually been codified, with emphasis on problems of conduct more or less peculiar to the profession in question. For the medical profession in the United States, codification of this sort has been undertaken by the Judicial Committee of the American Medical Association, which convenes periodically to revise and modernize the *Principles of Medical Ethics.*

This sort of codification is useful as a reference when debate arises over some moot point of conduct, and the current opinion of "organized" medicine is desired. However, it may be said, in general, that the individual who has mastered the science of ethics *in toto,* without specific reference to medical, legal, or any other specialized branch of the science, will rarely if ever be obliged to consult a written code covering his particular profession. An ethical man will be ethical no matter what field of endeavor he chooses. His courses of action will automatically be right ones, whether or not he consults the "official code."

The foundation of the official *Principles of Medical Ethics* is set forth in Chapter 1, Section 1 of the June 7, 1958 edition of the *Journal of the American Medical Association:*

> The prime object of the medical profession is to render service to humanity; reward or financial gain is a subordinate consideration. Whoever chooses this profession assumes the obligation to conduct himself in accord with its ideals. A physician should be an upright man, instructed in the art of healing. He must keep himself pure in character and be diligent and conscientious in caring for the sick. As was said by Hippocrates, "He should also be modest, sober, patient, prompt to do his whole duty without anxiety; pious without going so far as superstition, conducting himself with propriety in his profession and in all the actions of his life."

The doctor who has grasped the above-stated broad principles relating to his profession and who is familiar with the general science of ethics will have no need of a particularized code of medical ethics in order to behave as an ethical physician. Each member of the humanological team must master the principles of ethics *in toto.*

Epistemology

The third discipline, epistemology, is the science that seeks to find the answer to Pontius Pilate's haunting question: "What is truth?" Epistemology may also be called critical philosophy. Sages in ivory towers may spin wholly logical solutions to all problems. The epistemologist takes these solutions apart in the light of practical experience, isolates and analyzes each of the premises upon which the given philosophical utterance is based, and seeks to determine how much of the total is true, how much in doubt, and how much false. His standards and criteria for making his determinations are forever open to question and criticism themselves; epistemology can never be a closed book.

The potential medical humanologist is drawn to epistemology because he has an inquiring mind and is dubious about accepting whatever is told him as truth, unless he has full confidence in the scientific probity of his informant and is satisfied that the so-called truth has been intelligently put to the test.

René Descartes, the great seventeenth-century French philosopher, was one of the most devastating of epistemologists. Descartes lived at a time when any man of the world with a philosophic bent was constantly involved in controversy. Religious conflict between Catholics and Huguenots and reviving interest in ancient science, philosophy, and art kept inquiring minds in continuous turmoil and stirred up alarming questions that the medieval mind had looked upon as settled.

Descartes decided to make assiduous investigations of the many hoary truths that had been considered almost sacrosanct owing to their antiquity. He came to the conclusion that all ordinary knowledge, even that supported by the evidence of the senses, as well as scientific and mathematical information, is open to challenge. For purposes of argument, he proposed a philosophical test: assume that the world is a carefully contrived illusion commanded by a deity who *deceives* humanity. Once such a possibility is admitted, everything becomes dubious. Two plus two may not equal four except in the dream world of the individual making such a statement. The problem then was either to prove the assumption that the world is created and ordered by a deceiver and to doubt the validity of all so-called knowledge. Descartes' solution was as follows:

> I was convinced that there was nothing in the entire world, that there was no heaven, no earth, that there were no minds, nor any bodies. Was I not then also convinced that I did not exist? Not in the least. It was certain that I myself existed since I convinced myself of something (or just because I thought of something). But there is some kind of a deceiver who is very powerful and very cunning, and who always uses his ingenuity in order to deceive me. Then, for certain, I exist also if he is deceiving me, and let him deceive me as much as he wishes, he can never make me be nothing as long as I think that I am something. So that, after having considered this well, and having carefully examined everything, we have to arrive at the definite conclusion that this proposition: "I am, I exist," has to be true every time that I utter it, or that I mentally think about it.

There is no need here to present a full exposition of Descartes' profound answer to his own philosophical dilemma. The point to be considered is that the dilemma was one that he faced voluntarily. Many who call themselves philosophers have made it a practice to avoid such dilemmas by refusing to pose questions which may strike at the very roots of their impressive intellectual edifices. By posing the most devastating of questions, Descartes showed himself to be one of the greatest of epistemologists, and spurred tremendous advances in philosophy and science. Many of his conclusions have been debated as well they should, but his epistemological approach is the approach of all scientifically and philosophically minded individuals in our society today.

Sociology

Sociology, the fourth discipline, may be defined not so much as a science in itself but as a body of scientifically accumulated knowledge about human relationships. Sociology is not concerned with man's biology, history, or accomplishments *except* as they are a part of, affect, or result from human interrelationships.

Sociology is not, on the one hand, an exact and wholly abstract science like mathematics, nor is it, on the other hand, a search for a course of action in which the seeker will become subjectively involved. It has to do with relationships as they exist, not as they *should be*. It strives to be as precise and as complete as possible; yet it must incorporate into its fabric a considerable amount of "knowledge" that is imprecise and incomplete, inasmuch as a complete and precise understanding of all the human motives that influence social relationships has not been arrived at yet, and quite possibly will *never* be achieved. As far as personal involvement in a course of action is concerned, this is outside the scope of sociology proper. Sociology has to do with a search for truth, which may be translated into a positive program under another heading, as, for example, politics, public health, or personnel management.

Sociological factors are so important to the occupational environment that it is difficult to imagine how anyone having to do with the humanological end of business, professional or otherwise, can function at all effectively without a profound interest in and understanding of sociology. Certainly this interest and understanding must characterize the humanological physician. It is an important point of difference between him and those of his colleagues in the medical profession whose concern is with specific medical problems or specialties rather than the total health of man and how health is affected by his social environment.

Economics

The study of economics is the fifth discipline. One finds, among practical businessmen, a regrettable tendency to look upon economics as an "ivory-tower subject," for the diversion of theoreticians, not for the use of down-to-earth go-getters.

These are the same "practical" people who, for the sake of expediency, committed their companies to total medical care "package" deals in the tropics, who, for similar reasons, voted to cut down on corporate funds earmarked for "Utopian" fundamental research and increase the amounts intended for immediate tangible causes having to do with hospital beds and cure of specific diseases, and who, by the same "logic," populated whole communities with cheap labor drawn from underdeveloped countries abroad or depressed rural areas in the United States. An elementary background of "ivory-tower" economics might have spared them the disastrous results of these "practical" expediency measures.

It is not only the thoroughgoing down-to-earth businessman who tends to look upon academic economics with some suspicion. Many scientific scholars, including physicians, tend to remain aloof from a discipline that is not and cannot ever be an exact science, with all its premises and conclusions subject to experimental verification. This is because it is impossible to make economic observations under controlled experimental conditions characteristic of scientific laboratories. A scientist who wants to test the results of a certain drug on certain organs may be able to create ideal environmental conditions and use two groups of rats which differ only in that one group received the drug and the other did not. Obviously this sort of experimentation is beyond the economist. He cannot test rats or guinea pigs to see what they will bid for what goods under what conditions. He cannot persuade scientifically minded entrepreneurs, bankers, manufacturers, or retailers to embark upon courses leading to bankruptcy simply in order to permit him to study the early signs and symptoms of that particular brand of business failure.

Certain scientific persons may feel that they can afford to remain aloof from academic economics, but this surely cannot be true of the physician, who must deal with people that work for a living and whose major decisions must be made with an eye to possible economic consequences. The medical humanologist, in particular, must have an appreciation of economics. His advice regarding humanological policy can be worthless if it is not in keeping with the facts of the economic world in which this advice is to be put into effect. Any positive measures he advocates regarding the health of the individual, the employed group, or the environment as a whole will have to be conceived in terms of prices, wages, interest rates, stocks and bonds, banks and credit, taxes, and expenditure, with respect to future possibilities as well as present actualities. Otherwise, the real, economic world will prove an unhospitable arena for these measures, and they will inevitably end up as did Don Quixote's nobly designed sally against the windmills.

Human Ecology

The sixth discipline the doctor on the humanology team must understand is human ecology. This is the study of the effects of the environment on human beings. It covers by far the majority of the problems with which the medical humanologist will be confronted in the carrying out of his work. For this reason, it makes up a large part of the subject matter of this text. It has been discussed in general in prior chapters and will be discussed in detail in subsequent chapters.

When one thinks of environmental influences, one may find oneself thinking in terms of purely psychiatric problems, but there is considerable evidence to indicate that stress, an environmental stimulus, can be at the root of almost any problem heretofore considered either physiological or psychological. Stress factors have been found in relation even to the common cold. Harold G. Wolff, M.D., a specialist in this aspect of medicine points out that this connection between emotional cause and "physical" effect has been recognized by folklore long before the concept of psychosomatic disease was ever conceived. He says:

Our language bears witness to the fact that men have long known that certain life experiences, feelings, and bodily changes are connected. Let me remind you by quoting a few phrases picked out of everyday experience that indicate this general knowledge:

He was red in the face; he was hot under the collar; he was blushing; he was pale with rage; he was in the pink of condition; it brought tears to his eyes; his eyes popped out of his head; his tongue clove to the roof of his mouth; his mouth was filled with cotton wool; it took one's breath away; he breathed heavily with passion; he got into a cold sweat; cold hands, warm heart; cold feet; dripping with suspense; it was a nauseating experience; it makes me sick; it turns my stomach; he had a lump in his throat; he got a weight off his chest; he's a pain in the neck; he's a gripe; he trembled with fear; he shook with rage; he has the jitters; he's a stiff-necked fellow; he's weak with laughter; keep a stiff upper lip; and — faint heart ne'er won fair lady.

To complicate the problem, environments refuse to remain static. Once an individual has learned in one way or another to adjust to the stresses produced by one environment, it is superseded by a more complex, more stress-producing one. Ulcers for example, predominantly a woman's disease 100 years ago (about 3 to 1 in favor of women), now favors men to the extent of more than 10 to 1.

Deploring the advent of industrialization with its wake of situational and functional ailments (to be discussed in detail later) is not becoming to the true humanologist. The industrial environment is here to stay, and it will grow increasingly more complex. To plead a return to a simpler civilization is to be unrealistic. The challenge to the humanologist is to recognize that ecological problems are a price that must be paid for a way of life which few are willing to abandon. These problems cannot be evaded; the humanologist must familiarize himself with them and learn how to face them.

The Human Being

It is fairly obvious that, in addition to the sciences or neosciences listed above, the potential humanological medical man will have made it a point, long before his medical college days, to seek out courses of education having to do with such sciences as biology, chemistry, zoology, and comparative anatomy. The potential doctor in any branch of medicine will have been fascinated in his youth by all living things, such as birds, insects, frogs, fishes, and mammals. It is not until he has become fairly sure of his true calling that this interest becomes focused upon one particular creature, *the human being.*

It is obvious, too, that the interest shown in the above-mentioned fields of knowledge must be profound and backed by intelligence, inasmuch as only about one out of eight candidates makes the grade for admittance to a well-accredited medical school.

There is no need here to elaborate upon the years of hard work and the range of subjects involved in formal medical education. This is the common experience of every aspirant to the title of Medical Doctor. No matter what branch of medicine the finished doctor will enter, his background of medical education will include, first, studies of the normal human body, such as human anatomy, physiology, body chemistry, and psychology; next, studies of the abnormal or diseased body, such as bacteriology, pathology, psychiatry; next, studies of living patients; and finally, internship, during which period he fortifies his learning with actual practice.

In spite of the fact that this arduous course tends to "separate the men from the boys," all who complete it are not equally qualified in all respects. Nevertheless, it is not fair to judge each of these purely as an individual. He represents a profession. A physician's diagnosis is not an individual achievement. It stands for a rich inheritance of specialized experience, and whenever any doctor expresses his judgment regarding someone's health and well-being, he is expressing the aggregate of all that he has inherited together with what he has shared and learned to understand in his daily association with his professional associates.

The Five Potential Types of Patients

One important knack that is acquired during this training by the doctor who plans to work with people is the ability to see every individual with whom he deals as *five* potential types of patient.[1,2]

1. He sees the individual through the eyes of *preventive medicine*. This means that he sees the individual as someone who is in danger of encountering one or more specific health hazards against which he can and should be shielded. This is particularly true of the vertical patient, but even the horizontal patient should be viewed first from a preventive point of view and shielded from complications that might negate the curative steps taken against the disabling disease.

2. The doctor sees the patient from the standpoint of *diagnosis and/or early detection* of incipient illness.

3. He sees him from the point of view of *curative medicine*. To many laymen the doctor has no other role, and as a result an unfortunate relationship is created in which maintenance of health for the future becomes a problem the doctor cannot

---

[1] ROBERT COLLIER PAGE, M.D. "Correlation of Case-Find Methods with Medical Service in Industry." An address at the Tenth Annual Congress on Industrial Health Council on Industrial Health-American Medical Association and Medical Society of the State of New York, February 20-21, 1950.

[2] ROBERT COLLIER PAGE, M.D. An address at the Industrial Medicine Association, Buffalo (New York), April 27, 1955.

help the patient solve; the patient simply has no use for the doctor except as an after-the-fact repairman. This situation can be corrected as shown in 4.

4. The doctor sees the patient through the eyes of *educative medicine*. This means not only teaching the individual patient how to comport himself in such a manner as to remain vertical, but also explaining what the ideal doctor-patient relationship should be.

5. The doctor sees the patient from the point of view of *habilitative medicine;* i.e., as someone who through illness may have lost his ability to perform his job and who cannot be considered well until he is once more as productive as his physical and mental limitations permit.

These five viewpoints have to do with the patient who is either horizontal or is in danger of becoming horizontal. Many of these people appear in the office of the physician.

Constructive Medicine

The humanological physician, however, has little to do with such patients. The majority of the persons with whom he deals, directly or otherwise, are not only not ill, but definitely not in search of advice related to illness. Many of these may have capable family doctors and may be participants in wisely conceived programs of periodic examination. In other words, their problems of prevention, diagnosis, therapy, and habilitation are seen to and are of no immediate concern to the humanological physician.

If he has learned his specialty thoroughly, this latter physician will view these people, who can hardly be called patients, from an entirely different standpoint: that of *constructive medicine.*

One of the leading proponents of constructive medicine was Edward Stieglitz, M.D. As he has pointed out, this term is not to be confused with preventive medicine, nor is *construction* of health to be confused with health *maintenance*. The individual who follows all of the precepts of preventive medicine and who takes the proper steps to maintain his health, may be technically free from disease, but he may still not be *positively healthy* in the fullest sense of the word. He may, for example, be carrying around a debilitating load of excess weight. He may be unhappy or bored, he may lack pep or appetite, he may awaken every morning with a "brown taste" — the result of too many cigarettes or scotch highballs. He is not ill. He has no desire to go to a doctor's office. Technically he may be classed as healthy; a routine physical examination without too many probing questions would undoubtedly certify him as such.

This is where constructive medicine comes in. Its object is to *construct* health as a positive state of well-being, not merely to ward off, diagnose, or cure disease. To the humanologist imbued with the principles of constructive medicine, health is not just the absence of disease; it is a positive condition of *wellness* and happiness. It is the condition under which the employed person is most productive, and hence it becomes the paramount concern of the doctor who has to do with the business environment.

The medical student who has mastered the above fields of knowledge is now entitled to be called a Doctor of Medicine. He is not, however, prepared to call himself a specialist in occupational humanology. Before electing to be a specialist in this or any other field of medicine he will do well to ruminate on these words:

"Physician, Know Thyself"

First, he must admit his limitations. No doctor, however thoroughly trained, can be a topflight combination diagnostician, heart specialist, pediatrician, roentgenologist, urologist, gynecologist, orthopedist, neurologist, psychiatrist, obstetrician, and medical administrator, all rolled up into one. There are, after all, at least 34 different kinds of medical doctors, each with a specific role to play. In addition, there are nurses, hygienists, technicians, and the like. The physician must learn his own role, and should call for help when he is faced with a challenge outside his particular limitations. He may feed his ego by setting himself up as a thoroughly rounded physician who will undertake personally to handle every medical problem that is placed before him without consulting the recognized experts in connection with each given problem, but he will not be assuring his patients of top-quality service.

Second, he must resolve his own conflicts. A noted psychiatrist, Hyman S. Rubinstein, M.D., recently explained what this injunction means in terms of his specialty:

> The physician who has failed to . . . come to terms with his own impulses, who has refused to accept that he too is at times beset by urges which if acted out would lower his own sense of worth as a person — too often covers these all-too-human stirrings with a delusional cloak of self-righteousness. Such a physician when confronted by a patient who is troubled by similar unresolved conflicts, often presents such an outward air of piety that the patient is actually stifled. Both the patient and the physician are then on guard against sensing a lowering of self-esteem. The patient who may have been originally willing to bare his soul for the sake of peace of mind, now finds himself being interviewed by a doctor who struggles to keep from his own awareness any reminder of his own unacceptable self.

Dr. Rubinstein is speaking of psychiatry, but his words are true of all fields of medicine having to do with human contact.

The decision to specialize is not an easy one. The doctor who goes into occupational humanology has a difficult road ahead of him. It is not difficult for him to gain favor with the boss by playing golf with him or by becoming his personal professional attendant, but to gain the respect of the rank and file is another matter altogether. Appreciation will be rare, but the need and challenge exist nevertheless, and will be met by the doctor who has what it takes to meet it. The words of former New York Senator Irving M. Ives are pertinent here:

Always remember your first obligation is to your conscience. Often during my boyhood, I heard these words repeated by my grandfather, who was a clergyman. It was his private sermon for me — his personal version of Shakespeare's "To thine own self be true."

Who among us is not from time to time driven to make a decision between conflicting demands? A physician as a full-time employee of any group is a constant target for pressure groups, some pushing this way, some that way. All of us are too often tempted to compromise on the easy pleasant course. By doing one thing it may be possible to sidestep embarrassment, by doing another it is often possible to gain immediate applause; by doing a third the loudest voices may be quieted. How then is it ever possible to arrive at the right decision in the total interest of the human problem at hand?

"Remember your first obligation is to your conscience." If you have made a difficult decision, ask yourself how your conscience will react to it tomorrow, the next day, and the next year. If you have any qualms, then that decision is wrong. Change it.

The Humanologist as a Specialist[3]

There need be no question that the doctor on the humanological team, who is imbued with this philosophy, is as truly a specialist as the doctor whose major interest is disease of the heart or lungs, urology, or gynecology. Recognition of this fact is implied in the establishment of the Specialty Board of Occupational Medicine of the American Medical Association.[4] The physician on the humanological team may

---

[3] LEO WADE, M.D. "The Problem of Industrial Health as Seen by the Medical Administrator." An address at Conference on Education of Physicians for Industry, Mellon Institute, Pittsburgh, December 6, 1955.

[4] Application for Certification in Occupational Medicine, The American Board of Preventive Medicine Incorporated.

now be certified as a spokesman in his own right for the health needs of those employed. He is not just a medical doctor who happens to find himself in industry.

Recognition as a specialist in no way entitles him to be narrow in his concepts. He must wrestle with any tendency to limit his field of vision to the occupational arena. It must include social responsibilities. The employed individual is a part of the community at large; the employer wields a potent influence on the community at large. Hence the humanologist for tomorrow's needs will think in terms of the whole community.

This is summed up by Dr. Willard C. Rappleye, retired Dean of Columbia University's College of Physicians and Surgeons, as follows:

> The rapid evolution of this concept (i.e., community responsibility of the medical profession) in the United States has placed a new obligation upon the medical profession and the educational institutions of the country to produce sufficient numbers of necessary personnel of all kinds to make possible the maintenance and the improvement of health for everyone in the country.

The Doctor for the Needs of Industry

Industry, organized labor, and government, including the armed forces, have need for the advice and guidance of the trained specialist in occupational health. This need will increase automatically as the scope of humanology increases.

In order to make the best and most thorough use of his training and experience, the specialist whose interests are focused on the health needs of those employed will do well to consider one of the following possible careers:

1. Act as a free-lance consultant (*a*) to industry as a whole, large or small, (*b*) to management, and/or (*c*) to labor organizations or professional guilds
2. Be an associate in a group-practice enterprise in which representatives of other pertinent branches of medicine

are joined or in some way made readily available when needed
3. Be a member of a humanological team on a full-time career basis or as a part-time employee of an industrial, governmental, or labor organization
4. Be a member of a teaching faculty on a full- or part-time basis

In the opinion of the writer, the first two career possibilities offer the greatest challenge to the trained and fully competent physician. In each of these careers, he is at liberty to follow and put into practice the dictates of his own judgment, training, and experience.

The fourth career is the most important. There is a recognized need for top-notch educators in the field of humanology. Despite this need, however, there are few challenging openings at present.

As for the third career, there will always be a demand for medically trained humanologists who will commit themselves to a salaried position on an organizational chart. Before embarking upon such a career, the doctor should give careful thought to his personal goals. He must be prepared to make a periodic honest self-analysis and be aware of what he is doing and where he is going, before he reaches that "point of no return" where decisions as to his possible courses of action are no longer his to make. It may be noted in passing that there have been few distinguished physicians who have embarked upon full-time salaried careers who have been privileged to live long enough, with security and happiness, to reminisce upon past accomplishments.

The Registered Nurse

If the doctor is the captain of the professional arm of the humanological team, it may be said that the registered nurse is his executive officer. She is the one, ordinarily, who will be charged with seeing to it that his advice with regard to general principles of action is carried out properly in each specific case. While she is not, nor should be, expected to practice medicine beyond the practice of first aid within definitely established limits, she may be the only visible representative of the medical profession within a given occupational environment.

She is, in fact, the eyes and ears of the medical humanologist. To risk overworking the military analogy, it may be suggested that in addition to her functions as executive officer, in which she translates recommendations into concrete individual courses of action, she is expected to serve as reconnaissance officer for the medical profession in general. In industry, she is typically the first member of the humanological team to encounter the employed individual who is actually ill or is on the verge of illness, physiological or psychological. The maladjusted, the disturbed, the hypochondriac, the potentially senile, the mildly ill or injured, as well as the acutely ill, come to her immediate attention either voluntarily or thanks to the alert eye of the able supervisor, and it is up to her to see that these people obtain whatever medical attention they truly need, whether it be off-the-premises diagnosis, therapy, consultation, or limited attention at dispensary level.

Qualifications and Background

The actual functions and limitations of the nurse in industry belong properly in the next chapter. Here it is appropriate to limit discussion to the qualifications and background of this strategic member of the humanological team.

Plainly, the mere fact that she has been painstakingly trained is insufficient evidence that she will be a competent occupational health nurse. Since she carries out her most important functions in face-to-face contacts with all kinds of people, ranging from geniuses to semiliterate "brawn" workers, she must be broad enough to establish communication with these different kinds of people. In other words, she must be able to talk a variety of "languages." She must not put the unschooled loadcarrier on his guard by seeming to be superior to him, nor must she let herself seem to be on *her* guard in the presence of the rather conceited young recruit just out of college who has decided to grace the company with his august presence. She must have the knack of being at ease, friendly, and intelligent with all.

This knack is not wholly inborn. Anyone who likes people and who can get along with other people without acute discomfort can learn to develop this talent with experience. This is why the value of the humanological nurse increases in proportion to the time actually spent in an occupational environment practicing this skill.

Her Place in the Scheme of Things

Other factors which have to do with the value of the nurse in industry, in addition to training, aptitude, and ability to communicate are:

1. *Her position in the organizational plan.* — It goes without saying that she must be given the managerial authority as well as responsibility to carry out those functions which are hers by definition, though these may vary from industry to industry, and to make the decisions which fall within her professional scope. Moreover, inasmuch as her value increases with experience, she should have opportunity to advance not only financially but in organizational authority, provided she does not exceed her professional limitations.

2. *Liaison with other members of the professional arm of the humanological team.* — Liaison with the physician at the head of the team may be taken for granted. The other members, to be discussed more fully below, include the consulting psychologist who has to do with certain aspects of

humanological assessment in connection with placing and keeping square pegs in square holes, and the several specialists in plant safety and industrial hygiene whose special concern is with the elimination or control of specific occupational hazards. These team members are as much in need of the cooperation of the nurse as is the physician. Day-to-day experience in the particular occupational environment assigned to her gives her a down-to-earth understanding of that environment beyond the reach of the occupational psychologist or industrial hygienist. She is therefore in a position to interpret this environment so as to provide a realistic foundation for all humanological measures.

3. *Liaison with resources in the community at large.* — Not only must she be in communication with representatives of all significant branches of medicine in order to send those in need of medical help where they can get quality service at reasonable cost; she needs to be able as well to call upon appropriate nonmedical agencies, such as welfare organizations, clergymen, benevolent associations, and governmental bodies, when such are indicated as possible factors in the solution of a given individual problem.

Fundamental Principles of Occupational Nursing

A list of the fundamental principles of occupational nursing may be in order here to shed some additional light on the kind of person such a nurse needs to be. Discussion of these principles, however, must be reserved for the following chapter. These principles are as follows:

1. Though the humanological nurse may exercise initiative and professional judgment compatible with the organizational authority granted her, all nursing care, in the last analysis, is under the direction of a doctor who can be reached if necessary and to whom detailed reports will be submitted. Actually, this frees the nurse

by relieving her of the burden of responsibility for decisions outside her scope.
2. The primary interest of the humanological nurse is the vertical, gainfully employed individual; hence her principal concern is with prevention of injury and disease, or, after disease has occurred and responded to medical treatment, with rehabilitation for maximum productivity.
3. The nurse must be fully aware of her limitations. She may offer first aid for mild disorders which would not ordinarily be presented for medical treatment or which are easily and completely corrected.
4. Health assessment programs are not to be considered properly within the scope of the humanological nurse. If, through misunderstanding on the part of the employer, she is asked to institute such a program, she will turn the problem over to the physician to whom she is required to report.
5. The nurse will take advantage of every opportunity to disseminate effective health education, through displays, exhibits, counseling, etc., throughout her plant. With the cooperation of the consulting physician she will keep this material up to date. Dissemination may be done according to her own judgment, or in cooperation with managerial or supervisory lay personnel.

The nurse is first and foremost a nurse, and secondarily a specialist in humanology. Therefore it seems appropriate here to touch briefly on some basic factors having to do with nursing in general.

As to training, the nurse, like any specialist, will have to devote a certain amount of time to theory, rules, and similar "rote" work, but by and large, her education will be practical and empirical.

Like a good soldier, she will learn best by doing, and her education will consist of considerably more than the acquisition of knowledge. A nurse, in any branch of nursing, needs to acquire certain attitudes toward life and her role in it; she needs a certain toughness of spirit, modified by sympathetic understanding and flexibility of mind; she needs to have developed a "feel" for nursing as well as an understanding of it.

For this reason, it is difficult to understand how certain commercial schools of nursing have been able seriously to advance the claim that they have produced graduate nurses by means of home correspondence courses. Certainly the doctor on the humanological team will regard such claims with extreme caution. The matter need not be explored in greater detail here. The situation is being combatted by the major nonprofit associations having to do with the nursing profession, and is being corrected in practice partly by tightening of sound nursing practice acts, and partly by increased caution on the part of those who must weigh the qualifications of supposedly "trained" nurses.

Code for Professional Nurses

The American Journal of Nursing published, as long ago as 1950, a "Code for Professional Nurses," which is still applicable, as it will no doubt continue to be as long as illness exists in the world, as an outline of the absolutely essential qualifications of the nurse, in or out of industry. The code, somewhat condensed, is as follows:

1. The fundamental responsibility of the nurse is to conserve life and to promote health.

2. In addition to being prepared to practice, the professional nurse must maintain her status through continued study.

3. The nurse must remain with the patient until assured that he can safely be left or that adequate relief is available.

4. The religious beliefs of a patient must be respected.

5. All personal information entrusted to the nurse must be considered confidential.

6. A nurse recommends or gives medical treatment only upon specific physician's orders or in emergency, and reports to a qualified physician as soon as possible.

7. The nurse avoids misunderstandings or inaccuracies in carrying out doctors' orders by obtaining verification; she refuses to participate in unethical procedures.

8. The nurse sustains confidence in the doctor and other members of the health team; incompetence or unethical conduct of associates must be reported, but *only* to proper authority.

9. The nurse is obligated to give conscientious service and is entitled to just remuneration.

10. The nurse accepts only such compensation as the contract (actual or implied) provides. She does not accept tips or bribes.

11. The nurse does not permit her name to be used in connection with testimonials in product-advertising.

12. The nurse is guided by the Golden Rule in all professional relationships.

13. The nurse in private life adheres to standards of personal ethics which reflect credit upon the profession.

14. The nurse will not knowingly disregard the accepted patterns of behavior of the community in which she lives and works.

15. The nurse upholds the laws of her country and community and is fully conversant with those laws having to do with medicine and nursing.

16. The nurse is interested in the health needs of the public at large, and makes an effort to participate in local, state, national, and international health-improvement efforts.

17. A nurse performs the duties of citizenship, voting and holding office when eligible, and has an appreciation of the social, economic, and political factors which promote desirable patterns of community living.

The Occupational Psychologist

Another extremely important member of the professional arm of the humanological team is the occupational psychologist. He is by no means a psychiatrist, who has no place in the occupational environment except on a consultative basis in the event of specific need. The latter has to do with so-called *mental* illness. He is a *medical* specialist, who should be available in the community and called upon when mental illness is present or in the offing.

A Student of Behavior

The psychologist, on the other hand, is a student of behavior in *general*. He is not concerned with abnormal behavior except insofar as it is a factor in his measurement of a given individual's intellectual potential, his personality and emotional psychodynamics, his achievement and interest patterns, and his special aptitudes and disabilities. In industry, his interest lies necessarily with the vertical individual who, after being measured, is capable of being assigned to gainful employment. The sick individual who is in need of psychiatric consultation is automatically taken out of his hands.

The reason for making a point of this distinction here, obvious as it may be even to the elementary student of humanology, is that there are concrete instances in which management policy has failed to make such a distinction. There is no need to point the finger. More than one so-called executive development program has included the service of a full-time psychiatrist, a luxury difficult to justify on grounds of actual need or of elementary economics. A business which produces enough bona fide cases to keep a career psychiatrist busy is, one hopes, an unusual business, to say the least.

The branch of psychology which has to do with occupational health is *clinical* psychology as opposed to *applied* psychology. Clinical psychology consists of the systematic inquiry into the behavior of a given individual through the application of psychological tests and procedures. The test and procedures have been tried and accepted as valid by qualified authorities in the field of psychology.

The Role of Psychologists in Industry

The major role of psychologists in industry has been limited to such items as (1) selection and placement procedures; (2) job evaluation and merit rating; (3) accident reduction; (4) measurement and improvement of employee morale; (5) marketing and advertising research; (6) opinion research; and (7) efficiency measurements.

Mental Hygiene

In recent years, the concept of *mental hygiene* has come to the fore in humanological circles, pointing the way to new uses of the tool of clinical psychology. It is coming to be understood that in order to be effective, a mental hygiene program applied to the occupational environment must be comprehensive and all-embracing.

A Positive Human Engineering Program

An outstanding example of such a program, soundly tested and offering great promise for the future of occupational mental hygiene, is the thoroughgoing human engineering program which was conceived as long ago as 1944 at the Caterpillar Tractor Company, Peoria, Illinois, and launched in 1946, under a personnel consultant attached to the company's medical division staff. Since its inception, the program has been periodically kept up to date with the help of exhaustive follow-up studies.

The Eight Principal Facets

The program has eight important facets. They include (1) selection and placement, (2) induction, (3) interviewing and counseling, (4) training of interviewers, (5) training of supervisors, (6) mental hygiene education, (7) social service, and (8) research.

Selection and Placement

1. *Routine pre-employment test battery.* — This consists of four standard placement tests. Employees are assured that the procedures are never intended to be used as a means for rejection or discrimination against any prospective employee. The tests are the following:
   a) *Wonderlic Personnel Test,* designed to ascertain the applicant's mental ability to learn.
   b) *Bennett Mechanical Comprehension Test.*
   c) *Cornell Word Form,* for the detection of personality disturbances. It consists of a list of stimulus words next to each of which are two other words, the subject is asked to choose the one he thinks fits the stimulus word.
   d) *Cornell Index,* for measuring emotional maladjustment and facilitating diagnosis. This test consists of questions concerning complaints caused by emotional disturbances, e.g., insomnia. Certain questions have to do with crucial symptoms, e.g., problem drinking. These are called "stop questions," and indicate the necessity of immediate corrective action.

2. *Caterpillar psycho-graph profile.* — This is based upon the realization that no psychological test battery serves any useful purpose *per se* unless the results are properly interpreted by a *competent* authority, applied to actual industrial situations, and made comprehensible to persons who must actually make use of such tools. The profile shows, in graphic form, the applicant's test results and becomes a part of the medical as well as the personnel record. The confidential psychological nature of the data is available upon consultation with the medical division only. The data obtained from the selection tests are incorporated with all other pertinent data, medical and otherwise, to give a total picture of a human being. Those who have poor scores on the personality

tests are fortified against poor work adjustment by means of brief preventive psychotherapy or are referred to a community agency. Those with poor scores on the intelligence and/or mechanical aptitude tests are placed in jobs commensurate with their limitations.

3. *Special test batteries.* — The services of the personnel consultant's offices are often sought by other divisions in connection with special selection problems. These usually have to do with the appraisal of employees to be placed in responsible positions, e.g., trainees for sales representatives.

Induction

Basic employee orientation consists of an illustrated lecture explaining the various parts of the plant and giving information on such practical matters as first aid, methods of calculating wages, safety measures, etc. The mental hygiene aspect of induction is embraced in sections of the orientation course dealing with emotional adjustment (*a*) to make employees realize the potential emotional stresses of the new situation in which they find themselves, and (*b*) to make them feel that their work is worthwhile and a contribution to the betterment of themselves as well as society.

Interviewing and Counseling

An interview to study emotional health is ordinarily given as a matter of routine to all employees during their pre-employment medical examination. During this interview, which precedes counseling, the personnel consultant has at his fingertips the scores on the intelligence, mechanical aptitude, and emotional stability tests. Emphasis is placed upon the presence or absence of personality and psychosomatic symptoms. It also aims to uncover histories of problem drinking or criminal behavior, as well as to ascertain shut-in tendencies, excitability and irritability, impulsive behavior, worrisomeness, lack of interest, and sexual difficulties. The interview is informal and nondirective; the employee is allowed to direct the conversation into channels that interest him, while the interviewer merely acts as a guide.

Counseling, on the other hand, must be to some extent directive; a stand must be taken by the counselor. There are five primary aims of counseling:

1. The extraction of information for the use of management and/or labor
2. The interpretation of company policies to confused employees
3. The determination of available facilities or specialists for treatment of emotional problems
4. The proper administration of recreational activities or other effective morale measures
5. The salvaging of valuable employees "on the way out"

Evidence has been accumulated to substantiate the observation that counseling helps to make the employee aware of true conflicts that may underlie his surface complaints. The result may be that he is relieved and the surface problem is either diminished or altogether eliminated.

The counseling service is part of the medical division, with the result that employees feel secure that confidences will not be betrayed; the same feeling does not seem to exist where the counseling psychologist is attached to a lay department lacking the medical tradition of discretion regarding confidential information.

Training of Interviewers

1. *Lectures on personality and applied psychology.*—Some of the major lecture topics for the indoctrination of employment interviewers have been the following:

   a) The nature of personality, (physical, endocrinological, psychological, and environmental determinants) *b*) The development of personality, description of traits, concept of introversion, extroversion, role of drives and motivation

c) Adjustment mechanisms, defense, withdrawing, repression, rationalization, compensation, etc.

d) Relationship of personality to the intellect: classification of intelligence, definition of terms, concepts of amentia and dementia; effects of emotional maladjustment upon intellectual efficiency

e) Causes and symptoms of maladjusted personalities, "nervous breakdown," the psychoneuroses, and the major psychoses

f) Psychosomatic relationships, mechanisms, and symptoms, role of anxiety

g) Selective placement, placing the emotionally and intellectually handicapped in industry

h) Mental hygiene, prevention of emotional difficulties, therapeutic methods

i) Rehabilitation of the returning veteran, understanding the nervous veteran, recommending helpful literature for veterans and families

Emphasis in each of the above lectures, and in similar presentations, has been on *employee needs in occupational environmental situations.*

2. *Supervised interviewing.* — Having been indoctrinated in the theory of interrogation and pre-employment interviewing, new members of the humanological team perfect their interviewing techniques under a two-stage scheme which affords practical experience. The stages are:

a) The learners "sit in" on actual interviews conducted by the personnel consultant;

b) The learners conduct their own interviews under the supervision of the personnel consultant.

3. *"Headache clinics."* — Once weekly it has been customary for the personnel consultant to convene with the

interviewers for the purpose of discussing difficult cases. The case-conference method in the "clinical" sense has been used. Such an exchange of ideas enables the interviewer to glean worthwhile cues, and, if nothing else, they tend to bolster his confidence by the knowledge that others have had similar "headaches" in interviewing.

4. *Reading assignments.* — In addition to required readings in standard textbooks dealing with clinical psychology and interviewing techniques in industry, a number of other books of potential interest to interviewers have been made available in the library of the personnel consultant.

5. *Coordinating all data on applicant.* — This subject is covered in a series of lectures designed to show the interviewer how information obtained from the personal interview, test scores, employment application blank, and from the confidential merchants' report could be utilized to hire the employee so that maximum satisfaction and effectiveness could be achieved.

6. *Individual conference with the personnel consultant.* — Such time is utilized by the interviewer for clarification concerning decisions of hiring. Occasionally other interesting points which might not otherwise be brought to light are uncovered. Interviewers, it develops, often have problems of their own calling for "catharsis" if not concrete help.

Training of Supervisors

A series of lectures has been designed for foremen to instruct them in the handling of employees, particularly with reference to their emotional problems, in order to foster good productive attitudes. The content of these lectures is similar to the series of lectures given the interviewers, but detail is kept to a minimum. Actual case material is used to illustrate the dynamics involved in the development of significant symptoms. The aim of the lectures is to incorporate "humanology" into the supervisor's sphere of influence in addition to mechanics in which he has already achieved standing. (See chapter on "Supervision," page 493.)

As an integral part of this program, a manual has been written designed to train the foreman in the recognition and management of emotional problems that occur in persons under his authority.

Mental Hygiene Education

1. *Articles in employees' magazine.* — A systematic effort has been made to educate employees through short articles in the biweekly employees' magazine. These articles cover psychological principles, behavior, emotions, etc., in relation to the work environment, home life, and interpersonal relationships in general. Following each article, it has been standard operating procedure to insert a paragraph entitled "What to Do About It," containing brief advice concerning some steps that can be taken in the direction of improving behavior, acquiring new insight, and/or understanding mental hygiene.

2. *Bibliotherapy.* — The personnel consultant has made available to employees some 200 selected books. These contain discussions of various psychological topics. While any employee who wishes may make use of these books, it is standard procedure to suggest certain titles to those who are being counseled, so that counseling may be reinforced by supplementary reading. Extreme caution is exercised in prescribing specific books, in order not to confuse the reader with material beyond his particular grasp or momentary level of insight or to suggest additional symptoms to susceptible readers.

3. *Visuotherapy.* — In addition to use of the printed word as an educational medium, visual aids in the form of 2″ by 2″ slides have proved valuable. Pictures have been chosen that show how anxiety, frustration, resentment, aggression,

and the like can influence bodily dysfunction. Other slides show the central nervous system network and the paths of neural impulses.

Social Service

In a number of situations, it is necessary for the personnel consultant to have recourse to outside community agencies in order to help an employee effect a more complete vocational and emotional adjustment. In Peoria, veterans may be directed to the Veterans Administration offices for specific psychiatric and psychological attention. Other federal, state, and local agencies, including the local state hospital and some community social agencies, have proved to be of concrete value, as have employees' close relatives and local clergymen.

Research

The program makes provision for constant reappraisal of tools and methods. Cooperation, not only with other arms of the humanological team but with management in general, has broadened the clinical psychologist's understanding of this particular aspect of human ecology.

Broad Goals of Mental Hygiene

To summarize, it may be said that the broad goals of mental hygiene in industry are (*a*) to maintain the mental health of employees by insuring them happiness, pride of achievement, and good work adjustment, and (*b*) to avoid the waste of human resources brought about by faulty emotional adjustments.

The program described above, which may be trimmed down in size to fit any given organization offers procedures designed to help achieve these broad goals by accomplishing the following four goals:

1. Aiding in placing industrial personnel in positions which make possible the fullest use of their capabilities with a minimum likelihood of personality disturbances.
2. Providing machinery for combating personality disturbances when they do occur, (counseling, bibliotherapy, visuotherapy, etc.). It has been learned that the efficiency and satisfaction of employees can be maintained by making available help for their emotional problems. Moreover it has been clearly determined that persons with moderate psychopathology may be superior employees if properly placed.
3. Extending principles of mental hygiene to interviewers and foremen, thereby bringing these key people closer to understanding human elements at play in industrial-personal relationships.
4. Broadening the scope of the clinical psychologist in industry to the point where he can do something about the human relations and efficiency of the great mass of "normal" employees, not only psychoneurotic and other emotionally maladjusted persons.

The Industrial Hygiene Engineer and His Co-workers

The very fact that it is possible to talk at all about such matters as constructive medicine, total health, employee productivity, or control of nonoccupational illness in connection with the work of the professional arm of the humanological team is due almost wholly to the emergence of the *industrial hygiene engineer* and his co-workers.

In the early days of occupational medicine, the doctor in industry had his hands full wrestling with the problems of diseases and injuries arising directly out of occupation. His sole function was to provide treatment for, and seek to determine measures for the control of, the following general types of ailments, which are by-products of an era of dramatically expanding industrialization: (1) Injuries, ordinarily having to do with machinery; (2) Poisoning or other internal disease owing to exposure (*a*) to metals or other inorganic substances, (*b*) volatile solvents, (*c*) radiation, (*d*) organic chemical substances.

The physician, in such circumstances, simply did not have time to focus his attention on such matters as, for example, circulatory problems in older workers, even though he might be aware that these were more costly in terms of productivity than specific on-the-job hazards.

Technological progress increased the number of these hazards; at the same time it high-lighted the doctor's need for skilled help. As facilities for training them developed, the industrial hygienist, the toxicologist, the biochemist, the physicist, the engineer, the physical and analytical chemists, and others joined forces with the doctor to investigate the nature, prevention, and therapy of occupational disease processes.

Out of this teamwork came control techniques which enabled the doctor to devote more and more of his time to thinking in terms of the *total* health, happiness, and productivity of the employed human being. The strictly occupational aspect of employee health problems has shrunk to its proper proportionate size in the scheme of things. After the medical profession has played its proper role in the development of a control program for a particular hazard, that hazard becomes primarily a problem for the industrial hygiene engineer.

This is not to say that the medical consultant can turn over the entire responsibility for occupational health hazards to the industrial hygienist and then proceed to forget about them. For one thing, there seems to be no end to the development of new hazards as technology finds new products to be made in new ways with new substances and new machinery. For another thing, as Dr. Robert A. Kehoe has put it:

In connection with all environmental studies in which an expert is brought in to make observations, one never deals wholly with an engineering problem and never wholly with a medical problem, but with the combination of medical, physiological, and technological problems.

Furthermore, the doctor is at all times the captain of the professional arm of the humanological team, for a number of reasons which need not be argued here (note that in the clinical psychology project described in this chapter, it was found most effective for the psychologists and helpers to work under the aegis of the medical department). Another aspect of industrial hygiene that calls for the personal intervention of the doctor is that of bringing the need for control measures to the attention of top management. Here again, Dr. Kehoe, may be quoted. In an address as far back as 1951 he said:

> I have worked on the problem of industrial lead absorption for many years .... The information that is available at the present time should enable us to eliminate lead poisoning from American industry, except in the most unusual situations. However, this has not been accomplished. Even though, according to statistics, the number of cases is decreasing, the improvement is more apparent than real. The reason for this, in many cases, is that *management does not know the problem and has never been properly informed of it by physicians.*

What this indicates is not necessarily a lack of competence on the part of anyone concerned, but rather a lack of teamwork. It is axiomatic that any team captain, to be able to do a job with any degree of self-confidence, needs to have a full appreciation of the problems of each of his teammates. Medical consultants will do a more thorough job of captaining the industrial hygiene phase of the humanological team's endeavor if they familiarize themselves with the job of the industrial hygienist. This cannot be done in the hospital, dispensary, or the library of the Academy of Medicine. It must be done in the actual arena of the industrial hygienist, in the actual places where people work.

If one is to gain a constructive understanding of potentially hazardous working conditions, one must take time, watch how things are done, become acquainted with men and women on the job, examine and assess all operating facilities and their maintenance, inspect washrooms, shower rooms, and eating places, considering their comfort, cleanliness, and convenience, and measure, where possible, environmental factors with respect to light, noise, temperatures, chemical contamination of atmosphere, and so forth.

The medical consultant who does this from time to time acquires a new respect for his teammates in the field of industrial hygiene. He develops a broadened understanding of just what kind of person or persons he needs on his team to carry out this part of the total humanological endeavor.

The Qualifications of an Industrial Hygiene Engineer

A cold, clear, "brass-tacks" presentation of the qualifications of an industrial hygienist is set forth in a routine job-specification sheet prepared in 1951 by N. V. Hendricks, a recognized leader in the field. The following are the "minimum qualifications" for a "finished industrial hygienist":

1. *Education and experience*
   a) Graduation from a college or university of recognized standing, with major study in chemical, sanitary, or public health engineering or chemistry
   b) One year of graduate study in industrial hygiene or its equivalent
   c) Seven years of full-time paid professional experience, within the past 10 years, in the field of industrial hygiene
2. *Knowledge, skills, and abilities*
   a) Thorough knowledge of the principles and practices of industrial hygiene
   b) Extensive knowledge of public health or sanitary engineering
   c) Extensive knowledge of the techniques required for industrial-hygiene surveys and studies

- d) Good working knowledge of chemistry and the analytical procedures required in the field of industrial hygiene
- e) Good knowledge of industrial exhaust ventilation
- f) Some knowledge of toxicology, with particular reference to the toxicity of common industrial materials, and the ability to interpret toxicological data
- g) Exceptional ability to plan work effectively
- h) Exceptional ability to get along well with others, to secure cooperation, and to avoid antagonisms

Industrial Hygiene in Action

A classic example of industrial hygiene in action is afforded by a project in which the author participated personally involving the establishment of a cancer-control program for personnel engaged in processing high-boiling catalytically cracked oils. This project was classic in one sense in that it was carried out by the full panoply of specialists needed to round out a major effort in the field of industrial hygiene. In addition to the hygienist, the team called for the cooperation of the chemist, the toxicologist (experimental biologist), the nurse, the epidermologist, the biostatistician, and the safety engineer.

A Classical Example of Fundamental Research

The project was classic in another sense, in that it substantiated the remarks of Dr. Kehoe, quoted above, with reference to lack of communication with management. This particular classic aspect may be discussed here, in passing, before consideration of the project itself. Anything that has to do with cancer, it seems, can do things to the imagination of the most hard-boiled man of management. When the top-line officers of the industry in question were informed that a cancer hazard existed in certain operations of the business, logic flew out the window. The legal team became frantic with the thought of potential million-dollar law suits. Everything must be done hush-hush.

In spite of these obstacles, the medical department managed finally to round up a full research team as described above plus adequate university cooperation. However, despite what seemed to be a green light to finish the project and a more than adequate initial allocation of funds, innumerable obstacles appeared almost from the start. Certain low-echelon officers expressed disapproval of the project and put annoying stumbling blocks in the way. Management meetings were held, largely, it seemed, for the purpose of quibbling over definitions. One vice-president created a tempest in a teapot by demanding the definition of an industrial hygienist.

The real excitement came after the hazard had been determined and measured, and effective control measures recommended. Meeting after meeting was held to decide on ways and means of telling the exposed workers just what the hazard was and what to do about it. Elaborate circumlocutions were worked out by public relations people. Some representatives of management expressed fears that there would be demands for extra pay, civil suits, and the like, following revelation of the hazard to "exposed" employees.

What actually happened, when the employees were told by a member of the health team the true nature of the problem in language they could understand, was that full cooperation prevailed. All control recommendations were carried out with neither complaint or confusion. No increase in pay was demanded. The long-range result was that no cases of occupational cancer occurred, despite the proven hazard. The only harm, in fact, that seems to have been done by the project was that it left in its wake a few somewhat strained relations at top-echelon level. The strained relations could have been avoided with real understanding of the problem.

History and Highlights of the Team Approach

The project, despite petty obstacles, was a thorough and in many ways a conclusive one. Its history is as follows:

In the year 1942, the first commercial unit for fluid catalytic cracking was put into operation in a company affiliated with the Standard Oil Company (New Jersey). In this operation, certain petroleum fractions are heated in the presence of a catalyst, and large petroleum molecules are broken down into small molecules such as those found in gasoline, kerosene, home-heating oils, and diesel oils. There is a residual material which contains an appreciable concentration of high-boiling polycyclic aromatic hydrocarbons. This material is known as high-boiling catalytically cracked oil, slurry oil, or, after removal of the catalyst, clarified oil. Since certain high-boiling polycyclic aromatic hydrocarbons are known to be carcinogenic, it was suspected that this residual material might constitute a potential cancer hazard.

Occupational cancer had been reported as resulting from contact with soot, coal tar, shale oils, certain petroleum oils, and other related materials. However, catalytic cracking of petroleum was a new operation, and no information was available relative to the carcinogenicity of products obtained from this process. However, it was known that coal tar, containing certain compounds in common with catalytically cracked oils, had caused occupational cancer.

In 1942, samples of catalytically cracked oils had been studied at a St. Louis cancer hospital and found to be productive of carcinogens on the skin of mice. Because of these findings, certain humanologists in the oil industry, including the author, felt that there was a need to investigate the matter in detail, with the intention of determining just what materials were actually hazardous and to just what extent. In 1945, arrangements were made with the late Dr. C. P.

Rhoads of Memorial Hospital for the Treatment of Cancer and Allied Diseases, New York, to carry out experimental studies, using several species of animals. The studies were carried out under a grant from the Standard Oil Company (New Jersey) and certain affiliates, in cooperation with the medical department of the company. The Standard Oil Development Company prepared and supplied all samples, fractions, and blends of oils, except those of shale oil, which were obtained through the courtesy of the United States Department of the Interior, Bureau of Mines.

Studies were exhaustive. Mice, rabbits, rats, guinea pigs, and rhesus monkeys were subjected to repeated applications of oils blended in various ways, distilled at different temperatures, and variously treated. They were also subjected to applications of eight so-called "slack waxes," obtained from pressing operations, and containing large quantities of oil. Petroleum tars and American shale oil were likewise tested.

The reports of the tests, presented jointly by scientists of the New York University-Bellevue Medical Center and the Sloan-Kettering Institute for Cancer Research, showed that high-boiling catalytically cracked oils and fractions of these oils boiling above 700° F. were highly carcinogenic to the subject animals. Slack waxes and shale oils containing "guilty" substances were found to be equally carcinogenic; petroleum tars were found to be somewhat carcinogenetic, depending upon the processes of extracting them, but were less carcinogenetic than coal tars which are produced at higher temperatures.

Scientists, including engineers, biostatisticians, and chemists at the Esso Laboratories of the Standard Oil Development Company, reviewed these findings and carried on supplementary research based upon them in an endeavor to pinpoint the chemical nature of the carcinogenetic products, to develop relatively simple laboratory assay methods for predicting the "tumor potency" of high-boiling refinery products (since processes are forever changing and last year's elaborate research project may not be accurate with respect to this year's product) and to determine processing steps that can be employed to reduce or eliminate the potential hazards of these oils.

In order to ascertain the extent and degree of contact of employees with these oils in companies affiliated with Jersey Standard, an industrial hygiene survey was inaugurated in 1947 and carried out in six refineries where the catalytic cracking process was in operation. This study, carried out by N. V. Hendricks, indicated that slightly over 4,000 employees were engaged in work in which they might have skin contact with these oils or blends of these oils daily, every few days, or occasionally. It was learned that relatively few workers were exposed to the oil in the operation of the fluid catalytic cracking unit, but that after the oil left the cracking unit numerous employees were apt to come in contact with it in varying degrees.

Specific Industrial Hygiene Recommendations

After correlation of all data obtained in this team endeavor, the following industrial hygiene recommendations were made:

1. *Restrict the number of employees exposed.* — This is made possible by assigning *selected* groups of workers to jobs involving contact with these oils.

2. *Make medical examinations every three to six months.* — Exposed men should be examined every three months. The examination should consist of a complete physical examination, with special attention to the presence of warts, papillomas, tumors, cancers, dermatitis, or other abnormalities of the skin.

3. *Select the employees.* — The selection of employees who work with these oils should be made by the medical

department in order to eliminate individuals who have an excessive number of warts, precancerous lesions, keratoses, or other skin diseases. It is preferable to select swarthy or dark-skinned men with cleanly habits.

4. *Remove employee from exposure if lesion develops.* — If an employee after working with these oils presents warts or precancerous lesions on the surface of the body, he should be permanently removed from any employment in which he might again be exposed to these oils.

5. *Treat lesions promptly.* — Any employee working with these oils who has a wart or a cancerous or a precancerous lesion developing on the surface of the body should be given prompt treatment. This should be carried out at company expense.

6. *Wear protective clothing.* — Employees working with these oils should wear whatever protective clothing is necessary to prevent skin contact with these oils. The type of clothing would be dependent on the nature of the individual job at hand.

7. *Do not touch scrotum or face with oil-soaked hands.* — Since a high percentage of occupational cancers caused by coal tars, shale oil, and unrefined waxes are located on either the face or scrotum, it is felt that contact of the oil with these parts of the body should be avoided.

8. *Remove oil from the skin immediately.* — Employees who work with these oils should be instructed to remove oil from the skin immediately by thorough washing with soap and water.

9. *Take a shower bath daily before leaving the plant.* — All employees who work with these oils in such a manner that there is a reasonable possibility that they will have skin contact with the oil should be required to take a shower bath before leaving the plant. Obviously, in certain cases where the possibility of skin cancer is very remote, for

example, in the case of certain laboratory workers, the requirement of a daily shower bath would not appear necessary.

10. *Put on clean work clothes daily.* — Employees who work with these oils should wear clean work clothes daily. If underwear is worn, this should be clean.

11. *Provide special facilities for cleaning work clothes.* — Because of the potential hazard associated with these oils, it is felt that it would be undesirable to have an ordinary commercial laundry or dry-cleaning establishment clean the clothes. Therefore, in order to ensure clean work clothes, special cleaning facilities should be established. The cleaning of the clothes should be carried out under supervised conditions in order to prevent contamination of the cleaning agent.

12. *Reduce exposure points in plant to a minimum.* — In any plant where these oils are handled, the necessary changes should be made in order to reduce to a minimum the places where the employees are exposed to physical contact with these oils. This could be done by restricting tanks and lines having contact with these oils.

13. *Identify equipment containing these oils.* — All equipment containing these oils should be identified with the statements that the oil is hazardous, skin contact should be prevented, and in cases of skin contact the oil should be immediately removed by washing with soap and water. (Follow-up action on this recommendation was that all lines, tanks, and other pieces of equipment containing the oils in question were painted orange for identification.)

14. *Inform employees of potential hazard.* — Despite management qualms about employee reaction to such information, it actually had a beneficial side effect in that it made employees conscious of skin cancer. Whether or not employees had contact with these oils, they began coming voluntarily to the medical department with any occurring skin lesion.

15. *Supervise precautionary measures in each plant.* — In addition to precautionary measures already established,

arrangements should be made for a representative of the medical department to keep in touch with the precautionary measures in operation. This can be accomplished by making a survey of these precautionary measures every week or every few weeks. A health problem such as this can be properly handled only by ensuring adequate medical supervision.

16. *Make a periodic survey of plants handling these oils.* — A periodic survey of this problem should be made in each plant where this problem exists by representatives of the medical department at approximately six-month intervals.

17. *Eliminate possibility of inhalation of these oils.* — In all plants where these oils are handled, precautionary measures should be taken to eliminate the possibility of inhalation of droplets or vapors of these oils by means of special hoods and careful maintenance to prevent pump leaks.

18. *Provide the medical and industrial hygiene personnel and equipment necessary to carry out these recommendations.*

19. *Investigate blends of these oils.* — Studies were inaugurated to determine whether or not blends of these oils are less carcinogenic than the unblended oil. It was originally suggested that blends of these oils contain no more than 25 per cent of high-boiling catalytically cracked oil, slurry oil, or clarified oil.

It is suggested that blends of these oils contain no more than 25 per cent of heavy catalytic cycle gas oil, slurry oil, or clarified oil. This limit was based on the disclosure of biological experiments with blends of a catalytically cracked oil containing 40 per cent of 700° F.-plus fractions and a biologically inactive oil base stock. When it was subsequently recognized that the active components occurred predominantly in the 700-plus fractions, it was recommended that blends of these oils containing more than 10 per cent of catalytically cracked oil boiling above 700° F. be considered carcinogenic. It was recommended that such blends should not be sold to the ultimate consumer, who notoriously disregards recommended precautionary measures, and should only be sold to concerns intending to process the material further. A set of precautionary measures is given to each customer who purchases blends which are considered potentially carcinogenic, and every assistance is offered to these customers in helping them establish precautionary measures within their own plants.

20. *Make reports.* — It is requested that complete reports covering the industrial hygiene engineering and clinical phases of the program be submitted to the medical department every six months.

The Broader Aspects of Industrial Hygiene Peculiar to the Petroleum Worker

The above project illustrates the meaning of industrial hygiene with respect to the team approach to a specific health hazard, peculiar to a particular type of worker. Industrial hygiene has its broader aspects, as revealed in a study, captained by the author, of 66 areas in 14 refineries, three producing camps, and seven marketing divisions. It was evident from this study that the industrial hygienist has responsibilities embracing the entire work environment of a great many different types of employees potentially faced with a wide variety of health hazards.

These 606 areas represent those most suspected of being associated with a health hazard. In other words, each of these refineries, producing camps, and marketing divisions was first inspected by a trained industrial hygienist, and the environmental areas most likely to have health hazards were selected. In 181 of these areas (column V), on detailed survey, *no* health hazards were found to be associated with them. In 81 areas (column IV), simple mechanical remedies were found sufficient to eliminate the hazard. An example of this type would be increasing the capacity of a fan so as

TABLE I*

	TOTAL	I Definitely Hazardous or Unsatisfactory	II Potentially Hazardous Needing Periodic Review	III Suspect Needs Medical Review	IV Suspect Needs Mechanical Remedies	V Satisfactory
14 Refineries	488	62	154	94	61	117
3 Producing Locations	60	9	10	6	3	30
7 Marketing Locations	58	2	7	0	17	34
	606	73	171	100	81	181

*Surveys conducted by: Medical Research Division, Standard Oil Development Company

Table I — Results of Survey

to make already available ventilation completely effective, or designing a cover for an open vessel. In 100 of the areas (column III), the toxicological data or acceptable exposure levels were inadequately established to permit full evaluation, so that it was recommended that the men be brought under periodic medical supervision to develop this information. An example of this would be the cancer-control program, previously referred to, where certain oils have been found to be carcinogenic to mice, but it is not known whether or not they truly represent a cancer hazard to workers. The men were nonetheless provided with protective equipment and brought under periodic medical surveys.

In 171 of these areas (column II), it is known that the work is potentially hazardous, even though the controls at the time of the survey were perfectly adequate. In these circumstances, periodic study of the environment by an industrial hygienist and/or of the men by the medical department seemed warranted. An example of this type of situation is lead burning or sandblasting operations. In the first situation, the air that the workers breathe and the urine of the men engaged in this work should be analyzed for their lead content periodically. If either of these is found abnormal, the industrial hygienist can quickly devise the necessary controls. In the case of sandblasting, the air should be analyzed for its free (crystalline) silica content, and chest x-rays should be taken periodically of the men. Again, the purpose of this work is to insure correction of any unsatisfactory conditions early and before any irreparable damage has been done. This is prevention of disease rather than cure.

Finally, in 73 areas, conditions were found unsatisfactory. In these situations, it might be necessary to design adequate exhaust systems to rectify the situation, substitute less toxic materials for those being used, or modify the working habits of the men in order to eliminate the hazard. In all 73 such areas, the necessary corrective measures can be installed or instituted. Each area must be resurveyed in the future to insure that the controls recommended are adequate. Therefore, they are recorded under column II of the table. In no case, however, was it necessary to correct the situation by reducing the hours of work.

It was observed that the majority of the potentially hazardous occupations were not among process men but rather among the maintenance crews, such as potential lead exposures among painters, lead burners and bonders, welding-fume exposures among welders, etc. Such hazards or potentially hazardous occupations are common to many industries as well as the petroleum industry. Their control in all industries makes use of the same basic principles of sound industrial hygiene engineering.

With proper engineering, safety, and medical controls, no job in industry is too hazardous to undertake for health reasons.

Summary of Study

In summary, this study established the following facts:
1. In the four major subdivisions of the petroleum industry, namely, producing, refining, transportation, and marketing, some 500 basic occupations are represented. These extend all the way from laborers to pipefitters, boilermakers, process men, research chemists, etc., to office workers, including officers.
2. Of these 500 basic occupations, about one-fifth, or 100, entail exposures which are potentially hazardous if uncontrolled. Of these 100, not more than 50 are confined to the petroleum industry alone.
3. The remaining 50, including welders, machinists, painters, etc., are found in other industries too, such as the steel, rubber, and chemical industries.
4. It is recognized by all authorities in the field of industrial health that where specific occupational expo-

sures occur, be they in the petroleum or any other industry, the accepted method of prevention is to control the exposure at its source by standard industrial hygiene engineering techniques.
5. Statistical studies of actual findings made upon complete industrial hygiene surveys have clearly established the effectiveness of this approach wherever undertaken.
6. This positive approach to the health needs of the petroleum worker implies that there is mutual understanding and appreciation of the problems involved by all concerned: management, the humanology team, and the worker himself.

Hygiene Problems of Other Industries

Considerable space has been devoted to industrial hygiene problems of the petroleum industry. The author's experiences with this industry has made this a logical approach. The reader who is interested may find a number of excellent works covering industrial hygiene problems in other industries. The *Diseases of Occupation* by Donald Hunter, M.D., is a first-rate reference work for specific chemical compositions of material used in various occupations. For an excellent study of industrial hygiene in the chemical industry, the reader is referred to pertinent sections in a work entitled *Modern Occupational Medicine,* compiled by A. J. Fleming, M.D.; C. A. D'Alonzo, M.D., and J. A. Zapp, Ph.D., all connected with the humanological team of E. I. duPont de Nemours and Company (Lea & Febiger, Philadelphia, 1954). Other excellent reference works covering industrial hygiene in other industrial areas include the following: (1) *Occupational Medicine and Industrial Hygiene* by R. T. Johnstone (C. V. Mosby Company, St. Louis, 1948), (2) *Industrial Hygiene* by A. J. Lanza and J. A. Goldberg, (Oxford University Press, New York, 1939), (3) *Anthro*

Silicosis Among Hard Coal Miners, (Public Health Service Bulletin 221, United States Government Printing Office, Washington, District of Columbia, 1936), (4) "The Hazards of Coal Mining," by J. M. Regan, in *The Practice of Industrial Medicine* by J. A. Lloyd Davies, (J & A Churchill Limited, London, 1957, Pages 239-253), (5) *Pneumoconiosis, Beryllium, Bauxite Fumes, Compensation* by A. J. Vorwald, (Paul B. Hoeber, New York, 1950).

Hygienic standards. — Since 1955, a committee of recognized experts in the field of industrial hygiene and toxicology prepared under the aegis of the American Industrial Hygiene Association, Hygienic Guides which may be obtained from the Executive Secretary, American Hygiene Association, 14125 Prevost, Detroit 27, Michigan for 25 cents per copy. The pertinent facts of interest to professional and nonprofessional industrial hygiene personnel having to do with the following chemical substances are available at this source:

Acetaldehyde
Acetic Acid
Acetone
Acrylonitrile
Amyl Acetate
Anhydrous Ammonia
Aniline
Arsine
Benzol
Beryllium
Butyl Alcohol
Cadmium
Carbon Disulfide
Carbon Monoxide
Carbon Tetrachloride
Chromic Acid
1,2-Dichloroethane
Ethyl Alcohol
Fluoride-Bearing Dusts
    and Fumes
Fluorine
Formaldehyde
Hydrogen Cyanide
Hydrogen Fluoride
Hydrogen Sulfide
Hydrozine
Mercury
Methyl Ethyl Ketone
Methylene Dichloride
Nickel Carbonyl
Nitrogen Dioxide
Ozone
Sulfur Dioxide
Trichloroethylene
1,1,1,-Trichloroethane
Zinc Oxide

REFERENCES

AMERICAN MEDICAL ASSOCIATION. "Principles of Medical Ethics." A Special Edition. *The Journal of the American Medical Association*, (June 5, 1958).

Behavioral Sciences at Harvard. A report of a Faculty Committee, Harvard University, June 1954.

BERRY, CLYDE M., PH.D. "Selling Industrial Hygiene to Management." The *Medical Bulletin*, The Standard Oil Company (New Jersey), (October, 1951).

BRADLEY, WILLIAM R. "Industrial Hygiene Considerations in Plant Location and Design." *Chemical and Engineering News*, XXIX (March 26, 1951), pp. 1198-1200.

BRIDGES, CLARK D. *Job Placement of the Physically Handicapped*. New York: McGraw-Hill Book Company, 1946.

BURWELL, C. SIDNEY. "Evolution of the Doctor." *The Physician and His Practice*. Boston: Edited by Joseph Garland, M.D. Little, Brown and Company, 1954, pp. 3-13.

CALDERONE, FRANK A., M.D., M.P.H. "The Economics of Preventive Medicine — A Synthesis of Opinions." *Industrial Medicine and Surgery*, XXV (March, 1956), pp. 101-108.

CHAPMAN, LORING F.; THETFORD, WILLIAM N.; BERLIN, LOUIS; GUTHRIE, THOMAS C.; WOLFF, HAROLD G., M.D. "Highest Integrative Functions in Man During Stress." *The Brain and Human Behavior*. Baltimore: Vol. XXXVI, The Williams and Wilkins Company, 1958.

"A Code for Professional Nurses." *The American Journal of Nursing*, L (July, 1950), p. 392.

COOK, WARREN A. "Engineering Phases of Plant Health Control." *Chemical and Engineering News*, XXIX (April 16, 1961), pp. 1517-1518.

COUNCIL ON INDUSTRIAL HEALTH, AMERICAN MEDICAL ASSOCIATION. *Standing Orders for Nurses in Industry*. Chicago: American Medical Association, 1943.

CRUICKSHANK, W. H., M.D. "Mental Hygiene in Industry." *Industrial Medicine and Surgery*, XXVI (October, 1957), pp. 477-481.

CUBER, JOHN F. *Sociology: A Synopsis of Principles*. New York: 3rd edition, Appleton-Century-Crofts Incorporated, 1955, Chapter 1.

D'ALONZO, C. A., M.D. AND FLEMING, ALLAN J., M.D. "Occupational Psychiatry Through the Medical Periscope." *Industrial Medicine and Surgery*, XXV (October, 1956), pp. 466-469.

DEICHMANN, WILLIAM B., PH.D. *Signs, Symptoms and Treatment of Certain Acute Intoxications.* Miami: University of Miami Printing Arts Department, 1955.

DESCARTES, RENE. *Meditations on the First Philosophy.* 1640.

DUBOS, RENE J. "Health and Disease." The *Journal of the American Medical Association*, CLXX (No. 5, October 1, 1960), p. 105.

ECKERT, ROBERT E., M.D., PH.D. *Industrial Carcinogens.* Grune and Stratten, 1958.

ELLER, JOSEPH JORDAN, M.D. AND ELLER, WILLIAM DOUGLAS, M.D. *Tumors of the Skin, Benign and Malignant.* Philadelphia: Lea and Febiger, 1951.

FARNSWORTH, DANA L. *Mental Health in College and University.* Cambridge: Harvard University Press, 1957, Chapters 5 and 7.

GOODMAN, KENNETH E., and MOORE, WILLIAM L. *Economics in Everyday Life*, New York: Gin & Co., 1952, pp. 3-56.

HARDY, HARRIETT L., M.D. "Common Industrial Exposure and Diseases and Medical Programs for Their Control." A Symposium on Industrial Medicine, Harvard School of Public Health, April 3-4, 1953.

HENDERSON, R. M., and BACON, S. D. "Problem Drinking." *Yale Plan for Business and Industry Quarterly Journal Study of Alcohol*, XIV (June, 1953), pp. 247-262.

HINKLE, LAWRENCE E., JR., M.D., and WOLFF, HAROLD G., M.D. "The Nature of Man's Adaptation to his Total Environment and the Relation of this to Illness." The *American Medical Association's Archives of Internal Medicine*, XCIX (No. 4, March, 1957).

HINKLE, LAWRENCE E., JR., M.D., et al. "Studies in Human Ecology." The *American Journal of Psychiatry*, CXIV (No. 3, September, 1957).

JORDAN, EDWIN P. "Group Practice." *The Physician and His Practice.* Boston: Edited by Joseph Garland, M.D. Little, Brown and Company, 1954, pp. 78-88.

JOURNAL OF INDUSTRIAL MEDICINE AND SURGERY. Editorial "Another Kind of Corporate Medical Practice." *Industrial Medicine and Surgery*, XXIV (May, 1955), p. 237.

KEHOE, ROBERT A., M.D.; FOULGER, JOHN H., M.D.; ALBERT, ROY E., M.D.; ZAPP, JOHN A., JR., PH.D. "Criteria for the Diagnosis of Occupational Illness." *Industrial Medicine and Surgery*, XXIV (October, 1955), pp. 427-442.

KEMP, HARDY A., R.N. "The Role of Nurses in Industry." *Industrial Medicine and Surgery*, XXI (February, 1952), p. 62.

KILLIAN, JAMES R., JR. *Education for the Age of Science,* Department of Health, Education & Welfare (Washington, D.C.), (May, 1959).

LEVEN, AARON SAMUEL, M.D., D.P.H. "The Art of Human Relations in Industrial Medicine." *Industrial Medicine and Surgery,* XXV (November, 1956), pp. 519-522.

MCCORMICK, JOHN F., S.J. *Scholastic Metaphysics.* Chicago: The Loyola University Press, 1928, Part I, Chapter I.

MCGEE, LEMUEL C., M.D., PH.D. *Manual of Industrial Medicine.* Philadelphia: University of Pennsylvania Press, 1956.

MENNINGER, WILLIAM C., M.D., AND LEVINSON, HARRY, PH.D. "Industrial Mental Health." *Menninger Quarterly,* Topeka, (Fall, 1954).

OESTERLE, JOHN A., PH.D. *Ethics: The Introduction to Moral Science.* New York: Prentice-Hall, 1957, Chapter 1.

PAGE, ROBERT COLLIER, M.D. "Health of the Petroleum Worker." *American Medical Association's Archives of Industrial Health,* XI (February, 1955), pp. 126-131.

──────────. "The Doctor For Tomorrow's Needs." The *Journal of Medical Education,* XXVII (March, 1952), pp. 91-98.

──────────. "A Medical Control Plan for the Prevention of Occupational Disease." *Industrial Medicine and Surgery,* VI (September, 1937), pp. 481-496.

──────────. "The Doctor of Tomorrow and Federal Medical Services." The *New England Journal of Medicine,* 246, (May 15, 1952), pp. 765-771.

──────────. "The General Practitioner in an Industrial Era." *Ontario Medical Review,* XXIV (No. 6, June, 1957), pp. 415-420.

PAGE, ROBERT COLLIER, M.D., and HOLT, J. P., M.D., *et al.* "Symposium on a Cancer Control Program for High-Boiling Catalytically Cracked Oils." *American Medical Association's Archives of Industrial Hygiene and Occupational Medicine,* IV (October, 1951), pp. 287-297.

PAYNE WHITNEY PSYCHIATRIC CLINIC. "The Medical Value of Psychological Testing." A 1957 Report to the Board of Governors of New York Hospital.

RAPPLEYE, W. C., M.D. "Education Component of Medical Licensure." The *Journal of the American Medical Association,* CLIV (April 3, 1954), p. 1212.

ROGAN, J. M., M.D., F.R.C.P. "The Hazards of Coal Mining." The *Practice of Industrial Medicine* written by J. A. Lloyd Davis. London: J. and A. Churchill Limited, pp. 239-253.

RUBINSTEIN, HYMAN S., M.D. "Physician Know Thyself." *Current Medical Digest*, XXV (May, 1958), pp. 53-58.

RUTSTEIN, DAVID D., M.D. "The Influenza Epidemic." *Harpers' Magazine*, (August, 1957).

SAMUELSON, PAUL A. *Economics: An Introductory Analysis.* New York: McGraw-Hill Book Company, 1958, Introduction.

SCHOTTSTAEDT, WILLIAM W., M.D.; GRACE, WILLIAM J., M.D.; WOLFF, HAROLD G., M.D. "Life Situation, Behavior Patterns and Renal Excretion of Fluids and Electrolytes." The *Journal of the American Medical Association*, CLVII (April 23, 1955), pp. 1485-1488.

SMYTH, HENRY F., JR. "The Field of Chemical Hygiene." *Chemical and Engineering News*, XXIX (April 9, 1951), pp. 1196-1197.

SNYDER, JOHN C., M.D. "Ethics & Public Health." *Harvard Alumni Bulletin*, April 4, 1959.

STERNER, JAMES H., M.D. "The Medical Aspects of the Industrial Hygiene Program." *Chemical and Engineering News*, XXIX (April 9, 1951), pp. 1399-1401.

STIEGLITZ, EDWARD JULIUS, M.D. *The Future for Preventive Medicine.* New York: The Commonwealth Fund, 1945.

TERHUNE, WILLIAM B., M.D. "Psychiatry as a Career." The *Connecticut State Medical Journal*, XXI (February, 1957), pp. 93-98.

WADE, LEO J., M.D. "What Is Industrial Medicine?" *Industrial Medicine and Surgery*, XXIX (No. 2, February, 1960), p. 54.

WAGNER, SARA P., R.N. "Nursing Trends." The *Medical Bulletin*, the Standard Oil Company (New Jersey), XII (February, 1952), pp. 184-186.

WATTS, ISAAC, D.D. *Logic, or the Right Use of Reason in the Inquiry after Truth.* London: 1724.

WEIDER, ARTHUR, PH.D. "Mental Hygiene in Industry — A Clinical Psychologist's Contribution." The *Journal of Clinical Psychology*, XLV (October, 1947).

——————. "Some Aspects of an Industrial Mental Hygiene Program." The *Journal of Applied Psychology*, XXXV (December, 1951), pp. 383-385.

WHITTAKER, PAUL J., M.D. "Graduate Training in Occupational Medicine." "The Inplant Component." *Journal of Occupational Medicine*, II (No. 7, July, 1960), p. 312.

WOLFF, HAROLD G., M.D. "Stress, Emotions and Bodily Disease." The New York Academy of Medicine Lectures to the Laity. Published by the International Universities Press, New York, 1954.

WORLD HEALTH ORGANIZATIONS, *Technical Report Number 135*. Washington, D.C.: Pan American Sanitary Bureau, 1957.

ZIMMERMAN, CHARLES J. "Management's Role In Mental Health." *Advanced Management*, (September, 1960), p. 5.

8

A Positive Program of Occupational Health

A POSITIVE PROGRAM of occupational health which lies within the province of the medical and paramedical members of the humanology team begins where the strictly engineering aspect of occupational health leaves off. Otherwise, the potential assets of the medical department are likely to be frittered away in purely negative pursuits.

A Company Safety Program

An example of the latter alternative occurs when management confuses the functions of the medical department with the company safety program. Medical service in industry, according to present-day concepts, has nothing whatever to do with safety. Worker safety is an engineering problem and a management obligation; all the medical profession can do is help rectify unfortunate accidents after they have happened.

In this connection, the concept of safety may be expanded to include not only the engineering of measures to prevent so-called injuries but the entire objective of industrial hygiene. This takes in all organic illnesses resulting directly from *physiological* properties of the working environment. Properties such as fatigue or tension are largely psychological and do not, therefore, fall properly within the scope of industrial hygiene engineering; one man's tension is another man's stimulating challenge.

Physiological Properties of the Work Environment

Accurate criteria are needed to enable the medical consultant to determine which health problems are wholly attributable to such physiological properties of the work environment as are susceptible to industrial hygiene engineering and which are attributable at least in part to such properties of the nonoccupational environment or individual make-up as to require medical or paramedical attention. There are occasionally borderline cases which make a clear-cut determination difficult. The patient's unsupported statement that his illness is the result of working with some injurious material is not always reliable, however honest it may be from the patient's point of view. The prerequisite information for a determination of the extent of the occupational nature of an illness must be sought in an appropriate investigation of the environment, preferably at firsthand, and, when feasible, in clinical and laboratory investigation of the patient himself. The latter involves the exercise of meticulous care in the elicitation of a complete medical history, supplemented by a thorough medical examination, calling in specialists as required. For this type of study, a comprehensive background of physiological, toxicological, and clinical knowledge is essential. The medical consultant on the humanological team may be expected to have such a background. He has a right to expect that management will provide him with facilities for the fundamental physiological, toxicological, pathological, clinical, and epidemiological investigations required to make a determination of occupational illness and to map out a course of preventive action that may be put into effect by the industrial hygiene engineer. From this point on, the doctor and his medical and paramedical colleagues step out of the picture of occupational disease, except insofar as engineering or other human errors occur and after-the-fact therapy becomes necessary.

The Three Levels of an Occupational Health Program

An occupational health program,[1] assuming that all of the engineering problems have been solved, is carried out by medical and paramedical personnel at three principal levels: (1) the dispensary level; (2) the diagnostic level; and (3) the community-relations level.

The Dispensary Level[1]

It is at the first, the dispensary level, that the majority of employed persons may be seen by some member of the professional arm of the humanological team in most cases while they are still vertical.

The Registered Nurse Is the Key Person

At this level, the nurse is the key person. Every day she is faced with individuals who have seemingly minor complaints which have not previously been brought to the attention of any other professional person. Some of these complaints are truly insignificant. A word of common-sense reassurance may be proper medicine. In other cases, treatment of the problem may lie well within the professional limits of the graduate nurse. Occasionally, symptoms brought to the dispensary may be quite obviously indicative of a clear-cut disease entity or injury, and the nurse has no difficulty in making an immediate referral to the appropriate medical specialist or other type of practitioner. If the above-mentioned types of cases were all that were dealt with at the dispensary level, the nurse would need only to be ordinarily intelligent and awake on the job with no other qualifications. As it happens, however, an imposing number of

---

[1] ROBERT COLLIER PAGE, M.D. "Constructive Medicine." (The Answer to Industry's Medical Problem) An address before the International Medical Conference of the Standard Oil Company (New Jersey) and affiliates, New York, March 15-22, 1946.

the *seemingly insignificant* problems that are brought to the dispensary are actually highly *significant*. A recurring pattern of mild respiratory problems in a given individual, may, for example, prove to be the warning signs of a serious emotional upset. The capable nurse will learn how to make "horseback appraisals" of apparently minor problems, relating them to the individual's total work environment, and see to it that the proper corrective action is inaugurated and the proper specialized help called in when necessary.

Standard Orders for Nurses in Industry

The average dispensary, according to the recommendations set forth by the Occupational Health Institute, has a nurse in charge. She may comprise the entire staff in a medium- to small-sized business. In a large company, she may head a staff composed of one or more paramedical assistants. Her functions will be clearly defined in a standard procedure manual, such as *Standing Orders for Nurses in Industry,* prepared by the Council on Industrial Health of the American Medical Association (Chicago, 1943). A minimum of equipment will be available, and the accent will be on first aid. (See discussion of "Types and Scopes of Present-Day Service Units" in Chapter 2.)

The nurse and her associates at dispensary level should have an appreciation of all aspects of preventive medicine, even those aspects which are the responsibility of government and call for no active participation on the part of the dispensary staff. Industry has a stake in public-health measures, and is often in the position of helping to write out the prescription for such measures, even though the actual carrying out of the measures is wholly a governmental problem. A governmental problem which must be solved at federal, state or local level.

Airs, Waters and Places

Away back in the fifth century B.C., a Hippocratic work entitled *Airs, Waters and Places* was published, largely for the benefit of mercantile colonists who set up trading posts in remote places. This work was one of the first to deal with public health to an exhaustive degree, and it is interesting to note that at this early date, public health, occupational health, and total environment were all considered together.

The writer of this important work recognized that certain diseases were always present in a given population. He called these *endemic*. Other diseases, he noted, became excessively frequent only at certain times.

Although *Airs, Waters and Places* was the basic *epidemiological* text until the nineteenth century, when the nature of epidemics began to be understood in terms of bacteriology, the work was less concerned from a practical point of view with epidemics than with endemic disease caused by ever-present local factors. The ancient author summarized these essential factors of endemicity as climate, soil, water, mode of life, and nutrition. Practical advice was offered regarding the appraisal of these factors in selecting a site for a new community.

Control of these factors, by means of swamp drainage, adequate sewage disposal, etc., was not recognized at this stage. Later, in Roman times, the Latin flair for organization and large-scale engineering projects brought about a certain degree of control; aqueducts made pure water available in out-of-the-way places, swamps were drained, public baths were erected and carefully supervised for cleanliness, and the better sections of most up-to-date cities in the Roman sphere of influence had elaborate sewage disposal systems.

Medical knowledge, however, was primitive. The lowly bacteria were unthought of. Hence, public-health measures, however elaborately conceived, were still groping in the dark.

Public Health

In truth, even as late as 1900, despite tremendous nineteenth-century advances in sanitary science, public health had by no means come into its own. One reason for this seems to have been that no basis for teamwork between physician (and/or nurse) and the government engineer had been established. The field of public health seemed to be unrelated to the practice of medicine, and the practicing physician had little interest in public health. His only points of contact had to do with compulsory reporting of vital statistics and contagious diseases that happened to fall within his ken. Community activities going beyond sanitation and elementary epidemiology, such as hygiene education in the schools, infant-health protection, maternal hygiene, and occupational-health measures, seemed to encroach upon the practice of medicine. Considerable dissension was stirred up between practicing physicians and health authorities.

Some of the blame for this state of affairs may fairly be laid on the philosophy of medical education of half a century ago, which still persists in ever-diminishing areas. This philosophy was based on the concept that the sum total of medical service was diagnosis and therapy; that prevention was of no interest to the real physician. Since prevention *was* of considerable interest to the layman, preventive medicine was bound to make many of its most significant early advances *without* the blessing of the medical profession at large. The result: jealousy, misunderstanding, and lack of teamwork.

Within the last 50 years, this philosophy has been undergoing drastic revision, spurred considerably by the mushroom growth of occupational medicine. In addition to diagnosis and therapy, the practicing physician has assumed responsibility for prevention and rehabilitation. It has also come to be realized that medical care is part of a team effort, involving not only physicians, but also nurses, dentists, hospitals, convalescent homes, official health services, nursing homes, voluntary health agencies, etc., all of which need to be coordinated to provide a unified comprehensive medical care, available to all.

Public Health and Preventive Medicine

Dr. Wilson G. Smillie,[2] former Professor of Preventive Medicine at Cornell University Medical College, points out that some difficulties still exist regarding proper delineation of responsibilities in this coordinated plan. The fault lies, Dr. Smillie has said, with the fact that some confusion persists with regard to the distinction between the terms "public health" and "preventive medicine," which are closely interrelated but nonetheless distinct. Dr. Smillie has drawn the distinction as follows:

Public health is essentially the responsibility of the community for the protection and promotion of the health of the people.... *Preventive medicine*... is a term that is often used interchangeably with public health, but is quite a different thing... (It) is the responsibility of the individual, or head of a family, for the promotion of health of himself and his family. This can be done effectively only if the family follows the advice of an expert counsellor in these matters, namely, the family physician. Thus the ultimate responsibility for preventive medicine rests with the practicing physician.

One reason for confusion of public health with preventive medicine is that each of them may use the same tools to accomplish their purpose. A good example is diphtheria immunization. The primary purpose of public health in diphtheria control is to protect the community against invasion of the disease. To do this, it is not necessary that all susceptibles shall be immunized.... If only half the pre-school children are protected against diphtheria, then the disease may enter a local area, but will not spread.

On the other hand, a physician in practicing preventive medicine, must be sure that every susceptible child under his care is well immunized against diphtheria. He may even utilize the Schick test

---

[2] WILSON G. SMILLIE, M.D. "Incorporation of Preventive Medicine in Clinical Practice." An address before Health League of Canada, Toronto, February 14, 1958.

to be sure that the child is immunized, whereas the health department will seldom use a Schick test, since a *high proportion* of children are known to respond to toxoid. In this example, the same techniques are employed, but one is a public health procedure — the other is preventive medicine.

However, as Dr. Smillie points out, there has been considerable progress in the right direction in recent years, which will be accelerated by making the study of formal courses in public health and preventive medicine prerequisite for graduation from medical school.

Epidemiology

From the point of view of occupational medicine at the dispensary level, the topic of *epidemiology* is a combination of public health and preventive medicine. The relative responsibility of governmental agencies and occupational-medical consultant (aided by nurses) in the control and/or prevention of epidemics will vary from region to region. There are two principal phases of epidemiology: (1) environmental control (swamp drainage, DDT, etc.) and (2) immunization of susceptible individuals.

Environmental control in most developed areas is taken for granted as strictly a public-health matter. The humanological professional in a given industry may complain long and loudly if local garbage disposal facilities are so inadequate as to create employee illness, but his complaints will be addressed to some echelon of government, not of business, and it will be assumed that if any corrective action is taken, it will be taken by government.

There are still occasions, when industry moves into a remote, economically underdeveloped area, where it is necessary to treat control of environment as a problem of prevention for the medical department. It may be found, for example, that the health of the working population is adversely affected by pollution of the water supply. There are no government facilities. The medical department is asked to make a determination as to the exact relationship of the pollution to the extent of illness, the major causes of pollution, and the corrective steps to be taken. It may be necessary for such steps to be taken under the supervision of a medically trained person who can determine whether a satisfactory correction has been made. This is preventive medicine.

Such a situation is exceptional, and is becoming more so. Even where local governments are unable to inaugurate public-health measures, such organizations as the World Health Organization and others are making trained personnel available who know how to organize environmental-control engineering measures on a public-health basis, leaving medical and paramedical personnel free to concentrate on the properly medical aspects of prevention.

Immunization

Immunization, on the other hand, is usually under the heading of preventive medicine. Occasionally it is necessary, in a time of emergency, to carry out immunization on a large scale, as a public-health measure. Certain types of inoculation are standard and are prescribed by law in some countries. Certain other types, i.e., gamma-globulin to reduce susceptibility to measles, etc., or influenza vaccine, to be discussed in detail later, are not standard. Their use or nonuse depends wholly upon the judgment of the physician in the individual case. In the United States, as in other developed nations of the western world, most immunizations are carried out as matters concerning the doctor and the individual patient. They are voluntary on the part of the patient, who is simply persuaded as a sensible individual to take steps to protect himself against serious disease. Governmental or private welfare agencies or industries may make it possible for him to receive certain immunizations at little or no cost, but the voluntary aspect remains. It is a matter between the individual and his doctor. He may be fined or locked up if he pollutes the water supply, but not if he chooses to run the risk of contracting smallpox in his own country. This is an important illustration of the distinction between public health and preventive measures.

Despite the voluntary aspect and the absence of legal "teeth," preventive medicine in this country has chalked up some notable accomplishments. In the last half century, the death rate in the United States has dropped from 17 to considerably under 10 per 1,000. Life expectancy has been increased about 20 years. The combined death rate for measles, scarlet fever, whooping cough, diphtheria, and meningitis has been reduced by more than 97 per cent since 1900. BCG vaccinations for tuberculosis and modern antibiotic therapy have helped bring that once formidable killer nearer every year to its ultimate and inevitable extinction.

Requirements for Overseas

The voluntary approach is not always workable, however, in the case of individuals going to certain overseas areas where specific diseases are endemic and/or potentially epidemic. In such cases, where there are not specific laws regarding immunizations for stated diseases, the medical department may have to exercise judgment based on the seriousness of the disease involved, its contagiousness, its current incidence, the susceptibility of the individual worker, and the availability and effectiveness of the vaccine or vaccines. The principal of the greatest good for the greatest number will be brought into play here. If the disease is not extremely serious or of long duration, and if supplies of vaccine are short, it may be found necessary to restrict vaccination to those in key posts.

The following facts of inoculation requirements not only record many procedures that are standard in the United

States, but are also useful in the case of companies who send an appreciable number of individuals to certain indicated "potential disease" areas. Appreciable inconveniences for travelers can be avoided if medical departments take care to learn *routes* being traversed, not just destinations. Airlines maintain current data on such matters. The epidemiological reports of the World Health Organization and the United States Public Health Service are useful references, as are several armed-service publications which are readily obtainable.

The International Certificates of Vaccination of the World Health Organization are recommended for the use of employed individuals whose travel routes and/or destinations are outside of the United States or Canada. In many countries, especially in the Eastern Hemisphere, these forms are required legally, and in all countries they are acceptable.

Certain South American countries require individuals entering as residents to possess a certificate of good health, or, at least, of freedom from communicable disease, some based upon examination with a stated number of days or weeks. No one form for such a certification has become widely used. It is contemplated that an International Certificate of Health will be designed and used in the future.

Immunization Against Various Specific Diseases

Smallpox. — Vaccination against smallpox has, since its discovery in the eighteenth century, made this disease so rare in some parts of the world that doctors in the United States and Canada have been known to experience acute difficulties in diagnosing actual cases of the disease brought in from areas where it is still no rarity.

Vaccination against smallpox is not complete or acceptable until a definite reaction has been obtained, observed, and recorded together with the dates of observation. "No reaction" means no vaccination, and the procedure must be repeated

INTERNATIONAL CERTIFICATES OF VACCINATION

AS APPROVED BY
THE WORLD HEALTH ORGANIZATION

CERTIFICATS INTERNATIONAUX DE VACCINATION
APPROUVÉS PAR
L'ORGANISATION MONDIALE DE LA SANTE

TRAVELER'S NAME—Nom du voyageur

ADDRESS (Number—Numéro) (Street—Rue)
ADRESSE

(City—Ville)

(County—Département) (State—État)

U. S. DEPARTMENT OF HEALTH, EDUCATION, AND WELFARE
PUBLIC HEALTH SERVICE

FOR SALE BY SUPERINTENDENT OF DOCUMENTS
GOVERNMENT PRINTING OFFICE—WASHINGTON 25, D. C.
$2.50 PER 100

PHS 731
Rev. 1-57

Fig. 10 — INTERNATIONAL CERTIFICATES OF VACCINATION

until a reaction is obtained. Repeated failure to obtain any reaction may result from the use of vaccine which has become impotent because of age, improper shipment or storage, or from improper technique.

In most areas, successful vaccination against smallpox is considered to result in comparatively satisfactory protection for three years. Accordingly it should be repeated every three years except where medical indications or legal requirements necessitate its repetition at more frequent intervals.

Newborn children may be vaccinated immediately if necessary. The usual procedure in the United States is to vaccinate children within three months after date of birth.

Typhoid fever. — It is assumed that immunization against this disease is to go hand in hand with public-health measures against the water pollution at its root.

Inoculation[3] of adults against typhoid fever should consist of a series of three injections, subcutaneously at 7 to 28 day intervals, of 0.5cc of typhoid-para-typhoid vaccine. Protection can be maintained by yearly single injections of 0.5cc of the same type of vaccine at any time thereafter regardless of length of time elapsed.

Once a complete series of typhoid inoculations has been received, a booster inoculation is considered sufficient for one, possibly two years' protection, dependent upon the length of time since the original series. An individual who has *had* typhoid fever may be assumed to have acquired a lifetime immunity.

Children over 1 and under 2 should receive three injections of 0.12cc of typhoid vaccine; those 2 to 6 years of age, three injections of 0.25cc vaccine; those over 6 should receive an adult dosage.

---

[3] Dosages recorded are those recommended by one pharmaceutical house and are included here to illustrate accepted procedures either presently or formerly not necessarily the specific dose to be given in a given instance. The practice and method of immunization is constantly subject to changes i.e., (triple vaccine to quadruple vaccine — oral route instead of intramuscular route, etc.).

Typhus fever. — Adequate protection for adults against epidemic typhus may be obtained by two subcutaneous injections of epidemic-typhus vaccine, separated by a 7-day interval. Some cross-immunity to endemic typhus will be afforded by this procedure. A single booster injection of 1.0cc of epidemic-typhus vaccine subcutaneously within 6 months prior to departure will suffice for persons proceeding to typhus-potential areas, provided a series of inoculations has been completed within the preceding 3 years. Residents in such areas should receive booster inoculations at 6-month intervals.

Children 6 months to 2 years of age should receive three injections at weekly intervals of 0.12cc of typhus vaccine. Those 2 to 6 years of age should receive three doses of 0.25cc; those 6 to 10 years of age three doses of 0.5cc. Children over 10 years of age should receive adult dosages.

Cholera. — Adequate protection against cholera can be obtained in adults by two subcutaneous injections of vaccine with a 7-day interval between. The first dose should be 0.5cc; the second 1.0cc. A single booster injection of 1.0cc of vaccine subcutaneously within the preceding 6 months will suffice for persons visiting cholera-danger areas, provided a series of inoculations has been completed within the preceding 3 years. Those residing in such areas should receive booster inoculations at 6-month intervals. Children 6 months to 2 years of age should receive initial doses of 0.06cc, and second and third doses of 0.12cc. Those 2 to 6 years of age should receive a 0.12cc initial dose with second and third doses of 0.25cc. Children 6 to 10 years of age should receive 0.25cc initially; and 0.5cc for second and third doses. Those over 10 may receive adult doses.

Yellow fever. — The two largest foci of yellow fever are in South America and in Africa. In South America, Colombia, Venezuela, the Guianas, Western Brazil, Bolivia, and the parts of Peru and Ecuador east of the Andes are considered within the Yellow Fever Belt. Inland areas of Panama are also included. In Africa, roughly that area between 15 degrees latitude north of the equator and 10 degrees latitude south of the equator is considered within the belt. Eritrea and Northern Rhodesia, outside those latitudes, are also included within the zone.

For adults, a single injection of 0.5cc of yellow fever vaccine at least 15 days prior to arrival in an endemic area will provide immunity for 6 years. Booster injections of 0.5cc of vaccine should probably be given every 5 years. Children aged 3 months to 1 year should receive half the adult dose; all others should receive the full adult dose. Since young children do not immunize as well as adults, they should be reimmunized on return trips to endemic zones.

Many countries in which yellow fever is no real medical problem legally require that all arrivals who originated in or traversed yellow fever zone be inoculated.

Plague. — Inoculation is not recommended except when an epidemic is present. Immunization is not required by any country for entrance, but where an epidemic is present, inoculation should be received at the time of arrival in the area. Plague has been present in the interior of Bolivia, Burma, Ecuador, India, Indonesia, and Peru. Two subcutaneous inoculations of plague vaccine at a one-week interval, the first 0.5cc and the second 1.0cc, will suffice for relative protection if given within 6-months of proceeding to the plague area. Residents in such areas should receive a 1.0cc booster inoculation every 4 to 6 months. Children aged 6 months to 2 years should receive a series of three injections spaced one week apart; the first 0.06cc, the second 0.12cc, and the third 0.12cc. Children 2 to 6 should receive three injections similarly spaced; the first 0.12cc, the others, 0.25cc. Children 2 to 10 also should receive three injections, a week apart; the first 0.25cc, the others, 0.50cc. Children over 10 may receive adult dosages.

Poliomyelitis. — Inoculation against poliomyelitis is now recommended for all persons. Each of the inoculations consists of 1.0cc. The second inoculation may be given 2 to 6 weeks following the first, and the final inoculation may be given from 7 to 12 months after the second. This method is now being superseded by the oral route.

Tetanus. — Medical departments may advise inoculations against tetanus whenever they feel that such are indicated. Employees whose work might expose them appreciably to tetanus should be offered protection against this disease. Two intramuscular injections of alum-precipitated tetanus toxoid in 0.5cc quantities at intervals of at least 4 to 6 weeks suffice for an initial series. One stimulating dose of 0.5cc intramuscularly 1 year after the original series results in immunity for approximately 5 years. Booster inoculations should be given thereafter at 4-year intervals.

The majority of children in the United States receive their immunization against tetanus with the combined triple vaccine (pertussis, diphtheria, and tetanus) which is given them after the first 3 months of life. It is routine to repeat these vaccinations every 3 years until the age of 12, after which a stimulating dose of 0.5cc is used in case of infection due to exposure. Children who have been immunized against tetanus and have not received the triple vaccine should be immunized following the adult schedule of dosage.[4]

Lesser diseases. — 1. *The common cold:* As yet no vaccine has been produced which will provide sure or lasting immunity to the so-called common cold. As for environmental control measures, adequate ventilation engineering, and provision for normal thermal comfort may be of some use in that the former may control certain air-borne allergy factors which cause nasal or respiratory conditions that can be mistaken for common colds and that the latter may make a minor cold at least endurable, and perhaps barely noticeable, as it

---

[4] Quadruple vaccine (pertussis, diphtheria, tetanus, and poliomyelitis) is now recommended.

runs its course. Otherwise, prevention of colds by means of environmental control is like trying to keep dust off the windowsills of a home located near the railroad freight yards.

Vaccines have been promoted that are said to build up resistance to colds, or to eliminate the allergy factor. Many laymen believe that a shot of some kind of antibiotic, administered at the dawn of a cold, will somehow diminish its effects. This of course depends upon the unanswerable question: what would the effects have been? Certain infectious complications will react to antibiotics; if the incipient cold-sufferer was about to become involved with one of these complications, then his shot of antibiotic undoubtedly saved him, even if he did suffer miserably from the uncomplicated common cold. One will never know. The repeated use of antibiotics in the case of mere "gambles" such as this is frowned upon by most reputable physicians. They deem it wiser to wait and see if the complication develops and attack it then, rather than play with the risks of immoderate unnecessary use of antibiotics.

It is possible that a true immunizing vaccine against the common cold will be developed someday, and medical departments will be called upon to formulate plans as to what to do about it. Undoubtedly great pressure will be brought to bear by those who have a financial interest in the matter, and this will be augmented by members of management with individual opinions on the subject. It will be incumbent upon the professional humanologist to remain calm and exercise careful judgment, knowing that his decision may establish a precedent.

He will be well-advised to maintain the strict limitations that humanological tradition has established as to how far business may go in the practice of medicine. When the workers of a given industrial enterprise are threatened by an epidemic of a serious disease that may cause death or permanent disability, and when local private-medical facilities for immunization are inadequate, then the humanologist may legitimately take emergency steps. It is difficult to envision an epidemic of the common cold necessitating emergency measures. In any event, the humanologist will be called upon to assess each individual case on its own merits, and render his decision only after careful study — not only of the immediate situation but of the future implications of any action he may decide upon.

2. *Influenza:* The pandemic or worldwide epidemic of Asian Influenza in 1957-58 offered a number of excellent illustrations of the multitudinous factors and hairsplitting decisions that must be weighed by the humanologist in industry when a disease is dramatic and highly publicized.

Despite its speed of movement, the so-called Asian Influenza epidemic was expected and well understood before it began to cut a swath through the United States. The last pandemic with serious side-effects had occurred in 1918-19, and although there had been no way at that time of observing the virus causing the disease, the germs which came as *secondary* invaders were readily detectable by the microscope. These germs were, most frequently, the streptococcus which ordinarily causes scarlet fever; the staphylococcus, commonly causing boils and abscesses; the pneumococcus which usually causes lobar pneumonia; and a germ called the "influenza baccillus" which sometimes causes pneumonia. Discovery of these secondary invaders was only of limited practical value in 1918-19, but proved extremely useful in the 1957-58 pandemic, as there were a number of antibiotics effective against most of these germs. It was these secondary invaders that were responsible for most of the 300,000 deaths in the United States in 1918.

In the course of studies of lesser influenza epidemics in the 1920's and 1930's, three types of influenza viruses were successfully isolated. These were labeled A, B, and C. The A virus is the only one of these that is considered likely to produce severe pandemics, owing to its ability to change its form radically, i.e., to mutate. Ordinarily, an individual suffering from influenza will develop an immunity to the same type of virus, but this is not true if the virus has changed its form radically enough before returning in a second wave. It was the *second* wave of influenza in 1918 that was the most widespread and devastating.

In 1942, an effective "dead" vaccine was developed against a Puerto Rican type influenza A. However in 1947, this virus was replaced by a mutant labeled "A-prime" against which the vaccine was no longer effective. From that time on, the vaccine was kept up to date as new mutants developed. It was expected that the existing vaccine would probably be ineffective in the event of a new pandemic, but techniques had been developed to get vaccine production under way with considerable dispatch as soon as the new mutant of virus A made its appearance.

Thus, when the 1957 epidemic appeared in April, probably out of Communist China, prostrating one-tenth of the population of Hong Kong in less than one week and spreading rapidly into Singapore, Formosa, Indonesia, the Philippines, Australia, Japan, and India, the rest of the world was able to draw up its battle lines in advance.

In the United States, specimens of the 1957 virus were distributed speedily to commercial-vaccine manufacturers throughout the nation. The United States Public Health Service, through its Division of Foreign Quarantine, patrolled all ports of entry to the country, not so much in the expectation of keeping influenza (not a quarantinable disease) out of the country as to pinpoint the movement of the disease. The American Medical Association established a committee to study and report weekly to the profession at large (through the *Journal of the American Medical Association*) new developments in the "flu" situation.

The approaching epidemic was well publicized in the United States. In one leading magazine, an eminent professor of preventive medicine of a leading university discussed the forthcoming problem in terms comprehensible to the literate layman. The Surgeon General of the United States Public Health Service, was interviewed and his statements on the subject of Asian influenza was given a wide press distribution. What both of these authorities had to say that was of particular interest to the man in the street was that vaccine supplies would probably not be sufficient for total distribution and should be saved for "key" people in the community, such as doctors, nurses, policemen, vital transportation workers, and the like. Through these, and other sources, the general public was warned against the possible complications such as penumonia, that might be brought about by "secondary invaders." The man in the street was urged to seek medical attention at the first symptoms of illness; he was reassured that antibiotics were available which were effective against the major secondary complications. The commercial manufacturers of vaccine sent promotional and educational material out to practicing physicians throughout the country.

Judgment Requisite

This mass-alert was laudable in many respects, and undoubtedly saved some lives, but it posed some knotty problems for a number of company doctors and medical consultants who found themselves being pressed from certain quarters to set up programs to attack the disease at dispensary level. Specific vaccination programs, either for specially selected key employees or for all employees, were discussed in some instances. There were, however, a number of fine points to be considered. It was known, for example, that sufficient supplies for civilian use would not be available until after the epidemic had made its first strike in the United States and perhaps not until after the epidemic had already spent itself. Under such circumstances, it was apparent to any true humanologist that community needs would have to come ahead of occupational needs. Moreover, it was evident that even if spurred production made it possible to vaccinate 100 per cent of a company's employees with complete success, only part of the problem was attacked. The employee himself might escape the disease, but if his whole family became ill he would lose time from work anyway, since he would have to stay home to care for his family. A vaccination program, in order to be wholly effective then, would have to take in the entire community, and even if vaccine supplies proved adequate, which seemed unlikely, a community-wide program might well prove to be outside the scope of the industrial humanologist.

Another problem was that people allergic to eggs, feathers, or chickens were likely to be made more ill by the vaccine, which is egg-cultured, than by the influenza itself, which ruled out the possibility of 100 per cent protection in many areas, whether the vaccine was available in quantity or not. For these reasons, most companies felt that the wisest program was to disseminate lively, illustrated, educational material explaining the nature of the disease and urging employees to seek medical care upon its appearance. Vaccination was generally left to the combined discretion of the individual and his family physician. In many cases, however, where pressure was great, vaccination programs were inaugurated under company auspices, and batches of vaccine ordered. Inevitably many of these programs were delayed owing to late shipment of vaccine, which in some cases did not arrive until the peak of the pandemic was past.

A November 6, 1957, *Informational Bulletin* of the New York Hospital is illustrative of a typical constructive program of a major community facility faced with an epidemic. This bulletin, going out to all department heads and staff members, established clearcut procedures and policies as to how to deal with the epidemic. It recommended that all

patients for whom elective admission to the hospital was planned be vaccinated prior to admission, that patients with uncomplicated influenza not be admitted to the hospital except under unusual circumstances, that visitors be limited vigorously. It recommended further that hospital personnel be vaccinated according to the following priority system:
1. All medical and hospital personnel responsible for patient care
2. Essential nonmedical hospital and college personnel
3. Remaining hospital personnel

The priority system for patients was:
1. Patients with serious pulmonary or cardiac disease
2. Pregnant women
3. Others

The bulletin urged that physicians in the community make a major attempt to treat patients with influenza *in the home*, where there was considerably less risk of encountering secondary infections. Further, it was announced that if the epidemic reached serious proportions the hospital might have to eliminate all elective admissions during the emergency.

Owing to the fact that in most cases there was widespread cooperation between the medical profession, industry, and the community — with supplies of vaccine being allocated as they were needed and with no attempts on the part of any significant group to "hog the show," the epidemic ran its course without becoming a disaster. Similar epidemics may be expected in the future, and the wise professional member of the humanological team, from nurse to medical consultant, will play a cooperative role, exercising judgment based upon the merits of the individual situation, and envisioning business not as an entity in itself but as a part of the total community. Unless all implications are clearly understood and appreciated, negative instead of positive results are achieved.

Malaria

The story of malaria is one that every humanologist should learn by heart as an example of an all-out war on an international scale against a disease, in which communities, international agencies, businesses, and the medical profession all played important cooperative roles.

Malaria is now considered a major problem principally in tropical areas, but in 1855 it was endemic throughout the entire United States, with hyperendemic areas in the southeastern states. Its spread was associated with careless clearing of forests, and impounding of water for storage and other purposes. The United States Sanitary Commission of 1849 reported that the "most common disease of all was malaria."

By 1870, agricultural reclamation of swamps, marshes, and lowlands reduced the incidence of malaria—accidentally. In 1880, it was discovered by Alphonse Laveran that malaria is caused by protozoan organisms of the genus *Plasmodium*. There are many species of *plasmodia*, but only four of these produce naturally-occurring malaria in man, and only three are important: *P. falciparum*, *P. vivax*, and *P. malariae*. In 1897, Ronald Ross, a member of the British Medical Corps in India, demonstrated the parasites in the stomach wall of the *Culex* mosquitoes. In 1900, William G. Gorgas made the first large-scale demonstration of malaria control through mosquito abatement in Havana, Cuba. His main effort against the *Anopheles* mosquito (of which there are more than 70 species that carry the parasite from one human host to another) was directed to the drainage of swamps. Crude oil and kerosene were mixed to cover bodies of water which could not be drained. He also stocked bodies of water with minnows which feed on mosquito larvae. He also began the study of the different species of *Anopheles*, which require different environments and may be attacked in different manners.

The most striking example of the eradication of a species of mosquito from a limited area is the elimination of

Anopheles gambiae from South America. This mosquito was unwittingly introduced from Dakar, Africa, into Natal, Brazil, by a fast French destroyer in 1930. It found a suitable environment and soon caused a devastating malaria epidemic. *A. gambiae* proceeded to spread northward and in 1933 caused an epidemic with 100,000 cases and 20,000 deaths. A campaign was initiated by the Brazilian government in cooperation with the Rockefeller Foundation to eliminate this species completely from the Western Hemisphere. The attack consisted of the elimination of breeding places and the simultaneous use of larvicides and spray insecticides. Paris Green was the larvicide, used either as a dust or an oil solution. Pyrethrum powder was used against the adult mosquitoes. This insecticide, obtained from chrysanthemum flowers, acts on the nervous systems of adult mosquitoes. Experience since that campaign has shown that malaria control is more effective and more economical when directed against the adult insects by means of sprays rather than against the larvae.

The next important advance was the practice of *residual* spraying by means of sprays which maintain their lethal effect for a considerable period of time after application to a surface. D.D.T. has been especially important since its lethal effect lingers from three to six months after a single application. The mosquitoes which later come to rest on the surface are poisoned and die. If an adult mosquito rests but once on such a sprayed surface and takes up a lethal dose of D.D.T. at any time during the interval required for the development of sporozoites following an infecting blood meal, transmission of malaria by that mosquito will be prevented. If the D.D.T. is not quickly lethal it may still adversely affect the mosquito's ability to lay eggs and reduce its biting rate. The aim is not to destroy all *Anopheles* mosquitoes but only those which are infected with parasites obtained from biting man, and therefore are potential transmitters.

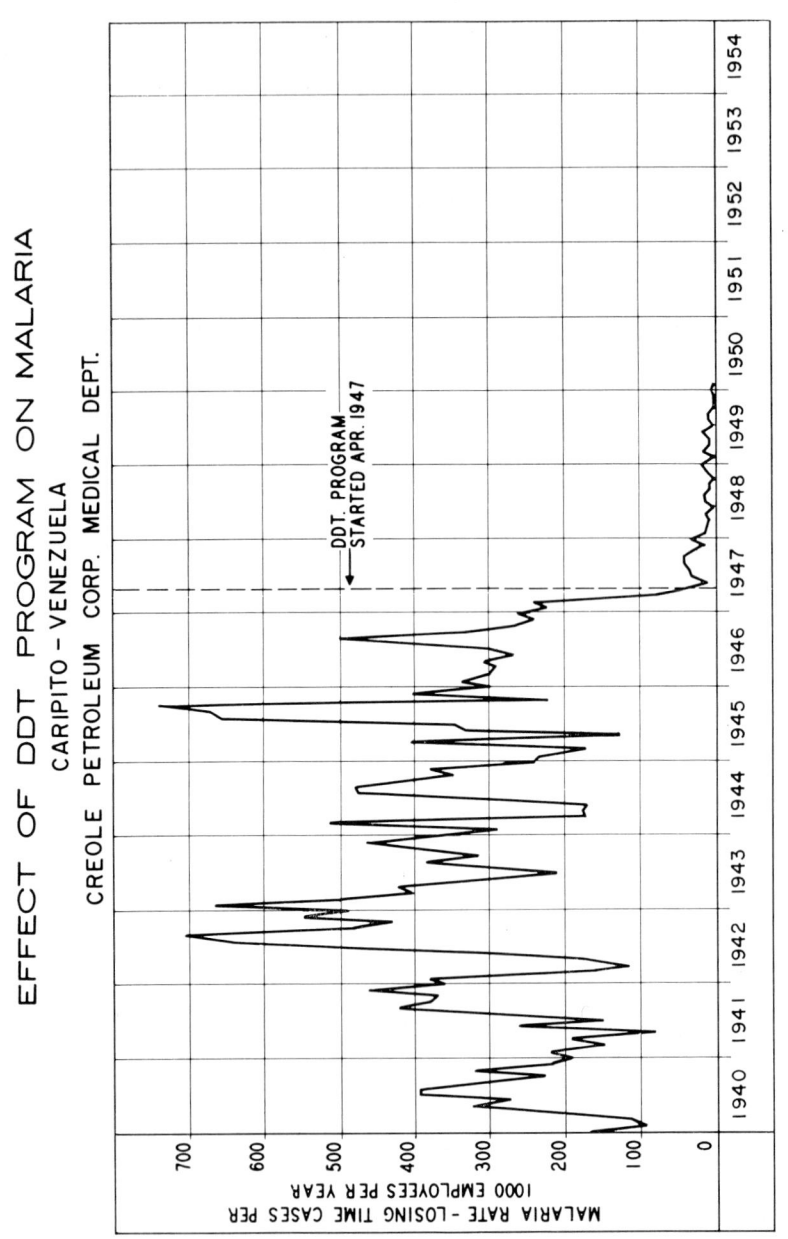

294   OCCUPATIONAL HEALTH AND MANTALENT DEVELOPMENT

Graph X — Effect of DDT Program on Malaria

Other insect-adulticides having residual action are benzine hexachlorine, chlordane, and dieldrin.

It has been noted that the present generations of house flies, *Culex* and even *Anopheles* mosquitoes are not so readily killed by these insecticides as were earlier ones before exposure had occurred. It is believed, however, that by effective residual spraying over wide areas, interruption of the transmission of malaria could result in the elimination of malaria parasites from a region before resistance to the insecticide could develop.

The accent now is on total eradication rather than indefinite control of malaria. This involves reduction of cases to such a small number that the disease cannot be maintained in a given area. This requires complete coverage of an adequately extended area by residual spraying for four years followed by a surveillance system for the following four to find and eliminate any remaining areas of malaria as well as possible newly imported cases.

Malaria has been eliminated entirely from large areas in Corsica, Cyprus, Italy, Ceylon, Mauritius, Argentina, Chile, Venezuela, British Guiana, Panama Canal Zone, Puerto Rico, Barbados, and Martinique. It is believed that eradication can be accomplished in a period of eight to ten years with no more than four to six years of spraying. However, an efficient surveillance system must be maintained, with simultaneous control measures over vast areas. Programs of this nature have virtually eradicated endemic malaria from the United States and many other areas of the world. This eradication may well be worldwide in the near future; the only obstacles are social and administrative rather than technical, biological, or economic. The humanologist in industry can play a major role in helping to remove these obstacles by presenting authoritative facts in terms of lay language as they appear in the literature.

Yaws

Another problem that is now gradually succumbing to concerted effort of many groups over wide areas is that of yaws, a contagious disease endemic in the regions lying between the Tropics of Cancer and Capricorn. An example of what constructive action on an international scale can do to combat this problem is seen in the Yaws Eradication Campaign carried out in Haiti as a joint effort of the Government of Haiti,[5] the World Health Organization, and UNICEF. The Yaws Eradication Service (SERPIAN) was established in Haiti in 1950. The personnel is composed of an international team including foreign consultants, administrative staff, (director, chief accountant, secretaries, and statisticians) and technical personnel. Among the latter are chauffeurs who have received training in elementary techniques of asepsis and penicillin injection.

Data, such as population figures and prevalence rates, was lacking. In the southern part of the country, the daily clinic method was employed, through which from 15 to 25 per cent of the population was treated.

After completion of a national census, it was possible to change to a house-to-house system. Inspectors assigned to different parts of sectors, went from house to house to inject 2.0 cc. of penicillin with aluminum monostearate in oil to each yaws case, and 1.0 cc. to each contact. Careful tabs were kept as to who was present and examined in each house visited, and those absent were scrupulously sought out and examined. This method permitted contacts with from 95 to 100 per cent of the population of a region.

With the assistance of four inspectors, a medical officer has been making regular surveys throughout the Republic. Secretions collected in house-to-house canvasses have been

---
[5] Francois Duvalier, himself a physician, as president spearheaded this effort.

regularly examined, and in the majority of cases a blood sample is also taken.

In undertaking this campaign, the Government has spent approximately $1,000,000, UNICEF, $630,000 in material and drugs, and the World Health Organization, $282,000 for payment of salaries of consultants and cost of certain materials. It has been reported that yaws is virtually eliminated in Haiti today. Methods successful in this campaign are being adopted throughout the tropical belt.

It should be noted that the success of these campaigns has sociological implications that will become more and more important to the humanologist with interests in the areas where such diseases were formerly hyperendemic. Victories over these diseases tend to increase the average life span by leaps and bounds. At the same time, the population increases, often at a rate too rapid to be absorbed by the resources of the community unless intelligent economic and social planning goes hand in hand with the war against the disease and is undertaken by those in a position to take the necessary steps. The industrial humanologist needs to be in a position to know if his company can absorb this increased population or if it can reach cooperative agreements with other agencies to see to it that economic growth keeps pace with this increase.

Mental Health

To return to the dispensary proper, another vital *preventive* role of the properly qualified company nurse is in the field of mental health. This is not to imply that the nurse ought to attempt to be an amateur psychiatrist. There are more than enough of these at large in any sizeable company, armed with half-knowledge gleaned from popular journalism; it is doubtful that they can have any but a complicating effect on the general mental health problem. The wise dispensary nurse will realize that the less she knows or pretends to know about the professional aspects of psychiatry the better she will do her job with regard to preventive psychiatry.

Considerable explanation is in order here.

First, one may ask why this role is not allotted to the psychologist on the humanological team. To a certain extent the latter plays a preventive role by helping to see to it that square pegs do not get into round holes and by carrying out a large-scale program of human engineering with regard to job environment. Preventive *psychiatry,* has to do with *individual* personal problems that tend to elude the large-scale approach, and that may arise in any but the absolute Utopia, which will probably never be engineered by any mortal psychologist. These are problems that arise in the home and complicate the job environment, or arise on the job and complicate the home environment, despite the best-laid plans of the entire humanological team.

These problems, when they manifest themselves in their first trivial-seeming symptoms, will inevitably come first to the attention of the nurse, or to the attention of some colleague or supervisor who will bring them to the nurse, and here is where preventive psychiatry begins.

These problems, in their correctible stages, tend to reveal themselves in the following atypical patterns of (1) drinking, (2) accidents, and (3) short-term absences.

Absenteeism

Absenteeism, estimated to cost industry about $10,000,000,000 annually, is actually caused by a small percentage of workers. Studies of a number of large companies over a number of years indicate that it can be predicted in a given year that three-fourths of all the absences will be accounted for by one-third of the workers. Moreover, it may safely be predicted that the majority of those employees with a high absence record in their first year of service will continue to rate high on the absentee rolls throughout their periods of employment.

The psychiatric implications of a typical, regular pattern of absences are highlighted by the fact derived from a study of women employees in the New York Telephone Company in the early 1950's. Women in the high absence-group were markedly more susceptible to the *functional diseases or disorders of feeling, state, thought, or behavior*.

The dispensary nurse with a seasoned appreciation of her role will not treat each cause of absence, where a definite pattern exists, as a separate entity; she will see each individual cold, headache, stomach upset, and the like as part of a chronic and repetitive disease. Without overstepping her professional limitations, she will seek the underlying personal problem of the employee.

Social Adjustment of the Absentee

Social adjustment of the absentee is something else the skilled nurse will observe. If the absence pattern is in any way psychogenic, there is usually a poor social adjustment on the part of the individual in question.

Perhaps even more damage is done to the industrial effort by the "half man" who reports to work regularly and promptly but is able to perform only at a low level of efficiency. It is not as easy to measure the magnitude of this problem in terms of dollars and cents as it is in the case of absenteeism, where the yardstick is simply the number of dollars paid out in sick pay for hours not worked.

"The Half Man"

The "half man" who stays off the absentee rolls is not only a sick man himself, but may be in a sense "contagious." As the quality of his work decreases, he will develop emotional defenses to protect himself against having to face the fact of his decline. Those around him may become, in one way or another, sympathetically involved, particularly if his defenses are plausible. The army "gold brick," for example, was usually able to persuade a number of his easily-led fellows that it was somehow more manly to shirk than to work. Those who were willing to work diligently were stigmatized as "eager beavers."

Accidents are an important clue in the detection of these "half men." Accidents tend to conform to patterns and they are frequently predictable. Some persons tend to repeat accidents on certain days of the week or month, particularly on days that have a special meaning to the individual. Some people seem to have a tendency to suffer repeated injuries to certain parts of the body.

Problem drinking may augment the absentee problem, the "half man" problem, or both. Or it may not manifest itself in anyway directly connected with the job until it is so deeply rooted that little or nothing can be done about it on an occupational level.

It is estimated that one out of every 50 workers is a problem drinker. The majority of problem drinkers have been working for more than 10 years. By this time, industry already has a heavy investment in them. Many are men of valuable experience, considerable responsibility, and proven ability. Unfortunately the employee with a drinking problem is frequently protected by a cover-up on the part of his co-workers, until his problem is past correction. Early detection of this problem is, in general a supervisory responsibility; hence, problem drinking is discussed in detail in a subsequent chapter on Supervision. However, when the drinking pattern is indicative of an incipient psychiatric difficulty, this can be determined at dispensary level, and the initial corrective steps can be taken there by counseling and preparing the employee for subsequent interviews with his supervisor and a medical consultant.

The Role of the Supervisor

It will be reiterated under the heading of supervision, but it may as well be stated emphatically here that continuous liaison between the supervisor at all echelons and the nurse at dispensary level is an absolute must in any positive occupational-health program. This is for the benefit of supervisor as well as nurse. The supervisor has been having serious problems with regard to his proper status and functions in the recent development of humanological concepts in business. His role has been changing, in many cases faster than he has been able to change to keep up.

Despite the fact that he is necessarily looked to by the worker for leadership and security, he has been often a confused individual himself. Much of his technical leadership has been assumed by the union steward and business agent. He frequently finds himself in the position of having to give orders he is not entirely sure he is authorized to give. In addition to these technical problems, he finds that he is being asked, or so it seems to him, to function as a nursemaid, a one-man prohibition squad, a kindly counsellor, and a "head-shrinker." He is expected to ferret out all of his employees' emotional problems, and treat them accordingly. For this he is not trained. It is inconceivable that he ever will be, except in a superficial manner. If he had the training requisite for all the duties that seem to be expected of him, he would not be a supervisor; he would be a professional.

This is the reason that many supervisors seem to do poorer jobs after supervisory training courses than before. They are suddenly appalled at finding themselves apparently responsible for matters entirely outside the context of any training they have ever had.

The supervisor who is sympathetic with the total understanding of humanology knows exactly when and where to turn for help.

The Nurse and the Supervisor

The seeming complexity resolves itself as soon as the nurse is brought into the picture. With this liaison properly established, the supervisor can come to understand that he is not expected to have x-ray eyes. By knowing those under him, he will have no difficulty in spotting atypical behavior. It is not up to him to try to determine whether or not such behavior has an emotional or psychogenic basis. It is up to him, when faced with the problem of an accident-prone worker, a lackadaisical worker, a pathologically pugnacious or mendacious worker, a perpetual absentee, or a worker who is clearly unable to do his job, to seek the prompt assistance of the nurse, who, through common-sense and her particular brand of training, can determine whether the problem is one that can be talked out, can be solved by simple changes at a low-echelon organizational level, or needs specialized intervention in terms of outside psychiatric treatment, home welfare service, etc.

It goes without saying that the liaison between nurse and supervisor should work both ways. The nurse needs the help of the supervisor too in the interpretation of any pattern brought to her attention. A considerable part of her decision as to the disposition of any given case will have to be based upon a thorough knowledge of the individual work environment, which cannot always be gained in full from the individual in question. In this connection, it is interesting to note how difficult it is, in many institutions for the mentally ill, to obtain complete data regarding the occupational background of inmates. Whether this is owing to a lack of cooperation on the part of industry or a lack of appreciation of the importance of the occupational background on the part of institutional officials is not known. In any event, the dispensary nurse is obliged to look upon this aspect of the individual background as extremely important; more important in many cases than the home aspect. It is to be hoped that in the not too distant future, preventive psychiatry programs on a large scale can be developed, utilizing representatives from government, community institutions, and business who have a full appreciation of the importance of collecting all data, occupational or otherwise, pertaining to the total environment of each individual studied.

In 1958, the author made an effort to obtain statistical data relative to the environmental background of persons newly admitted to mental institutions in the United States. The National Association was approached first, and this organization furnished a considerable body of statistical material which rather whetted than satisfied the curiosity. From this material certain illuminating facts emerged.

1. Out of a total of 96,731 males and females admitted in a given year (1954), the majority had mental disorders owing to
   a) Psychoses (33,105),
   b) Chronic brain syndromes associated with diseases or prenatal influence (33,547), or
   c) Acute brain syndrome associated with alcoholic intoxication (2,412) and other (996).

2. The majority of those admitted, 53 per cent, was in the 25-64 age group (the age of gainful employment). Only 11.09 per cent were under 25, and 35.9 per cent over 64.

Next, the author addressed a query to the Department of Health Education and Welfare of the Public Health Service of the National Institute of Health, whence the statistical information had originated. This had to do with previous environmental history. "In this respect," the letter stated "my interest lies in the realm of occupations. . . I wonder if any research has been done . . . on the 25-34, 35-44, 45-54 and 55-64 age groups with respect to what their given occupations were, assuming, of course, that they were gainfully employed some time prior to their need for institutional medical care."

In reply, the author was referred to statisticians in 20 states, forming the Institution's Model Reporting Area, since the central offices of the Institution announced "we do not systematically collect such data nor do we publish any data of this type."

The author's next letter to the 20 statisticians was as follows:

"In March of this year I wrote to the National Institute of Health, Bethesda, Md., asking certain explanations of data presented in Part II of their publication, *Patients in Mental Hospitals*, 1954.

"My questions were twofold.

"1. Admission diagnosis of all patients — broken into the following age groups:

>   22 - 32
>   32 - 45
>   45 - 55
>   55 - 65

"2. The occupational or environmental history prior to date of admission.

"Any help, facts or thoughts that you may be able to give me will be sincerely appreciated. My interest lies in the realm of preventive psychiatry. The role that occupation plays in this respect is a particular challenge to me."

Of those who replied to this letter, most said that it would be possible to obtain certain occupational data, under such broad categories as "Professional, Semiprofessional, etc." upon special request, but that such data was not easily tabulated, as it was not incorporated into routine admission statistics. Some reported that such information was doubtful. The state of Kansas furnished elaborate data as to occupation and age prior to admission, but, with certain exceptions, it was difficult to determine where there might be an industrial tieup which might be useful in developing a blueprint for industrial liaison in preventive psychiatry. The two broad categories in which the largest numbers of patients were represented were: (1) General Farmer, and (2) Housewife. Of course there would be no industrial tieup with either of these categories.

Obviously the above study will have to be expanded in order to prove anything from the point of view of preventive psychiatry in business. No doubt, when they are tabulated and correlated, statistics will show that the majority of persons admitted to mental institutions were gainfully employed at sometime prior to the first admission. (This would be truer in a state such as Massachusetts or New Jersey than in a largely agricultural state such as Kansas.)

If the above assumption is borne out, it may also be safely assumed that facts pertinent to employment may have had something to do with the eventual mental disorder. While it may prove difficult to seek a *cause* in employment, it is not illogical to seek *aggravation*. Certain facts which could be ascertained only through the cooperation of the employer might add up to a picture of a square peg in a round hole i.e., an individual who was pushed past his emotional-end point. Besides being of use in diagnosis or therapy of the incarcerated individual such facts could undeniably be used to prevent the same thing happening again.

The liaison implied above would of course have to work both ways to be most effective. It may be assumed that an up-to-date firm with an appreciation of humanology will have exhaustive records on each employee which should offer clues as to the degree of mental health and stability possessed by the individual at the time of employment and at such subsequent stages of employment that it was deemed necessary to bring his particular file up to date.

Correlation of all such data with other medical or institutional data at the time of institutionalization would add another dimension to the so often unknown and misunderstood mental-hospital patient, whose background, prior to the obvious onset of his illness and his isolation from the world of reality, is ordinarily shrouded in mystery.

Such procedures would not only give the individual's institutional doctors and nurses a picture of him at a time when he was deemed normal enough to play a role in the day-to-day occupational world, but it will offer clues to the industrial humanologist which will enable him to isolate the time, the place, and perhaps the specific incident at which the processes leading to mental ill-health were set in motion.

Without correlation of such data, mental hygiene must still grope in the dark. Here is a challenge for all interested in the mental-health problems of the nation at large — from the preventive and the rehabilitative as well as from the institutional point of view.

Space Medicine

An extremely highly developed form of preventive medicine at the dispensary level is encompassed in the science of *aviation medicine,* which was the forerunner of the science of *space medicine.*

This science is based upon the almost insurmountable human problems posed by modern spacecraft. Scientists of many disciplines have striven to make it possible for the human being to adapt to the conditions created by the greater speed, higher altitude, extended range, and increased complexity of operation which characterize the spacecraft of today and tomorrow. The ultimate worth of these efforts can only be as great as the degree of successful application of fruits of these efforts at dispensary level in this particular instance, at space surgeon level. The principal objective is maintenance of the flyer, and tomorrow, the space traveler, in the highest possible state of effectiveness under all circumstances.

Selection of the astronaut still remains one of the primary missions of space medicine. After selection, the most critical problems are encountered in the area concerned with the protection of the astronaut against the hazards and stresses of flight. The complexity of these problems is recognized when one considers the variety of these hazards and stresses: (1) from high altitudes: hypoxia, dysbarism, temperature extremes, ozone, cosmic radiation, and visual problems, and (2) from high speeds: stress through the application of acceleration forces, linear, angular, and radial — also extremely high temperatures and important visual problems.

Travel in outer space ushers in a new series of medical problems for which preventive medicine must be prepared. The human body can stand any rate of speed provided that it is constant. What hurts is a too-abrupt change in speed or direction. The standard of measurement is the g, representing the acceleration produced by the earth's pull at sea-level. Unprotected and supine, the body cannot stand more than $3\frac{1}{2}$ g's for more than about 15 seconds. In blast-off or re-entry, g forces increase, not only with respect to acceleration but to rate of change. This must be taken into consideration in the building of the space machine, but it must particularly be taken into consideration in the selection of the man to pilot that machine. He must be fit to withstand considerable punishment. Then there is the problem of weightlessness. Quick movement of an arm would turn a pilot head-over-heels if he were not strapped. Many astronauts who have experienced weightlessness have suffered nausea, vomiting, or vertigo to the point where they cannot function. One medically trained astronaut observed an illusion occurring during weightlessness that objects were higher than they really were. Armed forces cardiologists expressed doubts as to whether prolonged weightlessness would have a salubrious effect on circulatory and respiratory systems.

Other space problems are as follows: food and drink will have to be sucked or squeezed into the mouth, owing to the lack of gravity during much of space flight; personal hygiene and sanitation will be persistent headaches; a danger zone of radiation storms containing cosmic particles that may be injurious, particularly to the reproductive system, will have to be rushed through at top speed; isolation and loneliness can create considerable emotional stress, cause hallucinations, and trigger incipient psychiatric problems.

All of these problems point to a high level of health maintenance, plus selection of top-grade people, geared to stand up to a certain amount of punishment. It presents a worth-while challenge for students of humanology.

Space travel points up the potential of mantalent development and epitomizes the significance of occupational health.

In addition to its highly important preventive function, and its strictly limited therapeutic function (aspirins, etc. for headaches and common colds, first aid equipment, and such essential emergency equipment as a resuscitator which can be operated by a nurse or first aid man in cases of asphyxiation, shock, or severe asthmatic attacks), the dispensary has a circumscribed role to play in the allied fields of constructive medicine and educative medicine.

Counseling

The nurse fulfills her constructive medicine role through *counseling*. Some distinctions are in order here. First, the type of *constructive* counseling that falls within the province of the dispensary nurse should not be confused with the type of *therapeutic* counseling that might be offered by an internist or psychiatrist. The function of the nurse is not to try to advise truly sick people. These are not in her province, and should be referred with dispatch to the appropriate outside agencies.

What the nurse can do most effectively, within certain limits, is counsel people who are vertical and more or less healthy in order to help them attain to a better condition of health, happiness, and productivity.

Constructive Versus Educative Medicine

The other distinction that needs to be made here is between constructive and educative medicine. As counselor, the nurse is an educator, but in this particular function she is dealing with individual employees, telling them how to avoid or get rid of this or that individual personal problem. Educative medicine has to do with educating employees in general, through methods of dissemination that reach large groups of people.

Owing not only to her training but to her status, the nurse is able to offer common-sense counseling in a number of areas which the tactful supervisor would do well to avoid. She has to do with a number of people with minor emotional problems, or problems due to ignorance or inability to adapt to certain circumstances or mistaken attitudes; people who are by no means in need of psychiatric or other medical treatment, but who are in need of trustworthy guidance before their problems grow into major ones calling for expert and costly professional therapy. Many of these people are not truly aware of their growing problem, obvious as it may be to their co-workers, until it is explained to them by the nurse.

The supervisor is in a position to take steps to curb early problem drinking by availing himself of disciplinary techniques. Such techniques however cannot be of much value where the problem is not measurably affecting work performance and therefore not legitimately subject to discipline. The individual with halitosis, for example, or body odor, may actually be making the job acutely unpleasant for those who must work near him, yet if "even his best friend won't tell him," neither will his supervisor. Disciplinary action on such grounds would stir up damaging criticism of company policy.

Personal Hygiene — Obesity

The nurse, taking a friendly but impersonal clinical approach, is not resented when she discusses personal hygiene with the offender, helps him to understand the cause of his hitherto unsuspected-social problem, and offers advice as to how to correct it. The same holds true in the case of obesity, which in its early stages is from the point of view of the individual a cosmetic and social problem, rather than an organic or functional one. The perennially undernourished individual, the hustler who bolts a pickle sandwich in lieu of lunch after a breakfast of black coffee, the bored individual who is irritable with his colleagues, the disorganized, over-eager person who is always injuring himself in his haste to do too many things (the wrong way) at once — all of these will take advice from the nurse that would be hotly resented if it came from a supervisor or fellow worker (though the ultimate solution of the particular problem may call for some cooperation on the part of supervisors and co-workers).

Educative Medicine at the Dispensary Level

As for educative medicine at the dispensary level, many tools are available to the enterprising nurse who knows how to make use of them. Education will cover, on a general scale, all of the problems which are met with in counseling; moreover it may have a preventive function, by promoting and explaining health-maintenance measures. Education may also have to do with the adjustment of handicapped workers such as epileptics, paraplegics, injured persons, etc., by means of disseminating educational material that is available and that offers specific advice in such cases.

A health-educational calendar of one large company, reproduced on page 311, illustrates the normal functions of educative medicine at the dispensary level.

Moonlighting

Medical service at the diagnostic level is necessarily beyond the limitations of the dispensary. In some respects, however, as it has been indicated above, the duties of the nurse tend to bridge the gap between the dispensary and the diagnostic level. The subject of *moonlighting* is a case in point. In recent years, there has come into being a widespread trend for workers on the hourly-wage level to hold down an additional full-time or part-time job in addition to the regular one. As this generally involves a certain amount of night work, the practice has been called moonlighting.

Under normal circumstances there need be no objection to this practice from a health point of view. Health problems may occur if the extra hours cause exhaustion, owing to the fact that: (a) too many of the daily work-hours are tedious, or (b) physical reserves are used up before the end of the work-day, or (c) either job calls for a full commitment of mental faculties.

Most of the jobs available to moonlighters are neither too taxing nor too unpleasantly monotonous for the normal, healthy worker, who will tend to be buoyed up in any case by the comparative financial freedom he is earning. He may even be saving money.

Occasional signs of fatigue on the part of one of these diligent individuals may be brought to the attention of the

Month	Aims	Bulletin Boards	Racks	Company Paper	Other
January	To help workers learn about the common cold	Cold posters using characters from film. Announce film showing	Pamphlets: 1. The common cold 2. Drink fruit juice 3. Rest	Article, "The Common Cold," by physician	Show film "How to Catch a Cold," during lunch hour 2 days
February	To help workers learn about heart disease	Posters from the Heart Association	Pamphlets: 1. Heart Disease 2. Overweight 3. Rheumatic Fever	Reprint Article from American Heart Association	Display "The Heart of the Home" in the lobby
March	To stimulate interest and active participation of workers in chest x-ray programs	Posters on T.B. and lung cancer. Announcements of time, place of x-ray unit	Booklets from T.B. Association and Cancer Society	Article by union rep, "Let's All Take Part." Outline of program — (how, when, where workers can take part) —by nurse	Chest x-ray survey, 1-day
April	To help workers learn the Cancer 7 Danger Signals	Lucky Seven—play up idea that the 7 danger signals can be lucky if people recognize signs and go to physician immediately. Announce film showing	Booklets from Cancer Society: 1. It's Later Than You Think 2. How your Doctor Detects Cancer	List 7 Danger Signals. Article by nurse: "What is Early Detection"?	Show films during lunch hour: "Breast Self-Examination," "Man Alive"

Table II — A Health Education Calendar

Fig. 11 — Example of an Open Health Letter

company nurse. The moonlighter himself may seek advice (or "pep" pills, or something of the sort), or his supervisor may ask the nurse whether he should put his foot down regarding the individual's extra hours of work.

Ordinarily a short talk with the individual regarding his health and the nature of his work will provide the nurse with a common-sense answer. The signs of fatigue may be patently insignificant; they may be owing to merely temporary problems of adjusting to a new sleeping schedule; they may derive from too much party the night before.

On the other hand, the nurse may suspect that the individual is actually pushing himself beyond his limits, and in this case she cannot make a clear decision on her own. In order to determine what the individual ought to do for the sake of his own health, it is necessary to assess the individual's true capabilities with relation to the work assignments he has taken on, and this can be done only at the diagnostic level.

Dermatoses

In some instances, the nurse may find herself bridging two gaps: that between industrial hygiene and the dispensary, and that between the dispensary and diagnostic level. Dermatitis is an example of this. Skin problems, dermatoses, are in the forefront of strictly occupational diseases, and as such are of great interest to the industrial hygienist. Numerous "engineering" methods may be adopted to control this problem. Protective clothing may be issued, and a regimen of personal hygiene may be inaugurated, where industrial exposures are clearly responsible for the condition. Yet in many cases, it seems to transcend mere engineering techniques. Some individuals continue to have dermatitis problems from exposure to industrial materials which are not generally considered to be responsible for skin diseases. Unexplained allergies appear in individuals at unexpected times and places. Moreover, the line between strictly occupational dermatitis and ivy dermatitis is not always easy to draw. Even if it is obvious that an individual's skin condition was contracted during a Sunday picnic in a sumac grove, the relation between his condition and the materials with which he must work after he has contracted it is of concrete concern to industry. Exacerbation of his originally nonindustrial condition is strictly an on-the-job problem.

The dispensary nurse, in any event, must be prepared not only to act for the industrial hygienist in seeing to it that individual precautions are taken against dermatitis, but to act as the hand, eyes, and ears of the physician, both in providing first-aid relief for cases of dermatitis, however contracted, and in picking out those individuals whose affinity for dermatitis seems to imply a need for further examination at the diagnostic level.

Medical Services at Diagnostic Level

At the diagnostic level, the first major function is the placement examination. The term formerly used for this function was "pre-employment examination," but since its purpose is not so much to determine whether an individual should be employed as where he should be placed, the newer term is preferable.

The Placement Examination

It is well to remember here that, in the majority of cases, the individual undergoing the placement examination is young. He has undoubtedly undergone numerous physical examinations in connection with acceptance or rejection by the armed forces. Thus, the physical factors are usually known, and this is therefore not a major interest of humanology at the time of the placement examination.

Environmental History

Environmental history may be more difficult to obtain from the average individual than details of physical condition. As a result, the placement examination takes considerably more time than a routine physical. However, the extra time is well spent. The placement examination, for example, should include the following:

Basic Data	Name and address
	Date of Birth — Sex
	Name and address of private physician
Family History	Mother — Father — Siblings
	Background and medical history of family
Past Medical History	Operations
	Injuries
	Diseases
	Systems review
Environmental History	Birthplace
	Home
	Schooling — extent and attitude towards it
	Activities during school vacations
	Work record and attitude
	Social adjustment
	Emotional status
	Relations with family
	wife — children — home
	Ambitions
	Estimate of self
	Habits:
	tobacco—drugs—alcohol—diet—weight
	Extra-curricular interests
Military Service	Branch
	Rank
	Location
	Injuries
	Compensation
Physical Examination	Complete, including examination of
	fundi, vision, and hearing
	Rectal examination in males, at
	discretion of examiner

	Rectal or pelvic in females not required as routine
Laboratory	Hemoglobin (RBC when hemoglobin is abnormal)
	White blood count
	Differential
	Serology
	Urinalysis — chemical and microscopic
	X-ray — PA plate of chest
Evaluation of	(ECG not usually necessary)
Health Status	The examining doctor evaluates the health status of the individual in relation to the job that is being considered for him
Orientation Regarding Periodic Health Inventories and Medical Counseling	The employee is advised of the periodic health-inventory program, and scheduled for his first periodic examination, the date of which is fixed by his present health status

The examiner must state his opinions definitely and concisely so that management will have the information necessary to place the individual in work suited to his particular *medical* ability. Confidential information must not be divulged in these statements. It is recommended that statements such as one of the following be made for each individual with the appropriate explanatory comments:

1. *Able to Perform any Duties.* — Exception: Unless an examining doctor is so qualified, he should not give any opinion regarding the physical qualifications of airplane pilots.

2. *Able to Perform Limited Duties.* — The limitations should be concisely and clearly stated.

3. *Able to Perform any Duties When Defects Have Been Corrected.* — The following defects are examples: (a) hernia, (b) poor visual acuity which is correctible, or (c) acute bronchitis.

4. *Able to Perform Only Limited Duties When Defects Are Corrected, if Correction Is Possible.* — Examples would include: (a) organic heart disease, (b) persistent arterial hypertension, (c) history of recent hematemesis or hemoptysis, or (d) positive Romberg's sign.

Logic as to Place of Performance

There is some controversy as to whether the placement examination should take place on the company premises or in the offices of an outside consultant. More and more companies seem to find considerable logic in the latter course. One reason for this is of course the expense, particularly to smaller industries, of maintaining a full-time staff and equipment to carry out procedures that are by no means continuous. In a smaller company, there are bound to be a great many days during the year when the staff and equipment for the placement examination is either idle or engaged in make-work activities that are really not essential. This is true some of the time even in the larger business concerns.

Another reason for the current trend away from the in-plant examination is that some men, during recruiting season, are examined dozens of times. Young men out of college find themselves going through the same routine in one prospective company after another, until they are becoming somewhat cynical about the whole process. When it becomes general industrial practice to engage outside consultants, this shunting from pillar to post will be cut to a minimum. Doubtless in the same area, as many as 20 businesses may be served by as few as three or four firms of consultants, who may use the benefit of data gained from any given individual at recruiting time for the benefit of all firms possibly interested in placing that individual.

Placing the Medically Handicapped

An important humanological problem after the placement examination has determined and measured specific physical handicaps in an individual is finding the job which that individual can perform satisfactorily without straining his limitations. This problem is not solved simply by setting up a centralized shop for handicapped workers to carry out special assignments away from the critical looks of their nonhandicapped colleagues. Such segregation may cause considerable psychological damage, take away basic job satisfactions, and entail painful social readjustments.

A frequent solution to the problem is the "light duty" slip signed by an examining physician. The success or failure of such an approach to the problem will necessarily depend upon the ability of the line foreman or supervisor to understand precisely what such a slip means. To be on the conservative side, many a supervisor will place more restrictions on the worker than the physician intended; this can be devastating to morale. Sometimes the instructions on the light duty slip are in medical language that the handicapped worker himself cannot understand.

A system that has begun to achieve some success in recent years involves the use of two forms. One of these forms is prepared by the physician in extremely objective terms. It explains in details just what the physical capacities of the individual are in terms of some 50-odd job factors, ranging from sensory perception, through body position and environment. The other form is prepared by investigators with experience in job analysis. It deals with the same job factors from the point of view of the demands of the job itself. Correlation of the two forms in the case of an individual worker may prove most successful as a placement tool, particularly if the full cooperation of the line foreman or supervisor is obtained.

Such a procedure may be useful not only in the determination of the proper niche for the prosthetically disabled or more or less permanently limited person, who is at least understood as a personality by the average foreman, but also for such former "unemployables" as epileptics, who were at one time more or less frightening to the impressionable worker. Through a combination of the above, placement procedure and education of the epileptic himself as well as his colleagues and superiors, it is possible for the great majority of patients with seizures to live normal productive lives. Some, such as Julius Caesar, have transcended even this. Patients need to be protected from discouragement as well as from over-protection. Their seizures should be looked upon not as catastrophes, but as temporary disabling conditions such as migraine headaches, diabetic acidosis, or dysmenorrhea. Therapy for these conditions, which has become increasingly successful in recent years, need not be discussed here. From the humanological point of view, the important things are courage and understanding.

It is at this stage that the psychologist plays his major role in industry. For details regarding the fulfillment of this role the reader is referred to Dr. Weider's report on the clinical psychological program for industry, set forth in Chapter 7.

The Periodic Health Inventory

After placement, all employees should be scheduled for a periodic health inventory. Of all the contributions of humanology to the employed individual, the periodic health inventory is perhaps the greatest and the most deserving of gratitude. It is impossible to estimate how many employed individuals owe not only the prolongation of their usefulness span, but their lives and happiness as well to this procedure, which is still in its infancy.

The periodic-health inventory embodies far more than the routine clinical physical examination. It implies a thorough and exhaustive evaluation of the total individual, tempered with flexibility and individualization.

As long ago as 1919, the periodic health inventory as a measure of health maintenance was recognized, and its possible extension to the general population was envisioned. In 1931, the American College of Surgeons began an appraisal of industrial-health services. Since the end of World War II, the periodic health inventory has become an integral part of a constructive medical program.

The scope of the health inventory varies widely between organizations. Some depend upon mass screening techniques, personnel questionnaires, and brief interviews. Others may apply the public-health service device of multiphasic screening processes. Then there is the "dragnet" method based upon a wide scope of laboratory studies and the "personalized techniques," selecting certain laboratory studies as the result of a clinical opinion based upon a careful history and medical examination. All of the above methods are "after the fact." Little or no thought is given to the total individual in terms of health maintenance to say nothing of health *construction* for the employee.

The most valuable type of health inventory includes the sociologic history of the individual, all emotional factors, his reaction to stress and strain, the character of his home life, and facts in relation to his extra-curricular activities; all of which, when coupled with the actual physical findings, give the examining physician all thoughts necessary for him to consider the employee from a total environmental point of view.

Iatrogenic Diseases

For the prevention of iatrogenic diseases of all kinds, the employee must not be told to see the doctor and insist upon a yearly checkup unless the doctor specifically recommends it. Between periodic inventories the employee is encouraged to seek medical counsel only when he is ill or in trouble. During the first 10 years of employment, if the above procedures are carried out, it is possible to make a total medical evaluation or "profile" for each individual. This profile is of great value in considering the future health and welfare of each employee. During this entire period, the employee is schooled with respect to his own capacities and limitations. (See Chapter 18, Page 640.)

Positive Health Versus Negative Health

In order to do his job well, the examining physician needs to be imbued with the importance of positive health in contradistinction to negative health. His primary concern is not so much with the prevention or correction of disease as with the promotion of a healthier, happier, and better-adjusted individual. He explains with simplicity and authority physiological processes compatible with the aging process. He considers minor conditions such as, for example: alopecia (baldness), presbyopia (farsightedness), questionable elevations of the blood pressure and the like, as slight deviations from the normal instead of early symptoms of impending disease.

By so doing, he is considering the employee as an individual. The promotion of health in this fashion not only increases the employee's value as a worker but also as a total person who, in turn, should become a useful citizen in all aspects of community life.

Scope of the Periodic Health Inventory

An example of the recommended scope for the periodic health inventory is as follows:

Basic data
  The following information is recorded and the necessary changes made:
    Company
    Department
    Position
    Company service
    Private physician
  Summary of interval record
    Dispensary record

How many clinic visits and why, interim medical problems
Lost time owing to certified illness

Environmental history
Résumé of outside consultations
Present environmental status, including: Statement as to how employee likes his job — his supervisors — his fellow workers. Any changes that may have occurred in his estimate of himself, his ambition, home life, or relations with his family. Any specific hazards in relation to his job.

Physical examination
This is thorough and individualized. Previous laboratory examinations are repeated, plus any additional procedures considered necessary.

Consultations
As indicated

Impression and discussion
Impressions of the examiner as to the employee's medical status are recorded, and the employee advised in detail regarding his health status. This, of course, would include information that would help him in social adjustment, dietary habits, etc.

A statement is made on the medical record as to whether the individual is suited for his present job and what, if anything, needs to be done in the future.

Disposition
On basis of current health status, a reminder card is filed for follow-up and for next inventory. It is strongly recommended to the employee that he authorize sending a copy of the periodic examination to his private physician, so that both physicians may help him in health maintenance.

One of the reasons that the periodic health inventory accomplishes such great results in the prolongation of life and the usefulness span, is that it makes possible the early detection of a number of diseases which may not produce appreciable symptoms until they have reached an irreversible or nearly irreversible stage. One of the most important of these diseases is, of course, cancer.

By these health audits and individual health programs, which will add *life to years*, the employee's total health can be developed.

Cancer — Significance of Early Detection

All competent physicians routinely perform a rectal examination on all male patients who have reached 40 years of age. A pelvic examination* is routine for all women seeking competent advice. In the case of males, cancer of the prostate and rectum is most amenable to cure when detected early. In the case of females, cancer of the cervix and related areas responds to curative treatment when discovered in its incipiency.

These particular examinations are mentioned because they have become so much a matter of routine. There are exceptional cases, but there seems to be hardly any reason today why people should die of cancer of the rectum, prostate, or cervical areas, except from failure to undergo periodic medical examination.

Other forms of cancer have not been so readily diagnosed, but in general, as the incidence of these forms has increased, medical science has kept pace by learning more and more about the art of detecting them early. For these reasons, the frightening predictions of statisticians must be taken with a grain of salt. The statisticians tell us, for example, that 31 out of every 100 males born in a given year will eventually die of cancer, as will 36 out of every 100 females. Because the life expectancy of the adult American is now 69 years, it is predicted that the number of cancer patients will increase by more than 50 per cent during the next 25 years.

The statisticians, in noting increases in population, life expectancy, and in the incidence of certain forms of cancer, have managed to overlook the growing increase in constructive medical programs which will tend to bring the figures down — not up.

The recorded incidence of cancer of the lung and bronchus apparently did double in a reported period of 10 years, but it should be remembered that the records are a reflection of

---
*To include a Papanicolaou Smear.

improvements in diagnostic procedure and equipment, including X-ray, bronchoscopic, and pathological study of suspected lung tissue.

At present, 5 out of every 100 cases of lung cancer are being cured, and 50 out of every 100 could be cured if detected early.

Before the age of 65 in women, cancer usually originates in the reproductive organs (breasts and genitals). Today 30 out of every 100 cases of breast cancer are being cured, whereas 70 out of every 100 could be cured if detected early. Thirty-five out of 100 cases of uterine cancer are cured; 70 out of 100 could be cured if detected early.

Today only 15 per cent of the cancers of the rectum are diagnosed early enough for corrective action. Early detection in all cases of rectal cancer would result in at least 70 per cent of the cases being cured. Even cancer of the mouth will respond satisfactorily in 65 per cent of the cases. Today only 35 per cent are detected early enough to prevent distressing complications.

Of all cases of skin cancer, 95 per cent will respond to known forms of treatment. Today 85 per cent come to the attention of competent therapists in time.

Out of every 10,000 individuals between the ages of 35 and 45, there are 39 who acquire some form of cancer. Of these, 26 are females and 13 are males.

Out of every 10,000 between the ages of 45 and 55, there are 95 who acquire cancer; 56 females and 39 males.

Until the age of 55 the odds are in favor of the males. Between the ages of 55 and 65, the odds are 50-50. Of the 192 out of 10,000 persons who acquire cancer at this stage, 96 are males and 96 females.

After this the odds change in favor of the females. Of the 306 out of 10,000 who suffer cancer between 65 and 75, there are 171 males and 135 females.

The purpose of the above presentation is not to frighten the reader but simply to point up the almost epidemic proportions of the disease. If business and other allied groups will join to use the tools at their command to combat this "epidemic" as they have used the suitable tools for the virtual obliteration of smallpox, malaria, and the like, they can bring the incidence of incurable cancer down to barely significant proportions, even without any new development in cancer research. The tools at their command are periodic health inventories, scrupulously carried out at company expense for all personnel.

Other diseases frequently detected in their incipiency in a thoroughgoing inventory which might otherwise develop into serious, irreversible problems, include glaucoma, etc.

Glaucoma

Early detection of glaucoma is vitally important in the prevention of blindness. Industrial-hygiene measures having to do with eye protection from work hazards, proper optical aids, illumination and the like, are valuable guards against many types of possible sight damage, but they are of no avail against glaucoma. This disease has to do with a building up of pressure within the eyeball. The insidious pathological processes that accompany this rise of intraocular tension destroy sight without warning and without sign — until too late. Pressure on the optic nerve kills the fragile fibers, starting in the periphery. Central vision usually remains perfect until just before the end — which is blindness. Detection means there is hope of halting further progress of the disease — not of reversing the progress. There is rarely any restoration of the visual field that is already lost.

Formerly, it was difficult to include truly effective techniques for glaucoma detection in a routine health inventory. Instruments have been complicated, time-consuming and in some cases, inaccurate. There have been developed in recent years, however, techniques which give accurate findings with simple foolproof equipment within a few minutes, allowing the patient time to institute proper management, if necessary, with the aid of a qualified ophthalmologist.

Diverticulitis and Mucous Colitis

The relationship between diverticulitis and mucous colitis is more than superficial. Both take place in the colon (bowel), both *may* be associated with carcinoma, both *may* be symptomatic of an emotional reaction to stress, and both can become extremely serious if not given early attention.

In diverticulitis, one or more abnormal sacs (pouches) are created in the wall of the bowel, owing to pressure of fecal matter within and weakening of the bowel wall. Then these sacs become inflamed. When the inflammation is acute, signs of peritoneal irritation develop, such as localized tenderness and muscle spasm. Perforation may occur with localized peritonitis and abscess formation or generalized peritonitis. Neoplasm, sometimes listed as a sequel of diverticulitis, may be merely a coincidence.

As for mucous colitis, the most benign type is that in which the only symptom is the passage of red blood with each bowel movement, indicating that the inflammation is not too widespread throughout the colon. Diarrhea usually means the inflammation has made greater headway; though there may be no pain or abdominal tenderness. In the severe cases, cramps, fever, loss of appetite, nausea and vomiting occur, leading to weakness, anemia, and general debility. The prognosis at this stage cannot be optimistic; though in the less severe, early stages, complete recovery is not unusual.

Periodic evaluation permits early detection and allows for specific therapy and total understanding of the problem.

Tuberculosis (Pulmonary)

In 1955, 15,000 individuals died of tuberculosis in the United States and an estimated 250,000 were ill with it. There were 1,000 recorded deaths in New York City, a total of over 12,000 known cases and an estimated 10,000 persons who probably did not, and probably do not, know they have this disease. The dramatic fall in mortality from tuberculosis which began in 1947 in no way precludes the importance of early detection.

Adults still evidence infection by the presence of active tuberculosis with *tubercle bacilli* in their sputum or by the presence of tuberculosis in their children. Eighteen of every 1,000 youngsters in a group of preschool children seen in New York City Child Health Stations in 1955 showed positive skin tests for tuberculosis. Such findings in children are evidence usually not only of recent infection, but of an open active case of tuberculosis in some adult in the child's immediate environment.

Consequently, the chest x-ray continues to be an important part of all constructive placement and periodic health evaluations. It is an important and dependable tool in the early diagnosis of unsuspected chest disease.

Judgment, however, in its repeated use year after year becomes a question for the medical consultant and not the policy makers to decide. The biological effects of atomic radiation as it effects the human body, its reproduction organs, and the question of leukemia as a delayed effect has pointed up this significance.

The American Trudeau Society has emphasized that chest roentgenograms are only justified if they lead to the detection of previously unsuspected or clinically significant curable lung disease. Therefore, it is essential for those engaged in the practice of occupational health to evaluate their yields. Among infants, children, young adults, prenatal employees, and especially young diabetics, the tuberculin test should be used as the preliminary screening technique whenever possible and the tuberculin reactors should have x-ray examination of the lungs.

The frontier is the effective control of tuberculosis. To reach it, all the knowledge of the past concerning the relationship of all the concomitants of a faulty environment must be constructively applied.

Bronchitis and Bronchiectasis

The difference between bronchitis and bronchiectasis is that the latter is chronic and progressive, while the former is of limited duration. Both involve the bronchial tubes.

In bronchitis, there is an inflammation of the bronchial tubes, and in addition the trachea is usually inflamed. Acute bronchitis tends to be secondary, following the postnasal drip connected with a cold, or such predisposing ailments as measles, whooping cough, or influenza. A frequent complication of bronchitis is pneumonia, and the disease has been known to trigger latent infections of tuberculosis.

Bronchiectasis is characterized by dilatations of the bronchial tubes and an inflammatory reaction in the tube walls. Bronchitis is the factor most frequently responsible for this damage to the bronchial wall. Other causes are whooping cough, measles, influenza, and inhalation of irritating fumes or inorganic particles such as dusts.

If treated effectively in the early stages the chances for recovery from these diseases are good. Later, the process becomes irreversible, and it may lead to chronic invalidism or death from some complication. Lung surgery may be necessary here.

Unless the true cause is detected and scientifically understood, the so-called bad actors tobacco, smog, etc., may be credited erroneously.

Arthritis

Arthritis occurs in a number of forms. Some 1,300,000 Americans are said to be suffering from arthritis in one form or another. Early diagnosis and treatment can prevent it from becoming seriously painful and disabling in at least 70 per cent of the cases.

The two most common types are rheumatoid arthritis and osteoarthritis.

Rheumatoid arthritis has a tendency to attack people at the height of their employable period between the ages of 20 and 50. Those most commonly afflicted are physical as opposed to mental workers, including farmers and hourly-wage-rated employees.

This problem usually gives some warning before it strikes. The individual who stands at work all day may begin to notice slight pains in his hips and knees. The typist or linotypist may become gradually conscious of a tingling of the fingers and swelling of the middle joints. Persistent stiffness of neck and shoulders may be another warning sign.

Osteoarthritis usually shows up in people in their late years or late middle years. The early symptoms are mild. Often they consist of little more than a tingling sensation in the fingertips, though x-rays at this stage may show metabolic changes in the hips, knees, and spine. Overweight can be a contributing factor. Early detection and arrest of the progress of this condition is the only alternative to some degree of disability.

Deafness

This problem of deafness can be divided into four main classifications:
1. Central nervous system deafness, due to an atherosclerotic or aneurysmal brain hemorrhage; to hysterical or neurotic aphasia; or to epilepsy

2. Nerve or perceptive deafness, due to a skull injury such as fracture of the temporal bone; electric shock; caisson disease (the bends); toxic chemicals; infection due to cerebrospinal meningitis, mumps, tuberculosis, typhus fever, or scarlet fever; Meniere's disease; allergy; acoustic neuroma; pulsations of carotid artery; multiple sclerosis; rhesus factor sensitization; anemia; periarteritis nodosa; measles; congenital syphilis or german measles; tetanus antitoxin; presbycusis; or noise
3. Conductive deafness, due to otosclerosis; tumors such as adenoids, epitheliomata, osteomata; fungus growths; nonchromaffin paraganglioma; calcium deficiency; infection of the middle ear; blast injury; foreign bodies; or congenital defects
4. Malingering and neurosis

Early detection through the health inventory may permit arrest of certain of these conditions before total deafness occurs. Perhaps more importantly, it will enable the humanologist to make vital decisions as to placement. An employee with one of these conditions may be in a situation where normal work noises are aggravating his problem or where he is unable to function since the nature of his work demands acute hearing.

Back Pain

Back pain is an extremely common symptom which often poses delicate problems for the professional humanologist. Its etiology is varied and sometimes clouded in obscurity. There was a saying in the armed forces that if one wished to malinger and possibly evade service altogether his best bet was to simulate back pain, since it was so difficult to disprove that he was really suffering. Service in the armed forces has produced a certain number of tough-minded medics who proceed on the assumption that all or nearly all undiagnosable back pain has its roots in malingering. This is an unwarranted assumption. Even if psychogenic, the pain often has its roots in a genuine problem which is capable of constructive solution. Moreover, upon investigation, many cases of undiagnosable back pain, have proved readily diagnosable when proper techniques were brought into play. One known cause of back pain is allergy with a muscular reaction. Back pain may be "transferred" from genuinely injured areas in other parts of the body. Twinges of pain in the back mean, more often than not, that serious complications (crippling complications) can be in store if thorough diagnostic action is not taken quickly. The hasty diagnosis of malingering can invariably be eliminated by a total humanological evaluation of the work assignment.

Proctologic Disease

There is little that needs to be said about proctologic diseases here, as the importance of the proctologic examination has already been elaborated upon under the heading of Cancer. It should be mentioned in passing that such diseases, occurring in the area of the rectum, are not always serious. However, illness or injury in that area *may* be symptomatic of a malignancy or other progressive disease, and therefore scrupulous attention is essential at the first sign of a proctologic problem.

The Concept of Disease

The concepts of the essential nature of health and disease have changed. Hormones, enzymes, electrolytes, and other chemicals of the body are now being examined with the interest formerly reserved for anatomy, physical diagnosis, and microscopy giving us new ideas about common denominators in disease. It is becoming clearer that disease is not limited to an anatomical part such as an organ, a tissue, or a system, but affects the entire organism.

A disease or symptom complex which, from all objective complaints, may be limited morphologically to a tissue, actually may affect not only the normal function of the entire organism but also may set into motion a subtle progression of reverberations psychologic, sociologic, and economic, extending far beyond the sick patient. We are constantly reminded that a human being lives in equilibrium with his environment and that, in some degrees, he affects, and is affected by, all the conditions that surround him.

With acknowledgement of these facts as applied to a health-maintenance program, it becomes a relatively easy task to outline the essentials of an ideal periodic health inventory.

It is imperative for every individual to know what his health is and to learn to live within the budget nature has given him. This issue can never be dodged. Meeting it head on, the earlier the better, is the only approach to positive health.

When applied to industry, as previously explained, it begins with the placement examination which goes far beyond the former so-called pre-employment examination. The new employee is considered from a total environment point of view, with the physical examination *per se* taking its rightful place in the medical analysis of the total individual. The next step is the performance of a comprehensive periodic inventory at suitable intervals as prescribed by the examining physician. This inventory includes the sociologic history of the individual, all emotional factors, his reaction to stress and strain, the character of his home life, and pertinent information in regard to his extracurricular activities.

The Benefits to Be Derived

The benefits to be derived from such a program, from an employee as well as an employer point of view, are in direct proportion to the effort expended. The following profiles will substantiate this point of view.

PROGRAM OF OCCUPATIONAL HEALTH   333

Case 1: A long-service employee started with the company as an office boy. By attending night school, he not only graduated but also worked for his master's degree in accounting. He married and began to raise a family. Salary adjustments were in keeping with his particular assignments. He was ambitious and cooperative almost to a fault. Without regard for his own personal physical or mental well-being, the innate urge to go forward and become successful led him to accept working assignments which required not only a full working day at the office but evening and Saturday work as well. Prior to a medical setback, he frequented the medical department 374 times. These visits were always either before or after working hours or during the lunch-hour period. His complaints were always minor. Routine physicals revealed nothing unusual. He was always in a hurry. His one object was to succeed, and his impression of how to accomplish this was to please his superiors. At long last, the big opportunity came — a position to head up a department and to become a director as well. And then what happened? Fear and consternation got the best of him, and he suffered a complete physical and mental breakdown. Fortunately, he responded to prescribed therapy. A change in work assignment was obtained for him. A new pattern of life and work was decided upon at a time when something positive and constructive could be done about it. The problem now, however, is to handle this chap from a medical point of view so that he will not allow his natural aggressiveness to take him beyond his physical and mental ability to perform effectively and efficiently. Should he suffer another setback, it is quite probable he will be lost.

Case 2: This individual was born and raised in the New York metropolitan area. After completing high school, he worked for two years. He then decided on a teaching career and enrolled in a teacher's college. His studies were interrupted by three years' army service, after which he completed his work and ultimately received a master's degree in vocational guidance and occupational adjustment. He then accepted a teaching appointment in a private academy. However, the students were from wealthy families, indifferent to their studies, and he felt that they treated him like a servant. He felt that teaching had changed after the war, and, with the dissatisfaction from this position, decided to abandon that profession. He applied for several positions, and inasmuch as he had been recently married and needed a job, he accepted a position as a forecasting clerk. He continued in this for two years, although he gradually developed a dissatisfaction in his work which was reflected in lowered efficiency. Several attempts at transfer were unsuccessful and his discouragement progressed to the point where he was on the verge of resigning. Minor physical complaints developed which led to a periodic health inventory. On that examination physical findings were negligible. However, the discouragement was noted and an appointment was made with the employee relations people. Subsequently a transfer was negotiated to that department as a research assistant. He has remained in that position and has been quite pleased with the change. He finds the work interesting and his efficiency and morale have greatly improved. He is happily married, has one child, and apparently leads a modest home life. Most of his difficulties appear to have resulted from his indecision as to what type of career he should pursue. He now seems settled in this regard and apparently has found his niche. With a continued favorable environment, he should do well. This case is a simple one of misassignment. The personnel department could well have prevented this setback. A well-trained doctor with an inquiring mind detected the basic trouble at a time when something constructive could be, and was done about it.

Case 3: A 44-year old executive with 22 years' company service reported to the medical department for a routine periodic medical inventory. A profile was made. He was found to be in excellent health. A review of his previous medical history indicated that he had only a minimum number of minor illnesses. His record showed that he had been in the habit of having periodic physical inventories. He was a very affable, composed, and relaxed type of individual, and a review of his development was enlightening. He was born in a local town, attended public schools, and completed two and a half years of college. He married at this time and joined the company as a salesman. During his subsequent 22 years of service, he has had progressive promotions in regular stepwise manner at approximately two-to-three-year intervals. With each promotion, he had accepted greater responsibility; he has been able to adapt readily to new duties, and has never developed symptoms related to decisions associated with his work. On the contrary, he had developed the habit of putting aside problems associated with his work when off duty, allowing time for outside interests and his private family life. In addition to devoting a reasonable amount of time to his wife and four children, he has acted as a counselor for Boy Scouts, has been an active member of several business and social clubs, and has allowed time for hunting and fishing. His dietary habits have been excellent. He uses tobacco and alcohol in moderation, and allows at least seven hours for sleep nightly.

In brief, this 44-year old executive has led a well-integrated life. Admittedly, he has been endowed with certain assets which he has progressively developed to their fullest extent. Of importance, however, is the fact that he has done so with a measure of restraint and a sense of proportion. As a result, he is healthy, happy, and prosperous, and his chances for continued development in the company are excellent. It is possible for this individual to reach the top.

It is, therefore, apparent that periodic health inventories of the type described above should be available to all age groups of individuals. Many young people entering adulthood are burdened with anxiety, hostility, defensive attitudes towards themselves and others, feeling of guilt, inferiority, and other forms of self-disparagement and self-distrust. These symptoms, when coupled with an unsatisfactory working or home environment, may, if not clipped in their incipiency, develop into medical complaints beyond repair — the problem drinker being a typical example. Such a positive-health approach to the young will do much towards producing a happy, well-adjusted adult human being. Such a positive-health concept necessarily implies a high degree of cooperation with a paramedical staff. Where indicated, full use of medical and psychiatric social workers, psychologists, industrial hygienists, and public-health agencies must be utilized.

Individualization Essential

The periodic health inventory should be an individualized, thorough, and exhaustive evaluation of the total individual. Industry's basic motivation thus far for such examinations has been the protection of the employer. The value of this examination to the employer has been proved by increased production and marked financial savings. The individual as a person has been considered secondary to his value as a worker.

The many types of health inventories done in the past have stressed the physical examination and laboratory methods of detection. In line with the concept that disease affects the entire organism, periodic examinations should also be concerned with the individual as a psychologic, sociologic, and economic being. The job of the examining physician should be the promotion of a healthier, happier, and better adjusted individual. Periodic health inventories and the formulation of a total medical evaluation or "profile" should be available to all age groups of individuals. For these examinations to be of maximum value to employer and employee, personnel people, employee counselors, training groups, and administrators of insurance, benefit and retirement plans must act as a team. When all this comes about, a periodic inventory will assume a broad, positive, and constructive place in all health-maintenance programs.

Moreover, it may well point the way for a broader and more positive approach to our medical problems in the nation as a whole.

Educative Medicine at Diagnostic Level

Educative medicine at diagnostic level differs from that at dispensary level in that it deals not with preventive education for the healthy, but with supportive education for those with health problems that require the advice of a physician. It is at this stage that a diet and a way of life, is explained to the long-term obese person, the atherosclerotic individual, and the diabetic. The nurse at the dispensary level will be enlisted for follow-up action in such cases, but the medical counselor is indispensable in the beginning. He explains to each of these individuals the significance of his condition and the medical reasons why he must reorganize his life in order to come to terms with it.

Educative medicine also plays a role here with respect to preparation for promotion (see Chapter 13 on "Motivation and Promotion") and retirement (see Chapter 16 on "Retirement"). The importance of health-insight at these stages is elaborated upon elsewhere.

The stories of the following two deceased persons illustrate the difference between a company with a constructive policy and one with little or no policy as regards humanology.

Constructive Medicine Entirely Lacking

"Buddy" X is an example of the latter. A great many people must have read his obituary when he died a few years ago at the age of 43, for he had become something of a celebrity. A great athlete in his college days, and a professional football player for a few years after college, he had been hired by Corporation G largely for his publicity value. He continued to get into the papers, as representative of his company, by organizing junior athletic clubs in prominent cities and by making appearances on radio and TV. He was a great talker and an even greater eater and cigar smoker. By the time he had reached the age of 40, he weighed over 300 pounds. No one in his company would have considered suggesting that he reduce. His weight made him a *character*. When he waddled across the park to present the company's gold cup to the winning team in the local junior football league, he attracted all eyes, and as far as the newspaper photographers were concerned he was a natural. Those who knew him realized that he wasn't particularly happy. He felt that he wasn't really accomplishing anything worthwhile, and he worried about how long he would remain of value to his company. He couldn't understand why he was considered valuable at the time, since he was not aware that his function was more or less that of a clown or a character. He worked harder, worried more, and kept on eating. His death was surprising, but it was rumored that an official in his company said of it: "Not a bad deal. Look at the publicity we got."

Constructive Medicine at Its Best

The other example is David. David, as stated previously, is dead, but his medical life story illustrates how teamwork can bring understanding, happiness, and efficiency. For 15 years, David suffered from a chronic progressive, degenerative, noncurable disease of undetermined cause. This malady was first detected in its asymptomatic form when David received his first thorough health-maintenance inventory.

Twenty years had elapsed since his original *pre*-employment examination. In the interim, he had required only episodic care. His problem was one of high blood pressure with kidney stones, which were detected by x-ray study. At the time of his initial examination, he was suffering no symptoms in terms of pain or kidney discomfort.

Two years later in an effort to alleviate progressive symptoms which were having a most hazardous effect upon David's daily life, an eminent surgeon for a fee in keeping with David's ability to pay, performed a heretofore epoch reaching but unsubstantiated operation. David was given a 1 to 50 chance for survival. David knew all this. He was an understanding person. His superior at the office never forgot for a moment that David was an efficient employee. He personally encouraged him within a week after the operation to get well — that his job was there — that they needed him.

David's convalescence was slow but progressive. His company doctor and his company nurse cooperated with David's family and the operating surgeon to find David a general practitioner who would render essential periodic medical care.

After weeks of convalescence, David returned to work, first for a shortened day and a shortened week. Then only a shortened day. In record time, he was carrying his given load. His understanding family never added to the task by crowding activities in his way when not at work.

David was the breadwinner. He paid the bills. Years went by and his work efficiency remained high. His income increased. Upon each periodic medical examination, physiological and subsequent pathological changes compatible with his particular disease process were detected and explained to David.

He learned his limitations and courageously lived within them. Each year that David lived was actually equivalent to two for an individual in truly normal health.

Nevertheless, David was not a chronic nor an invalid. He was happy. He was employed. He was cognizant of his limitations. He was assigned to a job in keeping with his abilities to perform. He was efficient. He was proud. He was independent. He was a good citizen.

Management cooperated and the medical profession, the specialist, the general practitioner — each cared for David at a price that he could afford to pay. David died at the chronological age of 55. . . . Physiologically he was 68.

Community Relationship

The foregoing case is an illustration of constructive medicine at its best. It is also illustrative of the need for the humanological team to work hand in hand with the community at large, instead of trying to establish a miniature self-contained community within the confines of the plant proper. One ideal aspect of a cooperative industry-community plan is outlined in detail in Chapter 5 ("Corporate Giving"), where tomorrow's hospital is introduced. This need not be too Utopian for action *today*. Even if the building, as described, cannot be erected overnight, the facilities, particularly diagnostic facilities for the vertical man, can be found somewhere in the community today or can be encouraged to come to the community.

The Occupational Health Institute strongly recommends that outside consultants be on the list of the well-prepared company. There has been considerable controversy as to where the consultant should start. There are strong arguments in favor of keeping *all* features of diagnosis outside of and more or less independent of the company. In Chapter 6, it was shown that group-practice plans are emphatically recommended by labor. More and more companies will undoubtedly follow that route.

It is, of course, patently impossible for small companies to have diagnostic units, offering *complete* service, inside the plant. It is an intolerable situation when the larger companies insist upon setting precedents in this matter that the smaller ones are virtually forced to follow.

It is to be hoped that the larger companies, particularly those with a comparatively small percentage of employees in relation to the volume of business, will lead the way toward the practice of a type of constructive medicine in which industry as a whole can participate, and which will enlist the cooperation of every person and organization with a stake in the advance of humanology.

Consumer Effects

Another aspect of community relations which has to do with industries only of a certain type concerns the manufacture and distribution of products which may be injurious to health. Any product which is taken into the mouth, whether it is food, medicine, tobacco smoke, alcohol, or dentifrice is potentially a health menace. Any product which affects the air that is breathed in its vicinity, any chemical product, such as soap that comes into contact with sensitive skin, is potentially a health menace.

Humanology, since it embraces community-health facilities, and therefore community health in general, finds itself concerned even with the manufacturing end of business. Products which are potentially dangerous are by no means necessarily *dangerous*. Sweets which *can* contribute to obesity or tooth decay, alcohol which *can* promote problem drinking, medicine which *can* produce poisoning or drug addiction, detergents which *can* have an effect upon dermatitis — all belong to the everyday accouterments of the normal, healthy, civilized individual. The challenge of humanology is to keep constant tabs on such products, be assured that they *can* be intelligently used, and see to it that the public is informed as to their intelligent use. When unintelligent use of a given product can be curbed, the humanologically inclined manufacturer of that product can see to it that the proper curbing measures are put into effect.

Role of Humanologist

Once again, the question may be raised as to whether the humanologist responsible for this aspect of community relations should come from within or without the company. This can best be answered by pondering first whether in this particular capacity, the humanologist is representing the company or the community. There is no question that in the long run what benefits the community will benefit the company, but from an immediate point of view, it is best that the humanologist represent the community insofar as the health aspect of the manufactured product is concerned. Hence it will usually be deemed wisest to employ the services of an independent professional in the humanological field who is known to others in the community as one who has no special strings tied to him, in other words, one who can be counted upon to be objective, scrupulous, and impartial.

Costs of a Constructive Health Maintenance Program

Costs of health-maintenance services vary from industry to industry, from trade to trade, and from firm to firm within industries and trades. The clothing manufacturer's requirements may not be the same as those of the petroleum worker or the producer of chemicals. The requirements of none of these are exactly the same as those of businessmen engaged in the wholesale grocery trade, retail hardware business, or a dry cleaning service.

Variability of needs and costs. — Not only do requirements for health-maintenance services vary between industry and industry, but they vary, also, between company and company within a given industry. In general, smaller plants have higher injury frequency rates than larger plants. The variance in nonoccupational injuries and illnesses does not depend on the type of business but, nonetheless, may vary from company to company because of the location of the company and the conditions of the surrounding environment.

Some of the variables affecting medical needs are the number of employees in the business firm, the composition of the working group (percentage of women and over-age employees), the hazardous nature of operations (such as toxic exposures and danger of accidents), the workmen's compensation and group insurance loss ratios, and the absentee rate. Other variables affecting the cost of the health-maintenance services to be established are the type of service to be provided, the availability of medical personnel, and the degree to which community facilities may be utilized.

Some estimates of total costs. — There is a lack of reliable estimates which can be widely used as guides for calculating costs of in-plant health-maintenance services. However, for an idea of the total and initial capital requirements of some specific programs, the following estimates have been selected:

1. A cooperative industrial-health center established to serve 6 firms with an average total employment of 1,200 persons has operated at a cost of $3.24 per year per employee. About 60 per cent of the sum is for a nurse's salary, about 15 per cent for rent, and the balance for medical supplies and other items. A physician's services are not provided. However, an industrial nurse is employed and a well-equipped two-room clinic maintained.

2. In an experiment covering 7 plants in the Astoria, Long Island City area, the annual per capita costs of physicians' services, nurses' salaries, and medical supplies amounted to $13. This figure should be viewed with some caution, however, because the consultation and promotional services were provided without charge during the experiment. About 68 per cent of the $13 was spent for treatment and advice, 14 per cent for physical examinations, and 18 per cent for other services.

In Hartford, Connecticut, there is a Small Plant Group Service. Each company's expenditure varies according to the number of hours of the physician's service for which it contracts, and the cost of operating its own dispensary. Costs range from $15 up to about $30 per employee. The average over a period of years has been $20 per year.

In Atlanta, Georgia, a clinic was started with three participating companies. Each made an initial investment of $1 per employee. Space for the central clinic was provided rent-free by one of the companies and all of them helped to furnish it by donating or making cabinets, tables, and two beds. After the initial investment, the annual cost amounted to $9 per employee.

The cost per employee increases as the *quality* of the service to employees improves.

The small-plant programs referred to above, will give some idea as to cost which may be anticipated at decentralized areas, i.e., producing field, small marketing areas. The cost may be considered in terms of 0.5 per cent of the total payroll. For a headquarters office, it is recommended that a budget be allocated for basic capital expenditures to be spent over an 18-month period. And an amount not to exceed $50,000 be budgeted for operational and maintenance costs. This is a somewhat conservative estimate if the medical consultant be allowed to spend a good proportion of his time in the field; and the scope of periodic inventories covers field employees.

For the long pull, 1 per cent of the total payroll should provide enough funds for a well rounded constructive health-maintenance program.

Background philosophy for the foregoing remarks. — A decrease in expenditures is not necessarily synonymous with an increase in profits. A poorly conceived "saving" may well involve, at a later date, large and unconstructive expenditures. On the other hand, increased or new expenditures are often the shortest way to previously unrealized profits.

Medical services for years have been sold to management on such a basis. "Spread on medical services and the medical profession will provide an exuberantly happy and productive worker." This millennium has not arrived in spite of the provision of various amounts and qualities of medical services to employees — such as sickness disability pay, group insurance, retirement, and retention plans, etc.

Alert management and serious-minded medical consultants are on the lookout for objective evidences, i.e., some measurable indication of the effectiveness of occupational-medical service.

The average per capita cost, the number of medical examinations, the number of dispensary visits, the number of x-ray and other laboratory procedures, can be determined.

From a strictly monetary point of view these have little meaning for the following reasons:
1. There are variable norms for the desirable volume of medical care.
2. Volume and quality of services may be and often are inversely related.
3. Services and sickness benefits may be and often are provided without reference to employee needs.

One measurable quantity, sickness and absenteeism costs, has loomed large and there appears to be no end in sight. To some, so-called sickness absenteeism costs are an unmitigated evil which can be eliminated, or practically eliminated by an efficient company medical department.

Sickness absenteeism is not an adequate measure of a medical department's effectiveness. Figures alone often relate the fact that sickness costs go up with the implementation of medical services.

There are a number of reasons for this:
1. Sickness absenteeism is invariably inaccurately reported, particularly in units where medical control does not exist.
2. The availability of a doctor makes management and employees alike conscious of illness factors previously unrecognized or actively ignored.
3. The inauguration of a constructive health-maintenance program uncovers a backlog of previously unrecognized medical problems.

In the absence of such a diagnostic examination, the cost of or need for such an absence might well go unrecognized for 10, 15, 20, or more years.

In such instances the cost may be excessive for both the company and employee, i.e.:
1. Detection of a lung lesion — too late for effective therapy.
2. Detection of a rectal cancer in a potential annuitant — which was not amenable to surgery.

Such cases are costly to management and employee alike. A critical cost-minded manager considers this a new type of burden expense. They do not represent the kind of cases the "plans" people had in mind when present liberal-sickness absenteeism policy was conceived. They are the type of cases that the well-trained doctor of today wishes to find early so that a live patient rather than a premature death is the professional result.

These are not the type of cases to bear down upon in an effort to cut sickness-absenteeism costs. Regardless of so-called liberal-benefit plans, they rarely are liberal enough to fully cover such catastrophic illnesses.

Sickness absenteeism may cost from 3 to 10 per cent of the total payroll, with a relatively small percentage of employees being responsible for a major portion of sickness-absenteeism expense.

Abnormal sickness-absenteeism patterns are established during the first five years of employment. Patterns so developed are limited in frequency and duration only by the limits of a sickness-absenteeism policy.

Excessive sickness absenteeism has little or no relationship to the amount or quality of medical care. Many of the chronic offenders belong in a twilight zone, in which the strictly medical aspects are of secondary consideration to other aspects of the case which are primarily managerial in scope.

Such offenders are invariably rigid, uncooperative, self-centered, and inefficient. They create a morale problem wherever they happen to be assigned.

The high absentee employee has many stressful life situations both at home and on the job to which he has not adequately adjusted himself. This is a life long pattern. The low absentee employee on the other hand, has learned to adjust to similar stresses as they arrive.

Chronic excessive sickness absenteeism is therefore only one of a number of signs or symptoms of inadequacy on the part of an employee. Such problem cases are not amenable to orthodox-medical therapy.

Basic purpose of a constructive health maintenance program. — The only hope from a management point of view is early detection. This calls for teamwork and understanding on the part of all who have to do with employee welfare, i.e., the personnel man, the foreman, the first-line supervisor, the nurse, and the doctor.

The actual time lost from work in terms of cost is insignificant when employees while on the job do inefficient, nonproductive work. They may well be labelled "half men." Their adverse effect upon the morale and efficiency of fellow employees has also to be reckoned with.

For the long pull, 0.7 to 1 per cent of payroll is required to set up a constructive health-maintenance program.

REFERENCES

AMERICAN CANCER SOCIETY. *101 Answers to Your Questions About Cancer.* New York: American Cancer Society Incorporated, 1951.

AMERICAN TRUDEAU SOCIETY. "The Chest Roentgenogram and Chest Roentgenographic Surveys Related to X-ray Radiation Effects and Protection from Radiation Exposure." *American Review of Tuberculosis and Pulmonary Diseases,* (February, 1958).

"Asian Flu Round Up." *Medical Science,* (October 10, 1957).

BAYSEN, J. E. "Aviation Medicine at Crossroads." The *Journal of Aviation Medicine,* XXVI (April, 1955), pp. 124-129.

BONKALO, A., M.D. AND ARTHURS, R. G. S., M.D. *Facts About Epilepsy.* Toronto (Ontario, Canada): Canada Department of Health—Mental Health Service—Series 1, 1958.

BURNEY, LEROY. "What the Asian Flu Will Do to the United States or Us." The *United States News and World Report*, (August 30, 1957).

CAMPBELL, PAUL A. "Aviation Medicine on the Threshold of Space." *Annals of Internal Medicine*, L (No. 6, June, 1959), pp. 1542-1549.

CLARKE, ROBERT J., and EWING, DAVID W. "New Approach to Employee Health Programs." The *Harvard Business Review*, (July, 1950).

CONANT, ROGER G., M.D. "Causes of Deafness." The *Industrial Medicine and Surgery*, XXV (February, 1956), pp. 56-58.

DAY, EMERSON, M.D.; RIGNEY, THOMAS G., M.D.; BECK, DOROTHY FAHS, PH.D. "Cancer Detection: An Analysis and Evaluation of 2,111 Examinations." The *Medical Bulletin*, the Standard Oil Company (New Jersey), XIII (October, 1953), pp. 315-332.

U. S. DEPARTMENT OF THE AIR FORCE. *Flight Surgeon's Manual*. Air Force Manual 160-5, Sections 1 and 2, (October, 1955).

DICKINSON, FRANK G., PH.D. "Supply of Physicians Services." The *Journal of The American Medical Association*, (April 21, 1951).

ELSOM, KENDAL A., M.D.; SPOONT, STANLEY, M.D.; POTTER, H. PHELPS, M.D. "An Appraisal of the Periodic Health Examination." *Industrial Medicine and Surgery*, XXV (August, 1956), p. 236.

EPI-HAB. L.A. INC. 8962 Ellis Ave., Los Angeles 34, California. "Eradication of Malaria as an Endemic Disease in the United States." An Editorial. *Annals of Internal Medicine*, XLVIII (February, 1958), pp. 428-438.

FERLAINO, FRANK R., M.D. "The Common Cold Foundation." *Industrial Medicine and Surgery*, XXI (March, 1952), p. 139.

FOURCHER, K. R., M.D. *Constructive Medicine*. An address presented at Medical-Management Meeting on Medicine as Applied to Needs of Industry, Rome, September 8-10, 1954. Published by Standard Oil Company (New Jersey).

———————. "Diagnostic Care—An Economy." The *Medical Bulletin* of the Standard Oil Company (New Jersey), XII (August, 1952), pp. 399-404.

FRANCONE, MARIO PABLO, M.D. "Summary of a Presentation Made at the Second Argentine Congress of Labor Medicine, Mendoza, Argentina, April, 1953." Published in the *Medical Bulletin*, the Standard Oil Company (New Jersey), (October, 1953).

GAGE, ALFRED, M.D. "Applicability of Techniques of Military Malarial Control to Industrial Developments and Rural Population." The *Medical Bulletin*, IX (August, 1949), pp. 106-116.

GORDON, GERALD. "Industrial Psychiatry (Five Year Plant Experience)." The *Industrial Medicine and Surgery*, XXI (December, 1952), pp. 585-588.

GREGORY, S. "Brochitis-Occupational Factors in Incidence." *Tr. A. Industrial M.* V (April, 1955), pp. 2-9.

Guide for Industrial First-Aid Workers. Boston: Loss Prevention Service, Liberty Insurance Companies, 1958.

GUIDOTTI, FRANK P., M.D. "Periodic Health Examinations in the Hotel Industry." *Industrial Medicine and Surgery*, XXVI (November, 1957), pp. 506-910.

HEALTH INSURANCE PLAN OF GREATER NEW YORK. "Visiting Nurse Service in HIP." *Information Manual for Physicians*, 635 Madison Ave., New York.

HENDRICKS, N. V., B.E., CH.E. "Industrial Hygiene Today." The *Medical Bulletin* of the Standard Oil Company (New Jersey), XI (October, 1951), pp. 496-503.

HOLMAN, D. V., M.D. "Diagnostic Services at 30 Rockefeller Plaza." The *Medical Bulletin*, the Standard Oil Company (New Jersey), XI (No. 2, April, 1951).

HURTADO, ABERTO, M.D. "Some Clinical Aspects of Life at High Altitudes." *Annals of Internal Medicine*, LIII (No. 2, August, 1960), p. 247.

KAFKA, M. M. "What the General Physician Should Know About Aviation Medicine." The *Journal of Aviation Medicine*, XXV (December, 1954), pp. 689-693.

KRYNICKI, F. X. "The Chronic Low Backache." *Industrial Medicine and Surgery*, XXI (March, 1952), p. 111.

KUHN, HEDWIG S., M.D. "Glaucoma Detection in Industry." *Industrial Medicine and Surgery*, XXVI (July, 1957), pp. 327-330.

LIFE EXTENSION EXAMINERS. "How to Live." Published by Life Extension Examiners N.Y., 1947.

LUONGO, E. P., M.D. "The Relationship Between Industrial and Private Practice." *Industrial Medicine and Surgery*, XXIV (October, 1955), pp. 459-466.

MCCAHILL, WILLIAM P. "Nation's Handicapped Can Give America a Billion Man-Hours." Published in *The Crippled Child*. December, 1954.

MacDonald, G. E., m.d. "Dispensary Medical Service." The *Medical Bulletin* of the Standard Oil Company (New Jersey) XII (No. 1, February, 1952).

MacDonald, George. "Malaria as an Industrial Problem." *Industry and Tropical Health III*, published for the Industrial Council for Tropical Health by the Harvard School of Public Health 1957, pp. 27-39.

McLean, Allan O., and Taylor, Graham C. *Mental Health in Industry*. New York: McGraw-Hill Book Company, 1958.

McMurray, Robert N. "Mental Illness — Industry's Three Billion Dollar Burden." *Advanced Management*, (September, 1960), p. 18.

Metcalf, Wendell O. "Health Maintenance for Greater Efficiency." *Small Business Management Series No. 16*. Washington, D.C.: U. S. Small Business Administration, 1954.

Morhous, Eugene J., m.d.; Baker, James P., m.d.; Ballou, Charles, m.d.; Crumpacker, Edgar L., m.d. "Periodic Health Examinations." *Annals of Internal Medicine*, XLVI (No. 4, April, 1957).

Muller, John N., m.d. "Some New Frontiers in Adult Health 1956." The *New York State Journal of Medicine*, (April, 1957).

National Industrial Conference Board Inc. "Company Health Programs for Executives." Studies in Personnel Policy No. 147, N.I.C.B., New York, 1955.

The New York Hospital Information Bulletin (re Flu Policy), September 9, 1957; October 7, 1957; and November 6, 1957.

Newquist, M. N., m.d. "Inter-Relationships of General Practice and Industrial Medical Practice." A Symposium on Industrial Medicine, Harvard School of Public Health, April 3-4, 1953.

"Occupational Health Programs, Scope, Objectives, and Functions." (Editorial) *Journal of the American Medical Association*, CLXXIV (No. 5, October 1, 1960), pp. 1-6, 533.

Page, Robert Collier, m.d. "Importance of Pre-Employment and Periodical Medical Examinations: Summary of Medical Findings in 500 Cases." The *Illinois Medical Journal*, LXXIII (April, 1938), pp. 343-348.

—————————. "An Objective Appraisal of Periodic Health Examinations." An address at the Annual Meeting of the Public Health Association of New York City, February 18, 1953. Published in *Industrial Medicine and Surgery*, XXII (November, 1953), pp. 510-513.

Penalver, Rafael, m.d. "Manganese Poisoning." *Industrial Medicine and Surgery*, XXIV (January, 1955), pp. 1-7.

ROBERTS, H. S., M.D. *Difficult Diagnosis (Iatrogenic Illness)*. Philadelphia and London: W. B. Saunders Company, 1958, p. 384.

ROSEN, GEORGE, M.D., PH.D., M.P.H. *A History of Public Health*. New York: M.D. Publications Inc., 1958, Chapters 1 and 2.

SATALOFF, JOSEPH, M.D. "Occupational Hearing Loss." The *Eye, Ear, Nose and Throat Monthly*, III (February, 1955), pp. 105-107.

SCHEELE, LEONARD ANDREW, M.D. "Public Health in the International Scene." In American Association for the Advancement of Science, Centennial Collected Papers, Washington, 1950, pp. 134-138.

SCHRADER, V. D. L., M.D. AND ALBORNOZ, F. F., M.D. "Constructive Medicine." The *Medical Bulletin*, the Standard Oil Company (New Jersey), XII pp. 235-256.

SCHWARTZ, LOUIS; TULIPAN, LOUIS; BIRMINGHAM, DONALD J. *Occupational Diseases of the Skin*. Philadelphia: Lea and Febiger, 3rd edition, 1957.

SMILLIE, WILSON G., M.D. *Preventive Medicine and Public Health*. New York: MacMillan and Company, 1946.

STERNER, JAMES H. "Standards of Noise Tolerance." *Industrial Medicine and Surgery*, XXI (April, 1952), p. 165.

STOVALL, W. R. "Future of Aviation Medicine as a Specialty." The *Journal of Aviation Medicine*, XXVI (April, 1955), pp. 156-163.

"Sub-Committee on Noise in Industry." Annual report published by *Journal of Indiana State Medical Association*, XLVII (October, 1954), pp. 1170-1180.

SULLIVAN, J. D. "Psychiatric Factors in Low-Back Pain." The *New York Journal of Medicine*, LV (June, 1955), pp. 227-232.

TABERSHAW, IRVING R., M.D. "Industrial Medicine." The *New England Journal of Medicine*, 237 (August 28, 1947), pp. 313-320.

TAYLOR, CLIFFORD F. "Union Boss Bows to Private Medicine." *Medical Economics*, XXXV (March 3, 1958), pp. 102-118.

THAYER, W. R. "Matching Job Requirements and Residual Physical Abilities." The *Medical Bulletin* of the Standard Oil Company (New Jersey) XIV (1954), pp. 26-34.

WARSHAW, LEON J., M.D. AND TURELL, ROBERT, M.D. "Occupational Aspects of Proctologic Disease." The *New York State Journal of Medicine*, LVII (September 15, 1957), pp. 3006-3010.

WHITE, C. S., M.D., and BENSON, O., JR., M.D. *Physics and Medicine of the Upper Atmosphere*. Albuquerque,, New Mexico: University of New Mexico Press, 1952.

WILKINS, GEORGE G., M.D. "The Role of the General Practitioner in Industrial Medicine." The *Journal of the Indiana State Medical Association*, XLVIII (May, 1955), pp. 481-487.

WILSON, R. N. "Low Backache in Industry." *British Medical Journal*, II (September 10, 1955), pp. 649-652.

WOLFF, HAROLD G., M.D. AND GRACE, WILLIAM J. AND WOLF, STEWART. "Life Situations, Emotions and the Large Bowel." *Transactions of the Association of American Physicians*, XXVI, (1949), p. 192.

WORLD HEALTH ORGANIZATION, Field Office for the Caribbean. "Seminar on Treponematosis Eradication," 1957.

9
The Seven Cycles of Life

IT HAS NOT BEEN LONG since humanology became *three-dimensional.* Most of the seasoned men and women in the field can hark back without difficulty to the time when the employed individual was essentially a *two-dimensional* mark on a production chart or a serial number on an organizational table and when it was considered either radical or sentimentally unbusinesslike to view him as a total *three-dimensional* person, affecting and affected by his total environment.

The Fourth Dimension

Now that the battle for three-dimensional thinking appears to have been won, it becomes apparent to humanologists in the vanguard that three dimensions are still not enough. To evaluate the human being in his totality in terms of his total environment at any given time does not take into consideration that both he and his environment are constantly changing; that his past history of reactions to the changes in himself and in his relation to his environment is of the utmost significance today, and tomorrow, too. In other words, he exists not only in space but in *time;* his actions and reactions of today are intricately interlocked with conscious or subconscious memories of actions and reactions all the way back to his infancy. Just as a photograph of a split second of activity in a horse race does not convey a truly accurate conception of what a horse race is really like, so a here-and-now evaluation of a man based on present environment and personal aspects does not give a true picture of the constantly changing, constantly learning, constantly forgetting, constantly improving, constantly deteriorating, constantly acquiring, constantly relinquishing creature that is man. Time, then, is the fourth dimension, at any rate from the standpoint of humanology.

Evaluation of Man in the Fourth Dimension

In view of the fourth dimension, effective evaluation of human beings must be done in terms of both individual history and individual geography. Since man's reaction to the facts of geography, that is his place in the world, physically, socially, economically, etc., can be predicted with fair accuracy on the basis of his reactions to comparable facts in the past; history probably ought to be put ahead of geography, although a thorough-going evaluation will not emphasize one at the expense of the other.

History of the Individual

For the sake of orderliness in evaluating aspects of individual history it is helpful to divide the individual's total personal life into the following seven principal stages, each stage representing a cycle of life:

1. Period of parental dominance (birth to age 14)
2. Period of growing independence (15 to 20)
3. Period of adult experimentation (21 to 34)
4. Period of acquisition (35 to 45)
5. Period of peak attainment (46 to 55)
6. Period of substitution and compromise (55 to 65)
7. Golden period of reminiscence and leisure (66 to death)

The wise evaluator will view his subjects in terms of one of these *personal* stages before he narrows his historical microscope to concentrate on *occupational* stages in the indi-

vidual's history. Evaluation of occupational history may also be conveniently organized into *seven stages of progression*, as follows:
1. Recruitment
2. Selection and placement
3. Training
4. Motivation and promotion
5. Supervision
6. Direction
7. Deceleration and retirement

Each of these stages of occupational development will be discussed in detail in subsequent chapters. Here it is appropriate to dissect the *personal* stages of life as they affect the individual at any given time and as bases for prediction of reactions to conceivable future occurrences.

Period of Parental Dominance

The period of parental dominance is the most difficult to evaluate in the adult individual, owing to forgetfulness, misinterpretation of memories, or simple reluctance to discuss topics that may seem trivial in retrospect. Nevertheless, it is the most valuable stage from the standpoint of the evaluator. The more he knows about it, the more he will know about the essential and irreversible character of the individual being evaluated; for this is the period in which are molded the character and personality which a man or woman carries throughout life.

Superficially, this period is a period of battle with numerous infectious diseases, such as measles, mumps, and chicken pox. From a deeper, more constructive point of view, it is the period of total fortification against the more insidious problems of body and mind which are rarely thought of in connection with childhood. Insight and strength are rooted in this period.

During this period, the child is under the complete jurisdiction of adults. Since he is passive in such an environment and endowed with a capacity and willingness to absorb what is passed to him from above, his technique for living will be patterned on that of those who represent authority to him: parents, big brothers or sisters, patriarchal grandfather, or guardian. This leads to two important developments.

1. *Conditioned reflexes sensitized to given situations.* — The habits of eating instilled at this time, for example, may be lasting ones, inwardly if not outwardly. The individual reared to eat gargantuan dinners consisting of fats, starches, and sauces, may, on doctor's orders in later life, eschew food of this sort, but in the back of his mind it will always beckon to him; the smell of white gravy like mama made will always be the smell of emotional security.

2. *Duality.* — Significant to an understanding of this period is awareness of the *onset of duality*. At the beginning, the individual's life is entirely unified. He is the dead center of his world. Everything revolves around him. If any facet of his world conflicts with any other facet, he clings to the facet which gives him the most pleasure or security and simply ignores the other one. Then at some stage, ordinarily after the age of four, he is forced, in one way or another, not only to recognize, but to accept and come to terms with, a world away from mother's loving arms and tender solicitude. This may be a public playground, it may be kindergarten, it may be the back yard dominated by an older brother or sister who is unchecked by feelings of maternal solicitude for the little nuisance who gets in the way or keeps falling and hurting himself.

As he adapts to this other world, he learns to take strenuous steps to keep it separate from his home-and-mother world. He is not the same person at home as at school. He would not dream of discussing with his schoolmates his fears of the dark or his bed-wetting problems. At home, it is with great reluctance that he answers all the parental questions about the fascinating secret things he does with "the gang" Saturday afternoon. First, they wouldn't understand; second, it's none of their business.

For most of the rest of his life, barring the brief period of untrammeled adulthood between the end of school and the beginning of marriage, he must contend with some measure of duality. In later life, the conflict is between the job and the home and social life. From time to time, the essential conflict between these different worlds may be resolved, but the cessation of strife means merely armistice — not lasting peace. With the well-balanced individual, "cold war" might be a better term than "strife." Such an individual is well aware that the elements of overt conflict are always present in the double life he must necessarily lead. He knows that there is no final solution — no peace terms that will be wholly satisfactory. The claims of his job will never jive with the claims of home and society. Life is a series of compromises. How wisely he chooses his compromise measures determines whether he will be able to put the most into his job without sacrificing wife, children, and friends. The individual who serves on his local board of education and at the same time is ready and willing to work evenings at the office to execute an out-of-the-ordinary last-minute assignment is one who has learned to keep the cold war from becoming hot, though it may at times become uncomfortably warm.

The adult ability to contend with duality is learned in childhood. An important task for the thorough evaluator is to delve into the individual's past as deeply as possible to determine how well he learned the fundamentals of living two lives in early childhood. This will be a key to the amount of conflict between his present two worlds and to the individual's knack for keeping these worlds from crashing headlong into each other. The individual who lets the demands of home and society interfere greatly with the job will not

THE SEVEN CYCLES OF LIFE 357

reach the goals he may have set for himself. The one who lets the demands of the job take precedence over wife, children, and friends will find that the inevitable unhappiness he is creating for himself at home will cling to him even after he has passed within the sacred confines of the company, and in the end he will have defeated his own purpose by being as useless as a jobholder as he is as "head" of a family.

Period of Growing Independence

The period of growing independence is a painful one for both parents and child, but it is not altogether as crucial as the preceding period. By the time the adolescent begins straining at the apron strings, the foundations of his character will have been laid, for better or for worse.

It is at this stage that so-called juvenile delinquency occurs. Delinquency is, of course, relative. In its more virulent aspects, where major robbery, mayhem, and/or flagrant sex offenses are concerned, it is evidence of parental failure in the early stages of life — usually complicated further by an unwholesome environment where defiant misbehavior is glorified. There are, however, petty aspects which probably do not deserve the appellation of "delinquency." These include fast driving; flippancy toward tradition and other adult "sacred cows"; tentative and inconclusive experiments with sex, tobacco, and alcohol; bizarre tastes in clothing, music, and slang. All of these tendencies add up to "delinquency" in the minds of completely baffled adults who have forgotten or romanticized the days when they, too, were intent on shocking their elders. Actually it is a part of growing up. A wisely reared individual will come in time to realize that many of the ways of his elders are good ways; however, in this agonizing transition period it is psychologically necessary for him to reject those ways as he rejects the apron strings they seem to represent.

A wise parent will not altogether relax parental discipline at this time on the assumption that his job of moulding is complete. Even in his moments of rebellion, the adolescent senses that he is not wholly capable of going it alone, and he is confused by the absence of authority, however he may chafe at it when it is present. Over-exercise of authority, which leaves no room for initiative or independence of thought on the part of the youngster, may do insidious damage. Everyone is familiar with the person who was either completely cowed by a domineering parent or smothered by an over-solicitous one to the point where in adult life he is never quite able to strike out for himself, in thought or in deed. Such a person may be brilliant and charming and attain to positions of leadership on these counts alone. Inevitably, however, a time will come when he must make a vital decision unaided or initiate some forthright course of action that is not in the books. At such a time, when the chips are down, he will tend to vacillate and ultimately do too little too late.

The alert humanological evaluator has a responsibility to spot such persons before their other undeniable good qualities jockey them into positions requiring more maturity than they can muster. Knowledge of the individual's parental background during his years of groping toward self-reliance will furnish the clues for effective evaluation of such individuals.

A wise parent will find a happy medium between offering, on the one hand, the hard core of parental authority and, on the other hand, love which the puzzled adolescent needs, whether he thinks so or not, by encouraging the youth to take as much initiative and responsibility as he can carry. Such a parent will stimulate his teen-age child to work out problems for himself. He will not attempt to pry into every aspect of the youngster's away-from-home life. Respect for his youngster's privacy is highly important; it is the first recognition of his quasi-adult status. The perpetual mama's boy or daddy's girl is one who never had any privacy, hence developed no real individuality.

As this stage draws to a close, the individual who has learned successfully, i.e., with a minimum of violence, bitterness, or lasting antipathy toward parents, to stand on his own two feet and think for himself is rewarded by a temporary respite from duality. For a brief period he comes near being a unified person in all major respects. He is one person living in one world. There is no tug of war, with his family at one end and his schoolmates at the other. At this stage, free from apron strings, he *is* his family. Wherever he is, is home. For the time being, he has no overpowering ties. The past is meaningless to him, at least for now, and he is able to project himself wholly into the future. He may have had vague aspirations before regarding his role in life, but considerations of home and school were paramount then. Now these aspirations begin to jell. If he has successfully weathered the breakthrough into self-reliance, he is now capable of coming to some pretty concrete conclusions as to what he wants to be in the occupational world. He may not know now, or for many years to come, just how to get there, but the mere fact of knowing where he wants to go gives him an edge over the perpetually unweaned individual who must always vacillate, or the totally undisciplined juvenile delinquent whose only goal in life is to "live it up." The trained evaluator will take this into account. A study of the adolescent "weaning" experience of the individual being evaluated will indicate just how seriously that individual's aspirations are to be taken, and what significance is to be derived therefrom.

Period of Adult Experimentation

The period of adult experimentation is when the newly independent individual begins to look around to determine which of many possible routes lead to the goal he has picked for himself. If he is imperfectly "weaned," he may settle for any adequately salaried "security" job, regardless of whether it leads to his goal, but if he has been properly reared, he is unafraid at this stage of his life. When he takes a job, he is merely "trying it on for size." He knows there are other jobs. There are no strings attached to him.

This is as it should be. An individual who is in pursuit of a tangible and attainable goal is a happy and productive individual. An individual who has sacrificed his goal for the sake of security or who is in pursuit of a goal he will eventually come to learn does not really exist is in danger of becoming a disgruntled, frustrated individual, seriously handicapped by a myriad of real health problems, both psychological and physiological.

This fact is all too rarely taken into consideration in company recruiting policies. There is often a tendency to try to get and hold onto capable people, regardless of the fact that the company may not be able to offer these people the means of attaining their true goals in life. Consider the case of the dedicated young engineer now chewing his pencil in boredom in an organization overloaded with engineers whom it has not quite discovered how to keep busy. Consider the dynamic young organizer and administrator, president of his class in college, now debating his role in a family-controlled company where all of the top-line posts are held by the sons of the founders, all of whom have sons or sons-in-law of their own.

The unafraid young experimenter will back out of such blind alleys as soon as he has had an opportunity to size them up. But before long he will perhaps cease to be unafraid. Some ill-advised recruiter or placement counselor will throw a scare into him by pointing out to him that his record of job-hopping may prejudice potential employers against him. Such a record, he will be told, smacks of instability. If this warning is pointed enough and if the young seeker has grown a little weary of backing out of blind alleys (perhaps there is a potential spouse in the offing who is more than a *little* weary of all this seeming "backing and filling"), he will perhaps say, "Goodbye, goal," and settle for the tangible job at hand. If the company thinks it has served its interests by "hog-tying" this able individual despite the promptings of his true aspirations, it will find out to its puzzlement some years later that it is stuck with a rather cynical individual who is bored with his work, is a bad influence on his fellows, and is perhaps less easily gotten rid of than he was acquired in the first place. Such a person can be more dangerous than an overtly rebellious employee. Cynicism is infectious. No one wishes to be thought gullible; the clever cynic has a way of making those who do not share his views appear to be credulous louts.

Therefore, in the best interests not only of the job-seeker but ultimately of the company itself, the astute evaluator will endeavor to ascertain what the seeker hopes to become in the occupational world, and *why* he has these aspirations. Perhaps the individual's aspirations are wholly unreasonable. Perhaps he is a person of low intelligence who has somehow picked up the fixed idea that he is qualified to become at the very least a second vice-president of a complex financial house. In such a case, the counselor is justified in making an effort to point out to the individual the disparity between his aspirations and his true prospects. If, on the other hand, it is apparent that in some other company the aspirant may have an opportunity to realize his ambitions, it is the counselor's duty to all concerned to steer him in the proper direction. He may say: "We like your background. You are obviously a competent man, the kind we are always looking for; but we do not have what you want here. If you stay with us, you will eventually become embittered since we cannot offer what you really want out of life. May we suggest that you try Company A, or Company X, and see what they have along your line."

The First Ten Years of Employment

If it is plain that there is at least a possibility of the individual's reaching his goal within the company where he is being evaluated, the evaluator will endeavor to see that he is placed where his goal is at least in sight. This is not analogous to the carrot on the stick held in front of the donkey to make it move forward. The donkey only thinks he is progressing toward the carrot. No human being can be successfully kidded in such a manner for any appreciable period of time. Sooner or later there will be a rude awakening, not only for him but for his immediate superiors in the company. The individual should be told precisely what the chances are of attaining his goal, and he should be kept posted as to where he stands. If he fails to make it, he will know that there was a legitimate reason; nobody was pulling his leg. From an occupational health point of view these years of adult experimentation, the first ten years of employment, are the most important years of all. See Chapters 10, 11, and 12.

Period of Acquisition

As the individual goes into his next phase, the period of acquisition, at about the age of 34, he is once again afflicted with the problems of a dual life. This time it is duality with a vengeance. He hasn't quite identified himself with the job to the extent that he finds it more enjoyable than, say, an evening at the country club or a week end of hunting. Nevertheless, he is aware that at this stage his occupational life is the most important. He wants to get ahead in his chosen occupation, and though it may cost him pain, he will forego a jolly evening out if something pressing has come up at the shop or office. His prowess as a hunter, fisherman, golfer, raconteur, elbow-bender, or card-player is not taken very seriously by anyone. If he comes back empty-handed from a duck hunt, he is immediately forgiven. On the other hand, if he fizzles out in his occupational specialty — if, for example, he aims to become head chef in a hotel and winds up as third assistant salad boy — he is not forgiven, nor can he forgive himself without the aid of the bottle or a similar prop.

The employed person must therefore now take a stern line with his nonoccupational self and, if he is truly diligent and ambitious, consider the probable result. By the time he is 40, he will have pushed his nonoccupational self wholly into the background. Whether his working career is normal, unusually triumphant, or a series of disappointments, it has become his principal world. His nonoccupational self, despite the strictures of family and well meaning friends, has learned to stay humbly in the background, reasonably content with what small scraps were thrown its way — an occasional social evening out, a newspaper to read while commuting, small doses of radio or television, perhaps a beer or Martini after work, or 18 holes of golf on Saturday.

It is at this stage that normal physiological changes having to do with the aging process begin to appear. Over-absorption in the occupational side of life often allows these changes to make considerable headway before they are observed and dealt with. The major responsibility of the humanology team at this juncture is to see to it that each individual is thoroughly examined at appropriate intervals and alerted to the true state of his health. As new physiological limitations appear, the individual's capacity to carry out his assignment is reassessed, and proper steps are taken. Every effort is made to encourage those who are capable of doing so, to acquire a philosophy of life that is not exclusively job-centered.

The individual who learns to make sensible adjustments, who is not only a job-holder, but a proper spouse and parent and a good citizen of his community, comes at length to enjoy his job in every sense of the word. He needs to be prodded from time to time to be reminded that there are other aspects of life, but he responds intelligently to such prodding.

Period of Peak Attainment

For the average person who has managed not to fall by the wayside, the most satisfactory years are those between 45 and 55, when he is at his peak of attainment. If he has not attempted to push too far beyond the limits of his inborn capabilities, he is now thoroughly at home in his job. Off the job he has passed the trial-and-error period of being a parent to incomprehensible small children. His older children are either undergoing their own problems of growing independence or are approaching the threshold of the stage of adult experimentation. In either case, he is now able to see the results of parental leadership in the far more perplexing and crucial, if not so painful, period when they were under 12.

This is a period of self-assurance and calm. The inner security that comes from knowledge of his own competence in his chosen field enables him to think coolly and clearly. The frantic rush of yesterday seems a little absurd now. He realizes that there *is* time to do many of the things he has wanted. This is a good time for someone on the humanology team to persuade him to reflect on his past, present, and future and see where he stands. The danger in his self-assurance is that he may not be aware how quickly the foundation can be snatched from under it. The very fact that he has reached his pinnacle may weigh against him when the chips are down. An unexpected merger wherein there are more candidates for certain posts than there are vacancies may pull the rug from under him. "Let's give the position to young X," someone will say. "Old A has had his day in the sun; it's time for him to move over and give some up-and-comer a shot at it."

Another thing to be borne in mind is that the average individual who has reached his peak, whether he be a skilled laborer or a line officer, will, if he stays there long enough, be bypassed by one or more persons who were at one time junior to him in rank as well as tenure. The individual on the rise may carry with him a little-known grudge against his former superior. Some real or imagined grievance may have caused him to say: "Just wait till I get in a position to pull rank on that guy. He'll have wished he'd never tried to give me a bad time."

Then, again, the person who has reached his peak must take into consideration that the chronic diseases to which his advancing years have rendered him increasingly prone are not like the violent infectious diseases of his younger days. The individual who is beginning to have circulatory problems, for example, cannot knock off for a couple of days and come back to work as good as new. His problem is a progressive one and will inevitably affect his competence progressively. If he is in a position that many other younger men are competing for, his gradual deterioration will be taken more seriously than if he is in a "shelf" job that no one particularly wants.

The tendency of his superiors will be to give more consideration to the younger men on the way up than to the one at the peak who is showing signs of being on the way down. Someone will say: "Too bad about old A. He does a darn good job in spite of his health condition; but, after all, let's face it: He's bound to get worse as time goes on. Maybe we'd better kick him upstairs and give the job to one of those young eager beavers in the department."

If any of these things happen, the chances are that an unprepared individual who has been living on euphoria at his peak of attainment will suddenly find that he is totally without self-assurance and all of the other desirable attributes that go with it. He may change overnight from a calm, collected, capable job holder, parent, spouse, and citizen to a querulous, fussy, suspicious, embittered, and senile *problem child.*

The humanology team will do well to alert the person who is going along splendidly and self-confidently that the rug *can* be pulled out from under him and that he cannot and should not strive to avert this possibility by embarking upon a strenuous campaign of furtive office politics wherein all ambitious junior up-and-comers are promptly flattened before they can become dangerous. A progressive company with a well-entrenched and respected humanology setup will avoid rug-pulling tactics, knowing the human damage that can result. However, if the man at his peak is in a company which has proved itself capable of putting temporary expediency ahead of humanology, he will be on his balance at all times. He will guard his health and his self-respect so that when the rug *is* pulled, he will land on his feet. He will know and be prepared for the fact that the odds are against him because of his age. Therefore he will guard his most precious possession, his competence, which is his by the grace of good health, so that when the time comes to find another company, the odds will at least not be impossible.

Here is a challenge for humanology. The streets of the business world are thronged daily with competent men and women whose hair is a little on the grey side or the thin side. They are able, experienced, and — jobless. Some have been fired. Others were forced to escape from an intolerable occupational dead end. In any event they are, in nearly every case, far more valuable to any prospective employer than the bright-faced, brand-new college graduates who rub elbows with them in the same personnel offices and *get the jobs*. Humanology must learn precisely why this is so, and also find ways to explain precisely why it is wrong, before this challenge can be considered even half met.

The problem confronting a technological nation with an increase of senior citizens will be the development of under-employment not unemployment.

Period of Substitution and Compromise

Even if an individual at his peak of attainment manages to keep his footing and to remain in the position appropriate to one of his competence, there must inevitably come a time when, for his own good, he must face up to the necessity of relinquishing this position and slowing down. This is the beginning of the period of substitution and compromise. The age at which this period arrives varies with different individuals. The normal age is around 55.

Pride and bravado are complicating factors at this stage. The higher an individual has climbed in the occupational world, the more likely he is to be somewhat overendowed with pride. Pride has been the spur that has caused him to bear a bigger load of responsibility than most of his fellows. Moreover, he would be less than human not to have derived profound satisfaction from the very fact of success. Not only has he been more capable than the average, he has proved it to the world by virtue of his position and title in the business world.

It is not easy to ask such a person to adjust to a less demanding schedule and thus admit to himself and to the world that he is no longer as capable as he once was. This is an exceedingly delicate period. Mismanagement on the part of the humanology team can bring about lasting bitterness and needless inner suffering.

It is not enough here to indulge in a series of homilies on the foolishness of pride. The aging individual in question will not listen anyway. Pride has become a part of his fabric. Perhaps over the years, as he comes slowly but surely to terms with the realities of life and the aging process, he will learn to get rid of pride, or at least that much of pride that comes under the heading of mere vanity. At this stage, however, the humanology team can help him only if it takes his pride for granted and works out an adjustment which he can accept without seeming to humble himself.

This is not as impossible as it may seem at first glance. The individual's capacity to withstand psychological and physiological tension diminishes along with his strength, but his judgment, dependability, and know-how all tend to increase along with his experience. The humanological problem is not just to "get the old goat out of the way," but to place him where his existing capabilities can be put to valuable use so that he can continue to obtain prideful satisfaction from recognized accomplishment at the same time that he is learning to bask in reminiscence of past achievement.

An example of what humanology can accomplish in this area is the case of Mr. J., who, at the age of 58, was sales manager of a building-materials firm. Shortly after his 58th birthday Mr. J. suffered an acute heart attack. He made a good recovery, but was left with angina pectoris. An alert plant nurse noted that his on-the-job spells of pain regularly followed top-level business meetings. His company was faced at that time with an extraordinary sales problem. A heavily financed competing firm had recently made a powerful bid for the territories where Mr. J.'s company had seemed most firmly entrenched, using new techniques of selling and advertising which showed signs of being effective. Mr. J. had come up the hard way, through the depression years, and had proved his ability not only to sell goods in those difficult times, but to organize effective sales promotions and captain sales teams so large that many managers would have found them unwieldy. Now his methods, which had worked so well in times past, were being questioned by younger men with less experience. In effect, it seemed to him, his judgment, his competence, and his record were being doubted. Business meetings became stormy sessions. When it was suggested, logically, that the company had better meet its competition on its own ground and adjust itself to the more modern sales methods of the rival firm, Mr. J. felt that he was being criticized and challenged. To agree even to compromise measures, it seemed to him, would be to admit that he was incompetent. His back was to the wall, and he was fighting for his self-respect.

It became apparent that a younger man, more adaptable to new techniques, was needed to head up and reorganize the sales department. It became apparent, too, that Mr. J. needed to be placed where his judgment need not be open to doubt and where he might retain his self-respect and sense of accomplishment. There was no thought of saying to him: "J., you've had it. Your doctor says you've got to slow down, and, besides, we need new blood at the head of sales. Suppose we give you a fancy title and a cozy office out of anyone's way so you can write a history of the company, or put pins in maps, or cut out paper dolls."

A careful study of Mr. J. revealed that the secret of his past success in organizing large-scale sales campaigns and commanding a widespread but effective sales force lay not so much in any specific method or technique as in an innate ability to understand people and size them up in the course of a short interview. In his earlier days as a salesman in the field, he had used this knack to advantage by determining just what variety of "pitch" would induce any given individual prospect to pay attention and, perhaps, buy. Later, as organizer and leader, he had shown an uncanny ability not only to pick the right man for the right assignment, but to provide the appropriate motivation for each of a highly variegated assortment of individuals. His sales force included a number of temperamental and extremely individualistic "prima donnas" who would have defied the efforts of standard psychological testers or pedantic personnel administrators to classify them. Mr. J. knew the capabilities and hidden wants of each and was able to inspire them to put forth their best efforts in positions where they were able to shine.

The superficial operational techniques of selling had changed with the years, and Mr. J. was unable to change with them. However, human nature does not change. Age and experience had inevitably sharpened, not blunted, Mr. J.'s ability to understand people, particularly in terms of his specialized field of sales. This ability was still a valuable asset to the company and would continue to be.

The situation was put up to him as follows: "Mr. J., we have a tough battle ahead of us. It may mean that we will have to increase our personnel to meet it. At present, however, the problem is to improve the productivity and morale of the personnel as it stands. This means we are going to have to develop an "assessment team" to make a thorough study of our human resources in all departments. The team will include cost analysts, a psychologist, a medical consultant, together with a personnel expert. Each employed individual will be assessed in terms of his actual and potential performance, and each job will be assessed in terms of human beings. After the assessment has been made and policy has been established, the team will of course evaporate, but we will need someone full time to administer the policy. We want you to act as captain of the assessment team, and afterward to administer the policy. There is no one more appropriate for this job than you. No one else has the innate understanding of people plus the years of experience that you have. Of course, this will entail a sacrifice on your part. You will have to relinquish your power as sales manager and turn over the operational details of applied sales technique to someone else. However, in view of your present health status, you will agree that such a change is in order, will you not?"

Mr. J., who took pride in his understanding of people, was immediately aware that this new post was one in which he could express himself to the utmost. Secretly he had been unhappy with his present position and the interminable arguments caused by the stubborn stand from which his self-respect would not let him retreat.

Of course Mr. J. was not a run-of-the-mill individual. The proper humanological solution in his case would not be applicable to every conceivable successful individual at the threshold of his personal period of substitution and compromise. His case is presented merely as a clear-cut illustration of the general type of constructive action that can be taken, with individual modifications, in almost any imaginable case. The challenge is simply to find out wherein the particular individual is uniquely valuable to his company and wherein that value is enhanced by experience. This is not an impossible or even an extremely difficult task, considering that the kind of individual under discussion here is one who has risen in his occupational world and has held a comparatively important position long enough to give evidence of superior endowments.

Golden Period of Reminiscence and Leisure

The success with which an individual adapts to substitution and compromise will determine whether or not the subsequent years of retirement will be, as they can and should be, truly the "best of life, for which the first was made." As the average life span has increased, thanks to progress in medicine, the population of senior citizens has expanded enormously. In 1900, 4.1 per cent of the population was over 65. Based on the rate of increase since that date, it is estimated that in 1980, 14.4 per cent will be over 65. It may also be assumed, from the rate of progress in the field of constructive medicine, that a majority of these older citizens will be hale, mentally alert, and relatively free from disabling chronic disease. Even today, an individual who has passed what is commonly considered normal retirement age is less likely to be a candidate for the rocking chair in the corner than was his counterpart in 1900. Though officially classed as retirable, if not actually retired, he is probably an active individual, capable of carrying out innumerable tasks with vigor and competence.

Obviously, the problem of retirement has moved to a prominent place on the agenda of matters to be considered by humanology-conscious industry. At one time, it could be assumed that a large number of employees were bound to die in harness, and of those who didn't, the majority were either senile or on the verge of senility by retirement time. The few exceptions were too scattered to be particularly noticeable. Today, when any sizable firm is bound to operate in a community (or communities) in which a good part of the population is composed of active, influential, and vocal *ex*-employees, who are in a position to enhance or undermine the company's endeavors in the way of recruitment or general public relations, the postretirement adjustments of these people become a matter of pressing urgency.

When a person has devoted the better part of his life to his occupation, it becomes a part of him. It is not easy for anyone to step from a planned and organized routine of living offered in the occupational world into a sort of vacuum. This is what happens to the individual who has not planned for retirement. A company which has not given careful thought to its retirement-planning program may assume that many, perhaps most, of its retired employees are in this predicament. A rude awakening is in store for the management of such a company when it becomes apparent that these "lost souls" are its front-line recruitment officers and public relations media. One can imagine the kind of recruitment of public relations job they are accomplishing.

It is safe to assume that a company which has constructive policies regarding the early detection of incipient health problems is on solid ground insofar as the physiological health of its retired employees is concerned. In other words, health policy need not be considered a special aspect of retirement planning since it has to do with the individual at *all* stages from the time he is recruited.

The problem, then, is not a physiological one primarily; it is a problem of adaptability. A 65-year-old individual is not highly adaptable. He tends to be set in his ways. Habits of thinking and doing have become almost compulsive with him. If by this time his only meaningful pattern of life is occupational, one may confidently predict that after retirement he will wither on the vine if he does not find some other form of gainful employment.

From this it is clear that adaptation to living in a nonoccupational world must be accomplished while the individual is still young enough to be adaptable. This is why constructive retirement planning begins when the individual is 45 or, better, 40. Obviously, it is not economically feasible for an employed person at this demanding stage of his life to practice retirement literally. The humanological goal is to encourage him to develop his nonoccupational self.

First, it is explained to him that the person who concentrates solely on his occupational future during his climb to the peak of attainment does so at the expense of his nonoccupational self, which will be his *only* self after retirement. At the age of 40 the nonoccupational self is not really atrophied; it is simply overshadowed by the temporarily more important occupational self. It has learned to be shy, subservient, and retiring, and to speak only when spoken to. It must be brought, by degrees, out of the shadows and encouraged to stand on an equal footing with the dominant, bread-winning self.

A wise counselor will next seek to discover the truly worthwhile goals, dreams, aptitudes, and tastes that the employed individual has been forced to suppress for the sake of his occupational self. Somewhere buried in the individual is a stifled yen for something nonoccupational, something the individual would pursue avidly if it were brought to his attention and if he thought he had the leisure time for it: music, perhaps, or painting, politics, a history of baseball, a study of chess, travel, a cross index of *The Education of Henry Adams*, a thorough reading of *The Decline and Fall of the Roman Empire*, an amateur theater group, a poker club, rug-weaving, carpentry, electronics, or public service. Any human being has something of the sort hiding in him, and a competent counselor can help him take it out of hiding, dust off the cobwebs, and set it into motion.

The individual who, with help, has discovered his non-occupational self will in time come to anticipate retirement with pleasure because it will mark the beginning of a new life of meaningful leisure.

So much for four-dimensional thinking and the seven stages of life. The discussion thus far has dealt with the passage of time and its effect on human beings in general. The humanologist should also learn, at *any* given time, to identify types of employees.

Types of Employees

There are three main types of employed individuals: (1) the person who is capable with ease and promotable; (2) the person who is capable with difficulty and dangerous to promote; (3) the utter misfit. These are distinguishable at any of the stages and at all levels of organization.

The competent and promotable individual is well adjusted to his environment. He enjoys his job, and his off-the-job life is under control. In a position of command, he elicits the cooperation of his subordinates, making it a point to know the capabilities and motivating forces of each. Having made sure that he has given the right men the right assignments, he is not afraid to delegate authority as well as responsibility. This ability to share his power is the secret of the poised leader. Those under him, given an opportunity to participate in command, are aware of their importance in the scheme of things and are prepared to take on added responsibility when promotion time comes around.

The second type of employee, who is competent with difficulty, may be a first-rate lieutenant, but he is out of his

Fig. 12 — The Able Executive Who Surrounds Himself with Men of Similar Potential Abilities

depth in a position of command. Perhaps because of his genuine ability to carry out commands he wins the approval of his superiors, and when a vacancy occurs in a command position, he is tapped to fill it. Perhaps, statistically, his unit or department continues to show good results, but there is trouble in the offing. Unable to delegate effectively, he takes on his own shoulders a larger burden than is good for him over a long haul. He puts in unappreciated and uncompensated overtime. He carries a brief case home. He loses touch with his family, and his bored underlings, with no clearly delegated

376 OCCUPATIONAL HEALTH AND MANTALENT DEVELOPMENT

functions in the workings of what ought to be, but is not, a departmental team, find themselves at loose ends.

The humanology team will make every effort to prevent the unwarranted promotion of such individuals, and when a person of this type does find his way into a position of com-

Fig. 13 — The Executive Who Gets the Job Done For a While But Fails to Delegate Responsibility, to the Consternation of All as the Years Roll By

mand, he will be alerted to the danger of his position before the inevitable breakdown occurs. There need be no doubt that the breakdown is, indeed, inevitable, whether it comes via the several Martinis before the uneaten lunch, or the peptic ulcer, or the schizoid denial of hard reality.

The third type, the utter misfit, is not often found at the level of command; when he is, the situation is a perilous one. Sometimes, by virtue of nepotism, personal magnetism, or a large financial stake in the company, such a person gains a responsible post. Being utterly unable to carry out his assignment, even by dint of overwork, he is forced to rely on knowing subordinates who are willing to cover up for him in return for certain favors he is able to grant by virtue of his authority.

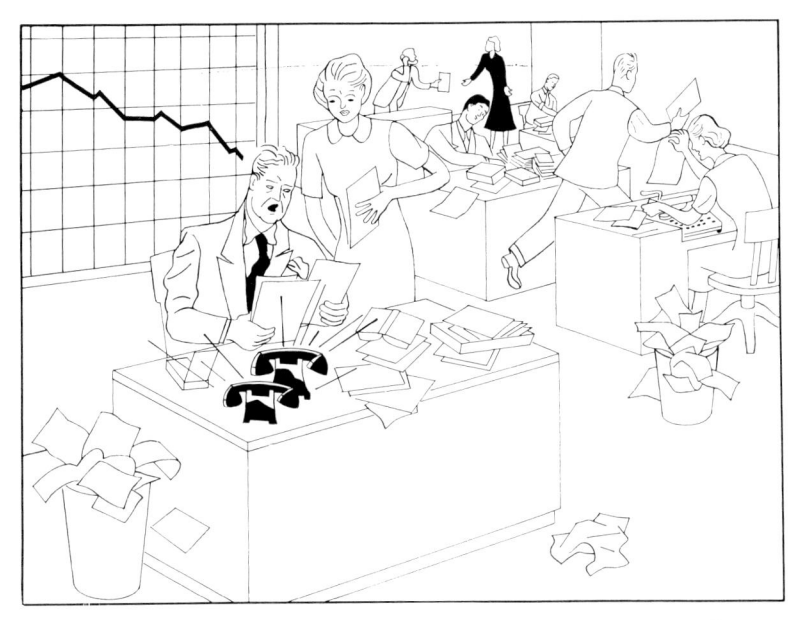

Fig. 14 — Everyone Apparently Working at Full Speed But Accomplishment in the Reverse

Eventually the lid is bound to blow off this hotbed of office politics. Sometimes it takes months to reorganize a department that has been sabotaged in this manner to the point where it is an effective and productive unit again.

Soundly based policy, administered with constant vigilance, will keep the misfit out of sensitive posts and will in the end make the misfit fit by placing him in the proper niche.

The foundation of an effective humanological program is the human raw material with which it has to deal. In this era of expanding technology, competition for the most desirable raw material is acute. Company recruiters must beat the bushes for quality manpower. Public relations programs, including fringe benefits, are designed to make job hunters think "Company X is the pleasantest place to work," or "Company Y offers a challenge for me," or "Company Z offers the most security in an uncertain era." Much time is wasted when the wrong men for a given company are drawn to its gates by a misguided recruitment campaign.

This leads to consideration of the first of the seven steps of occupational progression — recruitment.

REFERENCES

GREENHILL, STANLEY, M.D. "Mental Health and the Worker." *Journal of Occupational Medicine*, (December, 1960), pp. 567-568.

PAGE, ROBERT COLLIER, M.D. "The Seven Steps of Progression. An address at a General Management Conference of the American Management Association, New York, June 3-5, 1957.

─────────. "The Executive: What Health Means to Him and to His Company." An address at the Jersey Standard Coordination Committee Meeting, New Orleans, May, 1949.

─────────. "Maintaining the Health and Efficiency of All Levels of the Work Force." A Lecture given at course of Human Relations, Cornell University, February 1, 1954.

SELYE, HANS. *Stress of Life*. New York: McGraw-Hill, 1956.

10

Recruitment

THE WELL-ADVISED COMPANY keeps up some kind of recruitment effort at all times, even when there are no vacancies or when there is not even an imminent likelihood of vacancies. The keystone of recruitment, in all of its aspects, is an intelligently planned human relations policy, designed to make it widely known that a specific company is a good place to work. The effects of the most elaborate *direct* recruitment program, in which company representatives actually "go after" potential applicants, can be canceled out by the general bad reputation inevitably resulting from failure in the humanological setup.

Why Continuous Recruitment?

It may be argued that continuation of recruitment efforts through a period of depression, when a company is laying off employees instead of filling vacancies, is not only not necessary, but possibly cruel. Widespread advertisement of opportunities that do not exist is a bad practice at any time. Even when opportunities do exist, the scope of recruitment must be kept strictly within limits determined by the actual number and types of vacancies. The televised appeal that brings about a migration of 5,000 workers for 100 openings, the classified advertisement that attracts engineers of all kinds for a job that can be filled only by a specialist in electrical engineering — these leave a lasting bad taste in the mouth of the victims. Nevertheless, surprises are always happening. Key men quit or drop out of the picture owing to disability or death. The exigencies of a waning market or waxing competition may call for the addition of personnel specially equipped to deal with the emergency. First-rate salesmen are needed in a period of strong sales resistance. The installation of cost-cutting, labor-saving machinery creates a need for technically trained operators. Moreover, a period of layoffs is a sensitive one with regard to labor relations, and it sometimes becomes necessary to take on people specifically trained to cope with problems in this area.

Obviously, it is not feasible to halt operations while the personnel department waits for the appropriate applicant to answer its advertisement for a key vacancy or while a company scout is out beating the bushes for the indispensable man. If the company has not taken some pains to nurture its humanological reputation, this can be a long wait indeed. What is needed is a permanent waiting list of potential replacements or additions to personnel, and this list is obtained only through continuous recruitment.

Objects of Recruitment

The objects of recruitment efforts can be divided into two principal categories: (1) hourly-rated workers (unskilled, semiskilled, or skilled); (2) potential managers, line or staff officers, technical experts.

The alert company has on file at all times a list of candidates for all possible vacancies under each category. This list is gathered from a variety of sources. For good humanological reasons, priority is given, wherever feasible, to former employees who have been laid off or have been encouraged to leave for economic reasons. For example, during the 1960-61 recession it became necessary to lay off many employees but these individuals were the first to be employed when the recession was over.

Job Classification and Definition

A rather dry and nonhumanological but essential aspect of recruitment is that of job classification and definition. The personnel officer's bible in this respect is the *Dictionary of Occupational Titles,* which describes or defines some 21,000 jobs or occupations based on 100,000 job-analysis reports. Before humanology can effectively be brought into the picture, it is necessary for a company with a given operational goal to draw a blueprint of a hypothetical organization which is capable of achieving that goal. This entails breaking down the hypothetical operation into its component parts: the individual skills and aptitudes which, added together, will make the total operation work. For example, operation X will require supervision and operation of such-and-such a type of machinery, a sales manager and a sales force, a legal department, facilities for deliveries, a purchasing department, cost-accounting and bookkeeping personnel, filing personnel, top-line officers, a relations department, and so forth. The next step is breaking each of these various aptitudes and skills into individual jobs. How much work may be expected of one Y-machine operator in a standard work week? How many Y-machine operators will be required to achieve the level of production required for the projected operation?

This is where the *Dictionary of Occupational Titles* (hereafter referred to as *D.O.T.*) is useful. When the organizational planner is attempting to man a specific operation on paper, he can blueprint each of the individual assignments by finding out in *D.O.T.* what would be customarily expected of each potential assignee for how much pay and at what title. He can find out what background of training to look for in a man who might step immediately into a vacancy. Moreover, since *D.O.T.* lists job families, or jobs which are related to each other in skills or aptitudes required, he can tell what hypothetical individuals can be transferred from one job classification to another or can be upgraded with a minimum of additional training.

Limitations of Job Classifications

The job classification process is of course not humanological. Job classifications, however detailed, do not take into account human desires, needs, or true capabilities. The danger in relying too much on *D.O.T.* in prerecruitment planning is that too rigid classifications may be kept in effect even after postrecruitment selection and placement. Management may say in effect: "Look; according to *D.O.T.*, your background of experience shows that you can handle this job precisely the way we have it set up, at precisely the speed we demand, for precisely the pay you are getting. Presumably you are happy."

D.O.T., however, does not even pretend to take into account the fact that the background of experience may have been a background of misery owing to an original misassignment or error in judgment during the period of adult experimentation. Nor does it take into account that even the happily assigned individual may have hidden potentialities not even hinted at in his past job record, or hidden weak spots which have not yet had sufficient time to make themselves evident.

Job classification is indeed useful for *general* purposes in organizational planning, but when the blueprint is about to be translated into human reality, it should be taken with a smallish grain of salt. A table of organization should be humanologically flexible enough to allow the job to be altered within reason to fit the human being, not the human being to fit the job. Here a total health audit is of unlimited value.

Sources of Manpower

The humanological team certainly must be brought into the picture when sources of potential recruits are being determined, evaluated, or reappraised. There are 12 generally recognized sources of manpower; some are better than others for nearly all types of recruits, while some are of particular value for one kind of recruit, e.g., the potential technician, and relatively valueless for other types. A source that has proved excellent at a particular time or place may prove less worthy at a different time or locality. Many sources must be scrutinized with a wary eye at any time or place. There is no standard book procedure for evaluating these sources and keeping the evaluation up to date. This can only be done in the specific instance and at the specific time in question by a team of trained humanologists.

The accepted sources are described and discussed below.

1. *Employees within the organization itself.* — An individual who has been for any considerable period of time with a humanologically sound organization and who has carried out his work assignment with good grace and competence, as well as with recognition from above, will have one intangible asset that no recruit from outside can have: loyalty, *esprit de corps*. It is wise policy, therefore, when a vacancy occurs, to determine first if it can be filled by a present employee, even if it involves considerable upgrading and some extra training. This policy should not be carried to the point where loyal employees are promoted beyond their abilities to perform, but, in general, a member of the company family who can be made to fill the vacancy successfully ought to be considered to have a slight edge over the perhaps more glowingly recommended outsider. His promotion within the organization will have an inspiring effect upon the entire work force.

2. *Persons nominated by present employees.* — This source is good for filling vacancies in the lower echelons, and especially good in the case of raw recruits where a period of training is involved. It ensures a group of workers who are on good terms with one another. In the case of trainees, it minimizes the likelihood of transfer to another firm after

the company has already invested a considerable amount in training them. A worker tends to stay where his friends are, and on-the-job camaraderie adds considerable spice to any otherwise tedious chore. The total health potential of such individuals would normally be a matter of record.

In the upper echelon, this source is fraught with dangers. Cooperation at this level does not involve personal friendship so much as it does a mutual regard for different individual capabilities. An able vice-president can work smoothly and effectively with another vice-president whose business acumen he admires, even though in private life the two thrive in totally different social milieus and have nothing whatever to do with each other. Too close a personal or family friendship among top-line officers can lead to embarrassing situations, particularly when one officer finds himself obliged to question the competence of a lifelong pal, or cousin, or in-law. Of course, these problems are not insoluble. Many fine family firms exist and have learned to deal with such occasional delicate situations. However, most of the very large business organizations have extremely cautious policies regarding ties of family or friendship in high places. Some actually have rigid antinepotism rules, which make it impossible for any immediate relative (closer than first cousin) of any officer to be employed. This creates occasional hardships. The president of a New England manufacturing firm (who happened not to be a stockholder) found that his son was following in his footsteps and showed great aptitude in his particular field. He asked the board of directors to allow a relaxation of the antinepotism rule in this special case. His request was denied, and while the father was able to give the boy a good recommendation to his worthy rivals in the same field, he found that their attitude at first was: "Why is John trying to palm his son off on us? Why can't he take a chance on him in his own company? There must be something wrong somewhere."

Even at the lower echelon there is always at least some danger of cliques developing, involving groups of friends or relatives. A supervisor who is ordinarily effective and who knows how to be tough when the occasion arises can find himself totally at a loss when faced with a clique, the members of which cover up one for another, laugh at him behind his back, and cooperate in doing the minimum of work necessary to keep from being fired. Here again, humanology comes into play. A trained interviewer can usually spot the clique type, an insecure individual who draws his strength from the mob and who out of a conviction of inferiority feels that to survive he must enter into conspiracies to keep the entire group effort at an inferior level. This is the kind of person who protests to his union when he sees one of his fellows on the line putting forth a little more effort than he himself feels he can put forth. Such applicants can be weeded out early. Existing cliques can be broken up; the able individuals among them can be properly assessed, (including a health audit) and put into more responsible positions.

3. *Persons filing personnel applications.* — Such applications may be filled out in a standard manner, according to company policy, at the gate, or they may come in the mail and depart considerably from standard format. In times of depression, both types of applications may arrive in such numbers that it is difficult to keep them up to date or otherwise put them to constructive use. Many personnel managers can recall occasions at the depth of the depression when a vacancy occurred the very person for the job could be found at the plant gates. The odds were so often in favor of this happening that for many job classifications it seemed almost unnecessary to keep applications on file; the needed individual would show up in person when needed.

Such occurrences are infrequent in normal or moderately depressed times. However, there are always several people in any given area who are on the lookout for greener pastures or are temporarily adrift through no fault of their own and who would like to be remembered when a vacancy does come up. The standard cliché of the personnel interviewer: "Please don't call us; we'll call you," is not always an ironic joke. The trained interviewer is able to make an application blank serve as a true picture of a human being, with his significant abilities, limitations, aspirations, needs, and problems. These blanks are filed appropriately and are reviewed from time to time. The value of this source of recruits depends of course upon the skill of the interviewer. No standard form or test can do the trick.

A mailed application may be tantalizing. The applicant may show signs of real ability in his field, but, being unfamiliar with the wants of the company, he may leave out vital information. No responsible firm will fail to answer. *What* to answer may be a problem. Even if there are no vacancies, actual or imminent, it is desirable to maintain a good waiting list. If the individual resides at a distance from the company, it is unfair to ask him to put himself to expense and inconvenience when the vacancy does not exist. Even if the letter from the company makes clear that there is no vacancy but makes the suggestion: "Come anyway as we would like to see what kind of person you are," the applicant may be encouraged to build up false hopes. This will create an unpleasant public relations situation when he finds that there really is neither a vacancy nor prospect of a vacancy.

Delicacy and human perception can avoid misunderstandings of this nature. If there is doubt, it may be safest, humanologically speaking, to send a negative answer, thanking the applicant for his letter, and inviting him to drop in when he happens to be in the area. Should he appear, a total medical evaluation might well be part of the interviewing procedure. This commits no one and hurts no feelings.

4. *Immigrant workers.* — In the past, this was a source not only of cheap help for vastly expanding operations, but also of skilled technical help. In recent years, quotas on immigrants have lessened the significance of this source, although there is still an influx of lower-echelon labor from Central and South America and the West Indies. Occasional crises in the Eastern Hemisphere, such as the 1956 eruption in Hungary, bring about temporary relaxations of quotas. The majority of recruits from this source have lower standards of living than are the rule in this country; in many cases, they have considerably lower standards of health; many are illiterate. For these reasons there is a temptation to exploit them. "They never had it so good" is the exploiters' battle cry.

It becomes the duty of the humanologist to take whatever steps he can to prevent such exploitation, even when the immigrant concerned does not feel that he is being exploited and is more than happy with the low pay and lack of humanological consideration that his employer is inclined to offer him. Although physical fitness for a given assignment is of prime importance, the total sociologic significance is of equal concern.

If the immigrant has come to this country to stay, it is to the best interest of all concerned, that of his employers as well as that of his neighbors, co-workers, and citizen descendants, that he be absorbed as effectively as possible into his new community. The fact that he is happy with a dollar an hour, a verminous shanty without plumbing, shoes with holes in them, gunny sack clothes, and a diet of oatmeal for his rapidly growing family is altogether beside the point. He will not always be happy with this, particularly when his children, under the law, go to school and find out how America lives. However, even if one could predict that he would live out the rest of his days in blissful ignorance of his substandard existence, he is a threat to the health of his community as a whole — physiological, psychological, as well as *social*. Public-health officials may be able to take *some* steps in stemming the infectious diseases that are almost bound to emanate from a thoroughly poverty-stricken industrial environment. In the country, the problems are not so great in this respect. However, a brood of emaciated, perhaps embittered children, who are objects of contempt to most of their associates in school and who cannot help but look upon society as an enemy, will inevitably present the community with problems it cannot solve. The gang wars of the 1920's and the growth of the vast criminal organizations are still potent under the surface of life in the United States. These are a direct result of industry's past failure to try to help the "cheap labor" immigrants of pre-World War I days, to earn American citizenship and to become fully absorbed in American communities, with American rates of pay and American standards of housing, clothing, plumbing, sanitation, and schooling. The Puerto Rican influx of the 1950's offers a challenge to all humanologists to prevent a recurrence of the same social blight.

5. *Rural-urban migration.* — This is an increasingly significant source of recruits. Since World War II, about 900,000 individuals per year have been leaving the farms to enter the ranks of industry. There is no reason to believe that this trend will come to an end in the foreseeable future. The potential humanologist will therefore do well to ponder the implications of the farmer's economic plight, which is almost certain to grow worse. In certain areas of the country where the climate is predictable and water is provided by an abundant irrigation setup, a farmer can live moderately comfortable on 80 acres by growing special crops — notably seed crops or contracted legumes for freezing. However, such areas are rare. The weather must be just right, and the appropriate contracting companies must be interested and on hand. In most parts of the country, a farmer whose

grandfather was comfortable on 40 acres of land (which was about all he could handle with little or no machinery) must now consider himself a "marginal farmer" on 200 acres of land. Years of drought or glut add to his woes. From a large-scale economic point of view, there are altogether too many farmers producing altogether too much for the economy to support. Legislative relief measures can only prolong the agony for the small farmer, while the big farmer finds he can afford to become bigger and bigger with the aid of machinery. One small farm after another is being swallowed up by the huge corporate farming operations; one farm family after another moves sadly to the city. Technological progress in the future can only accelerate this trend, and the humanologist will be wise to familiarize himself with this increasingly significant source of recruits.

No two people are alike, and of course this holds true for farmers as well as urban folk; but there are a few legitimate generalizations that may be made about city-bound farm people. (1) Unlike many of the alien immigrant recruits, these people do not feel that they are substantially bettering themselves. For the most part, they feel that they are moving down in the scale of things. Most of them can remember better days on the farm; they do not adapt easily to the loss of independence, individuality, and family solidarity that seem to be involved in joining the industrial community. (2) They are trading what was to them an ordered, rooted existence for one that strikes them as utterly disordered and rootless. On the farm, life was based on the order of nature, the change of the seasons, the patterned life processes of plants and animals. Home and family were logical entities. There was no significant duality between home life and job life. Each was part of the other. Every member of the family had his share in the necessary work of the farm, which was at the same time home. Family discipline made sense. It was part of the order of things.

In the city, the "home" becomes just a place to sleep. Family members are not essential to each other. Dad has his job. Junior has his. Mom cooks and cleans. Sis goes to high school and runs around with a wild teen-age crowd. The only cement that can hold these people together is fondness one for another. When this is gone, there is nothing. As for leisure, it is a nightmare. No chores to do. No responsibilities. Nothing to do but sit and talk, or read, or — the neon lights of the corner tavern offer at least a partial substitute for a "filled-up" life. Perhaps Dad is proof against them by virtue of years of God-fearing self-discipline. Junior, on the other hand, might be less rocklike in character.

This move from a self-sustaining existence on the farm to a totally dependent one in industry is nearly always like a move over the brink into chaos. Where such migrations have been extensive over a relatively brief period, it has been possible to measure the results, particularly in the second generation. Over-simplification is dangerous. All families do not react badly to the disorganization of urbanization. Families, like individuals, may have inner resources which are passed down from generation to generation. Nor is a move to the city necessarily a break away from the stabilizing influence of religion. Nevertheless, the number of families which have adapted badly has proved to be considerable. This bad adaptation impinges on the community in the form of disrupted families, alcoholism, mental illness, divorces and desertions on an appreciable scale, with resultant loss of productivity on the part of those employed and an added burden on community welfare and public-health agencies.

Much of this grievous disorganization with its subsequent social results can be prevented by constructive measures on the part of the humanologist in dealing with these recruits. As former good citizens, temporarily beaten by circumstances beyond their control, they deserve special consideration and sympathy. All too often a once proud, former farmer, already tortured inwardly by loss of self-respect, is relegated casually to the bottom of the industrial heap ("after all, what industrial experience has he had?") where he becomes an object of contempt or derision or outright dislike owing to his ingrained independence of thought and habit, his individuality and religious piety made strong by years of lonely battle with the elements, and his awkwardness in speech and social give-and-take resulting from daily companionship with dumb animals or farm machinery.

These people need to be recruited with care and evaluated from a total-health point of view. After recruitment, they must be selected and placed (and adjusted to their new environment) with *minute* care.

6. *Women.* — Out of every three workers, one is a woman. This figure will increase. In 1890, the ratio of women workers to the total employed population was approximately 1 to 6.

From a recruitment point of view, the humanologist needs to know first why women work. There are a number of reasons, and it is a dangerous mistake to place all women workers into the same category.

The average woman seeking employment today is not so sure of herself as her grandmother was. Unless she is truly dedicated to a career, her life is full of ifs, ands, and buts. She wants a home, husband, and children, but she doesn't want to be an unpaid domestic *slavey*. She wants to do a day's work for a day's pay, but she doesn't want to be taken for granted by her employer. She is better educated than her grandmother and hence demands, and needs, more variety, more challenge, more food for the mind in her daily routine.

In a way, she is a product of her age. She was called into being by the vast modern corporation with its multifarious activities and its growing need for almost every conceivable kind of human skill — from brawn to technological know-how, from patience to finger dexterity. She is clearly a part of business. The economy would suffer were she to disappear suddenly from the scene. Yet she doesn't know just where she fits into the picture, and neither, it seems, do her employers.

When the invention of spinning and weaving machinery brought about the development of the first crude factory assembly lines, "womanpower" (presumably comprised of dependents who were happy to work cheap) assumed a new significance. Heretofore, weaving had been a man's job, calling for strength, skill, and technical know-how acquired in the course of a long apprenticeship. Now, a great many operations in the mass production of textile goods could be performed by a child.

Women, and children, too, could be expected to be grateful for any kind of wage at all in man's world. Man, being the breadwinner, was ordinarily on the lookout for jobs with at least some future, and employers gave him preference when such jobs were available. Only in times of economic depression, or when he was too old or disabled to get any kind of job easily, would a man willingly take a job which involved drudgery and no prospects of economic betterment.

As technology created more and more desirable jobs, more and more unwanted drudge jobs became available. The male file clerks found there was more money in selling; the male typists found that there was more challenge in operating typesetting machines; the male secretaries found they could escape the whims and tantrums of individual bosses by becoming certified public accountants. Business had two alternatives for filling unwanted jobs at times when the economic situation was good and vacancies plentiful: (a) raise the salary inducement to men; (b) fill the vacancies with women.

By now woman is solidly entrenched in business, but she is still, essentially, cheap labor. There are exceptions; women compete on an equal footing with men in a number of gainful occupations. They can be found in the top ranks of the arts and the professions; women own or manage important business establishments, notably in the field of fashion. But these are exceptions, and rather glaring ones at that. The average woman who works is expected to devote a larger percentage of her working time to plain drudgery than her male counterpart, and she is expected to do this for less money and fewer future prospects. Nevertheless, with rare exception, she is expendable; there are other candidates for her job, and if she doesn't put into it more enthusiasm than the returns would seem to warrant, she is subject to dismissal.

Not only is woman definitely a part of the business world, but the business world is definitely a part of woman's life as a whole. Though she had become important to business 50 years ago, the working woman represented only those of the impoverished strata. Today women graduates of colleges as well as nearly all noncollege women look forward to at least a taste of gainful occupation. Nor is marriage today always the end of a stint in business. No longer is there a stigma attached to the husband of the wife who works. Companies whose policies as recently as 10 years ago were to demand the resignations of women employees upon the occasion of their marriages have been obliged to revamp these policies and tender full recognition to the working *wife*. There is no longer a special type of person who can be called the working girl, whom heaven, it is presumed, will protect. The woman in business is, potentially, any woman in modern society.

Purely from a recruitment point of view, the humanologist in business will learn to assess a job requisition which can be filled by a woman in terms of the type of woman who should be sought to fill it, her motives for working, and her long-term occupational and marital expectations. An able, ambitious, and adaptable woman, married or not, should not be sought to fill a job designed for the budget-wise housewife in need of additional pin money. The time will come when it will be taken for granted that a married woman may rise as high in business as a married man, provided she shows the same competence and gains the same acceptance. This may not be the ideal picture of woman's place in the scheme of things. The average American still sees her through the eyes of his not too far distant rural or semirural forebears: as the center of home and hearth. However, home and hearth in this age have lost some of their former significance, however much one may regret it, and woman's place is wherever she chooses and is able to make it. The humanologist will face this fact. The able girl — the brilliant college graduate, for example — will not long allow herself to be earmarked for drudge jobs, just as the capable Negro has refused with success in recent years to let himself be classed solely as janitorial or domestic help.

The principal source for drudge jobs (and it will continue to be a considerable one) is that group of women — some married, some waiting to be married, some probably never to be married — who are not primarily interested in advancement or titles, but who need the money, either as a sole means of support or as a budget-relieving addition to the family income. The humanological challenge is to insure that job specifications tailored to these women do not make such demands that outside life is more or less pushed into the background. The woman who is in business for advancement may put her outside life in second place and still live a fairly full life. The working wife, the newly mature unmarried girl in an impoverished family, the young widow struggling to augment a meager inheritance and life insurance benefits to the point where it will feed a brood of children, the spinster whose only means of support comes from her job — these and others like them are often in danger of becoming wrapped up in their work to the extent that their private life becomes virtually a void. The results are tragic. Drudge jobs can never offer these women the satisfactions potentially available to them in their outside life.

Encouraging and helping women in this category to develop their nonoccupational selves is a humanological function that comes considerably later than recruitment. However, it is a good idea to anticipate this function in recruitment by seeing to it that job specifications are not unreasonable and ultimately ego-devouring. Total medical evaluation is pertinent to this end.

7. *Older workers.* — The increase in the proportion of people over 65 to the rest of the population is a constant one, which is discussed in detail elsewhere in this book. Suffice it to say here that this source is growing not only in numbers but in value. Measurement of the "average" potential of the 65-or-over worker has not been precisely made and probably cannot be made. Some firms have made it a matter of policy to find openings for their able "65-and-over" workers, who do not wish to retire. These have made an effort to assess individual capabilities and to correlate these findings into general principles and have learned how to set up a number of important job classifications for which recruits can be sought among persons retired from service with other firms.

In time, the fixed retirement age policy (to be discussed in a subsequent chapter) will have to be reassessed. Then the problem of fitting older workers into appropriate job classifications will come more properly under the heading of preretirement or deceleration policy, not recruitment. The majority of older workers not retained in industry will be considered more or less unemployable — for health reasons, perhaps — or because they are ready, able, and willing to be fully retired.

This is not the case at present, however. Every community has a large quota of older workers who are anxious to be on a payroll. One advantage of these individuals is that they have worked before and that their specialized job capacities can be determined by a humanological correlation of their background and of their physiological age. The humanological team that has taken accurate stock of the physiological versus the chronological age of those of its present employees who are aging in harness will have no difficulty in evaluating and placing recruits from the older age group outside the company.

8. *Schools and colleges.* — Except in the case of specific vocational schools, this source is considered to be one for "potentials"; in other words, for individuals who are sought not for what they are, but for what they may become. A comparative grading of such sources is imperative. Statistical school records are to be taken with a grain of salt. One school's standards in one subject may be high and in another subject comparatively low. Standards of individual schools must be carefully appraised by the appropriate expert on the humanological team in terms of the company standards for the type of employee sought.

Most companies have established a policy of interviewing college seniors for openings which may lead to advancement. Interviewers agree that today they often find themselves obliged to answer more questions than they are asked. College seniors are in demand, particularly when the college is noted for high scholastic standards, and they are in a position to interview their interviewers and pick the jobs that seem most promising to them. The interviewer should recognize this as a sign of potential worth. The overly humble, overly eager student who is all too anxious to sell himself to the first interviewer that comes around is a frightened man and risky material at this stage. No company can afford the time and expense involved in straightening him out. This must be accomplished at his own or his family's expense, and a sympathetic interviewer is doing him a service to point this out.

Most companies compete hardest for the so-called campus big shot, who is a class officer, a member in good standing of the best club or fraternity, a member of one or more honor societies, and undoubtedly a holder of a Phi Beta Kappa key, which he is close-mouthed about for fear he may not be accepted as a prince of good fellows. The humanologically sound company will stand back, in most cases of this sort, and let the other companies fight it out. This impressive senior expects, and may well get, a far better position than any properly cautious company will offer any rank beginner, however highly acclaimed. By the time he is 30, if he has not yet been trimmed down to size, this individual will have acquired a reputation (perhaps only partially deserved) for arrogance, for over-weening ambition, and for over-confidence in his own superiority. A kick downstairs is the only thing that can be surely counted upon to put him on the right track — if it doesn't come too late.

It is best to lay the cards on the table with recruits approached at this source and to tell them precisely what kind of person the company needs and what his prospects are if he proves to be this kind of person. Exaggerated claims of opportunities or conditions of advancement will tend to attract the something-for-nothing, emotionally immature type of individual who will become a chronic headache to the company which gets him.

Physical fitness is not a matter of great significance at this source. In many schools and in most colleges, physical examinations are a matter of course. Any deviation from the norm is duly noted and reported. Humanology is solely interested in goals and capabilities.

9. *Labor organizations.* — Humanology cannot enter effectively into this field of recruitment. It is one of the primary kinds of recruitment where a closed shop or strong union situation exists insofar as filling vacancies among the hourly rated ranks is concerned. The humanological problem begins after recruitment, at the stage of selection and placement.

10. *Employment services.* — There are two main types of employment services: (1) private; (2) governmental (in connection with unemployment compensation). The former,

mostly local in scope, vary widely, and each should be scrutinized closely as to its methods of evaluating the raw material on its waiting list. There has been in the past criticism regarding racketeering and "shakedowns" with regard to some of these organizations, but regulatory action on the part of industrial clients as well as government has changed this picture considerably. A really first-rate service can act as an outside extension of the arm of the company humanological team since it may interview many potential employees on a more relaxed basis than anyone within the fearful confines of the company can. Often the individual is so afraid he won't get the job that he is self-conscious and atypically nervous within the company gates, though he can be at his ease with an employment agent who can neither hire nor reject him.

Each of these private services sets itself up as an evaluator of the individual applicant and as an authority on job specifications. The humanologist can compare these services, partly by scrutiny of the actual practice of each and partly by comparison of the results achieved. Those services which have proved themselves to be shrewd in human appraisal and job analysis should be approached by the appropriate member of the humanological team to work out some method for establishing a clear-cut liaison as to techniques of human assessment in terms of job requisitions.

The government agencies were formerly avoided by employers as well as by the better types of recruits. The mortality of private services during the depression changed this situation considerably. Employers and job hunters both were forced to avail themselves of this service in the absence of others. Since the offices of these establishments now act as administrators of unemployment compensation, they offer a means for industry to reach a sizeable pool of individuals in search of employment. Perhaps the most important service these organizations have performed for humanology has been their intensive research in job analysis. Their findings are invaluable to the humanologist in determining how to make the best use of the raw material available to him through the various recruitment sources.

11. *Miscellaneous sources, such as lodges, churches, clubs, prison associations, and the like.* — These involve a liaison between some member of the humanological team and the leaders of these various organizations. By knowing the leader, he can get some idea of the reliability of such sources, which are secondary at best and to be used in special cases (perhaps where an unusual talent or skill is being sought and cannot be found through standard channels). It should be remembered that, in the main, the potentials that can be obtained from these sources are just as obtainable from other, more easily tapped and evaluated sources.

12. *Advertising.* — Most companies find the classified advertising sections of the local daily papers extremely useful in encouraging applicants for specific jobs to come and be interviewed. This usefulness varies from job to job and from paper to paper. In many large cities, classified advertising appeals may be made not only to rank and file recruits, but to candidates for managerial jobs and even top-line officers' positions. In smaller cities, an appeal for a potential top officer may go unread except by the wrong kind of person. Some papers, again, may have more influential advertising columns than others. If an advertisement for a certain general class of employee has never appeared in the classified columns of a given paper, the potential employer may safely assume that the potential filler of that type of vacancy has not acquired the habit of perusing those columns in search of a place and must be obtained from another recruitment source. Humanology cannot do much in this respect. It is a matter of following local customs and traditions, which are easily determined.

Radio and mass-circulation advertising which cover a large area, must be looked upon as potentially dangerous unless a company is seeking to fill a comparatively vast number of vacancies at the hourly rated level. Broadcast appeals of this sort have been known to cause mass migrations, with resultant social problems when the migrants found themselves unwanted in the area and unable to afford to leave the area.

The Positive Approach

By and large, the humanological team will approach all sources with an understanding of their comparative reliability, including the motives for working that may be found therein, and with scrupulous honesty as to what the company has to offer in return for what.

Sometimes there is a tendency to try to attract a great many more potential employees than can possibly be advanced in the foreseeable future. This is done on the assumption that you can't get too much of a good thing. If the recruiting process has been effective and a great many topnotch men have been attracted, considerable disgruntlement may result when a large number of these men have to be weeded out at the company gates. Worse disgruntlement may result if the extra men, the "supernumeraries," are not weeded out at the gates but allowed to rot in frustration in an organization which has no room for them in the posts they are suited for.

The company that wants a healthy, strong, courageous, and able complement of manpower will not exaggerate opportunities, but will seek to attract the individuals who will work long, hard, loyally, happily, and productively and who understand the meaning of responsibility insofar as it is applicable to their place in the scheme of things. Such a company will know how to identify and place the kind of men it needs, and it will do its recruiting with this end in mind.

REFERENCES

COLLEGE PLACEMENT OFFICERS. *Directory of Techniques of College Recruiting.* The Bureau of National Affairs Incorporated, Washington, D.C., 1951.

Dictionary of Occupational Titles. The United States Bureau of Manpower Utilization, Division of Occupational Analysis, 2nd edition, 1943.

FREEMAN, G. L., AND TAYLOR, E. K. *How to Pick Leaders.* New York: Funk and Wagnalls Incorporated, 1950.

LAPIERE, RICHARD T., AND FARNSWORTH, PAUL R. *Social Psychology.* New York: McGraw-Hill Book Company, 1949, pp. 323-327.

PAGE, ROBERT COLLIER, M.D. "Executive Health for Company Wealth." *Dun's Review,* (April, 1952) p. 16.

"Selective Program Cuts Executive Turnover: Studebaker Corporation Promotes from Within." *American Business,* (August, 1949) p. 12.

UNITED STATES DEPARTMENT OF LABOR. *Counseling and Placement Services for Older Workers.* Bureau of Employment Security, BES No. E152, September, 1956.

WAGNER, O. J. M., AND MURRAY, D. N. "Family Disorganization." *Social Medicine.* South Africa: Edited by E. H. Cluver, M.D., Central News Agency Limited, 1951.

WHYTE, WILLIAM H., JR. "The Organization Man." "Business Influence on Education." *The Pipeline.* New York: Simon and Schuster, 1956.

YODER, DALE, PH.D. *Personnel Management and Industrial Relations.* New York: Prentice-Hall, 1948, pp. 151-181.

11

Selection and Placement

WITHOUT A DOUBT, the most crucial occupational stage, from the point of view of both employer and employee, is that of selection and placement. Sound procedures here will make the necessary adjustments to the subsequent stages matters of mere routine. Unsound procedures may blight the entire working career of a promising individual and at the same time adversely affect his personal health potential.

Evidence of the importance of selection and placement is seen in the amount of space and attention given this problem in personnel manuals. Managements of individual firms and management organized in such bodies as the American Management Association have expended more energy and money in the search for a mechanical short cut to proper selection and placement than was ever expended three centuries ago in the search for a Northwest Passage. The result has been a vast array of psychological tests, personality tests, formal interview patterns, numerical classifications and subclassifications of past jobs and/or scholastic record, and algebraic formulae for reducing the elusive human element to a combination of letters and numbers that can be fed into a machine which will spew forth a card describing the perfect square hole for the presumably square peg in question.

No Mechanical Short Cut Will Do the Job

Like the Northwest Passage, the perfect mechanical short cut to ideal selection and placement has proved unfindable.

SELECTION AND PLACEMENT 403

Nevertheless, just as the search for the nonexistent Northwest Passage resulted in the development of a great continent and an extension of the frontiers of civilization, the search for the nonexistent short cut has sparked considerable valuable research into the nature of man and his occupational environment and has furnished humanology with tools which can be highly effective if properly used.

It has been learned — sometimes through bitter experience — that selection and placement are essentially humanological functions. Algebraic formulae and computing machines cannot do this job. It is theoretically conceivable that a machine could be devised which, if given (to the minutest detail) *all* the data about a given person, could determine with accuracy *whether* he should be selected and for *what* precise job. "All" data would include minutiae concerning the individual's hopes, his family life, his scholastic record, his childhood adjustments, his occupational record, and so forth. The machine would, of course, have no way of telling which data to linger over as particularly significant and which data to gloss over as nonpertinent and could therefore tolerate no omissions. Collection of such data would take days, weeks, perhaps even months of tedious questioning, requestioning, testing, and investigating. It is doubtful whether any business enterprise could justify so great an outlay of time and money even for so important a determination as whether and where to place its new recruits.

A Trained Humanologist Is Needed

A trained and skillful humanologist can, on the other hand, if backed by a first-rate humanological team, make a near accurate decision as to selection and placement in the course of one or, at the most, two interviews of timed duration. In a relatively short time, he can determine which data are significant and which are insignificant in the case of a given individual. Moreover, he can readily arrange significant data in logical order, placing proper emphasis on the more important items. A seasoned interviewer will often find the key to a given recruit within minutes, owing to his acquired ability to put first things first and dispense with pointless minutiae. Such first things are:

1. *The selectee's wants.* — The first thing the humanologist will want to know is what the potential selectee wants or thinks he wants out of life. A great deal of time can be wasted if this is not determined at the outset. At the same time the applicant is being selected for placement, he is experimentally selecting and placing himself in the company. If he is not given an assignment that leads in the direction he wants to go, i.e., if no such assignment is available or if the individual's ambitions exceed his true capabilities, he may decide to keep looking until he finds a suitable post, or he may simply grow stale and restive in the unsatisfactory post that is offered him. The interviewer must say to himself: "If I recommend X to such-and-such a job, will he want to stay with it, all other things being equal?" Very few companies can afford to place an individual in a job he will eventually quit, particularly if there is a costly training program involved.

2. *True evaluation of the selectee.* — The next step — the crucial one — is evaluation. Here the humanologist may avail himself of any number of mechanical aids, tests, formulae, and the like, but such devices must be considered as secondary. They only supplement the primary function of individual evaluation, which is based upon judgment and experience.

Interviewing is an art. It is not a cut-and-dried technique. In addition to know-how and a background of experience, the interviewer must be sensitive, understanding, and intelligent. A seasoned interviewer knows that each interview will be in some way different from all others and that it will, if properly conducted, lead in unexpected directions, perhaps revealing truths the interviewer had not intended to look for.

Presumably, at the beginning of the interview, the interviewer and the subjects are strangers to each other. The atmosphere is somewhat strained. Nothing of importance will be learned until an atmosphere of friendliness and mutual confidence has been established. Hence the skilled interviewer will make every effort to provide surroundings that are comfortable and private. His personal approach will be courteous and informal.

The opening conversation will be pleasant and general in nature. A good interviewer is one who knows almost instinctively whether he should carry the conversation himself or let the subject do it. As the ultimate goal is to induce the subject to "open up," the interviewer will endeavor to maneuver the conversation toward some not too personal topic involving the subject in some way or other. For example, if the subject has come from a considerable distance to the interview, perhaps he can be induced to talk about the complications of his journey. He may have had interesting personal experiences in connection with a recent spell of unusual weather — a snowstorm, a drought, or a heat wave. Anyone with a knack for social small talk will know how to handle such a situation.

Throughout the interview the give-and-take atmosphere of normal, unhurried, confidential conversation should prevail. Ideally, of course, the function of the interviewer is to be a listener, but in actual practice, if he seems to listen too attentively and impersonally without joining in the conversation himself, the interviewee tends to become self-conscious. The interviewer should appear to be a friendly, tolerant, interested, not too perfect but thoroughly trustworthy human being with desires and interests of his own. If he fails to project himself, temporarily, as a friend, the information he will obtain will be cautiously guarded and perhaps not altogether reliable. The interviewer has failed in his job.

Occupational Medicine and the Placement Interview

The connection between placement interviewing and occupational medicine may not be immediately apparent since it is not standard practice for the medical member of the humanology team to conduct such interviews or even to concern himself very much with job evaluation. This is considered wholly a lay function. All too often, professional counsel is not sought until errors in placement have begun to produce observable situational or physiological symptoms. This is often too late for constructive advice; it is nearly always too late for inexpensive action.

To correct this unsatisfactory situation, medical consultants in industry have been placing increasing stress on the occupational health aspects of selection and placement. The importance of close teamwork between medical consultant and lay personnel officer at this critical stage will undoubtedly be taken for granted in the progressive corporate enterprise of tomorrow.

Bridging the Gap

The present gap between the medical and lay members of the humanology team at the stage of selection can often be bridged by the enterprising consultant who is keenly aware of his true responsibilities to man, which exist whether or not they are written into company policy. Medical counsel is ordinarily brought into the placement picture in a rather nonconstructive manner in the form of the preplacement "physical" examination. However, the competent consultant will put this examination to more constructive use than may have been intended. Ordinarily, the young recruit is in good physical condition. With rare exceptions, if the young selectee has chronic problems, they are readily correctable or controllable. Decayed teeth can be filled; poor eyesight can be corrected by glasses; chronic respiratory trouble, such as sinusitis, responds to modern therapy. *Degenerative* chronic illness is highly unusual at this stage. Hence the purely "physical" part of the preplacement medical examination may be, to a reasonable degree, cursory. Although it may not be clearly stated in company policy, the consultant will be following his professional training by devoting a considerable part of the examination to history-taking and to determining what physiological and psychological resources the examinee brings to the job for which he is being selected. He will have had no difficulty acquainting himself with the requirements of the given job; this is information to which he is entitled and which he can obtain readily through the employing agency or which he possesses as a result of his personal knowledge and appreciation of innumerable work situations.

The consultant will make it clear to the examinee that he is acting in his, the examinee's, interests. A situation of the utmost delicacy may arise in the event that the medical consultant discovers that the individual in question is not suited, physiologically or temperamentally, for the job he is seeking. If company policy pointedly ignores the role of the consultant in placement evaluation, the consultant can do nothing for the examinee except warn him, for his own good, to seek placement elsewhere. In times of economic downswing, this is difficult advice to give and more difficult advice to take; the applicant may thank the consultant for his advice and take the job anyway since he cannot well afford not to. At best, however, the doctor's advice may at least put him on the alert. The first signals of frustration or physiological or emotional inadaptability to the job will not go unheeded — as they might in the absence of prior warning. The warned individual, though accepting the assignment "in the hand," may continue to keep his eyes open for a more desirable assignment. To this extent, the humanologically minded consultant may feel that he has done his part in the forestalling of a potential human disaster. The other alternative is, of course, to take the easy course, stay within the letter of company policy, and let each department head worry about his own misplacement problems.

The Significance of the First Ten Years of Employment

On the other hand, the management of a given business may have come to understand that the first 10 years of employment are *the* important years and seek competent advice and guidance in placement; for calling in the professional consultant after this period will not serve to correct initial errors in selection, placement, and early training. The professional consultant now finds himself in a position to function in a positive manner as a member of the placement team. His voice will be listened to not only in the placement of a given individual in a given job, but also in the tailoring of the specifications of a given job in terms of the given jobholder. Given such a status, the professional consultant can bring the principles of constructive medicine to bear at a time when they are most effective from the point of view of all concerned.

Handling the Unsuitable

The fact that an individual proves unsuitable for a given assignment need not automatically rule him out altogether. A *positive* approach on the part of all members of the humanology team will result in the determination of *positive* abilities on the part of those being evaluated. Therefore, at the same time that an individual may be classed as unsuited for one type of job, he will be classed as eminently suitable for some other type of work, unless he happens to be one of those extremely rare individuals who is wholly incompetent at any standard occupation. This does not necessarily mean that the individual who fails to qualify for a given placement opportunity will be automatically placed in another job for which he is properly qualified. An opening in the suitable job category may not exist, or someone with more seniority or better qualifications may be next in line for such an opening if it does exist. Nevertheless, the applicant's file will be kept alive, and he will continue to be regarded as a potential recruit, though under a different category.

Filling Out Forms

Policy will probably require the filling out of a certain number of cut-and-dried forms, even in situations where the role of the medical consultant is simply to determine physiologically whether or not the applicant is satisfactory. The conscientious interviewer will, however, keep his form-filling activities to a minimum. In the conversational ice-breaking prelude to history-taking, he will usually find it best not even to take informal notes. He has not yet won the confidence of the examinee, who feels perhaps that he ought to make, but is not making, a good impression. The sight of a pen or pencil taking down his rather guarded and possibly inept remarks will cause him to retreat further into his shell. Later on in the interview, once an atmosphere of confidence has been established, the exact opposite may be true. The interviewee who has been drawn out by an understanding physician whom he has come to regard as a friend will find that he is now enjoying himself, that he has an appreciative audience, and that it is flattering to have some of his remarks taken seriously and written down. He may not be aware that in some cases the interviewer is making notes not of what has been said, but of what has *not* been said. A sensitive interviewer knows that everyone has his own particular areas of privacy which may not be invaded even by a trusted friend. The overeager interviewer who tries to delve his way into one of these areas will find that all his efforts at establishing a confidential atmosphere are for nothing. The interviewee will begin to tighten up, and the interview may as well be concluded. The seasoned consultant will sense when he is heading toward an area of inviolable privacy and will hastily and tactfully beat a retreat before the hard-won *rapport* is lost. This is the explanation of occasional notes on "unspoken remarks" or significant silences.

Psychological Tests

The question of the practical value of psychological tests or personality tests is a hotly debated one. These tests have come in for more than their share of ridicule from independent-minded intellectuals, rugged individualists, and the like, who see in them a means of reducing the many-faceted human being to a combination of letters and numbers. These letters and numbers, when decoded, spell out either "desirable" or "undesirable." Suppose, for example, the code letters for "desirable" are J2, under a particular grading system. It follows that those whose tests are rated J2 must be more or less alike in personality. This puts a premium on sameness. Where, then, the critics ask, is business to find the creative, independent-minded individualists — the Fords, the Edisons, the Westinghouses, the Bells, without whom there can be no real business progress? The so-called "organization man," the "good Joe" who conforms and who dresses, looks, thinks, talks, and lives exactly like the rest of his colleagues may be a good administrator, they say. He may even be a good salesman or staff man. But he is not, they insist, a creator.

The designers of the tests insist that this reasoning is unfair. The purpose of the tests, they say, is not to establish a standard of sameness. There is not necessarily one rating which means "desirable" as against all others which indicate "undesirable." Scientific thinking, they point out, is impossible without classifications and general categories, and the purpose of these tests is to classify individuals into categories which will give evidence of their peculiar capabilities, limitations, likes, and dislikes. If a test shows that an individual is "different," i.e., socially awkward or exceptionally endowed in some notable respect, his rating does not penalize him as "undesirable," but it indicates to the realistic future employer that such an individual will not thrive in a supervisory or administrative slot where the prime requisites are to get along and be one of the boys in his relations with his equals, to be inoffensively correct in dealings with superiors, and to be looked upon by subordinates as "one of our kind, only a little bit better." This does put something of a premium on sameness in certain areas of endeavor, but only in areas such as sales, supervision, or administration, where the "different," "superior," or awkward person could jar upon the people with whom he would have to deal. There are other areas where the "different" person, if he possesses the proper qualifications, may be highly desirable. A good feature of the tests is that they indicate intelligence and ability as well as so-called personality factors; therefore they should be used for placing, not rejecting, the rugged individualist. As for the exceptional creative thinkers, the tests cannot be expected to find them or place them. The tests are based on norms, not on rarities. When the Edisons, Fords, and the like come along, they will find their place — or *create* their place in the scheme of things, tests or no tests!

There is some truth on both sides of the debate. Timid placement policies tend toward the rejection of those discomforting individuals who do not fit into reassuring patterns of safeness and sameness. This is negative placement; its deleterious results in the community as well as in business have been explored by countless sociologists and psychologists in recent years.

However, the designers of the tests are correct when they insist that these situations cannot be blamed on the tests themselves, but must be charged to managements that misuse them. If a company has a positive placement policy; if the purpose of all placement interviews throughout the process of selection is to discover valuable abilities in the interviewee — not merely to weed out the "odd sticks"; and if the tests are used properly as secondary tools simply to help round out the picture of the individual in question, then the criticisms of these tests are not valid. They are invaluable aids.

Significance of Character, Personality, Aptitudes, and Limitations

One thing that should be borne in mind at the placement stage is that, for better or for worse, the employee's character, his personality, and his essential aptitudes and limitations are now established for the rest of his life. From now on he can be trained to acquire and improve skills and know-how, but basically the kind of person he is now is the kind of person he will be 30 and even 40 years from now, except that *then* his limitations will be accentuated, whereas *now* they are camouflaged by the eagerness and stamina of youth. The trained professional consultant knows how to evaluate the bright-faced youngster in front of him in terms of middle age and even old age.

A good argument for the dollars and cents value of bringing a medically trained evaluator into the placement picture is seen in the case of Walter J., a young apprentice printer who was brought into a large metropolitan printing concern.

At selection time, it was determined by a lay evaluator that Walter was intelligent, had a flair for mechanics, and was high in manual dexterity. Had a health assessment been undertaken, it would have been discovered that during his adolescence a number of bouts with infectious diseases, complicated by undernourishment, had resulted in underdevelopment of his muscles. He was frail. Moreover, though he was inclined to be accurate in everything he did, he was not rapid in thought or action; he reacted badly to overvigorous pressure to speed up, and too much haste physically nauseated him.

SELECTION AND PLACEMENT

Walter was adept to learn. When allowed to feed a press or make up forms at his own speed, his passion for accuracy compensated for his comparative slowness. Everything came out right the first time, except in rare cases, which eliminated the necessity for costly and time-consuming corrections. It was noticed, too, that he was learning bit by bit to operate more rapidly. His foreman, not knowing his true physiological and psychological status, not only pressed him to work faster, but required him to carry heavy forms about the shop until Walter nearly collapsed.

Walter continued to do poor work when pushed beyond his limit, and did even poorer work when exhausted from carrying weights beyond his muscular capability. After two months of this he showed up on the absent list. It was ascertained that he was home "sick." He said he had the shakes.

Walter never returned to his old job. It was learned later that he had applied for a job at a printing plant which was more scrupulous in its policies relating to placement evaluation. An interview with a medical consultant had cast light on his basic difficulty. Management at his new plant had allowed him to develop his skill at a speed normal for him and had not required him to overtax his inadequate physique. He developed into a highly respected journeyman printer with supervisory responsibility and was worth many times the salary paid him by his company.

An analysis of the costs to the former company resulting from this error at the stage of placement would have looked like this:

Loss		Gain	
Total wages paid	$500.00	Salable work produced by Walter	$125.00
Foreman's time for training (80 hours at $2.25 per hour)	170.00		
Work damaged by haste	90.00		
Wages paid replacement to bring him to same stage of training	500.00	Salable work produced in same time by replacement	110.00

(Table continued on next page)

Loss	Gain
Foreman's time training new man 170.00	
Work damaged by new man 35.00	
Totals $1,465.00	................$235.00
NET LOSS $1,230.00	

In other words, because it could not see where the medical consultant fitted into the humanological team at placement time (though it had an able and conscientious lay interviewer), this company literally misspent $1,230.00 in two months — all on account of an employee at the lowest echelon. One can imagine how such waste multiplies in higher echelons.

Underplacement and Overplacement

In evaluating a candidate for placement, it is well to remember that *underplacement* can be as dangerous as *overplacement*. In other words, while it is generally recognized that it is bad business to put a man in a job that will overtax his abilities, it is less well known but equally true that harm can result from placing a man in a demeaning job that will in no way challenge his abilities.

The skillful evaluator will, however, take this truth, with a small grain of salt. In placing an able employee, it is not necessarily wisest to put him in *the* position commensurate with his abilities, but rather to put him in a more elementary job that will lead, stepwise and logically, toward the higher post. In the foregoing chapter on recruitment, there was a brief discussion of the problems involved in obtaining the highly qualified, often rather conceited, "big man on the campus" who expects to step right out of college into a highly paid managerial or supervisory position.

John P. McT. was just such a college graduate, and his case is illustrative.

John was one of the head men in his class socially as well as scholastically in a leading eastern university. His parents, who had undergone a number of financial struggles and who felt somewhat inferior socially, were most ambitious for John and had stimulated him to outshine all of his acquaintances in every way possible. As a result he was younger than the average at graduation time.

He was eagerly sought after by a number of firms because, though his exceptional background did not put him in the category of the "good Joe" type of conformist, he had shown evidence of definite leadership ability in various campus extracurricular activities, including athletics. His only apparent drawback, to some of the industrial recruiting people who interviewed him, was that he seemed to have a somewhat contemptuous attitude toward his intellectual or social "inferiors" and that he had a tendency to judge and find fault with others. A number of companies backed away from him for this reason, but a considerable number of major companies still felt he was worth bidding for. The best job offer was made by a large financial house, which offered to start him at $6,000 a year as right-hand man to one of the line officers whom he would, in due course, replace.

As it happened, John had a number of cultural and recreational interests in common with the line officer to whom he was assigned. This was not an "act" on John's part, but a genuine lucky break. John purchased a home in a suburb near the home of his senior officer, and the two became good friends. For this reason the officer was blind to an air of arrogance which John displayed in his dealings with subordinates and which was widely discussed and resented.

In due time, John's senior officer was promoted to a higher office which involved a transfer to a major branch office in another city. A junior vice-president was appointed to replace him. John, who had been confident that he would be the replacement, stayed where he was, as special assistant to the newly promoted junior vice-president, whom John looked upon as inferior mentally, culturally, and socially. While he made no attempts to express his feelings in this matter, he was not able completely to conceal them.

His new superior was under no illusions regarding John's arrogance, which was a disrupting influence among immediate subordinates and which had at one point caused a valuable secretary to burst into tears and resign. Moreover, he shared none of John's interests, a fact which afforded John a modicum of contemptuous amusement. No one could have been more surprised than John was when it was announced to him, four years almost to the day of his original employment, that he was expected to hand in his resignation.

"You are a valuable man, John," he was told. "With your grasp of financial matters and intelligence you will go far in some financial house once you have learned not to ride roughshod over people, but the disruption your attitude has caused here makes it impossible for us to keep you any longer."

John was staggered. He even made a special trip to the city where his former superior had been transferred in a last-straw attempt to obtain a reversal of management's decision. His old friend was embarrassed but unable to help.

John spent over a year making the weary rounds of all the companies that had bid for his talents during his senior year at college. The personnel staffs of each of these had had wiser second thoughts after their first try to secure his services. Perhaps a rumor regarding the circumstances of his resignation had gone the rounds, making John the victim of a polite blacklist. He will never know for certain. Finally he was able to persuade a smaller financial house in a midwestern city to take him on — conditionally — at a nominal salary and at a basic level which would lead him to a suitable position, provided he moved ahead at a reasonable pace and made an effort to fit in. Thoroughly humbled, John did not only put his nose to the grindstone, but made an effort, eventually successfully, to appreciate the true merits of people at other intellectual, moral, or cultural levels. He is now a line officer with this company, on his way to the top if some larger company doesn't pirate him away. He is thoroughly grateful to his former superior who had "fired" him and kicked some insight into him when he was still young enough to benefit.

Had his first company felt obliged to continue to put up with his attitude for the sake of his real abilities, he would have become hardened in his "sins" to the point where, perhaps at the age of 45, the disruption of his subordinates would have become impossible to overlook. His dismissal at such a late stage would be a gateway to nowhere. The early detection and interpretation of his limitations by a supervisor with authority would have prevented this dilemma.

Misassignment Produces Medical Symptoms

A law that should be as fundamental to the humanologist as the law of inertia is to the physicist is that *any gross misassignment will eventually produce medical symptoms.* This is also true of any unrelieved series of occupational disappointments, such as are experienced by the salesman who finds himself confusingly in an area of unusual sales resistance, the entrepreneur whose well-planned investments fail to yield a return one after the other, the writer whose frantic efforts earn him only a drawer full of rejection slips, or the loyal underpaid subordinate who sees one promised raise after another vanish into a jungle of plausible excuses. However, the subject of disappointment belongs properly in a subsequent chapter covering motivation.

In misassignment, a corollary to the above-mentioned law is that the least obvious cases of misassignment (those in which the individual appears to have made a good adaptation) are the ones in which medical symptoms are most likely to appear. In the obvious cases, either the misassignment is observed and corrected at an early stage, or the individual in question, being fully aware of his predicament, quits, blows his top, or discovers some more or less effective method of compensating. Occupational slang for this last procedure is "letting off steam" or "getting the poison out of one's system," which may involve sports, hobbies, politics, or other after-hours activities not necessarily having to do with the use of alcohol.

In the nonobvious case, however, where there is an appearance of smooth adaptation, the individual himself may not be aware of his misassignment. A veil of self-deception will hang between his conscious mind and the subtle damage he is doing to himself. If no one else happens to observe what is happening, this individual will continue to push himself blithely into the performance of tasks against which his entire inner being rebels. Eventually this inner being or subconscious will strike back in the only way it knows how, by producing a symptom of illness to warn the misassigned that it is time to stop, look, and listen. The same habits of self-deception which he has acquired in order to stick to his job will enable him to shrug off the first minor symptoms of potential functional illness. Even if he receives constructive medical advice at this point, he may refuse to accept it or act upon it, assuming that these symptoms, like the symptoms of his boyhood illnesses, will in time just go away. If by this time he is fairly well advanced into his middle years, this creeping loss of health will be progressive and, when he is finally forced to recognize his true predicament, irreversible.

The Significance of Stress

During the Korean War, a number of studies of stress under different types of combat action were studied by teams of scientists in cooperation with the armed forces. One group of men was thoroughly studied before and after an extremely violent but comparatively short-lived battle in the course of which a strongly fortified enemy hill position was captured, with a large number of casualties on both sides. Another group was studied before and after a long but comparatively less dangerous action during which it held a strongly fortified defensive position against a protracted siege by the enemy involving aerial and artillery bombardment augmented by a series of unsuccessful ground attacks. Casualties in the latter group were low.

Particular care was taken to note chemical changes in the blood, urine, and secretions of the ductless glands. It was noted that in the first group chemical changes were radically marked, particularly in the secretion of the adrenal glands. In the latter group, changes were present, though not quite so marked as in the first group; particularly noteworthy was a change in the blood count. Perhaps the most remarkable finding was that the extreme chemical aberrations in the first group corrected themselves in a relatively short time during a rest period behind the line, whereas some of the changes occurring in the group subjected to a less dangerous, less violent, but more protracted siege lingered for weeks. The Korean War terminated before it was possible to make extensive investigations into the psychosomatic significance of these changes in terms of actual performance or to study the physiological effects of a *series* of violent battles or protracted sieges.

Physiological Limit and Emotional End Point

The foregoing studies seem to point toward scientific verification of the clinically observed truth that while the average individual has an astonishing ability to get used to a variety of highly unpleasant situations, there is a physiological limit to how long he can *stay* used to any given situation.

Studies of employed individuals reveal further that the individual and his environment are constantly changing in their relation to each other. A job may be so planned and designed as to be a constant, although this is both rare and in most cases impracticable. No human being is a constant, as is made evident by the following examples:

Mary, who seemed to have been sent by heaven to the advertising department of Hooligan's Department Store 10 years ago, is now so fed up with writing the word "sale" in big block capitals that her attention continually wanders and she can't concentrate on the simplest assignment. She is in a rut. She has grown older and wiser, but the job hasn't grown with her.

Jack, on the other hand, who muffed his first sale of an insurance policy and who had butterflies in his stomach every time he rang a stranger's doorbell, is now the leading salesman in his division and expects to win the regional award this year. He has grown into his job.

Bertha G. had a reputation in her firm for being a fast and accurate typist; no one could dictate rapidly enough for her to get it down incorrectly in shorthand. Today she still likes her job and envisions herself as being as good as ever, but a series of crushing problems at home, complicated by the fact that she is going through her menopause, have virtually incapacitated her. Her employers are embarrassed about pointing out her innumerable mistakes to her and are wondering how best to retire her.

Bill Y. proved to be the perfect man to handle the Corporation X account for his advertising firm 10 years ago. It was rugged work involving strenuous travel, late hours, and an iron stomach for Martinis and the same old, tired-out musical comedies and night club floor shows. Half a year ago he suffered a mild coronary attack. He is by no means seriously incapacitated. He is as good an advertising man as ever, but the kind of work he was doing for Corporation X is entirely out of the question.

Evaluation and placement are therefore continuous processes. The prevention of illness owing to a change of relationship between employee and job can only be accomplished by a periodic evaluation of every job in any given business. Likewise, the job evaluation must be correlated with a complete audit of the individual performing the given job.

Fatigue

One of the first things to look for in assessing a job situation is fatigue. The factors producing fatigue will vary from individual to individual in identical job assignments, but in any event, whether the fatigue is the result of the individual's changed attitude toward the job or the result of something inherent in the given job or individual, the conclusions to be drawn are the same: fatigue is evidence that someone was not properly placed at the time of evaluation, and either a new placement or a new set of job specifications is usually indicated.

Fatigue of the central nervous system plays a much more important role in industry than does strictly muscular fatigue. The following aspects of fatigue were noted in the course of a series of industrial research projects focused on the problem:

1. An individual can feel extremely tired without having exerted himself.

2. Fatigue can disappear abruptly if something interesting occurs.
3. The mere thought of doing certain types of work creates a sensation of fatigue.
4. Fatigue can develop quickly in unpleasant social situations.
5. When tired, an individual makes mistakes more frequently; the mistakes in turn increase tiredness.
6. After a day of pleasant physical activity one may not feel tired but be ready for further activity.
7. In an emergency situation, one can undergo unusual emotional and physical strain without feeling correspondingly tired.

Studies made in stepped-up war-production industries during World War II revealed the following causes of fatigue:

1. *On-the-job factors*
    a) Excessive speed-up of work
    b) Boredom due to repetitive work
    c) Awkward movements
    d) Lack of properly spaced rest periods
    e) Improper posture
    f) Excessive noise
    g) Excessive heat
    h) Inadequate illumination
2. *Off-the-job factors*
    a) Loss of sleep
    b) Intemperance
    c) Emotional disturbances
    d) Inadequate nutrition
    e) Drugs
    f) Infections or other diseases

Fatigue shows itself in a number of ways. Many of these may be accurately interpreted by an alert supervisor and referred to the appropriate member of the humanology team for corrective action. This will be discussed in detail under the heading of "Supervision" in a subsequent chapter. Some of the signs of fatigue can be interpreted by both professionally and nonprofessionally trained individuals. The experienced plant nurse, for example, is expected to be skilled at recognizing the underlying fatigue factor in a number of physiologic and psychologic complaints which are dealt with at dispensary level.

The symptoms of fatigue are as follows:
1. Decreased quantity of work
2. Decreased quality of work
3. Increase of accidents
4. Physiologic and psychologic effects
    a) Digestive disturbances
    b) Loss of muscular control
    c) Insomnia
    d) Backache
    e) Vague debilities (possibly increased susceptibility to respiratory infections)
    f) Irritability and unstable emotions, stubbornness, faulty decisions

When a given individual shows signs of fatigue, the first function of the humanologist will be to determine whether it comes under the heading of "supervision" and is either temporary or readily correctable, or whether it comes under the heading of "placement" and calls for a complete change in work assignment or work environment. Only situations in the latter category need be discussed in the present chapter.

Poor Placement as a Cause of Fatigue

Correction of a situation in which poor placement is the cause of fatigue calls for considerable tact, particularly if the situation is one in which the individual was formerly highly competent at his particular assignment. The abovementioned case of Mary, who had become bored with her job in the advertising department of Hooligan's Department Store, may be examined in the light of two possible hypothetical methods of arriving at a solution, one wrong and one humanologically sound.

SELECTION AND PLACEMENT

To illustrate the first alternative we shall assume that Mary was approached by a tactless superior after a thorough study of her case had been made by a medical consultant in cooperation with her immediate supervisor. This superior, though tactless, was not an unkind person. He was truly sympathetic and meant to be helpful. He invited Mary to join him in a cup of coffee at an isolated booth in the store's lunchroom. On the table, he had laid out several of Mary's recent advertising layouts, each of which showed some glaring error or omission owing to inattention caused by fatigue due to boredom.

"Mary," he said, in as gentle a voice as he was able to muster, "we want to help you. We know exactly what you're up against . . . ."

Immediately Mary sensed that some criticism of her work was in the offing. She became tense and defensive.

"What do you mean, 'up against'?" she asked. "What's wrong?"

With a show of patient resignation the superior pointed to the mistakes, circled in red pencil.

"Look, Mary," he said. "Mistakes like these are not like you. You did better work than this when you first joined us, with no experience at all. Now, we know your problem. We know that you're getting stale in this position. It's beneath you, and it's boring, and as a result, of course, you . . . ."

Here Mary broke in tearfully:

"What you're trying to tell me is that I'm no good at my job. That's not true. It really isn't. I can explain every one of those mistakes. You think it's all carelessness, but it isn't."

She began to "explain" each error in terms of an incorrect order from her immediate supervisor, or a sudden emergency in the advertising department which drew her attention away from the drawing board at a crucial time, or a bad headache, or an out-and-out forgery on the part of someone who had access to her layouts after they had left her department.

Her superior raised his hand. "Look, Mary. Let's not quibble over these matters. What's done is done. What we have to think about is the future."

"I promise you I won't make any more mistakes like these. I'm on my guard now. Every time Harry (her supervisor) gives me instructions for a layout I'll double-check on every detail before I put anything on paper. If lightning strikes or if somebody falls downstairs, I'll keep my eyes glued to the drawing board, I promise."

"Of course, Mary. You say that and you mean it, but mistakes will go on happening unless we can get you away from that drawing board."

"I get it," she said. "You want me to resign." Her voice was hostile.

"Not at all. What you need, Mary, is a change of background. Actually you are capable of moving up to a higher position in your own department, and that would be the best solution. But unfortunately there's no opening there. We can't fire Harry just to make room for you. The second best solution is to move you to a challenging job in some other department, where you can display your abilities and at the same time benefit from a complete change of pace and scene. I think we have the ideal opening, and I'd like to talk it over with you."

"Sur-r-re you have," Mary said sarcastically.

The change in placement was made since it had to be made, and Mary accepted since she had to accept or resign, but it did not turn out to be the constructive change it had been intended to be. Mary's impression of the talk with her superior was that its sole purpose had been to scold her about her mistakes. She looked upon the change in assignment simply as punishment for poor performance, not as a boost to her morale and productivity. She was self-conscious since she felt that everyone in the store knew the reason for her transfer, and her attitude toward the store management was one of resentful defensiveness. She had passed the point where she could respond to constructive placement in this particular company.

To illustrate the second, constructive, alternative we shall assume that the officer who invited Mary to chat over a cup of coffee was an expert in humanological methods, seasoned in dealing with troubled individuals and liberally endowed with the essential tact. This officer had seen all of the red-circled layouts which had been the first symptoms of Mary's predicament, but he had no intention of bringing them to the table or even of mentioning them. They represented water over the dam.

"Mary," he said pleasantly and matter-of-factly, with no attempt at unctuousness, "you've outgrown your job. Anyone with eyes can see you've been far too good for it for some time past, and we feel that a promotion and a raise in pay is long past due."

Mary beamed. He was telling her what she knew in her heart: that her job was beneath her. He went on:

"By all rights you should be promoted in your own department, but unfortunately there is no opening for a promotion there now. Some day there will undoubtedly be one which you can and should fill if your interests haven't taken you on a totally different tack. In the meanwhile, there is a challenging position in one of the other departments which we feel you can handle capably with perhaps a month or so of orientation. This transfer of course amounts to a promotion, and there is a raise of pay involved."

Mary was extremely proud of the confidence shown in her ability and went to her new assignment with vigor and enthusiasm. As predicted, she had little difficulty orienting herself to a new type of work. It is predicted that she will become head of her present department before the top vacancy opens up in the advertising department. The department store and Mary have benefited.

Situational State

Correction of a simple fatigue situation involves little more than a modicum of tact, patience, and a willingness to be of help. However, if these are not brought into play quickly, at the first evidence of the problem, the fatigue symptoms will become more complex, more deeply rooted, and more pronounced. The individual in question may then be considered to be suffering from a situational state.

The term "situational state" is applied to the seemingly very real but undiagnosable complaints affecting persons who have reached their "end point." They have been subjected to as much stress and maladjustment as they can stand, and since they are powerless to change the condition of stress or maladjustment, they take refuge in sickness.

To a certain extent, everyone is subject to situational states, but some people are less so than others. Some have a capacity for enduring more stress than others or a talent for escaping from stress-producing situations before damage is done.

The average individual is not likely to be driven to the end of his rope by the usual doses of family friction, work pressure, or financial trouble. Although everyone is theoretically subject to a situational state, there is no danger unless one of two things happens: (1) Some of the usual day-to-day annoyances and irritations expand immoderately and unendurably. (2) The individual somehow has lost his usual ability to take his daily dose of friction in his stride.

Symptoms of Situational State

Any and all symptoms of organic disease may indicate a situational state.

The author recently participated in a study of 100 situational states discovered in an employee group served by one medical department in a leading United States corporation. Each case was studied under 28 different headings concerning

SELECTION AND PLACEMENT 427

past history, present history, working and home environment physical findings, and complete x-ray and clinical laboratory evaluation.

The common factors observed are indicated below in three graphs.

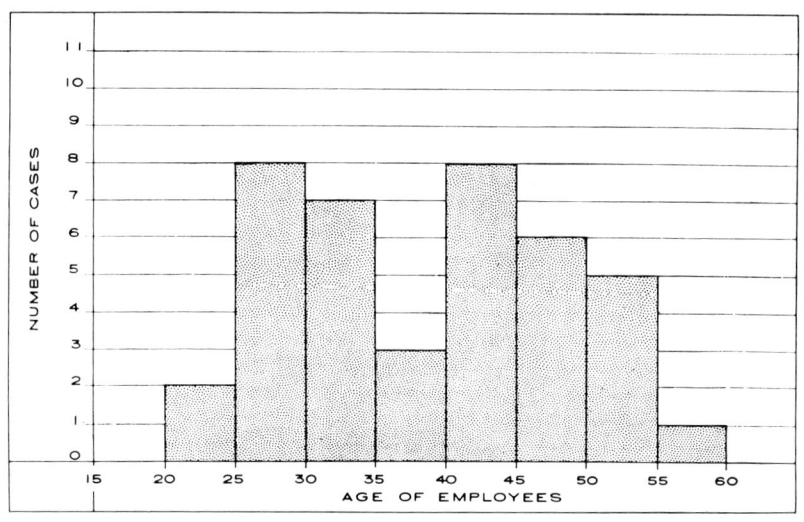

Graph XI — SITUATIONAL STATES—AGE OF EMPLOYEES

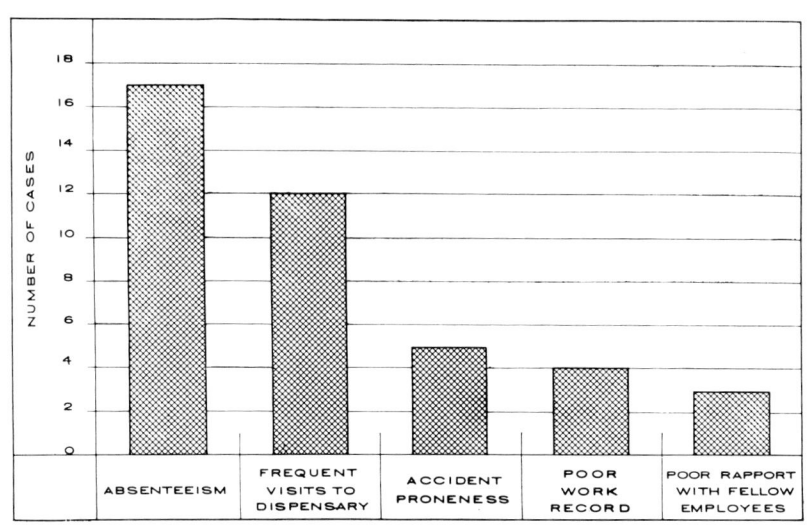

Graph XII — SITUATIONAL STATES—BY REASONS FOR REFERRAL

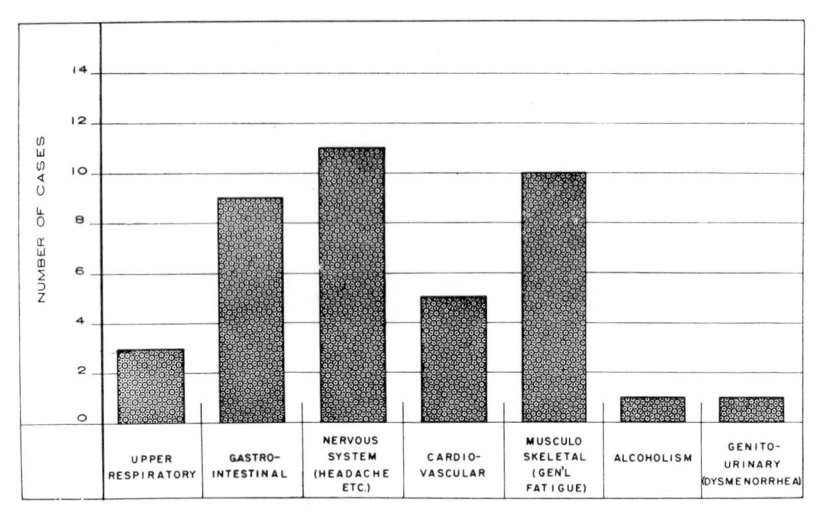

Graph XIII — SITUATIONAL STATES—BY AREA OF PRESENTING COMPLAINT

An analysis of the cases reveals the following:

1. While the group varied in age from 25 to 62 and in length of service from three to 37 years, *over 80 per cent of those studied* were within the age group between 25 and 40 and had from 3 to 15 years of company service. These facts indicated that the situational state is a problem of the young and relatively short-service employee who should still be salvageable and capable of many more years of useful, happy service after removal of the causative problem.

2. The past history of a significantly large number of these employees included the problem of a broken, or nearly broken, home in which the parents either had been divorced or lived together in a state of constant antagonism.

3. The presenting complaints, made by either patient or management, which brought these individuals to the attention of the medical department were most commonly (*a*) absenteeism and (*b*) frequency of visits to the dispensary.

The prime offender in the latter category was a 25-year-old girl with seven years of company service who had over 300 visits to the clinic during that time. Her complaints centered largely about her upper respiratory system. Complete work-up of her nose and throat, paranasal sinuses, and lung fields revealed no abnormalities or diseases in these areas. Once the patient's confidence had been gained, however, a free discussion of her home environment revealed the presence of an aggressively selfish mother who on two occasions had broken up this girl's engagements and thus prevented her from leading her own normal, happy life.

4. Ambition would seem to be common to most patients suffering from a situational state. Over 75 per cent of the situational problems were found in employees whose ambitions for rapid advancement in their jobs were well known to management. Unfortunately, in many cases this ambition was not accompanied by the mental capacity nor the ability to accept the hard work and sacrifice which is a prerequisite to major advancement.

5. In *none* of these cases, despite the fact that each was worked up completely, was there found an organic disease which could have accounted for the symptom or complaint which brought these patients to the attention of the medical department. Graph XIII, on the opposite page, shows the presenting symptoms, all lacking an organic base.

Case Histories

The following are actual case histories illustrative of the above-mentioned aspect of the situational state:

a) Male, age 40, with 20 years of company service, suffering from severe asthenia thought to be due to tuberculosis, but in reality based upon dissatisfaction with the part of the country in which he was currently working and a deep desire to return to another section.

b) Female, age 25, with seven years of company service. Made 300 clinic visits in seven years because of preoccupation with diseases of upper respiratory tract. Discussion revealed an unhappy home situation having to do with selfish mother. The patient had reached the point where she felt trapped by her mother's complete dependence upon her.

c) Male, age 25, with five years of company service. Complained of insomnia and increasing nervousness, owing to complete dissatisfaction with present position. He was unable to advance in his job because he lacked the education to go higher, yet he refused to take advantage of company policy, which would have aided him in obtaining his education.

d) Female, age 36, with eight years of company service. Presenting symptom: extreme nervousness and inability to concentrate on her job. The situational state was found to have stemmed from two factors: (1) an unhappy home situation; (2) an unfortunate job situation in which she was given a position of authority for which she was not suited.

e) Male, age 32, with six and one-half years of company service. Presenting symptom: hypertension found to be secondary to a situational state based upon the employee's presumed superiority to all working with him in his office.

f) Female, age 35, with three years of company service. Referred to the medical department because of physical attacks upon other members of her department. The situational state was found to be due to an unhappy home life involving the presence of a demanding mother and a parasitic older brother.

g) Male, age 42, with 19 years of company service. Presenting symptom: epigastric pain suggestive of duodenal ulcer. The symptoms were found to be related to a situational state caused by an extremely unhappy home problem; the patient was afraid to ask his wife for a divorce.

h) Female, age 31, with 10 years of company service. Presenting symptom: complete withdrawal suggestive of schizophrenia. A situational state was found which was based upon a secret love affair which this employee was having with a married man.

i) Male, age 41, with 18 years of company service. Presenting symptom: easy fatigue and loss of memory. A situational state was found based upon an unhappy working environment. This employee had been doing the same relatively uninteresting type of work for 18 years and for five years had been attempting to obtain a transfer to a position which he would find more stimulating.

j) Female, age 29, with six years of company service. Presenting symptom: frequent indigestion and epigastric pain. A situational state was found based upon the employee's deep resentment of the fact that her assignment called for more overtime than did the assignments of the other employees of her department.

SELECTION AND PLACEMENT 431

k) Female, age 40, with 14 years of company service. Presenting symptom: pain and "mass" in breast. A situational state was found based upon a home problem; her father had convinced her that all unmarried women of 40 or more died of cancer of the breast.

l) Female, age 52, with 13 years of company service. Presenting symptom: easy fatigue and tearfulness. A situational state was found based upon an unhappy working environment. This employee was secretary to an executive who in 13 years had never complimented her on her work.

m) Male, age 51, with 21 years of company service. Presenting symptom: easy fatigue, nausea, and vomiting. A situational state was found based upon the employee's having a completely domineering wife. This man was used by his wife as a household drudge in spite of the fact that he had a position which required long hours of work.

n) Male, age 39, with 21 years of company service. Presenting symptom: joint pains and easy fatigue. A situational state was found based upon an unhappy home environment. Due to his wife's tremendous ambition, this employee had gotten himself so far into debt that it was impossible for him to raise enough money to pay all of his bills.

o) Female, age 31, with five years of company service. Presenting symptom: chest pain and dyspnea. A situational state was found based upon the fact that this employee was single but had been keeping steady company with a married man for the preceding three years.

p) Male, age 56, with 31 years of company service. Presenting symptom: chest pain and dyspnea. A situational state was found based upon the fact that this man had a deep fear of retirement since so many of his friends had died within the first year after they had retired.

q) Male, age 44, with 17 years of company service. Presenting symptoms: headaches and insomnia. A situational state was found based upon an unhappy home and working environment. He was separated from his wife and unhappy with his job because he felt that he had wasted 17 years in which he could have been a teacher.

r) Male, age 47, with 17 years of company service. Presenting symptom: lower abdominal pain and diarrhea. A situational state was found based upon an unhappy working environment. This man had great potentialities as a public relations expert, but he had spent 17 years in a job in which his talents were completely wasted.

s) Male, age 33, with eight years of company service. Presenting symptom: headaches, insomnia, and tremor of hands. A situational state was found based upon an unhappy working environment. This man had a phobia concerning flying, but he was forced to fly many times a year because of his job.

t) Female, age 46, with 10 years of company service. Presenting symptom: diarrhea and lower abdominal pain. A situational state was found based upon an unfortunate working environment. This woman was convinced that for reasons of birth and breeding she was completely superior to all of her office associates.

u) Male, age 40, with 14 years of company service. Presenting symptom: frequent colds, nervousness, and irritability. A situational state was found based upon an unhappy home environment. On two occasions this employee had discovered his wife in adulterous relationships. He continued to live with this woman only in an attempt to keep his family, including three children, together.

Approach to Therapy

In none of the foregoing cases was an attempt made to outline a definite plan of therapy. In view of the variety of presenting symptoms and causes, it was concluded that in handling the situational state no clear-cut, routine textbook type of approach was feasible. However, the following approach is recommended for all situational problems:

1. The confidence of the patient must be gained through mutual trust and the clear understanding that information discovered by the professional team will remain confidential.

2. The professional consultant, physician, registered nurse, or clinical psychologist, when confronted with a situational state, must endeavor to aid the patient in bringing practical common sense to bear on the problem. This will be found effective in the great majority of such cases, and there will be little need for referral to outside consultants specializing in emotional disorders.

3. Where an unhappy working environment is found to be the cause of the problem, the cooperation of management should be sought in working out a solution, provided the patient's consent has been obtained for an open discussion of the problem.

Restoration to good emotional health may often take place completely within the office of the staff nurse, provided the proper cooperation is available from all members of the humanology team. A review of the cases cited above will make it obvious that the majority of situational problems are of the sort which the old family doctor would have solved with ease because of his intimate knowledge and appreciation of his patients' environment, both at home and at work.

It must be understood that the average situational state would not present a true occupational health problem if it were not for the fact that so many people have a tendency to kid themselves. They refuse to recognize the reasons for their symptoms and are often insulted when a physician or a nurse, in her role as counselor, attempts to explain that their symptoms reflect a situational state rather than a true organic disease which requires definitive medication.

The individual who cannot or will not analyze himself to find out what daily problems are causing his symptoms will eventually become considerably sicker. From time to time, as his occasional or intermittent spells of ill health cause him to retreat from his problems, he will enjoy temporary respites from illness, but as soon as he is feeling better, he will go back and subject himself to the same old stress again, and the process will repeat itself.

He will take patent medicine, go from doctor to doctor until he becomes thoroughly confused, and eventually, unless he acquires personal insight somewhere along the line, his spells of recurring illness which now cause him to be frequently absent from work may well develop into a true functional disorder and ultimately into organic disease; or they may lead him down that one-way street into a psychiatric state.

These complications will be explained and discussed in detail in their appropriate place. (See page 542.)

It must be understood here that the situational state must be detected early. By far the great majority of these problems can be solved soon after their onset at the simple medical or paramedical consultant-patient level. This implies, of course, that there is mutual appreciation and singleness of purpose among all persons in industry whose interests lay in the realm of humanology. The professional consultant in attempting this work will find it highly important to work on a team basis with the paramedical staff, supervisors, and members of personnel departments.

Nonoccupational Disability

The problem of nonoccupational disability (where the symptoms have a functional, organic, or otherwise tangible basis) does not come properly under the heading of selection and placement. This highly significant problem will be discussed in detail at a later stage. However, there are some aspects which relate to placement and which should be taken into consideration in the preplacement study of the new employee.

It should be emphasized that the pre*placement* examination is not a pre-*employment* examination. Presumably, the individual in question has been selected as an employee. The problem is to place him in a position in keeping with his total medical potential and mental capabilities.

The placement aspect of this problem is illuminated by a study of 132 applicants for positions in one branch of a major United States corporation.* Sixty of the applicants were found fully qualified for all types of work. Of the 72 others hired, 47 had conditions which necessitated periodic re-evaluation since there was a likelihood of a future need for treatment or adjustment of employment. The remaining 25 were

---
*Socony-Vacuum Oil Co., Inc., East River Plant, 1950.

hired in spite of the fact that each had at least one condition calling for immediate medical correction. These were given six months time to correct or control their medical problems. Several had carious teeth, uncared for owing to lack of funds. Others who had visual defects had been unable to afford eye examinations or glasses. The company making this study acted on the assumption that industry has a responsibility for taking on some of the social burden of employing individuals with substandard health, provided that the health problem can, with reasonable care, be controlled or corrected, and provided that a job can be found which will not aggravate the condition. However, it was felt that it would be financially impossible to be indiscriminate in this respect and employ individuals who might be expected to be absent frequently owing to illness. Moreover, it was felt that the majority of newly employed workers should provide the energy and production that would make possible the catering for older, long-term employees who were being slowed down in anticipation of retirement.

Importance of Placement Examinations

The placement examination is valuable in detecting and reporting such conditions as hernias, ankylosed joints, amputated digits, scars, and other deformities which, if not noted, may become the basis for unjust disability claims against the company, either as a primary or as an aggravating injury. More important, however, is the need to evaluate these defects in order to avoid appointing individuals to jobs where the disability may prove to be a handicap or a hazard.

Complexity of Nonoccupational Disability

Nonoccupational disability is a complex subject with a number of ramifications which will be illuminated further under the headings of "Supervision" and "Relinquishment of Power."

It may be mentioned in passing that, just as the individual may change with respect to his job, the job may change with respect to the individual. Technology is continually producing new tools and new techniques which can completely alter the nature of a job. Occasionally this creates a placement problem which is readily solved by the above-mentioned principles. In general, however, problems of this sort come under the heading of "Training" and will be discussed further in the appropriate chapter.

Summary

The humanologist, at the stage of selection and placement, needs to determine the wants and needs of the new employee. Having determined this, he will endeavor to evaluate the individual in terms of his present abilities and limitations, his hopes for the future, his staying power, his outside interests, his family and social adjustments, and his background, in addition to the determination of his exact physical status. He will treat each employee as an individual, using standard laboratory and psychological tests and formulae only as supplementary tools, and all decisions will be based on the following three principles: (1) The first 10 years of employment are the crucial ones. (2) Any gross misassignment will eventually produce medical symptoms. (3) No human being is a constant, and a good original assignment can become a misassignment with the passage of time.

REFERENCES

Davis, Stanley W. "Stress in Combat." *Scientific American,* CXCIV (March, 1956), pp. 31-35.

Ewalt, Jack R., m.d. "Emotional Problems In Industry." An Address at the Thirty-seventh Annual Meeting of the American Petroleum Institute, Chicago, November 11, 1957. Copy on file at American Petroleum Institute. 50 West 50th St., New York City.

FLINN, ROBERT H. "Fatigue and War Production." *Medical Clinics of North America*, XXVI (July, 1942), pp. 1121-1143.

HOAG, A. E., M.D., AND HOWARD, M. N., M.D. "The Value and Operation of an Industrial Medical Program." *The American Medical Association's Archives of Industrial Hygiene*, III (April, 1951), pp. 375-383.

"Personality Tests." *Industrial Relations News*, (April 7, 1956) Reprinted in *Industrial Medicine*, XXV (May, 1956), p. 204.

KAPLAN, HAROLD I., M.D., AND HELEN S., PH.D. "The Psychosomatic Approach in Medicine." *Annals of Internal Medicine*, XLVI (No. 6. June, 1957).

PAGE, ROBERT COLLIER, M.D. "Present Philosophy of Company Medical Services." An address to the European and North African Employee Relations Conference, Antwerp, Belgium, September 15-19, 1952.

—————. "The Situational State." *It Pays to Be Healthy*. New York: Prentice-Hall, 1957, p. 27.

PIGORS, PAUL, AND MEYER, CHARLES A. *Personality Administration*. New York: McGraw-Hill, 1956, pp. 71-89.

"Occupational Health on Radio." *Industrial Medicine and Surgery*, XXIII (August, 1954), pp. 336-369.

SINGER, HENRY A. "The Management of Stress." *Advanced Management*, (September, 1960), p. 11.

WARNER, NATHANIEL, M.D. "Psychiatric Problems in Industry-Viewpoint of the Consulting Neuro-Psychiatrist." *Medical Bulletin*, the Standard Oil Company (New Jersey), XII (February, 1952), pp. 83-90.

WILKINS, GEORGE G., M.D. "The Significance of Non-Occupational Disability in Industry." A Symposium on Industrial Medicine, Harvard School of Public Health, April 4, 1953.

12

Training

TRAINING IS A CONTINUOUS process from the date of first employment until the date of severance, and is applicable to individuals at all levels of employment from office boy to president. The purpose of training is to teach each employed individual to obtain the best possible results (in keeping with organizational goals) with all the tools available to him and with the physiological and psychological equipment he has.

The humanologist learns to view training from two standpoints simultaneously:

1. From the point of view of the employer i.e., the development of skills and abilities which will enable employed individuals to perform specific roles in the carrying out of the company's objectives;
2. From the point of view of the individual i.e., the improvement of the individual's ability to take pride in his achievement, the improvement or stabilization of his earning capacity, the development of a positive and meaningful role for him in the social structure.

These two points of view must be reconcilable at all times. This is true of the office boy who may one day become president, the file clerk who may become secretary to the chairman of the board, the eager beaver go-getter who marries the village sweetheart and unwittingly places her as well as himself in an untenable position when his employer sends him abroad to represent the firm.

In the prehumanological era of management theory, training was looked upon from a strictly economic or organizational point of view. The questions that were asked were: "Can John Doe be taught to do such-and-such a job in accordance with standard job specifications? What are John Doe's ultimate potentialities in serving this company? What shall be invested in him, and how soon can dividends on this investment be realized?

The humanologist will ask the above questions as a matter of routine, but he will reconcile the answers to these questions: "What does John Doe want to be? Can he be taught to like doing such-and-such a job, or to take pride in doing it well? What other aptitudes, hopes, needs, prejudices, or family pressures are acting upon John Doe."

If this is not done carefully, the end result of training in the case of John Doe may well be that:

1. He will move on from job to job, getting the benefit of costly training without returning anything to his employer in terms of quality performance;
2. He may adapt outwardly to the assignment for which he has been trained, but in time become incapable of quality performance owing to situationally-induced illness.

Training inevitably becomes more than mere education in job skills, even at the lowest echelon of employment, when it is taken into consideration that the good performer in any job is more than a good job holder; he is a good total person. His attitude is good, his health is good, his adjustment to all phases of his life environment is good. A breakdown in any part of his totality will eventually manifest itself in all the other parts of the whole. For example, an employee with an unhappy home life may for awhile turn out excellent work, since his job is temporarily an escape from home, but in time the home situation, if not corrected, will produce observable physiological or psychological symptoms that will affect not only his work, but perhaps his after-hours skill at the corner billiards table, and even his ability to share a good joke with the boys.

Hence training must in the long run become training for life as a whole. This will be discussed in detail at the appropriate point later in this chapter. It goes without saying that this concept of training embraces the concept of preparing square pegs for square holes. Square holes do not always remain square. Changes in technology mean not only changes in tools but changes in job specifications. No one is ever so fully trained that he does not require careful scrutiny to make sure that he and his job have not changed with relation to each other, and to retrain him for his present job or train him anew for a different job if it turns out that he is a square peg in a hole that has become round, or if he himself has become round in a hole that has remained square.

Training, of course, must be based on a thorough knowledge of the individual. In many cases, this will include his wife and his children, particularly if the nature of his job is such that lack of adaptability on the part of his family can dampen his ability or enthusiasm to perform.

All of these factors must be taken into consideration by the humanological team and communicated to the proper organizational levels if situational states are to be kept at an absolute minimum and hours spent by instructors or supervisors are not wasted. Waste at the initial training stage may be impossible to compensate for later, since the first 10 years of employment are what establish the entire occupational pattern of the employed individual.

This Age of Automation

Industry is moving inexorably toward the age of automation. It is doing this in spite of itself. Many present-day managers don't want it; it frightens them; it will demand acumen and speed of decision combined with technological know-how that is considerably less in evidence in the average present-day manager. Supervisors shy away from it; it means they will have to be reschooled in techniques of commanding work units with an entirely new relationship to their tools of production. Organized labor is prepared to fight it; it will do away with a large percentage of jobs for what are now called unskilled laborers; everyone will need some kind of skill since machinery will perform the routine "muscle" jobs.

In spite of this resistance, most of those who resist admit that the technological advance in the direction of automation is inevitable. A company which requires the services of 100 men to maintain a given production level in a given period of time, cannot hope to compete with a company which requires 10 men to maintain the same production level in one-fourth the time. The first function of a business is to stay in business, and if this can be accomplished only by laying off some unskilled men and installing machinery to replace them, this cold-blooded solution can scarcely be avoided.

Moreover, most of those who now tend to resist the advent of automation, will agree that its concomitant, vastly increased production, will improve living standards for everybody in the long run. What is feared is the terrible adjustment that must be made before that highly desirable goal can be reached.

Art of Training a Science

The "art" of training will have to become a "science" to meet this challenge. The old concepts of years of apprenticeship or gradual step-by-step absorption of techniques in order to become a journeyman, professional or technical expert, supervisor or manager, now calls for revision. Training methods to be effective must be quick enough to keep up with technological progress. The former holders of routine muscle jobs will have to be trained to the point where they can take their place in the ranks of skilled labor. The high productivity of a technological age demands a large population of gainfully employed potential buyers. An increase in productivity under technology, may mean a decrease in production workers, but it will mean an increase in jobs at the distribution and service levels.

Every Employed Person Requires Training Now

Everyone in the business world from top line officers to the packer in the shipping department requires training. From the point of view of top management, it will be a luxury rarely to be afforded to indulge in the kind of tentative thinking exemplified by the popular burlesque of so-called "Madison-Avenue-ese": "Let's put it on the train and see if it gets off at Westport."

When a top management decision involves investments in millions of dollars worth of machinery; when figuratively speaking, a mere tap on a push-button will call into being a vast, overflowing cornucopia of products that somehow have to find a market, then the man who makes the decisions has to *know* in advance whether or not "it" is going to get off "at Westport."

Significance of the Permissible Margin of Error

Of course free competitive enterprise will always contain an element of gamble, and occasional wrong decisions will have to be allowed, but the permissible margin for error having to do with both man and materials will grow necessarily smaller and smaller. The dangers of placing the brother of the boss' wife or the favorite drinking companion of the number 1 stockholder in a sensitive management position will become prohibitive unless the individual in question has proven himself thoroughly capable of tackling the complex problems of management in the age of automation.

Mantalent Versus Manpower

At the level of supervision, automation will mean the personal leadership of work groups composed not of so many replaceable units of manpower but of highly skilled individual experts. "Mantalent" then will be a better word for a work group than "manpower." Automation will make the humanological element more than less important at the level of the supervisor. Where the disciplinary methods of the old-time line sergeant or straw boss might have gotten results of a sort, they will become utterly futile. The supervisor will be dealing with specialists, some of whom may know more about some specific technological detail than he does. Moreover he must keep these individuals functioning against a continually shifting technological background, involving constant readjustment of skilled individuals to new techniques, improved machinery, and so forth. His units of mantalent are by no means readily replaceable; therefore he has a responsibility with regard to their motivation on the job as well as their physiological and psychological capacity to stay with it. He is no longer a whip-cracker; he is a highly important arm of the humanological team, and will need to be extensively trained in humanology as well as technology.

Minimum Standards of Employability

At the employee level, particularly at the level represented by organized labor, automation will no doubt call for a reassessment of minimum standards of employability.

The earnest and honest union representative of a large number of unskilled workers may say in all sincerity:

"It's all very well to talk of 'mantalent' replacing good old-fashioned manpower. It sounds quite Utopian. But I happen to represent quite a large number of people who couldn't be classed as anything but manpower-muscle-power, if you will. They may not represent any kind of talent, but they are human beings. They have to live. You can't just throw them on the scrap heap, and you can't blame them for bending every effort to resist a technological system that will automatically throw them there."

All Individuals with Few Exceptions Can Be Trained

There always have been and no doubt always will be some technological misfits: genetically subnormal individuals who are incapable of fitting into the occupational scheme of things and must be taken care of by private or public welfare agencies. However, the majority of these supposedly untalented muscle workers do not fit into this category. Training methods that are now being tried, tested, proved, and improved are capable of developing industrially valuable technological talents in any individual not classifiable as feeble-minded or psychotic. It was explained in a previous chapter that industry will find itself more and more obliged to take an interest in community medical services. By the same token, for humanologically related reasons, industry will find itself obliged to take a larger role in community educational facilities, at the primary as well as secondary school level. The training of the employed individual of tomorrow begins not the day he passes the plant gates, but the day he goes to school; perhaps even sooner.

What Was Learned in World War II

Although the future tense has been used in the foregoing discussion of the age of automation, the subject of training for this age will henceforth be in the present tense. It is going on at this moment. As far back as World War II, it was being learned that by the proper methods even semi-illiterate supposedly unintelligent individuals could be trained to become expert at operating relatively complex technological equipment such as artillery fuse-cutters, antiaircraft range finders, surveyors' transits, and the like. An uneducated private in a radar searchlight outfit possessed technical skills that would have awed a college-educated second lieutenant of infantry in World War I. Yet, by present day standards of training, not only in the armed forces but in automation-conscious industry, the training methods of World War II seem almost primitive. Moreover, private as well as governmental agencies are seeking, and beginning to find, methods of coordinating primary and secondary school education with adult training.

Training Research

Training research that has been carried on in the armed forces and in business has developed clues as to methods of teaching certain scholastic subjects in such a way that the students actually *learn* them, and not just learn to achieve passing marks. Adult training has benefited too from discoveries made at the school level. Another result of this coordination has been an agreement on goals. A common secondary-school attitude regarding a certain type of student has been in the past: "Nothing short of a miracle would pound any education into that youngster's head. As soon as he reaches the legal age, he'll just quit school and dig ditches for the rest of his life; so why waste effort?" Such an attitude today can only result in a social tragedy. If nothing short of a miracle will educate the youngster; then the school, with the help of government and industry, will have to determine some way of passing that miracle.

Technique for Living

It is an important function of the humanologist, whether he be medical consultant, psychologist, nurse, hygienist, or representative of management or labor, to see to it that a primary goal of training at all levels is a technique for living.

Nearly every time an individual becomes a grave problem to industry, and ultimately a problem to society-at-large, the tragedy can be traced back to some basic ignorance or self-deception regarding a basic fact of life. Despite the widespread literature on the subject of health, a questionnaire of employees of a given company regarding their knowledge of basic health facts might be unpleasantly enlightening. An almost indispensable vice-president of a highly competitive manufacturing company was put out of action for nearly a year following a surprise heart attack. He had had adequate warning in the form of not-too-severe attacks of angina pectoris the preceding year, but he had never heard of angina. He had been treating what he thought was heartburn with a proprietary medicine purchased through a mail order house. The private secretary to the president of a shipping firm created considerable confusion plus the cancellation of at least two important purchase orders that no one could find in her absence; she had collapsed suddenly at her desk and was in a semiconscious condition for several days as a result of a prolonged attempt to lose weight on a diet of coffee and vitamin pills. "But I was careful to take *all* my vitamins," she sobbed later in the hospital.

Health Safeguards

There is a limit to what any company can do to safeguard the health of its valuable, expensively trained, and difficult to replace employees, supervisors, technicians, and managers. It can offer periodic diagnostic services, it can participate in medical and hospital expense-indemnity plans, it can provide a financial base and establish standards for A-1 community medical facilities, it can subsidize medical education as well as fundamental research into the nature of man, and finally it can offer its employees education in health construction and maintenance. Beyond this it can hardly go. The rest is the responsibility of the individual.

Training Methods and Health Education

This is why the medical member of the humanology team has a responsibility to see to it that the most up-to-date and effective training methods are brought to the problem of health education. Except in rare cases, he will find himself in a position to inspect health education in the local schools and instigate corrective action where imperfections are found. This will prevent considerable grief at company level in the future.

At the plant, he must make it clear to all concerned that the mere dissemination of popularized health-information pamphlets does not constitute sound health training. Neither does a series of educational lectures or movies. This kind of material is useful as a supplement to health training, but as any seasoned educator is aware, no one will learn much from any kind of educational aid unless he first has a desire to learn. A training program must include the inculcation of a desire to learn on the part of each trainee.

This leads to a consideration of the truth that lies at the core of all training, whether the trainee is a budding physicist wrestling with the secrets of the shape and substance of the universe or a two-year-old child making the transition from diaper to chamber pot. This truth, put boldly, is: *All training is self-training.*

In other words, the best a trainer can do is stimulate an individual to train himself, offering what help is necessary along the way. The passive trainee, accepting what is handed to him by someone else, may acquire enough information to get him through an examination or informal on-the-job quiz. He does not absorb the information discriminatingly, nor will he retain it, since it was never wholly received by him. A case in point is the army private with his rifle. Long before he ever goes to the firing range with live ammunition he is shown diagrams illustrating the way the bull's-eye should sit on top of the front sight which rests in the middle and flush with the top of the notch of his rear sight. He is told how to correct his rear sight up or down, right or left, depending upon where his sights were actually pointing at the moment of the explosion and where the bullet actually landed. He spends hours in the hot sun dry-firing; that is: aiming his unloaded rifle at a target and squeezing the trigger in such a way that he is surprised when the click comes. Perhaps, he is not particularly interested in becoming a good marksman. Somebody, a noncom or one of the company officers may fire an embarrassing question at him; so he pays attention to the words. He doesn't make an attempt to try to imagine the real meaning of these words in terms of himself on the range. When he looks at the bull's-eye on the diagram, he doesn't try to see it in his mind's-eye as a real bull's-eye with the sights of his own rifle lined up on it. If someone asks him whether the trigger should be snapped or squeezed, he will answer correctly: "squee-ee-eezed." But he will have made no real effort to experiment with the trigger on his own rifle to determine the difference between snapping and squeezing. He won't *feel* it.

Then one clear morning he will be marched or trucked to the range with his colleagues to "fire for record." When his turn comes he will be told that he is entitled to a few practice shots to "get zeroed in," or to adjust his rear sight to compensate for structural errors, distance, wind velocity, and so forth. Confidently he will assume the prone position and look through his sights, remembering the diagram. But the diagram was only a plan, a blueprint. The real bull's eye, 200-yards away, seems microscopic. The effect of bright light seems to blur the outlines of his sights; so that they do not in any way resemble the clear-cut layout on the diagram. Finally he thinks he has it. The front sight seems like a great blunt, blurry-edged, shimmering club; the little microbe that is the bull's-eye is lost on it somewhere. He snaps the trigger, nervously. The kick startles him. It seems to him that when the explosion occurred he had his sights lined on a patch of blue sky, though he knew he had pointed at least in the general direction of the target before the blast. The marker pulls down his target and finds it pure and unsullied.

"How did you call your shot?" asks an officer.

"Well, sir, I-I-"

He is completely at a loss. Somebody down the line laughs.

"Okay," says the officer, kindly. "I guess you just haven't gotten the idea. Here, let me show you."

Remembering the laughter, he wants badly to learn now and not make an utter fool of himself. His training has just begun.

Pride a Prime Motivating Force

The above-mentioned case is not intended to prove that a trainer should play upon an individual's fear of ridicule before he can begin to train that individual effectively. Pride is of course a prime motivating force. People will do a great many seemingly irrational things for the sake of preserving status. Pride however can backfire. For example, it might have turned out that the private was literally untrainable as far as marksmanship was concerned. In such a case, the laughter of his comrades might well have undermined his self-confidence and competence in the areas where he *was* trainable. The trainer with an understanding of humanology, makes a careful study of each potential trainee, and will take care to avoid availing himself of the motivating force of pride if the foreseeable result will be the *shattering* of pride.

The case above simply illustrates the futility of training any individual who does not want to be trained. Therefore, a study of what motivates the individual is always in order before a program of training can be applied to him. This is true at every other stage of progression in an individual's occupational life, and the broader aspects of motivation will be dealt with in the following chapter.

Motivation and Training

An illustration of motivation with regard to training is seen in the case of an able employee in a butcher shop attached to a large manufacturing corporation whose employers decided to offer him further education which would equip him, eventually, to become head of the shop. The employee, who had been denied many educational advantages in his teens, showed himself an eager student, willing to put in extra hours of study in order to assimilate several courses not in the curriculum specifically designed for him. Upon completion of the course of training, he reported to management with the request that he be transferred to the training department. His employers were astonished. They told him he could never hope to earn the salary as a teacher that he would earn as head of the butcher shop. Besides, they said, he didn't have the educational qualifications. He asked them to check with the school where he had trained. Management learned there that this zealous young man had received top grades in every major course offered by the school and was eminently qualified for a training job. The question "why?" remained a puzzler until it was learned that among this young man's acquaintances, a teacher, however poorly paid, had a more acceptable social status than that of a butcher. It was suspected, in fact, though not proven, that the young man was having difficulty persuading the young lady of his choice to marry him unless he made the change in status. Irrational as it may have seemed to his employers (irrational in fact as it might come to seem to the young man himself in later years with more mouths to feed and a bigger mortgage to pay) it was, at the time, a legitimate and compelling motive. The appropriate transfer was made.

Training the Long Service Employee

Another motivating factor that must be taken into consideration is the resistance to the implications of criticism that may be read into the inauguration of a training program. Of course this doesn't apply to newly selected employees, many of whom actually look forward to the training period as a sort of painless introduction to the awesome mysteries of a new way of life. The long-service employee may look upon the suggestion that he needs training as a hint that his work performance is unsatisfactory. If management is not sensitized to this often unspoken attitude, and if the individual is pushed into a training program with no effort on the part of management to discover his feelings or to induce him to take a positive attitude toward training, the employee will tend to resist learning.

The old cliché "You can't teach an old dog new tricks" has been disproved in actual experiments with old dogs as well as old people, but the myth persists because of a general lack of understanding of the motives that so often underlie resistance to new ideas and new ways of doing things. The old-timer in the machine shop who is stubborn and set in his ways and eventually must be dumped because of his bellicose attitude toward his younger supervisors and trainers plus his refusal to adapt to new techniques and machinery, is, more likely than not, a victim of "meat-axe" training methods. At some stage in his career, no doubt, some manager or supervisor decided that his methods of production or his tools were out of date, and told him bluntly that he needed to be trained in new methods or adapted to new tools. This worker quite likely nursed a secret suspicion:

"They don't like my work. They're trying to ride me. They want to get rid of me. Well, I'll show them I can be stubborn."

His superiors undoubtedly meant well. It simply didn't occur to them that the worker might possibly take a negative attitude toward training. With patience, understanding and frankness, they could have induced this worker to *want* to adapt himself to new methods. An approach like the following might have been effective in this particular case:

"Have you heard about the newfangled stunt our competitors are trying out? If they think they can beat us that way they have another think coming, don't you agree? Of course it's a great idea on paper, but you can't make even a wonderful idea like that work unless you've got the right men — seasoned men with judgment, to carry it out. Tell you what! Just for fun, why don't you study up on their idea (I can get a trainer and all the material you need) and then show them how a real expert can make it work. Bet it will make them look sick."

In other words, you *can* teach an old dog new tricks if you can make him want to learn them.

When the medical consultant or the plant nurse is faced with unaccountable symptoms on the part of an employee who has for a considerable period of time been well adapted to his job, and if no undue pressure or stress is evident in home or job relations to form an apparent basis for a situational state, it will frequently pay to look for buried emotional wounds caused by maladroitly applied training methods. The employee himself may not be aware that his feelings were hurt when he was told to adapt to new ways. In this case, the inner disturbance is even more insidious than a recognized one, and that much more in need of exorcism by the application of tact, sympathy, and common-sense at the appropriate time.

Resentment

Resistance to learning is not necessarily always based on resentment. Man is a creature of habit. Having established one pattern of doing things and having become comfortable in that pattern, he will resist any attempt by an outsider to force a different pattern on him. He may not be conscious of his reluctance, but his inner equilibrium can be upset anyway, with possible medical results, unless he is skillfully made to become dissatisfied with his present pattern of life and earnestly desires a new pattern.

This is why any newly placed employee is receptive to training. The very fact that he has newly come to his present job is an indication that he was dissatisfied with his former condition of employment or unemployment. He is seeking a new pattern. This is helpful to the trainer in that it makes him acquiescent to almost any kind of training that does not lead to a loss of status in his society. From the point of view of the medical consultant, however, it must be looked upon as potentially dangerous. His very acquiescence at this stage, his willingness to adapt to almost any pattern as long as it is a new one, may lead him into paths contrary to his true desires, needs, and capabilities. He may agree to a course of training which will lead him into a branch of business where he will inevitably be unhappy, and become a knotty placement problem. Or he may plunge into a course of training which he is not equipped to complete satisfactorily, with the result that his company will be out of pocket and he will be a humiliated, sick individual.

The above disaster can be prevented by constructive humanological action at the placement stage.

How to Deal with Training Resistance

As for dealing with training resistance on the part of the established employee, the humanologist gains an advantage if he knows what questions are likely to be running through the mind of the individual who has been told he is in need of training. The most common ones are as follows:

1. What happens to the work I turn out?

2. How does one go about practicing up on this new skill (or with this new tool) without letting other people know that I am awkward at it?
3. What controls are exercised over my work?
4. Will this be like going to school?
5. What kind of people are they hiring to do this work?
6. What can I tell my friends about what I am learning?
7. How will this help my future in the company?

Even if none of these questions are asked one may accept them as implied and provide answers for them anyway. Answers in every case should be thoroughly frank. No good purpose is served by permitting the employee to infer that successful completion of a training course will lead to a promotion or raise in pay if such is not actually the case. Neither will any good be done by minimizing the importance of successful completion of the course, if failure in this respect will materially diminish the employee's value to his company.

A Training Program

Before plunging headlong into a training program, a company must answer the following questions: (1) What training needs exist? and (2) From what do these needs stem?

Training Needs

Training needs tend to stem from one or more nine basic causes:
1. abolition of "manpower" jobs owing to automation, necessitating a conversion of some manpower to mantalent;
2. anticipated turnover or retirement;
3. seasonal fluctuations in labor force;
4. preconceived expansion of individual aspects of the business;

TRAINING

5. new installation of machinery;
6. reorganization of departments for efficiency;
7. new specifications in labor contract;
8. complexity of required skills; or
9. changes in labor market.

Constant Aspects of a Given Job

In setting up training requirements for any given job, it is essential to determine what are the key points that make or break job performance; i.e., those aspects of a given job which are constant regardless of the individual performing the job.

To take a rather simplified example: a lifeguard at a club swimming pool may be required to have passed tests in resuscitation and life-saving. It goes without saying that he is expected to be able to swim. He must be free of infectious diseases that can be transmitted through the medium of the pool water. He may be required to perform maintenance tasks such as operating the chlorination apparatus, dragging the pool with chemicals to inhibit the growth of algae, draining and scrubbing the pool periodically, and keeping the locker rooms clean. He may have disciplinary tasks having to do with keeping boisterous youngsters from splashing and pushing people into the pool. Certainly he is required to keep track of everybody in the immediate vicinity of the pool, paying particular attention to the weak swimmers. All of the above may be called key points. Anyone who holds the job may be subject to every one of the above requirements.

However, there are many areas in which one job performance may differ from another, depending upon the individual. One lifeguard may feel he can perform his tasks best by patrolling the pool continuously. Another may keep the pool under surveillance from an elevated seat. One may keep the youngsters in order by organizing them into games. Another may prefer to lay down the law. Still another may be in and out of the pool frequently. These are not key points. They do not in any way affect the performance of the essential job, and depend upon individual preference.

Precedent and Individual Preference

It is the duty of the humanologist to determine and keep management conscious of the gaps between these key points and see to it that they remain gaps. There is a dangerous tendency among people who set up job specifications to rely upon precedent. "So-and-so did this job exactly such-and-such a way, and since he was a good producer, everyone should do that job in exactly that way." To escape this tendency, management must determine precisely what the individual job is supposed to accomplish in the over-all scheme of things; not precisely how the job is to be carried on. Having made this determination, management can decide just which methods of performance, and just which steps, precedented or unprecedented, are absolutely necessary for the effective performance of this job. This will establish the key points, which will define requirements for *any* given individual assigned to the job. All else should, if possible, be left to individual preference. This is a first-rate safety valve to guard against an excess of frustration and maladjustment. The trainer should make an effort to encourage the individual to find performance methods that are best suited to him in all except the key aspects of the job, which cannot vary.

Actual Training Areas

After establishing the need for training and the cause of that need, it is necessary to determine in which area of occupation the training is needed. The principal areas include: (1) orientation, (2) job skills, (3) human relations, a *must* for new supervisors, (4) technical and professional arts, (5) management and supervision, and (6) general education i.e.. techniques for living.

Training on the Job

The experience of most companies has been that it is most practical to conduct training on the job. About 75 per cent of all industrial training is so planned.

The trainee, particularly in the area of job skills responds best when he is given a feeling of accomplishment. This is best arranged if the training course is laid out in steps, each one not too far beyond the preceding one. Each time the employee completes a step successfully he has a feeling of having gotten somewhere. If the achieving of each step means the acquisition of an ability to do something new, the stimulus to carry on is even greater. The trainee, who becomes discouraged and hops from company to company before he has ever produced enough to justify the expense incurred in training him, is often put off by too great a distance between steps. His goal seems so far off that he begins to have doubts that he will ever attain to it. Such an individual is often in need of medical advice. Discouragement becomes chronic and may in time produce symptoms.

Habits and Age

The tendency of the nonmedical individual charged with training the individual who has passed his 45th birthday is to dismiss the grooves of habit as evidences of presenile cantankerousness. This is nearly always erroneous. Anyone, even a youngster, can become deeply rooted in a habit. Anyone, with proper guidance, can break a habit, regardless of age. This cannot be accomplished overnight. If an automobile manufacturer were to put out a big-selling model with the brake pedal to the *right* of the accelerator, a number of people would become involved in serious accidents shortly after the advent of the new model, and they would not, by any means, be all over 45.

Humanologically-oriented trainers often find it useful to remove old props when a new device is being introduced that is radically different from its predecessor. For example, if stenographers are to be trained to use an entirely new keyboard, they will be reminded that this is something new only if all the other old familiar props are radically altered. If everything from shiftlock to roller action is precisely and comfortably the same she will find herself forever slipping back into old patterns of typing, and bruising herself against a brutal wall of frustration.

Morale Factor

A morale builder in training is some form of concrete recognition of successful completion of a course. An award, diploma, or certificate will give the individual some feeling of having "arrived," even if training entailed neither a raise nor a promotion.

Purposes of Training

It must be re-emphasized that the purpose of training is to construct, not reconstruct; i.e., use the materials already in the individual's make-up, rather than remould him into a rigid preconceived pattern. It must further be underlined that the first duty of training is to help the individual develop a technique for living.

Finally, the humanologist will keep in mind and keep management alerted to the fact that training must be flexible and adapted to different varieties of individuals. To narrow training procedures is to create narrow employees, supervisors, and managers, who will not be able to fluctuate with the flux of times and conditions.

This is in actuality another way of restating the concept that square pegs must be placed in square holes, and must be restudied from time to time to determine whether there has been a change in "squareness" of either the peg or the hole. If the humanologist is scrupulous in this respect, and if training is continuous as jobs change or as people change in relation to jobs, situational states will not occur; the adaptation of the individual to his job will be sound, as will his adaptation to life in general. He will know his abilities and his limitations, and will obtain the deepest personal satisfactions in doing what he is cut out to do.

The trained individual will, in short, use his mind and body to the best advantage of all concerned: himself, his employer, and his family.

REFERENCES

FRYER, DOUGLAS H.; FEINBERG, MORTIMER R.; ZALKIND, SHELDON S. *Developing People in Industry.* New York: Harper and Brothers, 1956.

HALSEY, GEORGE D. *Training Employees.* New York: Harper and Brothers, 1949.

HAMILTON WATCH COMPANY. *Guide to Employee Award Planning.* Lancaster, Pennsylvania: Hamilton Watch Company, 1958.

Industrial Training Program. Industrial Development Corporation, Jamaica, W.I., February, 1958.

PLANTY, EARL G.; McCORD, W. S.; EFFERSON, CARLOS A. *Training Employees and Managers for Production and Teamwork.* New York: The Ronald Press, 1948.

STESSIN, LAWRENCE. "Managing Your Manpower." *Duns Review,* (September, 1960), p. 115.

"Training in Occupational Medicine." *Industrial Medicine and Surgery,* XXVI (No. 4, April, 1957), pp. 201-208.

ZALEZNIK, A. *Foreman Training in a Growing Enterprise.* Division of Research, Graduate School of Business Administration, Harvard University, Boston, Mass., 1951.

13

Motivation and Promotion

Every Stage of Employment Has Its Motivational Aspects

FROM A BROAD humanological point of view, every stage of employment has its motivational aspects. One can hardly discuss the subject of recruitment without considering the motives that underlie any given individual's desire to work for Company A in preference to Company B. At the final stage, during which the individual is being decelerated preparatory to retirement, the *opposite* of promotion, the humanological team must concern itself primarily with making him follow this course willingly, even happily; this is nothing more nor less than a motivational problem.

However, the only stage in which motivation is the *sole* factor, the only stage involving administrative as well as a medical problem which is wholly motivational in nature, is the stage at which the individual, having been recruited, selected, placed, and trained, wants to or is expected to improve his occupational status. This may not necessarily mean a raise in rank or in pay. It may mean increased security. It may mean the approval of his employer, his on-the-job colleagues, his family, or his off-the-job friends. In a sense, this stage may properly be considered as having to do with some form of *promotion*. Because of its readily measurable physiological, psychological, and sociological effects, all related to motivation, this stage is a sort of magnifying glass for the entire problem of motivation. Hence, while the motivational aspects of the other occupational stages are discussed as they arise in the appropriate chapters, it has been deemed logical to explore the *fundamentals* of all human motivation in terms of promotion, at which stage they are most clearly revealed.

Fundamentals of All Human Motivation

Motives have been defined as "anything that arouse, direct, and sustain behavior," by Fryer, Feinberg, and Zalkind in *Developing People in Industry.* First motives are physiological: the helpless infant cries out for food, water, and warmth. These simple motives become complicated in time. For example: the desire for food becomes a desire for specific kinds of food. Moreover, since it is a combination of desires in which many parts of the individual is brought into play, it may be broken down into its component parts; for example, the yearning of the immediate mouth area may be satisfied with a thumb to suck, or a teething ring.

Physiological and Social Motives

To complicate matters further, the satisfaction of the basic desires of the helpless infant is dependent upon the loving care of some adult, ordinarily a mother. In actuality, the person tending the infant may be doing so only out of a sense of responsibility or for some financial consideration, but to the infant, who has no comprehension of such things, it is *loving care* with emphasis on the *love,* a basic need. Because he needs to be loved as well as fed, clothed, and sheltered, his already complicated and devious *physiological* motives become almost inscrutably intertwined with *social* motives. It is difficult to disentangle these basic motives, even in early childhood before the so-called reproductive instinct has further beclouded the already murky picture, and to state with any finality whether such processes as honesty, ambition, cooperation, creativity, malice, and the like, originated primarily in the desire for food, for warmth, for maternal love, or for a simple change of diapers.

Motives develop from basic needs which are associated with certain *relievers, incentives,* and *satisfiers.* The responses which satisfy inner tension may not always appear logical. The individual who kicks a wastebasket across the room when suddenly faced with a frustrating situation is behaving wholly irrationally, *on the surface.* Inwardly, however, he may have, in a split second, transformed the wastebasket into the symbolic embodiment of the disturbing factors in the situation, and then wiped them out of existence with a kick. He is now able to face the frustrating situation coolheadedly and positively; the action was not so irrational after all.

These satisfying or relieving responses are, for the most part, learned; not inborn. People learn to behave in those ways which seem to satisfy their needs best.

Nine Basic Human Needs

A valuable humanological tool is the list of "Nine Basic Human Needs," developed by a New York firm of management consultants, and studied in terms of family, social, or on-the-job factors which have to do with each of these needs, as they affect the employed individual. These are as follows:

A. *Food and physical welfare needs,* influenced by these factors:
1. Working conditions (dust, light, arrangement of tools, noise, fumes, temperature, moisture)
2. Shelter
3. Transportation
4. Living wage (sufficient for minimum physical requirements)
5. Food during working hours

6. Facilities (lockers, etc.)
7. Clothing
8. Rest periods
9. Freedom of motion (opportunity to change physical position and relieve muscular fatigue)
10. "Housekeeping" at work (cleanliness of surroundings)

B. *Personal development needs,* influenced by:
1. Housing (opportunity for making a home)
2. Social recreation
3. Living wages (family requirements)
4. Working hours and schedule (hours at home)
5. Transportation (hours at home)
6. Community opportunities for children
    a) education
    b) social groups
    c) religious activities
7. Confidence in the future

C. *Desire for achievement,* influenced by:
1. Having a known goal
2. Knowing what is expected of one
3. Having some measure for individual accomplishment
4. Having available proper training to achieve what is expected of one
5. Having a clear understanding of the limits of the job
6. Having a proper reward as recognition for desired achievement
7. Guidance into activity appropriate to individual ability
8. Provision for steady progress

D. *Desire for activity, variety, and novelty,* influenced by:
1. On-the-job
    a) determination of proper job limits
    b) opportunity for physical movement
    c) relief from monotony

           d) rest periods
           e) recognition of individual differences
           f) change from job to job (within limits)
           g) change of job within group
           h) opportunity for initiative in thought
        2. Off-the-job
           a) having proper recreational facilities
           b) social activities
   E. *Need for release from emotional stress,* influenced by:
        1. Opportunity for physical activity
        2. Having proper grievance outlets
        3. Having clearly understood company policies, particularly regarding status
        4. Counseling (listening by others)
        5. Social contacts on and off job
        6. Religious activities
        7. Security
        8. Proper organization and selection of working group and supervision
        9. Enough properly trained supervisors who are sympathetic and easily available
       10. Home atmosphere
       11. Confidence in fairness of wages, promotion, and task assignment
       12. Belief in justice of management
       13. Smoking on rest periods
       14. Miscellaneous relaxers (i.e., gum or candy)
   F. *Need for security of status,* influenced by:
        1. Security of employment and home
        2. Pension systems
        3. Seniority
        4. All types of insurance
        5. Definite and understood promotion policy
        6. Company stability
        7. Company reputation

8. Good company discipline
9. Impartial treatment of employees
10. Freedom from undue coercion by unions or management
11. Stability of purchasing power
12. Confidence in management
13. Opportunity to maintain personal habits
14. Confidence in social order

G. *Need for worthy group membership,* influenced by:
1. Recognition of a good job (performance rating)
2. Belonging to a good company
3. Chance to join safety committee, fire brigades, first aid unit and the like
4. Acceptance by group
5. Union membership
6. Group pride (morale)
7. Intergroup competition
8. Congenial grouping
9. Company-wide occasions and symbols

H. *Need for sense of personal worth,* influenced by:
1. Democratic disciplinary system
2. Recognition of individual effort
3. Execution of suggestions made
4. Being consulted
5. Being treated as a person
6. Having one's opinion treated with respect
7. Being made to feel important to supervisor
8. Being made to feel important to members of the group
9. Being called by name
10. Sympathy toward individual problems of others
11. Recognition of religious feeling
12. Having absolute value as a unit of life
13. Having a feeling of worthwhile accomplishment

I. *Need for a sense of participation,* influenced by:
   1. Knowing what is going on
   2. Knowing why it is going on
   3. Knowing the use of end products
   4. Helping determine the conditions under which one works
   5. Knowing the relation of the job to the finished product
   6. Knowing the relations with fellow workers
   7. Knowing the relations of one's group to other groups
   8. Helping in the administration of policies
   9. Knowing the relation of the job to supervisor's job
   10. Having knowledge of future plans
   11. Knowing how outside conditions affect the job
   12. Sense of citizenship in the community
   13. Knowing the relation of one's job to others in the group
   14. Opportunity to participate in group recreation

Influencing Factors and Basic Motives

It will be noted that some of the influencing factors are repeated under two or more different basic motives.

These nine basic motives and the major influencing factors are well worth the careful scrutiny of every lay or professional member of the humanological team who is ever called upon either to find motives which will make individuals perform in a desired manner or to understand the motives for behavior that is undesirable or baffling.

In addition to knowing the basic motives, it is important to know the relative importance of the influencing factors. These factors might be termed secondary motives; their resolution points toward the resolution of one or more of the deeper, basic ones. Over simplification or lack of discrimination regarding these secondary motives has often lead to confused company policy. For example, it has often been assumed that the only really important motive is money, or happiness, or gratitude, or fear, depending upon who is doing the assuming. However, a careful analysis of these motives will reveal that they all have their shortcomings, though each may be partially valid in a given situation.

Money Is an Important Motive

Of course, money is an important motive. It is undoubtedly the main reason the average individual seeks a job in the first place. There are wealthy or otherwise subsidized individuals who may seek occupation as an escape from boredom. Retired people often seek gainful jobs to restore a lost sense of pride in belonging to a productive group. But such are the exception; not the rule.

Title a Motive

Title is another important motive. There is a bit of snob in nearly everyone. An individual would rather tell his friends that he is Chief of Sanitary Facilities than that he cleans out washrooms.

Fear as a Motive

Fear is not always as bad a motive as some critics have made it out to be. Fear of an individual supervisor will get out the work in a short haul, but in the long haul it will backfire. The disagreeable hypercritical "Little Hitler" will eventually turn his frightened underlings into hardened cynics, who, if they don't quit after relieving themselves of some choice bottled-up expletives, will find ways to evade work, sabotage production, take time off, launch whispering campaigns, and carry out a number of other elusive actions that will in the end undermine the supervisor.

Another kind of fear can be more productive for a longer period: that is fear of being thought ill of, which is a major motivating factor in the life of the ulcer-prone or headache-prone eager beaver. This kind of fear, though it produces results and must be considered in many respects laudable, must at all costs be kept in bounds at the level of supervision. The medical consequences, which will be discussed in detail later in this chapter, can be costly and irreversible.

Gratitude and Happiness as Motives

Gratitude or happiness are rarely ever adequate motives for increased on-the-job productivity. A company that makes every effort to keep its people happy will earn dividends in loyalty, but when it comes to spurring people to greater and greater efforts something else must be added. As for gratitude, it is a hard truth that employees do not produce in order to repay the paternal employer for the benefits and privileges he provides. On the contrary, benefits and privileges are quickly taken for granted.

On-the-job attitude surveys have shown that employees put the following goals ahead of money:

1. Security
2. Chance for advancement
3. Self-expression

The desire to be thought well of, to be loved and respected, is so strong that many individuals have been known to reject opportunities to move into more highly paid assignments which might cost them the friendship or the good opinion of those important to them. This is particularly true in countries or in areas where there are rigid and well-defined social strata to which an individual belongs. Such strata are becoming increasingly rare or ill-defined in modern America. The newcomer from the British Isles who says: "I'm working class and proud of it," is looked upon with bewilderment by his American or Canadian co-workers, who do not have any sense of kinship to any class whatever, and many of whom sincerely believe that they or their children will become company officers or professional men some day.

Goals Become More Complex

The higher an individual rises in business the more complex become his goals. He has points of resemblance to the professional soldier who dies not in battle merely for the sake of his monthly pittance, nor out of fear of his top sergeant, nor out of gratitude for the chevrons on his sleeve. Such a soldier is impelled by subtle motives, by a blend of pride in his outfit, love for his country, belief in his mission, self-identification with the goals of his regiment, division, army, or nation. As for the responsible officer in a business organization, he works extra hours, carries extra burdens, faces greater risks in terms of health as well as job security than does the lower-echelon employee. For this he is paid more, is driven more, is given more impressive titles, but above and beyond the salary, the demands of the stockholders, and the name in gold leaf on the frosted glass door, there is a more impelling motive. He has become identified with his company; he is imbued with its goals. When his company pursues a certain line, he says: *"We* are doing such and such . . ." *NOT:* "The *Company* is doing such and such. . ."

This is the kind of motivation that counts. *Esprit de corps* at all occupational levels will do more to keep people producing at peak capacity than a multitude of tangible rewards or threats.

Each Employee Must Have a Goal

One rock-bottom requirement for getting people into action is to give them a goal to shoot at. This is why promotion is widely considered, sometimes erroneously, to be the best means of motivating people. The perquisites of promotion are attractive, seductive, and sometimes, in the long run, destructive. There have always been a great number of people in the world who would rather die, literally, than be thought ill of. The orientals express this latter alternative effectively in the expression, "to lose face." A Roman centurion would have preferred the expression "honor," which he could not define but which he could preserve by falling on his sword whenever circumstances conspired to make him look ridiculous.

The centurion who fell on his sword might have a statue erected in his memory; the oriental who committed harikari after loss of face might be commemorated in a line or two of poetry penned by a worshipful descendant. However, the former fifth vice-president of Company X who perished in harness as *fourth* vice-president in an effort to maintain status vis-a-vis the Jones' could hope for little better than a few inches of type buried in the obituary page of the daily paper; the prominent obituaries are reserved for vice-presidents above the rank of second; the front stories are only for those who perish spectacularly.

Envy or Pride as a Motive

In other words, honor or face can lead into sordid blind alleys in this occupational world. The humanologist should be eternally on the alert to spot and redirect those avid promotion seekers whose primary motive is envy or pride.

The individual who should be encouraged to seek promotion is the one to whom promotion is a true fulfillment of self, a needed extra challenge to genuine talents.

In view of the fact that any company must have a large percentage of employees who are capable of improving their work performance but whose interests would not be served by promotion, a major challenge to the humanological team is to devise stimulating incentives that do not have to do with promotion. Perhaps, in many companies, the emphasis on promotion as a motivating force can be eliminated altogether on the assumption that those who can and should be promoted will be ready and willing when the time comes, without any advertisement of the arrangement.

Promotion and the Unpromotable

Those who are truly unpromotables should be made aware of this fact so that they may be cushioned against shock, envy, or chagrin every time a colleague or friend is moved up beyond them. However, elimination of the promotion incentive, means that some other way must be found to obtain the wholehearted cooperation of these people.

By consulting with and coming to a closer understanding with people on the job, the humanologist comes to realize that there are a number of ways of making sure that improved work performance can offer employees more direct satisfaction than minimum effort. The minimum output, on analysis, more often than not proves to be the result not so much of a lack of promotion incentive, but a lack of communication on the part of management, complicated by a lack of team spirit, originating at the supervisory level, and sometimes a feeling of occupational insecurity on the part of the worker.

Insecure people tend to seek out the company of other insecure people. Jack, an assembly-line worker who fears his supervisor is plotting to fire him because of his lack of speed will approach Tom, who has been reproved a number of times for clumsiness, and George, who has shown signs of being somewhat slovenly in housekeeping tasks. These three will decide to stick together and cover up for each other by turning in all-round substandard job performance. Dick, Harry, and Bill, who have been capable workers, will begin to feel self-conscious about being out of the gang. One by one they will join the conspiracy of their fellows against their common enemy: the boss.

Team Spirit and False Security

Here is a form of team spirit. All of these individuals have now achieved a sense of false security, and since they have discovered a common bond of friendship, they have begun to derive a sort of perverse social satisfaction out of their job situation. They might almost be called happy employees. But the team spirit evidenced here, and the resultant morale is wholly negative. The root problem is lack of communication. The supervisor was not able to convey to these men that his function was not to criticize them, but to help them turn out better work and to weld them into a happily productive unit.

Communications

Perhaps, this lack of communications was not the fault of the supervisor himself. Perhaps, he too was just a cog in a vast impersonal machine, organized in such a way that the production goals of each unit were visible only to someone at a rarefied level of management. The size and complexity of many modern corporations tend to make for disunity rather than for team spirit. The organization tables may look well on paper, particularly when framed and hung on the tapestried wall of a skyscraper management office several hundreds of miles from the actual scene of productive toil. Nevertheless, the humanologist has an obligation to point out to management that the most ingenious of organizational plans may still fail to stimulate maximum productivity at the level of the unpromotables unless it allows for team spirit, initiative, and a feeling of accomplishment among *small* units. Each employee must be made to feel that he has played a part in the desired effort and that the effort was worth while.

Mechanical Efficiency Does Not Guarantee Human Efficiency

James C. Worthy, the personnel director of Sears Roebuck and Company, pointed out as long ago as 1950 that an organizational plan designed for mechanical efficiency may fall down by failing to provide for team spirit. He said that:

Size and complexity may appear logical in terms of widely held theories of business organization, but in both cases improvements in mechanical efficiency are at some point overbalanced by losses in the willingness and ability of employees to cooperate in the system . . . . Intelligent planning on the part of management in setting up the formal structure of organization can do much to improve the quality of human relations in industry. Flatter, less complex structures, with a maximum of administrative decentralization, tend to create a potential for improved attitudes, more effective supervision, and greater individual responsibility and initiative among employees. Moreover, arrangements of this type encourage the development of individual self-expression and creativity which are an essential ingredient of the democratic way of life.

Unpromotables are not fundamentally different from people in other kinds of jobs. Anyone tends to work at less than peak performance if his emotional needs receive no satisfaction in the work situation itself.

Essential Factors for the Development of Team Spirit

The following factors are essential to the development of the team spirit:

1. An opportunity for workers in a unit to know each other instead of being hemmed in by impersonal administrative systems. This permits each member to see where and how his job fits into the total scheme of things.
2. A job for each worker that is meaningful in itself and that calls on the individual to use and develop potential capacities.

3. Frequent contact of workers with supervisors and line officers. Contact with the latter is particularly important to enable employees to judge them as human beings and not on the basis of rumors about "those guys in the front office" nor in terms of imperfectly communicated impersonal administrative actions.

Spontaneity of behavior is another consideration in the building up of an *esprit de corps*. This is another argument for the division of large organizations into small operating units. Having done this, the next humanological step is not to try to impose "good human relations" too arbitrarily from above. Employees, like other people, need to have some part in establishing their own human relations. An effective way to get cooperation is to build on what already exists. To disrupt spontaneously formed relationships in a small group creates disunity in that group. An individual who has lost his sense of loyalty to and oneness with his small operating unit will find it difficult if not impossible to transfer that loyalty and self-identification to the larger, more impersonal group.

The establishment of integrated units with team spirit actually offers the unpromotable what is, in effect, a promotion. It raises his status from that of insecure, or bored, or unimportant job holder to that of a liked and respected productive worker who is important to his co-workers, his supervisor, and his company. Moreover, though this new status may not involve a raise in pay, it offers him something that money cannot buy: a feeling of accomplishment and of psychological and physiological harmony with his environment. Too often this cannot be said of many of his better-paid, more elaborately titled superiors, who have been driven by some inscrutable compulsion to plunge deeper into a white-collar jungle which will inevitably defeat them, and for which they are totally unequipped.

This latter type of employed individual, who has, for various motives gone beyond his depth or committed himself to a disastrous course of action from which he cannot gracefully extricate himself is not unlike the unpromotable who has never been enabled to make the most of his place in the scheme of things. He has had, undoubtedly, a number of medical symptoms in the recent past (see discussion of "Situational States" in Chapter 10), but his ambition or his pride or the needling of his family or neighbors has caused him to overlook these symptoms, perhaps to mask them with proprietary medicines, barbiturates, artful self-deceptions, or alcohol. There is probably not much that could have been done about them anyway. These symptoms would have been found to have a more or less imaginary basis, and his doctor or nurse would have ordered him to put his situational "house" in order. He would have refused to accept the diagnosis or abide by the prescription.

It is inevitable that if drastic action of some sort is not taken with respect to such an individual, his comparatively easily shrugged-off situational state will develop into a full-blown functional disorder, in which the formerly imaginary symptoms become related to medical disturbances that are not only real but persistent.

Functional Disorders

Illnesses[1] that can ordinarily be classed as functional in nature are: tension headache, indigestion, obesity, alcohol addiction, colitis, insomnia, anxiety states, unexplainable fear, certain types of nondegenerative circulatory dysfunction including functional hypertension, and even some cases of sterility. The functionally ill person presents real symptoms of bodily dysfunction, but it is not "organic"; that is to say, there is no tangible evidence of any concrete change

[1] Aggravating symptoms for the most part.

or damage to any organ or portion of an organ. A sufferer from functional hypertension can, for example, live past 90, but as one such is reported to have told his physician: "It has seemed much longer."

Functionally ill individuals, though often appearing to be in the pink of health are among the most pathetic with whom the average doctor or nurse must deal, and this is particularly true when they are seen on the job where the distressing and elusive symptoms are complicated and increasingly aggravated by a continual decline in ability to perform. It is a vicious circle. The symptoms first have a marked effect on job performance; then anxiety over the latter aggravates the symptoms, which in turn cause a further decline in job performance. If the unhappy individual trapped in this merry-go-round is not discovered and extricated peremptorily, he may find ways to speed up his inevitable downfall: too many barbiturates, the morning whiskey bracer, escape into mental fantasies, continuous nibbling of rich foods, unauthorized and possibly harmful forms of medication, faith healers, and quack religions.

Functional Disorders Must Be Detected Early

Despite the fact that functional disorders are more easily diagnosed than situational states, they are more difficult to treat. Functional illness is still an enigma to the physician. Yet the average general practitioner is aware that sufferers from these health setbacks make up the great majority of his patients. Medication is hardly ever the answer, and the sufferer is doomed to a lifetime of disappointment, frustration, and unnecessary expense if he continues to seek a cure for his condition in this direction. The only hope for cure or alleviation lies in knowledge of the root problem, and constructive therapy based on this knowledge. This calls for close cooperation between the humanological team at the individual's company and the patient's family doctor. The former group will have special knowledge of the occupational environmental factors in the case, and can correlate this knowledge with the physician's awareness of family, social, and medically historical factors.

Because the inner emotional causes of functional illness are usually hidden even from the most sympathetic eye, sufferers from such disorders are generally misunderstood by their friends, relatives, and employers. They are rarely happy. They talk more of their ailments and of their doctors than any other group. They are forever seeking a cure, and each new miracle of medical advance increases their unhappiness, for it seems inapplicable to them.

These patients frequently, without meaning to do so, put obstacles in the way of the doctor who would like to establish communication. They are automatically resentful at the imagined psychosomatic label, and this is understandable, since the most baseless rumor involving the word "psycho" can put an innocent individual on a sort of invisible intangible occupational blacklist which can affect his vocational future. Moreover the doctor is faced with the difficulty of translating rather abstruse scientific ideas into terms the layman can understand. If the patient lacks a scientific background, which is the rule, he will tend to misunderstand significant aspects of what he is being told, and visualize what he cannot understand in terms of what he fears.

Here is another reason why teamwork between the company humanological team and the private physician is to be sought. If the sufferer knows that his company is earnestly endeavoring to help him solve the problems in his environmental background; if he is made clearly aware that the intention is not, negatively, to find what is wrong with him, but to determine, positively, what is right for him, his wall of resentment is certain to tumble down.

Moreover, such teamwork renders it unnecessary for the doctor to rub salt in old wounds which sometimes needs to be done in order to elicit painful information that could not otherwise be obtained. With the help of the humanological team, the doctor can be made aware of the embarrassing but significant negative facts regarding decline of work performance, misbehavior, carelessness, and the like. The patient, understandably, is reticent about discussing such things, and the doctor is fortunately in possession of the information he needs, without having to put an already sick person through unnecessary pain. He can skip the embarrassing external *results* of inner emotional distress and go directly to the root of the matter; *the distress itself, and its cause.*

The Distress Itself and Its Cause

In individuals past 40 who have already begun to exhibit actual physiological changes having to do with aging, the possibility of functional complications must be taken into consideration and watched for with minute care.

For example: Garry B, a 55-year-old bricklayer, had atherosclerosis. The condition was mild and not out of the ordinary for an individual his age. His doctor had explained to him how to keep it under control and live a full life. He had adjusted and seemingly all was well.

Then, like a bolt out of the blue, he was severely criticized by a contractor for a flawed job. A section of a wall had collapsed, and though it was proved later that the fault lay not in Garry's workmanship, but in the material ordered by the contractor, he was deeply wounded, self-critical, and certain that he was slipping.

He developed, at this time, what he described as a heart flutter. His first thought was to associate this new problem with his atherosclerosis. He became seriously alarmed, and this added to his inability to think clearly.

He went to his doctor, a capable but overworked general practitioner, who suspected that the so-called flutter might have a situational base, but who found himself unable to approach the problem constructively from this angle owing to Garry's hostility toward any line of questioning having to do with his occupational or home environment. This doctor decided to take a chance and proceed on the assumption that Garry was right and the flutter was related to the atherosclerosis for which a course of therapy had already been inaugurated. In the event the new problem was truly occupational, perhaps the situation would clear up of itself. Therefore, he simply stepped up the present course of therapy, adding a few dietary restrictions, rules as to rest, sleep and the like, and hoped for the best.

After a few weeks a bitterly disillusioned and discouraged Garry went to another doctor who told him that outside of mild atherosclerosis, there was nothing wrong with him. Garry became even more bewildered and depressed.

Finally the contractor who had unwittingly caused the trouble in the first place began to worry about Garry. Approaching him at the lunch hour he said to him in a bluff but kindly voice: "Look, Garry old boy, you seem to be knocking yourself out on this job. What's the matter, pal; you sick or something? Why don't you take off for the afternoon and go see a doctor?"

"I don't know," Garry said, sounding woe-begone. "I've been to a couple of doctors. They can't do anything for me. I've had hardening of the arteries, not bad, of course, but now I've got this flutter; it comes and goes. One Doc tells me I must be crazy; there's nothing wrong at all. Another Doc wants to know whether I beat my wife and how I like my job, and a lot of silly stuff like that."

The contractor laughed and patted Garry on the shoulder. "Don't kid yourself, Garry. There's a lot more to that stuff than you realize. This racket can give anybody *heart flutters* or a darn sight worse. Where do you think I got my ulcers? I took on some war contract work that I should have known I wasn't big enough to handle. Everything went wrong. When I started to get sick the Doc told me I was in over my head on the job and I almost punched him in the nose. Darned if it didn't turn out he was right. Look, Garry, I'll bet you're worked up over that caved-in wall deal. Maybe I should have apologized. I know it wasn't your fault."

Garry relaxed a little, but not entirely. In any event he agreed to go back to his family doctor, *with* his employer. This time he was ready to cooperate, and he was backed up by the feeling of security that resulted from his employers' expressed confidence in him and solicitude for him. The result, after much delving, was a happy one and Garry became, relatively speaking, a well man.

Functional Disorders May Affect any Part of the Body

Functional illness may affect *any* part of the human body. Functional disturbances of the lower gastrointestinal tract, i.e., colitis — are characterized by fleeting belly pain, bowel irregularity, and sometimes diarrhea, while the upper portion relates its feelings in terms of pain, gaseous eructation and bloating, and vague, generalized abdominal fullness.

Functional illness shows up in the cardiovascular system in the form of the extra heart beat, noncardiac chest pain, shortness of breath, and irregular cardiac rhythm. Certain functional departures from normality cause inordinate fright. These may include questionable hypertrophic arthritis of the spine, insomnia, and morning fatigue.

Anxiety attacks and tension headaches are also manifestations of functional disturbances of the body's nervous system. Recent research has shown that even the common cold may strike some people more frequently than others for reasons having to do with the emotions. Stress and the secretions of the nasal passages have a proven relationship.

Migraine Headache

The migraine-type headache nearly always indicates a need for painstaking scrutiny of the sufferer's relation to his environment. The headache-prone individual is often quite different from the victims of other types of functional illness. He tends to enjoy the kind of stress and strain that might upset another individual. Oddly enough, his headaches seem to appear in the *absence* of obvious stress, although boredom or guilt feelings over inactivity may be considered inverted forms of stress in themselves.

The interviewer should be on guard for certain clues that indicate the headache-prone personality. The majority of migraine-sufferers are perfectionists. They are often the children of perfectionist parents who have expected a great deal of them. As a result, they are worriers. If they feel that at any time for any reason they have done less than their best, it upsets them. When they are facing the challenge of overt stress, they are at peace with themselves because they are accomplishing something; they are living up to what mother or father always expected of them.

These people are much more likely to be at war with themselves than with the rest of the world, though they may be seriously exasperated when people in their circles, i.e., their wives, husbands, children, close friends, or close associates at work, fail to live up to certain standards of perfection.

Another clue to individuals with the headache tendency is that they find it difficult to be casual. They worry considerably about what other people think of them; their conversation, hence, is likely to have an artificial sparkle. They are often fussy about clothes and manners, and avoid people with a happy-go-lucky philosophy of life for fear of being thought like them. They live according to plan.

The migraine type of headache, which incidentally, includes the menstruation headache and the menopausal headache, is the immediate result of the stretching of the walls of one of the arteries in the head. Although usually on one side, it may become general, or shift from side to side. While it may begin at any hour, it often strikes early in the morning and awakens the sufferer. It is likely to be severe and pounding, with nausea and sometimes vomiting. Most people with migraine headache are depressed, snappish, and somewhat antisocial until the headache is over. Then a feeling of euphoria is not unusual.

Headaches May Be Nonfunctional

Headaches may be physiologically based and nonfunctional in nature. The high-blood-pressure headache is an example, but this is the exception and not the rule. A rise in blood pressure is rarely enough in itself to cause headache. A great many individuals have headaches in association with high blood pressure, but in most cases they are individuals who might be expected to have them *without* the high blood pressure.

In one study, it was found that nearly all patients with hypertensive headache had been troubled with headaches for years before the hypertension was discovered. High blood pressure merely precipitated a condition that was going to occur anyway.

Other factors which may trigger a situationally based headache are alcohol, smog, hot weather, a smoke-filled room, damp weather, or an argument, however, in nearly all such cases it may correctly be assumed that the patient was going to have a headache anyway.

Headaches with a Physical Cause

Among the types of headache ordinarily considered as having a purely physical cause, it is surprising how many are related to emotional stress.

One type of headache, not as common nor as intense as migraines, is the result of tightening of the jaw or other facial muscles. This headache is often the result of a state of fear or apprehension.

Headaches owing to eye disorders are sometimes the result of infection or muscular malfunction. The commonest form of glaucoma, which occurs when there is an increase of fluid pressure within the eyeball, is known to be a reaction to stress.

The well-known sinus headache is directly the result of inflammation of sensitive membranes during a head cold or sinusitis. This is caused immediately by infection, but, as noted above, it has been determined that some people are "nasal reactors," which means that they react to undue stress by being inordinately susceptible to head colds or to hay fever.

Treatment of Tension Headache

Treatment of the tension headache at first involves considerable questioning. The circumstances preceding each headache should be ascertained, through history-taking and, if necessary, through interviews with members of the patient's family and associates at work. The important thing here is to establish communication with and gain the confidence of the patient, who should be encouraged to analyze himself in an effort to discover the underlying cause of his problem. Sedatives may be prescribed as momentary pain-relievers, but none are specific for cure.

The average headache-prone individual must be made to face the fact that while he can learn to avoid or come to terms with most of the headache-producing situations, the essential "headache proneness" cannot be cured. He may escape headaches by learning to shrug his shoulders occasionally, to relax, to take life pretty much as it comes, but anytime he tightens up again and reverts to his old attitude toward life, the headache will catch up with him.

Obesity as a Functional Disorder

There may be some debate regarding the inclusion of obesity in the list of functional disorders. True, overweight, with few exceptions, is the direct result of overeating, and habits of overeating may begin in a number of ways totally unrelated to stress, for example, it may be a result of parental prodding in early childhood. However, once such a habit is firmly established, it takes on an emotional significance. A full stomach becomes a symbolic retreat to childhood, i.e., an escape from gnawing anxiety which masks itself as gnawing hunger.

Compulsive eating is probably a more common disorder among Americans than compulsive use of alcohol, narcotics, or tobacco. Its end result can be more disastrous. Statistics support the fact that the fat person dies of degenerative diseases before his time. Rarely does a notably overweight person reach the age of 90.

The three principal killers of the obese person, often at the height of his productive career, are: (1) heart disease, (2) high blood pressure, and (3) diabetes.

Frequently in the case of coronary thrombosis, the blood vessels have become so loaded with fat that one of the thickened sclerotic coronary arteries nourishing the heart muscle becomes clogged and blocked, causing a part of the heart muscle to die instantly from lack of nourishment.

As for blood pressure, it is known that a direct relationship exists between blood pressure levels and body weight in youngsters as well as adults.

In diabetes, frequently a by-product of obesity, one must constantly be on guard against infections or cardiovascular renal disease.

An illustrative case of a functionally obese, overambitious businessman is that of Claude S.

This man at age 38 was on the way to a successful career. At college, he studied engineering, but his flair for administration soon landed him in an executive capacity in the financial end of his business. He enjoyed good living, good food, and good wine.

A series of exasperating surface infections such as boils and carbuncles at this time, suggested to Claude's doctor that sugar studies might be in order. The results of these studies convinced the doctor that Claude was a *potential* diabetic. He explained this at length and told Claude his only hope was to keep this disorder under control by means of careful diet. Claude was not overimpressed. The years rolled by, during which he suffered minor medical disturbances and continued to put on weight.

At the same time, he kept on climbing that ladder of success, and, as many successful men do, began to think of himself as an expert in all matters, particularly including health.

At the age of 54, Claude suffered his first anterior coronary occlusion. This grounded him for several months, but his personal ambitions at the time stimulated him to make an effort to get well.

Three more years went by and he began to show signs of brain involvement. This was characterized by listlessness, falling asleep in meetings, lack of alertness, and forgetfulness. In his opinion, he was misunderstood by his fellow associates, particularly his superiors. He resented advice from anyone, including the medical profession.

His degenerative tendencies progressed. One day he suffered a severe pain in his left leg. All professional efforts were brought to bear; Claude's life was saved by amputation of his leg at the level of the thigh. His actual diabetes was not difficult to control, but the fat content of his circulatory system was a difficult medical problem to cope with.

He became morose and paranoid. For the sake of his ego, his employers did not retire him, as was normally the custom in cases like his, but retained him as a consultant. His judgment was poor, but this was not readily discernible by his immediate associates. Too late it was discovered that he had engineered the dismissal of one of his former associates who had befriended him.

In time, his premature senility progressed to the point where he was confined to his home. In conversation, he lacked sense and coherence. In time, his circle of friends dwindled to none.

His actual death was uneventful and came as a blessing to those in his immediate family circle.

What Experience Teaches

The humanologist in business will learn by experience that it is usually futile merely to prescribe a reducing diet for the obese individual and tell him to stick to it. First, he must induce the patient to put his mind and soul in order, so to speak, and make an effort to direct his thoughts to nongustatory pleasures. At the same time, the patient must be encouraged to indulge in as much introspection as may be necessary to determine just where hunger leaves off and emotion takes over in his yearning for food. What pressures or fears underlie appetite?

The next step is to establish a sane diet in which calories are cut down, but not essential nutrients. Starvation diets can take off pounds; they can also cause dangerous and perhaps irreversible manifestations of malnutrition. Reducing isn't supposed to be easy. The purpose is not only to lose pounds, sensibly, but to lose the desire to overeat; the diet that supposedly knocks off weight in a hurry without ever making one hungry will not alter inner compulsive eating patterns, which will return as soon as the regimen has come to an end.

There are a number of standard and proven diets which should be known to the properly qualified plant nurse. Such a nurse should also be adept in common-sense methods of establishing the attitudes prerequisite to the inauguration of any dietary regimen. If the overweight individual is otherwise in normal health, it is rarely necessary to call in a medical consultant to help lay out the proper course of therapy. The exception will be if there are indications of underlying physiological or psychological problems beyond the professional limitations of the registered nurse.

Problem Drinking and Personality Disorders

Two other functional disorders which may or may not be related, depending upon the individual and the circumstances, are problem drinking and personality disorders preliminary to the psychiatric state. Since both of these disorders invariably begin as supervisory problems, they are only mentioned here and will be dealt with in detail in the following chapters on Supervision.

Peptic Ulcer

Extremely pertinent here is the well-known *organic* disorder, with the *functional* background and the *situational* root, known as the peptic ulcer. Every person who has interested himself in the humanological aspects of occupation has had some sort of experience with this problem.

Among employed persons, the ulcer may be considered the elite of all functional illnesses. It is truly the service badge of competition. No longer is it the exclusive property of the male. Women who are usurping many onetime male positions in the business world are beginning now to enjoy the right to have ulcers.

Many kinds of anxiety or emotional distress can stimulate excessive activity of the intestinal muscles and excessive secretion of the acid-forming glands of the stomach, resulting in the formation of a peptic ulcer. In a detailed analysis of 201 ulcer cases which came to surgery, 64 per cent were in the duodenum. The remainder were located in various segments of the stomach.

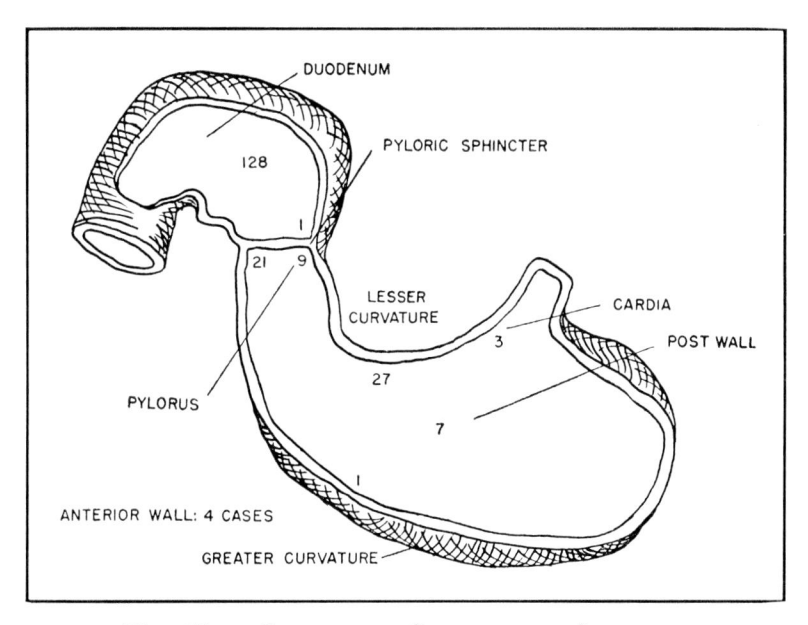

Fig. 15 — Segment of Stomach and Intestine

Causes of ulcer may range from excessive worry over detail to an inner sense of inadequacy. No matter what the cause, the ulcer, once arrived is good for a stay of from six to eight weeks, with recurrence in the offing unless the true cause is understood by the physician and patient.

The following cases are illustrative:

Harvey F., 38-year-old single man, started as a sailor as a boy, obtaining his officer's papers after 10 years at sea. Having served satisfactorily as a second mate for a number of years he was assigned to the next superior post. In this capacity, he was master of the ship during the captain's shore leaves.

The first time Harvey served as relieving master he developed the classic symptoms of a duodenal ulcer. He was referred to a specialist who saw to it that he was relieved of sea duty for a year. At his request, he then returned to sea duty as a second mate, and after two years was again elevated to first mate, and, in consequence, relieving master. At the end of the first voyage, the captain announced his intention of not returning to sea for the subsequent one. Four days before Harvey was to take the ship out he experienced a recurrence of his ulcer problem. A subsequent medical examination substantiated the opinion that his ulcer symptoms had to do with his added responsibilities.

Harvey has now come to face the fact of his limitations. He has returned to work and is now one of the company's best *second* mates.

Walter B., 34, is a college graduate with a degree in biology, who is now employed in the personnel department of a department store. He lives with an incompatible mother-in-law, his wife, and two children.

He has a medically confirmed ulcer (duodenal). He claims it is impossible for him to stay on an adequate diet. Lunch facilities at the store are not good. His mother-in-law pooh-poohs his complaints as nerves and has persuaded his wife that he is not in need of special diet or extra rest.

Recently an acute emergency occurred in the evening shortly after dinner when things had been going badly. Walter suffered a severe gastrointestinal hemorrhage. He was rushed to the hospital. Rest, and several quarts of blood brought him around. Six weeks later he was able to return to work.

Walter is aware of the exact nature of his problem but appears to be unable to do anything about resolving it. His actual working environment needs further humanological investigation. Walter is unwilling to face this fact or make a constructive effort to untangle his home problems with forcefulness. What he needs, as do many in his situation, is some coaching in techniques of honest self-analysis. Here is where the humanologist can be of inestimable value. Time, patient understanding and the facility to communicate are essential qualifications for the humanologist here.

Technique of Honest Self-Analysis

A device that can be used to encourage potential victims of the various types of functional illness is a "Self-Analysis Form" which can be easily and inexpensively prepared by the appropriate member of a given humanology team. This form, consisting of a card or sheet of paper to be given to the individual in question, may have printed across the top in big block letters:

WHERE DO I STAND HEALTH-WISE?

Below this should be two columns: one on the left headed "Situational" and on the right "Functional."

	Situational		*Functional*	
	Yes	No	Yes	No
1. Job				
2. Home				

Under "Situational" may be printed (1) Job. Under this in smaller type may be a number of pertinent questions, such as the following:

Am I fit for my job? Is it taking me where I want to go? Do I have that Monday-morning queasy feeling? Do I avoid pressure? Does the boss ride me? Am I sickness or accident prone? Do I get along with my fellow workers? Do I like my boss? Am I a clock-watcher? A calendar counter? Am I a perfectionist? Am I keyed up at the end of the day? Does after-work traffic exasperate me? Is my job a challenge or just a living? Other questions....

Next write (2) Home, and under this, questions like the following: Is home life moderately happy? Is my wife a teammate? Do we talk over all family problems together? Does she fully understand what I do for a living? Do I fully understand her day-to-day problems? Do I take part in child rearing? What are the names of my children's teachers? What marks did my children get in all subjects last semester? Do my children know what I do to earn a living? Do I need financial help from parents or in-laws? Do I have to design for living to mitigate in-law friction? Am I adult enough to be on friendly terms with my mother-in-law? Is home a bore? Do I need a drink or other diversion to get away from it all? Do I play with my children? Do we ever go places as a family? Do I have any hobbies? Do my wife and I wrangle over conflicting tastes? Money? Other pertinent questions. . . .

Explanation a Counseling Function

It should be explained to the individual in need of self-analysis that the purpose of these questions is not necessarily to find *solutions* to the problems uncovered, but simply to arrive at an awareness that the problems *do* exist and may have something to do with certain inexplicable symptoms. If a problem cannot be utterly solved, sometimes, it is possible for the individual to come to some kind of constructive terms with it once he is aware of its existence.

Under the right-hand column headed "Functional" it may be written: (1) Tension Headache. Underneath this in smaller type, some of the important identifying clues to the headache-prone individual may be written.

Next it may be written: (2) Obesity, followed by questions similar to these: Am I overweight? Am I in the habit of overeating? How much of this is due to genuine hunger? How much may possibly be emotional? Does food tide me over a period of anxiety? Am I pleased and proud when my children wolf down more food than is necessary?

Next may come: (3) Problem Drinking (If I am truly a problem drinker I will not answer these questions honestly). These questions may follow: Do hangovers ever interfere with the performance of my job? Have I ever mixed drinking with work? Do I need a drink for conviviality and fun or for medicinal effect? Do I ever *need* a drink to carry on? Do I invariably turn quarrelsome after heavy drinking? Is my wife a heavy drinker? Do we have family quarrels about drinking? Do I share nonalcoholic pleasures with my family on weekends? After an evening of moderate drinking do I lose my memory of what occurred? Do I gulp down the first two drinks at a party to get a quick edge? Do I sneak drinks in the pantry during parties, or fortify myself with a stiff one before leaving home?

Such devices will not be by any means 100 per cent successful. As was pointed out in the prior chapter on "Selection and Placement," all forms and formulae must be taken with a grain of salt and used only as supplementary tools. Moreover, as was explained in the chapter on "Training," you can lead a horse to water, but you can't make him drink. No one who does not truly wish to analyze himself will do so, no matter how many helpful devices are placed at his disposal.

Communication

The key to the problem is *communication* on the part of the appropriate humanologist. He must get through to the individual in question, make him understand the vital necessity of *knowing himself*. With *proper communication and thorough self-knowledge on the part of the employee* there need be no really significant health problems or cases of insipient maladjustment in the area of motivation and promotion. The responsibility for such an atmosphere falls on the shoulders of the front-line supervisor, who will be the subject of the two following chapters.

A Layman's Confirmation

In a recently published autobiography, the well-known inventor, Philip K. Saunders, strikingly confirms these chapters on Supervision.

Writing in a light and humorous vein, the author describes how prolonged frustration reduced him to a state of physical and mental debility. He developed most of the symptoms described in the preceding chapter.

His undoubted inventive and administrative abilities at last receiving recognition, Saunders promptly recovered. At 60+ he is a healthy and happy man, as the present writer can affirm.

That Saunders should so exactly have blue-printed the case for humanology makes his work "Dr. Pantofogo" published in 1959 by Prentice-Hall, recommended reading for all students of the subject.

REFERENCES

DIETZ, NICHOLAS, JR., PH.D. "The Common Cold and Stress Conditions." *Industrial Medicine and Surgery*, XXVI (No. 5, May, 1957), pp. 229-233.

GOODMAN, J. I. "Obesity-Relationship to Chronic Disease." *Geriatrics*, X (February, 1955), pp. 78-82.

NEW YORK BANKERS ASSOCIATION. *Scientific Morale Survey*. New York: New York Bankers Association, 1957.

PAGE, ROBERT COLLIER, M.D. "Alcoholism and Industrial Health." *Industrial Medicine and Surgery*, XXIII (No. 4, April, 1954), pp. 145-147.

——————. "Honest Self Analysis." *It Pays to Be Healthy*. New York: Prentice-Hall, 1957, p. 51.

PAGE, ROBERT COLLIER, M.D., AND RANKIN, L. M., M.D. "Review of Ulcer Surgery at Presbyterian Hospital 1921 to March 1935, Inclusive." *The American Journal of Surgery*, XXXVII (August, 1937), pp. 219-231.

PIGORS, PAUL, AND MYERS, CHARLES A. *Personnel Administration*. New York: McGraw-Hill, 1956, pp. 376-378.

RICHARDSON, BELLOWS, HENRY AND COMPANY. "Nine Basic Human Needs." *Developing People in Industry*. New York: Harper and Brothers, 1956.

WOLF, STEWART, M.D., AND WOLF, HAROLD, M.D. *Headaches; Their Nature & Treatment*. Boston: Little, Brown and Company, 1953.

WORTHY, JAMES C. "Organizational Structure and Employee Morale." *American Sociological Review*, XV (No. 2, April, 1950), pp. 169-179.

14

Supervision

(Part I)

Definition of a Supervisor

THE SUPERVISOR must be defined as one who is in a position of *direct command* over one or more employed persons, charged with the attainment of some specific goal or goals pre-established in accordance with company policy. He is the person who sets the pace of training, who gives orders, hands out discipline, inspires, or fails to inspire, quality performance, makes notes of individual strengths, weaknesses, and significant changes, recommends who should be promoted or demoted, and, in general, makes or breaks whatever policy may be handed down from above. If it is a question of "make" he may not always get the credit; someone over or under him will most often get the notice. But if it is a question of "break" he nearly always takes the blame. He is a key individual, and to survive he must be a thoroughly healthy individual as well; healthy in mind and body.

Meaning of "Supervisor"

There is a tendency to stretch the meaning of "supervisor" beyond its natural limits, and thus to confuse the actual functions of supervision. The manager of a branch of a department store with 10 or so employees can spend a considerable part of his time exercising supervisory functions. He is in and out of each of his departments several times a day, and he is possibly on a first-name basis with all of the employees. When an emotional crisis is brewing in "Ladies-Ready-to-Wear" or when there is an unaccountable fall in sales in "Yard Goods," he steps in and takes personal, direct action. He is captain of the team.

From the foregoing, it might be erroneously assumed that the manager of a branch in the same chain in a larger city, employing 100 or more employees, is also essentially a supervisor, though on a larger scale. The small store manager, promoted to command of such an operation, making such an assumption, would surely come to grief. He would find that he had neither the time nor the detailed knowledge to be in direct command of every operation of the store. In order to operate efficiently, he would have to delegate functions of supervision to assistants or to individual department heads, depending upon the size of the department.

Even in the small, 10-employee store, one may assume that there are times when direct supervision on the part of the manager is difficult if not out of the question. At certain times of the year, particularly in the Christmas season, when extra hands are employed and an extraordinary turnover of inventory occurs, it may become impossible for the manager to keep a supervisory eye on all that is happening, and intervene personally and directly in each problem: each bottleneck, each hysterical breakdown, each accident, each quarrel, each misunderstanding, or error. At such times, he must rely upon the supervisory abilities of certain seasoned and trusted subordinates.

Too often the supervisor is looked upon merely as the person assigned to obtain a certain predesignated performance norm from a group of employees. True, that is his economic, organizational function, but in order to fulfill this function he must be a great many things. It is taken for granted that he must be a leader; nowadays it is becoming increasingly evident that he must be something of a teacher, since there are fewer and fewer jobs which involve such simple commands as: "Here, take this shovel and go over there and dig," or "Load up this wheelbarrow; take it over there and dump it, and bring it back for another load." Today's supervisor is more and more likely to be in charge of operations which are complex enough to require a continuing training effort on his part to meet the requirements of his assignment.

Supervisor Must Be a Vanguard

What will come to be accepted as humanology progresses is the fact that the supervisor must also be a vanguard or reconnaissance unit for the psychologist, the medical consultant, the nurse, and the hygienist on the humanological team. He is dealing with human beings whose ability to stay with the team and carry out the duties delegated by him, happily and enthusiastically, depend upon their physiological and psychological health. He is in a position to be the first to notice when a change in attitude toward work or in patterns of productivity indicates a questionable health factor or an ominous development somewhere in the individual's environment. At such times, he can act as a counselor; he can offer common-sense; he can establish communication; he can even disseminate a certain amount of elementary health education. Properly trained for his supervisor's role, he will know, too, when to refrain from personal action in the problem and move quickly to seek qualified help.

In this capacity, he will of course maintain a close liaison with the plant nurse and the plant hygienist, who will in turn be in a position to call in whatever professional help may be indicated. Under certain circumstances, particularly where large numbers of relatively low-paid workers are involved, the supervisor-nurse-hygienist team may find it necessary to cooperate with outside welfare agencies. Constructive action in such a case will ordinarily come into being thanks to the sharp but sympathetic eye of the well-trained supervisor, who is sensitive to the early symptoms of environmental derangement.

Health Requirements of the Supervisor Himself

The concrete responsibilities of the ideal supervisor with regard to the health of his subordinates will be discussed in detail later in the next chapter. First it seems logical to analyze the health requirements of the supervisor himself, management's responsibilities with regard to his health, and his own responsibilities with regard to health maintenance.

Obviously, a supervisor must be in A-1 health psychologically. A person beset with emotional problems of his own, job-based or otherwise, cannot be expected to be fully sensitive to often barely detectable signs of emotional distress in others. His personal health habits must be exemplary. The gluttonous machine shop foreman is in a poor position to give advice to the subordinate who has allowed himself to become too obese to perform certain maintenance functions involving crawling under pieces of machinery. The hangover-prone supervisor who needs a restorative pick-up at lunch time is limited in what he can say with effect to the employee with an incipient alcohol problem.

The foregoing is not to be taken to mean that every supervisor must necessarily be free of health defects. Such a policy would be not only cruel; it would keep a number of highly valuable individuals from serving in the responsible posts for which they are fitted. It has been noted a number of times in foregoing chapters that health is relative; one man's optimum health is another man's health problem. It is largely a matter of personal adaptation to the facts of life, plus philosophical insight. Following are two illustrative cases of supervisors with varying health handicaps. It will be noted that one had learned the secret of turning liabilities into assets, and was truly healthy despite readily measurable health problems; the other failed in this, for reasons deeper than the diagnosable problems itself, and was ineffective as a supervisor in the long run.

Burt P., 48, was an old hand in the timber business. He was in charge of a crew of lumbermen in the Pacific Northwest when he suffered his first diabetic coma. Following recovery, Burt discussed the problems of getting back to work with a sympathetic employer and a medical consultant. The employer stated frankly that at first thought there seemed to be some grave objections to sending Burt out into the field in command of a crew. The "mess" facilities were not geared to cope with one man's special diet, for one thing. Burt was a "trencherman from away back," and even if it were worked out to provide him with the strict, carefully measured diet required, plus the necessary medicines and equipment, he would be required to exercise tremendous will power, particularly when joining the boys at their usual gargantuan repasts in the mess shack after a long, hard, invigorating day out-of-doors. The other problem, his employer said, was the fact that Burt would have to spend most of his days and nights at a considerable distance from any adequate medical help in the event of an acute emergency.

The medical consultant pointed out that as long as Burt was fully cognizant of his limitations and adhered strictly to his prescribed regimen there was no reason he could not carry on in his former capacity. He said it was his understanding that Burt was a highly reliable individual who could be depended upon not to take on any job he was not sure he could handle. This was his choice and it was not an easy one. There was a less demanding job open for Burt at one of the company mills near a fair-sized town, where adequate medical facilities were available. If he chose this, he could perhaps afford to take a few chances with his health problem. If he chose to go back into the field with a crew, he could not under any circumstances take chances. He must make the most rigorous adjustment possible. If an emergency occurred owing to his carelessness, it might be possible to save his life; the crew was always in communication with the head office, by radio or telephone, and a helicopter was available in case of any severe medical emergency. But it would not be possible to save his job. There would be no second chance.

Burt carefully considered all of the ramifications and after a time said:

"I'll take the crew then. Trust me. I'm not saying: I'll try it and see how it works out. I realize that wouldn't be fair; you can't afford to send someone along as an understudy just in case the experiment doesn't work. What I mean is: I'll *do* it. If the day ever comes that I think the going is beginning to get rough, I'll give you fair notice in plenty of time so that we can break someone in to take over for me."

Burt made the transition successfully, learned to live within his limitations, and is still a first-rate supervisor. He will undoubtedly continue to fill an important post until retirement age.

George L., 52, managed a small suburban sales branch of a large insurance company. He had a good record as a manager, as a salesman, and as a trainer of new men on the sales force. A number of young hopefuls who were hired as potential officer material had been started under George because of his abilities to train and inspire those under him.

Changes in George's behavior had come about so gradually that no one could say accurately just when he had ceased to be the loyal, capable, and an admired supervisor of former days. When he had his first mild heart attack his doctor was shocked at the length of time it took him to recuperate. When he finally returned to his old position it soon became obvious that it would be necessary to replace him, perhaps to kick him upstairs to some honorary position where he would be safely out of the way.

His attitude was that of a thoroughly embittered man. He was suspicious of his sales force, accused them of talking behind his back and submitting false reports about him to the home office. Orders from the home office were angrily set aside. On one occasion, he wrote a lengthy letter to the president of the company, whom he had known from school days, describing how he, George, had been conspired against by a former subordinate, one of the young hopefuls who was now a junior vice-president of the company.

The company humanological team was called into action, too late, it seemed. It was learned that in actuality George had been an extremely sick individual for a long time prior to his circulatory accident. Organically, his problem had been atherosclerosis affecting most of his circulatory system, with some degree of brain involvement. Complicating this, however, was a severe situational state based on his feeling that he had been getting a raw deal from his company. It had never been explained to him clearly about the young hopefuls — that they had been hired specifically to be groomed for management positions. Every time one of these men was called to a higher post, "right out from under me," as George put it, he looked upon it as a reflection upon his own standing with the company. He could not understand why he was not the one to be chosen. Every time he heard of the subsequent promotions of his former boys, his feelings of inadequacy made life intolerable for awhile, and he spent a number of sleepless nights trying to readjust himself to the *here and now*. For a number of years, he was able to motivate himself with the delusion that any day now someone in an upper echelon position would note his long, faithful, competent service, and reward him with a boost into the ranks of high management. His heart attack blasted that delusion forever. He saw immediately that there was no chance for advancement, and he was no longer able to drive himself to loyal endeavor.

George is now retired, an embittered man. He was offered a *harmless* post but rejected it as a *cruel* joke. Recently he had another heart attack. The prognosis in his case is grave.

Undoubtedly his company was at fault for failing to communicate to George precisely where he stood in the general scheme and why it was necessary for certain previously picked subordinates to be promoted over his head. Perhaps it might also be argued that George lacked the philosophy or character to continue in a supervisory position if his petty pride was at stake. Whose fault it was is more or less beside the point here. What is more to the point is that a health problem, unreckoned with and uncontrolled, turned a good supervisor into a totally incompetent employee. All other considerations aside, if George's company had made it a policy to inquire periodically into the total health of its employees, and particularly its supervisors, George's problem would have been discovered at an early stage, and his case history would surely have had an ending more satisfactory for all concerned.

Meaning of Emotional Maturity

The first thing to be sought in determining whether an individual is supervisor material or not is — emotional maturity. This might almost be called the *sine qua non* of an effective supervisor. The above mentioned case of George illustrates a certain lack of emotional maturity. A really mature individual would not harbor feelings of envy or inadequacy. He would have made an effort to find out just where he really did stand, without waiting for someone from above to come and tap him on the shoulder. An emotionally mature person is proof against many situational problems, and can be expected to adjust intelligently to organic problems that arise.

Of course 100 per cent emotional maturity is extremely rare, if not utterly nonexistent. The individual who thinks himself wholly mature is an individual who merits careful watching when the chips are down. The noted psychologist, Dr. A. H. Maslow of Brandeis University, devoted years to the study of living, as well as historical individuals, and concluded that out of the thousands he had examined, only 43 could be described as meeting fairly rigorous tests of emotional maturity.

The potential supervisor, then, is not necessarily one who has attained the goal of maturity, but one who has a pretty good idea as to how far he has to go, and who is moving in the right direction.

Six Basic Principles of Maturity

The Research Institute of America lists "Six Basic Principles of Maturity," culled from a study of the opinions of the leading psychologists in the field. These are:

1. *Self-acceptance.* — The individual who is ready to assume a position of command is, ideally, one who has learned to appreciate himself without trying to be what he cannot possibly be. He is able to appraise himself and face up to his bad as well as his good traits. He is not torn by inner conflict and is thus in a position to direct his attention outward. These questions offer a handy yardstick to determine an individual's degree of mature self-acceptance:

a) Does he know how he operates? Is he aware of his strengths, weaknesses, motives for action?
b) Is he fair to himself? Does he make the most of his good points without brooding over his bad ones?
c) Is he making an effort to improve the bad points that *can* be improved, without wasting time following an impossible standard of perfection?

2. *Acceptance of others.* — The way an individual reacts to the weaknesses, aggressions, or malfeasances of other people is a judgment on himself; not on them. Anger over petty matters such as a clerk's carelessness in making change, a youth's thoughtlessness, a restaurant waitress' delay in bringing in the entree, this is not becoming to the sort of person who is charged with the job of leading a team. Certainly anger at real or apparent stupidity, a stupid reaction in itself, being emotional rather than logical, is wholly out of place in an individual who from time to time must be called upon to train, guide, and be an inspiration to a bunch of "dumb greenhorns." Every recruit is, after all, a "dumb greenhorn" until a good supervisor makes something else out of him.

Acceptance of others does not mean letting them get away with improper or questionable behavior. The good supervisor will know when to put his finger on error, because he does so without being emotionally involved. At the same time, he doesn't require the full approval of others in order to respect himself.

Acceptance of others also means striking a balance between over and under dependence on others. The seemingly totally self-contained individual is concealing a basic emotional disturbance. The individual needs the team as much as the team needs the individual; the one who outwardly denies this fact is more than likely nursing a deep wound which needs to be brought into the light. Over-dependence, on the other hand, is also a sign of immaturity. The following standards of mature dependency have been listed:

   a) *It is occasional; not full time.* The mature person leans on others in times of stress, but remains independent in the area of his strengths.

   b) *It is realistic.* A mature dependency is directed to those who are able and willing to fulfill the individual's needs. If the individual finds that those to whom he directs his dependence cannot or do not wish to give him the help he needs, he looks elsewhere with no feeling of hurt or frustration.

   c) *It is reciprocal.* A mature person knows when to depend on others; at the same time others know he may be depended upon in time of need.

Acceptance of others entails a sense of the appropriate in human relations. The mature person picks and chooses carefully the people in whom he confides for the sake of letting off steam. He does not confide in everyone. Reserve and dignity at appropriate times is part of the equipment of the healthy supervisor.

3. *Sense of humor.* — Sarcasm or ridicule must be ruled out of this category, since they are more often symptomatic of a desire to escape from a gnawing sense of personal inadequacy than from a mature ability to see things as they are. When an unpleasant situation cannot be changed, to shrug it off with a joke is a sign of maturity. The hostess at a dinner party who said with a twinkle in her eye, when the butler dropped the platter with the one and only turkey: "James, take that out and bring in the other turkey," was a real supervisor. James was already sufficiently mortified by his error; there was no need to rub it in. By carrying it off with good humor and making a fellow conspirator of James, she proved herself a thoroughbred. One may be sure that she allowed the incident to remain closed as of that moment, with no futile post-mortem recriminations or feelings of chagrin. Tact, it should be pointed out, goes hand

in hand with humor, as the above tale illustrates. A person deficient in humor can ordinarily be relied upon to be equally deficient in tact and vice versa.

4. *Appreciation of simple pleasures.* — Background may determine what simple pleasures are enjoyable; it is difficult to list any specific pleasures which will be enjoyed in equal measure by all emotionally mature individuals. Almost anyone with a happy background in a rural or semirural area will always derive genuine satisfaction from the sights, smells, or sounds of woods, meadows, or gardens in appropriate seasons. The city-bred person who is wholly mature may agree that such things are pleasant indeed but he may be totally at a loss to understand how anyone in his right mind can gain any significant inner satisfaction from them.

The inability to mature in this respect may often be blamed on the individual's parents. Many a person born with a silver spoon in his mouth finds it next to impossible to grow up and savor the basic things of life simply because as a child he was never allowed to be bored. His life was one of constant variety, with each new pleasure more elaborate than the last. Occasionally an individual finds the means to rise above such a background. Perhaps he is forced to it by adversity at an early enough stage so that the habits of world-weary pursuit of the ever-new, ever-exciting, have not become permanently fixed. Sometimes the other aspects of an expensive and good education, plus the example of friends of character, may combine to instill a constructive desire to become mature in this one respect. Otherwise, the result, when the pleasures run out, is almost invariably tragedy.

Ralph Waldo Emerson's definition of success offers the most all-embracing list of simple pleasures which will be appreciated by any truly mature individual:

a) To laugh often and love much;
b) To win the respect of intelligent persons and the affection of children;

c) To earn the approbation of honest critics and endure the betrayal of false friends;
d) To appreciate beauty;
e) To find the best in others;
f) To give one's self;
g) To leave the world a bit better, whether by a healthy child, a garden patch or a redeemed social condition;
h) To have played and laughed with enthusiasm and sung with exultation;
i) To know even one life has breathed easier because you have lived — *"this is to have succeeded."*

5. *Enjoyment of the present.* — It has been said that the best insurance for tomorrow is the effective use of today. The emotionally mature person does not exhaust himself with needless apprehension over that aspect of tomorrow which he cannot possibly change. A bird in the hand is worth two in the bush, and the individual who lets go the bird he has for the sake of the two he may capture in that seemingly nearby bush may find himself with no birds at all.

Moreover, the mature person does not kick himself about what has happened in the past. The past is a closed book. In George Bernard Shaw's play *Captain Brassbound's Conversion*, the childishness of fighting the past is skillfully depicted. This play puts the viewer's maturity to the test, by playing a typical Shavian practical joke on it. In the beginning, the apparent hero is a man with a grievance. The grievance is quite genuine, he has been badly treated, and he has dedicated his life, rather romantically, to obtaining revenge upon the person who hurt him. The viewer is invited to sympathize with this dashing avenger, and share in his desires for vengeance. The grievance is presented in the blackest possible light. Then suddenly toward the denouement, something unaccountable happens to the viewer's sympathies. A seemingly scatterbrained lady, one of the avenger's protagonists, suddenly begins to look more and more like a mature, humorous, and thoroughly delightful person, while the glamorous avenger looks more and more like a sulky, humorless little boy with his nose out of joint because of something that can't possibly be helped now. In the end, the avenger repents his foolishness and decides to live in the present, and it is hoped that an edified audience has done the same.

Immature people cannot tolerate the ambiguous. They want to know exactly how everything is going to come out. One suspects that they peek at the last page of the detective story they are reading. The immature supervisor, with a big challenging assignment, will fume and fret and push his subordinates beyond their normal capacities, simply because *he* cannot sleep for fear the job will not be done in time. As a result, he may get the job done ahead of time with a thoroughly demoralized crew, or demoralization may set in so early in the assignment that the job is never properly completed. "Sufficient unto the day is the evil thereof" is the motto of the competent supervisor, who does not try to live more than one day at a time, and who assumes that if his unit is working as a team, the work of tomorrow can be worried about when tomorrow comes. If, at this rate, the assignment is not completed in time, then it was obviously an improper assignment. No petty deadline is worth the destruction of a good team.

6. *Appreciation of work.* — A beachcomber is not an emotionally mature person. The ideal supervisor is the kind of person who, if he were placed in a position where he didn't *have* to do anything but lie on a beach and eat, drink, and sleep, would immediately *find* work to do. Such individuals as Henry Ford II, the grandsons of John D. Rockefeller, Averill Harriman, to name a few, are examples of the kind of emotional maturity that conduces to constructive endeavor even when there is no economic need for it.

There is a difference of course between the person who works because he has to, which is the case of the average supervisor, and the person who works in spite of not having to. The latter is in a position to do a certain amount of picking and choosing as to what kind of work he will do, while the former is usually severely limited in this respect. If he is truly of supervisor calibre, the latter individual will have avoided the types of jobs which run counter to his wants, needs, or capabilities, but often in the end he has been obliged to choose in a limited field of alternatives ranging from unpleasant to mildly unpleasant. This is unfortunately one of the economic facts of life, however the more idealistic of humanologists might wish it otherwise.

The answer in such a case is to learn not to fight work and decide to enjoy it. The supervisor in a field that might have seemed to him essentially unpleasant in the beginning will have learned the secret of taking pride in a "bang-up job." That is what makes him a supervisor. That is why those under him are happy to follow in an eager endeavor to keep the team's productivity record high. Enthusiasm is contagious, particularly when it produces results proving that we the team did a great job.

This final step to emotional maturity is the one that seems to be the most difficult for many people, and the most rewarding once it is made. The case of Eddie Arcaro, the jockey, is illuminating:

In 1956, when he was 40, an interviewer noted that after 19,000 races and nearly 3,700 first-place finishes, he was riding with the enthusiasm of a beginner, though there had seemed to be a falling off in some of the years immediately preceding. Arcaro said:

"I think I'm riding better this year than I have in the past two or three years. I'm feeling better than I have in four or five years. I've got more energy and my whole outlook is different."

He was asked why.

"Well," he said, "Last spring I was seriously thinking of quitting. I was lackadaisical. I didn't care if I rode or not . . . . I'm not a psychiatrist but I psychoanalyzed myself. I got to thinking what I would do if I quit, and I realized that I didn't know anything else I'd rather do. Psychoanalyzing myself gave me a different outlook .... I was fighting my work; now I want to go on with it .... I'm having fun. My wife even kids me about it."

A Check List of Emotional Growth

The Research Institute of America has printed a "Checklist of Emotional Growth," which might well serve as a useful supplementary tool in the periodic health evaluation of promising personnel. It contains the following questions to be answered by the individual himself. Presumably he is possessed of the insight to give honest and accurate answers and learn from the experience of analyzing himself:

1. Did you take formal training or instruction in the past year to further your progress?
2. Did you step up your reading?
3. Did you increase your participation in the last year in group activities, e.g., company teams, civic associations, church groups?
4. Did you improve your ability to handle routine and repetitive activities, e.g., correspondence, putting up storm windows?
5. Did you review past activities to determine which have been desirable? Which ought to be dropped?
6. Did you find it easier to deal with people?
7. Did you have fewer emotional flare-ups?
8. Did you get greater enjoyment out of periods of relaxation and recreation?
9. Did you devote more time to thinking about the reasons why other people behaved the way they did?
10. Were you more likely to concentrate on one activity until it was completed?
11. Did you devote more time to, and get greater satisfaction out of, helping others solve their problems?

12. Did you improve any of your skills or develop new talents?
13. Did you come up with some new conclusions about yourself, your personality, your habits?
14. Did you go in for new and more varied activities, develop new friends?
15. Did you find yourself making a larger number of independent decisions?
16. Did you find it easier to live with problems for which you had no immediate solutions?
17. Did you change some of your opinions and feelings about things?
18. Did you show a willingness to expose yourself to new experiences?
19. Did you gain a clearer conviction and a better understanding of the basic truths, religion, or philosophy in which you believe?

The Specific Value of Such Tests

It cannot be repeated often enough that tests of this nature are primarily for the benefit of the examinee, not the examiner, and this must be communicated clearly to all concerned. Units of organized labor have in a number of instances registered strong protests, and in some cases obtained strict prohibitions, against what has been deemed misuse of the periodic medical examination as a management tool to get rid of unwanted employees. Whether or not these protests have always been just is beside the point. The bitter and sometimes ineradicable ill-will that has been created is not beside the point. If an employee thinks, rightly or wrongly, that he is being examined by a professional person hired by the company for the purpose of railroading him, a grave humanological error has been committed on the part of management. This holds true at all echelons, from recruit to top-line officer.

Cult of the Positive

Finally, in assessing, or reassessing, the emotional maturity of the supervisor, the humanologist will do well not to be blinkered by the cult of the positive, which one hopes has just about run its course. For the last 20 or 30 years, popular magazines and books have disseminated a vast amount of material regarding the psychological, spiritual, and even economic values of thinking, living, and talking in a *positive* manner. While some of this material has been not altogether without value if taken with a grain of salt, the sum of it has tended to produce a breed of individuals who are often in need of being re-educated to common-sense before they can be wholly trusted with supervisory responsibility.

The perpetually positive, go-ahead-and-do-it, be constructive-not-destructive attitude is more often than not a cover up for inner uncertainty — for some known inadequacy that must not be revealed to the world. Just as every action has its equal and opposite reaction, so does every job with important positive aspects have its equally important negative aspects. By the same token, every constructive enterprise involves a certain amount of destruction. The emotionally mature supervisor strikes a proper balance between positive and negative; between construction and destruction. A major part of a job is learning *not* to do things. A major part of growing into responsibility is *destroying* old habits of thought and action that stand in the way.

State of Becoming

Supervision is not only a state of *being* but a state of *becoming*. In other words, having picked the right person, trained him correctly, and placed him in the perfect supervisory slot where he is performing with a high degree of competence, the humanological team will not assume that that is where the matter ends. Supervision is a continuing education. When a given supervisor ceases to find motives for learning and improving his supervisory techniques, one may suspect that he has grown above, or away from, his job, or that the assignment was improperly motivated in the first place. This often occurs when organizational blueprints prepared at top management level fail to provide for delegation of responsibility, authority, and initiative to lower-echelon supervisors. These latter become mere relayers of orders from above or complaints from below.

Translating Policies and Objectives in Terms of Individual People

The area in which the most competent of supervisors will continue to improve and increase his value to management is that of translating policies and objectives into terms of individual people. The enlightened policy-making officer who has become alerted to the importance of humanology and who has shed himself of the widespread myth that a no isn't as valuable as a yes, will make a point of seeking the opinion of his seasoned supervisors before writing a specific policy. He will also check with his supervisors from time to time to determine whether or not existing policy is effective as regards the individual human beings with which it has to do.

An example of this approach is seen in the case of Blank Company, a midwestern manufacturer of plastic novelties. This company had a serious community relations problem, owing to a strike which had involved violence on both sides. The local press had been unsympathetic to the company, which was not a user of local advertising space and hence at a disadvantage in obtaining space for publicity releases which would present its side in the dispute. After settlement of the strike in labor's favor, the ill will had persisted to the extent that a company officer serving as director of a fund drive for a local church was asked to step aside in favor of someone whose name would not hurt the campaign.

It was decided to hire a public-relations expert, to be attached to the already existing advertising department. The company, being a small one, had not given thought previously to public-relations problems.

The functions of the advertising department were not creative; that aspect was handled by an agency. Primarily, this department existed to process facts to be turned over to the agency, to check on results of agency campaigns, and to serve as liaison between the sales department and the agency. A considerable amount of routine work was involved, and this department had always operated on a rigid schedule; a time clock kept track of hours put in by each member of the department.

The new public-relations man was utterly frustrated by this routine. His title of "Director of Public Relations" did not exempt him from punching the time clock, even though he was supposedly to be rated according to ideas conceived and results produced instead of hours spent in the office.

He explained to the chief of the advertising department that he felt he should be out, questioning representatives of various groups in the town, acquainting himself with key people at the newspaper and the local radio station, sounding out workers and their union representatives as to wants and needs. After he had done this, he said, he might be in a position to work out some ideas regarding concrete steps to be taken by the company to re-establish sound community relations outside the plant and human relations inside it. However, even this working out of ideas was not going to be achieved on a time-clock basis, he said. It would take considerable floor-pacing, rechecking of facts, sleepless nights, and the like.

The advertising department chief was a sympathetic individual. He said he understood the problem exactly but it could not be explained to "Old Joe," the company's president, founder, and chief stockholder, who had lived all his life according to a rigid schedule, learned during a farm boyhood, and was utterly intolerant of anyone who was not precisely like him.

In order to justify his existence from punching in to punching out time, the new director had to content himself with sitting at his typewriter writing news releases and radio "spots" which, though approved at company level, were rarely used. He finally managed to persuade his employers that if they would purchase a modicum of advertising space in the local papers and time on the local radio station, the chances of placing material would be greatly improved. Since the company distributed its product nationally and had no need for local mediums to push sales, it relegated a small percentage of its budget for local goodwill advertising. This opened the gates for a trickle of free publicity, and the public-relations "director" was duly congratulated. He himself knew he was not doing his job of improving community relations to any significant extent and continued to feel frustrated. He reached the point where he woke up in the morning with an unexplainable array of aches and pains. It was all he could do to drag himself out of bed and force himself to go and punch that hated time clock.

His supervisor, who sympathized with him, covered up as well as he could for his frequent tardiness. As for his work, it became increasingly difficult for him to bring himself to type news releases which he knew were almost entirely ineffective, even when printed in the paper or sounded over the air. He developed a tendency to stare at a blank sheet of paper helplessly until he noticed he was being watched, perhaps by someone from the top echelon who happened to be walking by. Then he would type rapidly pure gibberish: "Now is the time for all good men to come to the aid of the party," just to look busy, until somehow by a feat of sheer will power he was able to bring himself to start a publicity release.

At length, the failure of the public-relations campaign became evident even to "Old Joe," the president, and this gentleman was prevailed upon to have a humanological or "mantalent" survey made by a consulting firm in the field of human relations, the object being to determine how great a proportion of the company's unsatisfactory community relations lay in unsatisfactory relationships at the plant itself. Among other things, the survey disclosed a tendency on the part of the top management to hold too tight a rein on the entire operation, with the results that supervisors were not being used as supervisors. It was recommended not only that supervisors be given more freedom in building and leading their teams, but that they be consulted occasionally to determine the effects of top-echelon policy in terms of people.

As a result of these recommendations, the chief of the advertising department was called into consultation with "Old Joe" and the vice-president in charge of personnel shortly after the completion of the survey. After outlining some ideas he had about welding the major part of his department into a team, he said:

"Now the rest of us don't mind this time clock. We don't like it much either, but since we have to conform to a certain routine based on time, we can put up with it. This public relations man is something else again. He's got ideas about what he could and should be doing, but he can't do it chained to a typewriter from 9 to 5 every day. He's got to get around town, making friends, learning things about the community and where we fit in, getting himself oriented. He's got to put in evenings, talking to people or just pacing the floor stirring up his brain. That time clock is preventing him from doing a job. I know it and he knows it, but up till now there hasn't been any way to put it across. I'm afraid the clock is actually making the young man sick, though he won't admit it. He hasn't been looking well — I won't go into all the details — anyway, I made him report to the nurse last week. I couldn't get out of him what she said to him; so I went down there myself and asked her if I ought to order him to go to a doctor. She said: 'He doesn't need a doctor. What he needs is to go get himself another job, but try to make him see it.'

"Another thing; why is he under me, when he has the title of 'director?' He doesn't fit in with my team. We're doing cut-and-dried work; he's doing creative brainwork. I may be wrong, but if he builds up any of the ideas that are in his head now, he'll make such a big thing out of community relations you'll have to get him an assistant or two. Why not make him a one-man department right now in anticipation of that, and give him an office of his own, where he can do his job without having people peer over his shoulder?"

"Old Joe" grumbled at first. He still had a basic distrust of anyone who was not as thoroughly enslaved by routine as he himself was. However the recent trouble had convinced him that he was not infallible. He agreed to exempt the public-relations director from the time clock and allow him to come and go as he pleased. This was to be done on a six-months' trial basis, to see if the change produced results. As for the suggestion to detach public relations from the advertising department as a separate department, he preferred, perhaps wisely, to be conservative, and refrain from making any drastic organizational changes until the exact scope of community relations could be determined.

The gradual result of the above-mentioned changes has been that the Blank Company has become a highly respected part of the community. This of course was not wholly due to the expanded functions of the public-relations director, who incidentally is head now of a growing department; the organizational changes recommended as a result of the humanological survey established channels of communication and a team spirit which greatly elevated employee relations. However the community studies made by the public-relations director opened up doors of good will that mere in-the-plant relations might not have done. He was able to determine community needs which the company was able to help fulfill: baseball uniforms, a park and coach for a group of underprivileged youngsters; a community diagnostic unit; remodelling and partial underwriting of the community amateur theater; instruments for the city band; leadership in numerous welfare fund drives; scholarships for promising high-school students.

The advertising department chief cited above had learned something that some excellent supervisors do not learn even after years of studying human beings: that the intelligent "maverick" may have worthwhile reasons for wanting to be treated differently, that to enforce conformity on him for the sake of expediency or "democracy" will solve no problems worth solving.

Responsibilities to Top Management Versus Subordinates

Perhaps the most delicate line any supervisor can learn to draw is that between his responsibilities to top management and his responsibilities to the members of the team he leads. A little bit of power is heady stuff, particularly to the young man getting his first taste of supervisory authority. Having passed his personal period of adult experimentation, he is well into his period of acquisition. Little things don't bother him. All that matters is the shining goal in the distance. He is on his way. He is rather full of himself. If the old-timers resent him and think he is stepping on too many toes; that's not his problem. He's going up the ladder of success while they are going down. By the time they are turned out to pasture he will be a big shot.

Then comes the terrible day when he is faced with an assignment that can make or break him, and he knows that the success of this assignment depends wholly upon the eager, willing, capable cooperation of his underlings upon whose toes he has so frequently trod. Bluster and threats won't accomplish it. On such a day, a supervisor who has always been strictly a management man is likely to acquire a lasting though momentarily painful education.

The opposite tack can be just as bad; perhaps worse. The supervisor who has learned how little he can really accomplish without a close understanding with his subordinates may decide to become a bit of a politician. He will bend over backward to be liked by the gang. Those under him will in time wise up to the fact that this supervisor can be taken advantage of, since he can be counted upon to refrain from taking a tough stand even when it is in the best interest of the company. His underlings may indeed become fond of him as a pal, but in his capacity as tribune of the people against the interests of his employer, he cannot, in the end, accomplish all of the unpleasant but necessary things that have to be accomplished on the way to an assigned goal. He will inevitably be revealed as one who is unfit to supervise.

A Static Impression of the Individual

The humanologist will avoid being too arbitrary in passing judgment on errors like the above. Everyone makes mistakes in the course of growing up. The thing to determine when a supervisor is found to have erred disastrously either on the side of being a management man or on the side of being a good Joe, is whether he is still capable of growing up and willing to make an effort to do so, or has leveled off in his growth. If the former is true, there is no reason why he cannot be expected to derive considerable benefit from his past errors and qualify later on for another chance at supervisory responsibility. One of the worst aspects of psychological tests, formal rating procedures, and the like, though not intentional on the part of their authors, is that they tend to create a static impression of an individual. Such tests should never be kept in the file for more than a year at the outside without being brought up to date. People continue to grow, to change their capacities and reactions, up to surprisingly ripe ages; a person who has been rated unfit or immature at a given time may prove to be eminently fit a year or so later, though he seldom is given the opportunity to prove that he has changed. The results of a test are taken as final all too often and as a result the individual becomes pigeonholed.

The Use of Gentle Reins

Management will be well advised to use gentle reins in breaking in the neophyte supervisor. This is his first real chance to get ahead in the Horatio Alger fashion, and it is normal for him to be a little overeager at the beginning. Understanding advice from a superior who has held a similar post in the past will do more to season the new supervisor, than meat-axe methods, which may make a cynic of him.

To a certain extent, a supervisor is required to be a disciplinarian, and as my parent, successful or otherwise, will testify, a disciplinarian is made not born. The new supervisor will not have had a great deal of experience in disciplining adult human beings, unless perhaps he has had the benefit of a position of rank while in the armed forces, where discipline is backed by considerably more authority than is acceptable in a private enterprise.

Cracking the Whip

Being human, his first tendency, if he does not duck his responsibility altogether by being lax and easy-going, trusting to luck, will be to exercise the simplest form of discipline, known familiarly as cracking the whip. The function of humanology at this stage will be to lead him into a more constructive course, before he not only creates an unhealthy situation in his unit, but completely undermines his own authority. The stern, whip-cracking type of supervisor will inevitably end up a lonely, frustrated individual, even if his methods obtain passable, though never brilliant, results and he is not removed from authority. There seems to be a streak of mulishness in human nature which makes employees balk at arrogant, authoritarian methods. Groups of employees, particularly those who are considered unpromotable and have no particular motive for putting up with any tinpot Napoleon, will gang up against an arbitrary supervisor, laugh at him behind his back, go over his head with petty complaints, and do just enough work to keep from being fired and no more.

Basic Principles of Good Discipline

The Olin Mathieson Chemical Corporation, with plants throughout the United States has developed a wallet-size reminder card for foremen, outlining the "Basic Principles of Good Discipline." Knowledge of these principles is at the foundation of real team spirit. On this card discipline is divided into three main categories: (1) Prevention, (2) Control, (3) Punishment. A similar card, or form might be an applicable tool for the indoctrination of supervisors at all levels in any type of industry, though it should be borne in mind that tools of this sort should not apply to the A-1 seasoned supervisor of long experience, who has gained intuitions and understandings that are considerably deeper than those that can be put into black and white on a form.

The Olin Mathieson reminder card, printed on both sides, is as follows:

Human relations
"Basic Principles of Good Discipline"
Discipline — the attainment of an objective by action which:
  Prevents — by training which directs, molds, strengthens or perfects
  Controls — by enforcing obedience or order
  Punishes — as a last resort
Discipline that prevents
  1. Proper training
  2. Set up practical rules
  3. Explain reasons for rules
  4. Set good example
  5. Establish consistency
  6. Encourage self-discipline
  7. Good clear communications
  8. Build right attitude
Discipline that controls
Day by day
  1. Check — as often as situation demands

2. Determine cause for breaking rule
   a) Misunderstanding
   b) Improper instructions
   c) Special pressures
   d) Poor working conditions
3. Correct the cause for breaking rules
4. Constructive criticism
5. Consider the individual
6. Public opinion
7. Precedent or experience
8. Keep good records

Discipline that punishes

   (Last Resort)

Types of Action
1. No action
2. Oral reprimand and warning
3. Written reprimand and warning
   a) Department head approval
   b) Union representative when presented
4. Time off without pay
   a) Division head approval
   b) Consult industrial relations
   c) Fix time for return to work
5. Discharge
   a) Division head approval
   b) Consult industrial relations

Considerations in deciding on penalty
1. Penalty fits offense
2. Company policies
3. Effect on individual
4. Effect on department
5. Consistent with past policies and possible future ones
6. Explain punishment
7. Future cooperation of individual
8. Make punishment acceptable if possible

Did action result in attaining desired objective?

One value of the principles cited above is that they establish clear-cut channels, enabling the supervisor to determine where to act on his own initiative and authority and where to share initiative and authority with higher echelon.

Supervisor's Responsibilities

This discussion is leading to a detailed analysis of the supervisor's responsibilities which have to do with medical problems of the members of his team. Before dealing with those responsibilities which have to do with health and which call for a liaison with properly constituted medical or paramedical people, it is appropriate to consider certain health problems which actually belong entirely under the heading of discipline. Too often these problems are erroneously dealt with as medical problems and medical problems only.

REFERENCES

Foreman's Letter. Published by National Foremen's Institute, New York, N.Y., June 4, 1956. An Interview with E. A. Babb, Training Manager for Olin Mathieson Chemical Corporation, regarding discipline.

MASLOW, ABRAHAM HAROLD. *Motivation and Personality.* New York: Harper and Brothers, 1954.

NILES, M. C. *Middle Management.* New York: Harper and Brothers, 1941.

PAGE, ROBERT COLLIER, M.D. "Possible Contributions of the Employee Relations Department to Better Utilization of Medical Services." The *Medical Bulletin,* the Standard Oil Company (New Jersey), XIV (March, 1954), pp. 2-11.

——————. "Medical Needs of Industry Abroad." Published by the Council for Tropical Health, Harvard School of Public Health, Boston, 1955, pp. 13-19.

——————. "Significance of Non-Occupational Disability." *A.M.A. Archives of Industrial Health and Occupational Medicine,* VI (No. 1, January, 1950), pp. 1-12.

PIGORS, PAUL, and MYERS, CHARLES A. *Personnel Administration.* New York: McGraw-Hill, 1956, pp. 60, 89, 138.

SULLIVAN, A. M. "Men, Morality and Management." *Duns Review and Modern Industry,* (October, 1960), p. 58.

TEAD, ORDWAY. *The Art of Leadership.* New York: McGraw-Hill, 1935.

RESEARCH INSTITUTE OF AMERICA, INC. *How to Handle Women.* 292 Madison Avenue, New York, N. Y., 1935.

RESEARCH INSTITUTE OF AMERICA, INC. *Marks of Maturity.* 589 Fifth Avenue, New York, N. Y., 1956.

15

Supervision
(Part II)

True Sickness Absenteeism

A MAJOR PROBLEM IS that of absenteeism. True sickness absenteeism as innumerable statistical studies have shown is a minority problem and highly amenable to constructive medicine (the supervisor's role with respect to the prevention of specific occupational diseases is dealt with in Chapter 7).

In one large industrial plant, studies over a six-month period showed that 50 per cent of the work days lost were accounted for by 5 per cent of the employees. Studies over longer periods show the same individuals reappearing on the lists of chronic absentees year after year. Granted that a small percentage of those chronic absentees were in that category owing to true, serious physiological illness, a large number, the majority as it turned out upon further study, were found to have been motivated by a variety of minor problems, which, it seemed likely were not shared by the majority of workers who did not present an absenteeism problem. The humanological team at the plant in question determined therefore to make an unusual approach to the problem and seek to determine *first* what were the motives underlying the good attendance record of the 95 per cent of workers. After this, the motives underlying chronic absenteeism were to be further explored.

Five Basic Motives for Attendance

The motives for attendance on the part of the loyal employees appeared to be as follows:
1. The basic need to sustain life. The majority of individuals must work in order to support their families and/or themselves;
2. The desire to be useful. Most human beings have an extra-materialistic urge to accomplish something worthy of pride, which will win the thanks, admiration, and/or appreciation of others;
3. Habit. The idea of work is ingrained into most people from childhood. This can be exaggerated, as in the case of the restive individual who simply cannot endure leisure and who dreads retirement;
4. Sociability. For a few people, the basic satisfactions to be obtained from a job arise from social contacts.
5. Group loyalty. The feeling of belonging to a team spurs many individuals to suprising levels of loyal and self-sacrificing accomplishment, to the extent that a mild indisposition such as a headache will be put out of mind if it entails letting the group down.

Five General Motives for Nonattendance

In the light of the above positive reasons for attendance, a study of the factors underlying the absenteeism of the minority group resolved itself into five general motives:
1. True physiological illness depriving the individual of the capacity to work;
2. Emotional illness that renders it impossible for the individual to give to his job the requisite amount of attention or concentration;
3. Unfounded concern about health that magnifies minor complaints to the point where they outweigh motives to work;

4. Actual dislike of work. Many of those in this group may be categorized as lazy, the root of the problem may be the result of improper supervision or poor employee placement;

5. Competition of other interests away from job. Home problems, other business enterprises; excessive involvement in leisure activities such as sports or the arts may distract an individual, and cause him to look upon his workaday job with extreme distaste or at best as a necessary evil.

A supervisor who has made an effort to become acquainted with the members of his team will soon learn to distinguish between those cases where medical help is needed and those in which morale or disciplinary factors are paramount. The latter, it is generally agreed, are in the majority, particularly in the case of short absences, though reliable statistics are difficult to obtain. By far the majority of these absences are listed as being due to respiratory disease. Among women who are categorized as chronic absentees, indispositions are often connected with real or imagined dysmenorrhea. The company doctor or nurse is not in a position to play policeman. When the individual in question returns to work, the stated indisposition has presumably run its course. The individual may or may not be telling the truth; there is little or nothing medical examination can do but assume the former. See Chapter 11 "Situational States." Thus, in such cases, it is up to the supervisor to keep a sharp eye out for patterns, and for other personal factors that may be affecting the absentee's desire to be regular in attendance. If illness cannot be fairly readily ruled out, the supervisor should seek the help of the humanological team. The humanological team will, in turn, refer the individual in question to the appropriate medical official so that he can answer the following questions:

Questions To Be Answered

1. Is the employee ill?
2. If so:
    a) Is his illness a justifiable basis for his absenteeism, or is it merely an alibi?
    b) Is he under a physician's care? Does he follow the doctor's advice?
    c) Is his condition likely to improve or worsen with passage of time?
    d) Are his health and life expectancy likely to be adversely affected by continued work on his job?
    e) Is his absenteeism in anywise related to an industrial illness or injury?

At the same time, the humanological team should ascertain the following facts from the supervisor who first called attention to the absentee:

1. Is the employee an effective and desirable employee despite his illness?
2. Is he in a job where his suboptimal performance owing to poor health may contribute to or cause substantial deviation from optimal productivity of a production unit or department?
3. Can he be used effectively in some other assignment more nearly compatible with his physical or emotional limitations?
4. Is the employee eligible for retirement under any of the company plans?

If, in the opinion of a qualified medical consultant, the individual in question is not ill, or if any actual illness discovered is deemed to be immaterial to the absenteeism problem, and if the supervisor cannot obtain sufficient information to form a basis for action, the humanology team will seek to determine the following:

1. Is the individual's absence typical of his department? If so, what environmental factors physical, social, or

psychic are influencing him and his associates' lack of enthusiasm for the job? Can these be corrected?
2. Will reprimanding or other disciplinary measures on one or more occasions change his attitudes?
3. Should the individual be discharged or retired for the good of the service?

Major Illness and the Family Doctor

Where major illness, physiological or psychological, is the chief factor, the family doctor becomes the only hope for solution.

Where hypochondria or excessive concern about health is a cause of inattention to work, common sense on the part of the nurse can usually solve the problem, if the supervisor himself fails to establish constructive communication. Reassurance will often help the anxious individual gain proper perspective. The supervisor's role in dealing with an individual of this sort is to try to help him gain an objective picture of himself and the team problems he is causing by his poor attendance.

The Job Factor

The supervisor assumes a major role when the poor attendance record reflects dislike for or disinterest in his job. This motive for absenteeism is frequently caused by faulty communication between supervisor and employee. The former will be grateful for the opportunity to discover this and take steps to correct matters through better understanding before the situation has become too deep-seated to be reversible. Sometimes it will turn out that the individual has been grossly misplaced, or that there is no possibility of adjustment between supervisor and employee. When this is the case, interdepartmental transfer may be the only solution.

Competing Interests Outside the Plant

The problem of competing interests outside the plant poses another kind of problem for the supervisor. At all times, he must make it clear to the employee that he is expected to pull his weight with the other members of the team, and that failure to do so, for whatever reason, will eventually call for concrete and perhaps drastic changes. Where the problem involves the worker's family, the supervisor may go only so far and no farther in taking action. If he has the confidence of the individual concerned he may offer commonsense advice, but as far as positive steps are concerned, he must limit himself to providing helpful leads, perhaps with the assistance of the humanological team, to the appropriate agencies equipped to handle the specific problem. Most communities have a number of such agencies. The supervisor should be able to obtain information as to the proper agency or agencies in any specific case. The employee relations department or plant nurse should have such information on file and kept up to date at all times.

Whatever the reason for the absence, legitimate or otherwise, the supervisor will obtain the best results if he makes it a point every time to remind the absentee in a good-humored, not a fault-finding manner that he is missed when he is away. A conversation similar to the following may often "cure" the chronic short-term absentee:

Foreman: "I'm certainly glad to see you back, Tom. How's the cold?"

Tom: "Oh, it's okay now. Feel like something the cat dragged in, but I'm a whole lot better off than I've been the last couple of days."

Foreman: "Well, don't overdo it and get yourself back on the sick list again. We've been in bad shape without you. I had to take Bill and Jack off their machines to fill in for you. It threw us all off schedule. You'd be surprised to realize just how indispensable you really are around here."

Problem Drinking

Another problem which is often erroneously considered a "medical" one, though it lies almost wholly within the realm of discipline or human relations at the supervisory level is that of problem drinking, popularly miscalled "alcoholism." To the average layman and to many physicians, the terms "alcoholism" and "the alcoholic" conjure up a picture of a hopelessly advanced condition with no recognition of a variety of causes and differing degrees of severity. The expression "problem drinker" is broader in scope and lacks many of the misleading connotations of "alcoholic."

Definition of the Problem Drinker

From the point of view of industry, an individual is considered a problem drinker when repeated or continued overindulgence interferes with the efficient performance of his work assignment. Clearly, such a "diagnosis" can only be made by the supervisor. The medical consultant may be able to measure the relationship between an individual's consumption of alcohol and the state of his digestion or the condition of his liver. A welfare worker may be able to determine the relation between an individual's drinking habits and the size of his bank account. But only the supervisor can say whether or not the individual's drinking habits constitute a problem in terms of performance. Moreover, only the supervisor can institute early corrective action at a time when it is inexpensive and sure. Late action, necessitated by the supervisor's failure to act preventively, will rarely be inexpensive and never be wholly sure.

Problem Drinking Is Still Essentially a Mystery

Laymen have in the past leaned far too heavily on the medical profession not only in the search for solutions to the problems posed by the consumption of ethyl alcohol, but for an understanding or a definition of the problem itself. As a result, problem drinking is still essentially a mystery. The results have been examined in medical laboratories. Psychologists have probed into causes and effects. Sociologists have scrutinized the use and misuse of alcohol in terms of environments, traditions, and beliefs. Volumes have been written. Outwardly it has looked as though considerable progress has been made. In actuality, however, the surface of the problem has barely been scratched for the reason that invisible walls of custom and usage have stood between the drinker, problem or nonproblem, and those who would study him. Experts on alcoholism must form their estimates of the total problem partly from what they see with their own eyes, often from a distance, and partly from whatever factual information can be gleaned from the tiny minority of problem drinkers who come to the attention of medical researchers or penal authorities.

Statistics Are of Questionable Value

Statistics on this subject must be taken with a large grain of salt. People who don't drink at all will gladly testify that they don't; people who have reached the point where they realize they need help will testify that they drink to excess. The vast majority of adults, however, ranging from true problem drinkers to near-teetotalers who drink small amounts of wine or beer socially, ritually or medicinally, will call themselves moderate drinkers who can take it or leave it alone.

Obviously then, reliable statistics are unobtainable from drinkers themselves. The patterns of problem drinking that have been measured have been the patterns of those individuals who have for various reasons sought help, been apprehended for overt public misbehavior, or created domestic or job problems too flagrant to be camouflaged. How typical and illustrative these patterns are cannot be judged until it has become possible to study them in the light of problem drinking patterns which are currently covered up. Industry, enlisting the help of the supervisor, can penetrate this wall of custom. Most of the light that now needs to be cast upon the problem will have to be provided by members of humanology teams in industry working cooperatively. A soundly-based humanological survey will tap another source of information that has long been neglected: the working bartender. A good bartender, whether in a public establishment or a private club, does not drink on the job, except perhaps at closing time. Over the years, he trains himself to spot the various kinds of problem drinkers, from trouble makers to moochers. At the same time, he acquires valuable insight into drinking patterns that rarely if ever come to the attention of the statistician or clinician.

A team approach to the problem, captained by business, and enlisting the aid of supervisors, bartenders, and police court officials or bridgemen who deal with problem drinkers at the legal level, will contribute immeasurably to understanding of problem drinking. In the meanwhile, most generalizations about problem drinking, however scientifically couched, must be scrutinized cautiously.

Four Categories of Problem Drinking

It has been possible to divide overt problem drinking into four broad categories:

1. Pathological intoxication. — A condition seen in people with a constitutional inability to handle alcohol. Small amounts cause profound changes in behavior and severe after-effects which adversely affect work performance.

2. Environmental habituation. — This condition can be seen in individuals who consume large quantities of alcohol for business or social reasons over a fairly long period.

3. Dipsomania. — A condition of uncontrollable periodic impulses to drink to excess, triggered by some type of periodic psychological disturbance within the individual.

4. A neuropsychiatric problem associated with an independent symptom of alcohol addiction. — This condition can be seen in an individual who has an underlying major psychiatric problem which breaks through from time to time to manifest itself in a number of ways including excessive drinking.

Three Types of Problem Drinkers

In each of the above categories except that of pathologic intoxication, there have been found three particular types of problem drinkers:

1. The chronic excessive drinker. — This individual is psychologically immature and egocentric, and, from the sociological point of view, desocialized or unsocialized, begins to rely more and more upon alcohol to solve his problems of maladjustment. The temporary solution of these problems is followed by the development of new problems raised by behavior while drinking.

2. The compulsive drinker. — The compulsive drinker develops clear-cut loss of control. For awhile, knowing that one drink leads to another, he can postpone or choose the occasion for taking that disastrous first drink. Gradually he loses even this ability. He knows that alcohol means trouble, but for a long time he continues to buoy himself up with the delusion that this time it will be different.

3. The chronic alcohol addict. — In addition to the basic maladjustment and the compulsive features of the first two groups, this individual develops readily diagnosed disease, such as an alcoholic psychosis, liver disease, peripheral neuritis, etc.

The above-cited types are the full-blown ones. Illustrative case histories abound in the annals of occupational medicine. An experienced plant nurse can cite such cases by the dozens. From the point of view of the supervisor and from the point of view of truly constructive action, such case histories are superfluous. Only at the early stages of chronic compulsive drinking, where case histories do not exist, since the problem has not become medically recognizable, can the most effective action be taken. Here is where the supervisor comes in.

Problem Acknowledged by Top Management

As pointed out by James F. Oates, Jr., President of the Equitable Life Assurance Society of the United States:

Alcoholism usually does not develop suddenly — most often it takes a number of years to come to full ugly maturity. Long before it reaches this stage, a good supervisor often has identified the trouble and watched it grow — but unfortunately has done nothing but watch. The supervisor should be taught the symptoms that may indicate early alcoholism and then turn to the Medical Department for confirmation — or denial — and, when indicated, for prompt, active, full appropriate treatment. The supervisor must learn that the industrial alcoholic is a special type of drinker — not a "skid-row" derelict but usually a mature family man with many years of relatively stable employment behind him.

Every supervisor should be trained in how to deal with that unfunny phenomenon, the hangover. There has been a great deal of widespread propaganda to the effect that problem drinking is just a "disease" to be treated with the same kind of sympathy that would be extended to the sufferer from pneumonia or a broken leg. Discipline, stern talk, reproof, it is said, are out of the question, since one cannot scold or punish an individual merely for having the bad fortune to be ill.

Talk of this sort may be applicable to the advanced-problem drinker, who has passed the point where he can hold a job or where he can be considered as a problem of industry. He is truly an ill person who cannot respond to ordinary common-sense measures. However, the competent worker who has only recently begun to come to work occasionally, usually on Monday mornings with a hangover, is a long way from being a chronic alcoholic addict. He is in the same position as the individual who eats spoiled food or who makes no effort to dress himself sensibly for inclement weather. In other words, while he cannot properly be considered a sick person, he is certainly flirting with sickness. One cannot treat him today for the sickness that may develop tomorrow; one can only take common-sense preventive action — meaning, in this case, discipline.

The supervisor will do well to remember two basic truths: (1) A hangover is not funny; and (2) the man with a hangover is not capable of doing a normal day's work.

Except in rare cases, the solution is not sending him home to sleep it off. This is dangling temptation in front of him; the chances are he will learn about that sovereign hangover cure: another drink.

On the other hand, little will be solved by coddling him on the job and giving him the impression that he can come in Monday after Monday unfit to do his job. An effective way to speak to such an individual may be as follows:

Foreman: "Get your hands off that machinery before you break something, Jim; go get a broom and sweep up. If you need aspirin, the nurse will fix you up. In any event, make yourself as useful as you can around here until quitting time without getting involved in anything dangerous or complicated. I won't expect you to do a full man's job today, and for this reason I think it's only fair that I recommend you be paid half a day's wages. Next time you come in so hung over you can't do a day's work, I'll recommend that you be docked a day's pay. The third time, you'd probably better not bother to come in at all. I don't want you on my team."

A man who comes to work regularly and gives no evidence of alcohol addiction outside of a Monday morning hangover is still amenable to reason and common-sense, and still capable of insight.

If he goes undisciplined for a long period of time and his hangovers begin to occur almost daily, common-sense will have difficulty in piercing the wall of alibis he is building about himself. Still later, when he checks in regularly after lunch with liquor on his breath, he has reached a point where the average foreman or supervisor or personnel man can be of little or no help.

The Result of Cover-Up

As mentioned briefly above, one of the most effective bars to solution of the alcohol problem in any given occupational environment is the cover-up. This is as much in evidence in the higher echelons as in the lower. The slipping line officer is shunted into jobs where presumably his drinking problem will be less conspicuous, and where his bungling can do the least harm. Elsewhere in the company, the Monday-morning hangover is reported as "flu," alcoholic disease is treated by the family physician as "enteritis" or "indigestion," the machinist with the fruity breath after lunch is sent home by the foreman with "a touch of sinus trouble."

The Problem Drinking Detection Unit

The "Problem Drinking Detection Unit" is one humanological approach to the problem in industry that has proved effective. A certain number of individuals from every occupational level who have solved personal alcohol problems can be of great value on such a team. Tact and diplomacy are of the essence. A balance should be taken between the theory that every man's home is his castle, and industry's fair request that each individual do a day's work for a day's pay. Confidential reports on known problem drinkers should be checked from time to time, but before the unit, representing management, approaches a problem drinker directly, the problem of effective discipline should be discussed with the drinker's supervisor. In some cases, it may be possible to get desirable action by approaching the individual through a trusted friend who can induce him to change his habits of living before need for drastic action becomes obvious.

At one time, an objection to detection units would have been that their investigations might bring up data concerning an individual which cause a whimsical teetotaling or hypocritical superior to be unfairly prejudiced against him. This could not happen in a humanologically sound organization. An up-to-date supervisor is well aware that his concern with alcohol begins *only* when it stands between a worker and the performance of his job. Detection should in no way be tied up with official disapproval. It should be done only with the welfare of the worker in mind.

A good supervisor knows his limits. The full-blown alcohol problem is out of his realm. He will not try to deal with it; neither will any other echelon of industry. There are many organizations outside industry, such as units of Alcoholics Anonymous, specialized private clinics, and the like, which sometimes achieve remarkable results. These cannot of course be brought effectively *into* industry. The individual who cannot do a job may fairly be asked to go elsewhere until he has solved his problem, though it may be expected that the humanological team will make every effort to see that the individual makes contact with whichever agency offers the most hope for his particular problem. In some cases, it may seem practical as well as humane to hold a position open for an individual who is about to make an honest effort to straighten himself out, but in cases where past history clearly indicates that such effort will be unsuccessful, industry will set common-sense bounds to its charitableness.

The wise supervisor will make every effort to avoid playing the role of the amateur psychiatrist. If the early patterns of unwise drinking seem to be owing not so much to a gradual sliding into bad living habits as to some real underlying psychological or physiological problem, the supervisor will seek help at once.

The Periodic Health Audit

In such a case, as in the case of all true medical problems which do not respond to discipline or job satisfactions, the supervisor's best friend is the periodic health audit. At one time, this was considered to be no particular concern of his. It might be argued that he had a marked responsibility with regard to occupational illness and on-the-job accidents, but nonoccupational illness was either the problem of the employee himself or, in more enlightened concerns, of top management, which after all controlled the purse strings and established policy.

The Supervisor Is Affected by Nonoccupational Illness

In practice, it has developed that the supervisor is most directly affected by nonoccupational illness, most immediately aware of it, and best able to take effective steps, on the spot, when it occurs. As far as he is concerned, a man absent owing to a head cold is just as crippling to his team as a man absent owing to an accident on the job. Moreover, the absence has immediate effects. He is the first to hear about it, and it is up to him to reorganize the general work load in order to compensate for the absent individual. Besides, he *knows* the individual personally. He knows things about the individual's absence patterns, problems, abilities, general stamina, and the like, that do not appear on a statistical chart drawn up for the edification of the vice-president in charge of health problems, a wholly mythical character, introduced for the purpose of illumination.

Nuisance Diseases

When it is taken into consideration that the great majority of absences are caused by off-the-job medical problems, the concern of the supervisor with the total health of his subordinates becomes even more plausible. In a company having

10,538 employees scattered throughout 25 plants, a typical experience for one calendar year was that well over half the absentees as well as man-days lost were accounted for by respiratory and gastrointestinal diseases obviously unconnected with plant hygiene or safety measures. The average length of absence in the case of such illnesses ran between four and seven days per person: long enough to throw a producing unit off schedule; not long enough to allow the supervisor to draw up reorganization plans. These might be classed as nuisance absences.

Absences Attributable to Degenerative Disease

Far more significant from a long-haul point of view was the showing on absences owing to so-called degenerative disease. While the number of individuals absent for this reason was only 279 out of a total of 10,719 absentees, the number of man-days lost amounted to 10,346, a little over 37 days for each single absence.

Altogether, these nonoccupational absences, the numerous short-term nuisance absences added to the relatively few long-term serious absences, accounted for a total of 52,726 man-days lost in a year's operation. The total of absences for all reasons was 89,799.

What a Careful Analysis Will Show

A careful analysis of these absences from a point of view of possible preventive action is revealing. The brief absences listed under respiratory disease, accounting for the greatest number (27,238) of man-days lost might be considered the least susceptible to preventive medical measures. The humanological team must devise ways and means of breaking this inclusive "respiratory" heading into a number of subheadings, indicative of different approaches to the problem.

Respiratory Diseases

The approach to clear cut pneumonia or pleurisy, for example, is well-defined, but the majority of the respiratory diseases accounting for this large proportion of absences are so-called common colds. The words "common cold" may mean a multitude of things. An exhaustive humanological survey of colds in a large industry or a number of industries will no doubt reveal several appropriate subclassifications of the common cold. It has been noted for example that a variety of cold can affect a large number of individuals subjected to unusual stress. Also some varieties of cold seem to be more infectious than others. It has been observed that sometimes, under normal circumstances, a carrier can infect a whole department. At other times, only a few, notably those with low morale or inadequate motivation will become seriously indisposed.

Further study of this subject in a responsive environment will no doubt render the name "common cold" obsolete, and will suggest preventive measures not necessarily medical. Surely constructive action can be taken to prevent those types of colds that arise from low morale, unhappiness, or unusual stress. While the highly infectious, short-term cold will have to be classed along with death and taxes as more or less unavoidable until our knowledge of this subject is more profound, plant hygiene measures may inhibit the spread of such colds within a department, but employed individuals will continue to catch colds at home, particularly if they have children attending public schools.

Gastrointestinal Disturbances

The absences listed under gastrointestinal, accounting for 15,142 man-days lost, likewise may be considered in a very high percentage of cases preventable. Here again, a further breakdown into subclassifications would have been revealing.

A nationwide industrial survey of absences from this cause would undoubtedly show that the majority were directly or indirectly situational in origin; hence preventable by supervisory understanding, tact, and common-sense. The balance, due to nutritional errors or to contamination, or perhaps to masked alcohol misuse, would surely respond to routine control measures.

Degenerative Disease and the Periodic Health Audit

It is with regard to the absences owing to degenerative disease that the periodic health audit shows its true value to the supervisor. It is these absences that are of the greatest significance to the supervisor's team. The five to seven-day respiratory or gastrointestinal absences are nuisances, but the team can adjust to them without undue difficulty; the absentee can step back into his old spot on his return to work without causing an upheaval. In the case of degenerative disease however, when the average absence lasts over 37 days, drastic adjustments must be made. Not only is the team itself thrown off balance by the necessity to compensate for a prolonged vacancy, but the returned ex-invalid poses a sensitive readjustment problem. The chances are that after each absence he is diminished a little bit more in his ability to perform, and the supervisor must know when and how to make the appropriate allowances and balances.

Experience has shown that a program of periodic examination, increasing in intensity for those employees who have reached their middle years, will reveal incipient degenerative disease at a time when it can be reversed, checked, or held at bay, *before* it has a chance to manifest itself in terms of prolonged absence or inability to perform.

At Consolidated Edison Company of New York, a total of 707 men over 40 availed themselves of the benefits of a company-sponsored periodic examination over a seven-year period, with tangible results. More than 85 per cent of these were in the 45- 65-year-old span. About half of these had conditions of major medical significance in themselves or as precursors of potentially serious disease. Approximately 15 per cent of the group knew of certain conditions they wished to have checked by the examining physician, but a sizeable number of those examined, 35 per cent, had conditions revealed to them for the first time! This was even more impressive when it was noted that more than half of these conditions were *without symptoms of any kind*. The cardiovascular system was most frequently involved. The author's experience in examining 290 executives is shown by the following chart and caption.

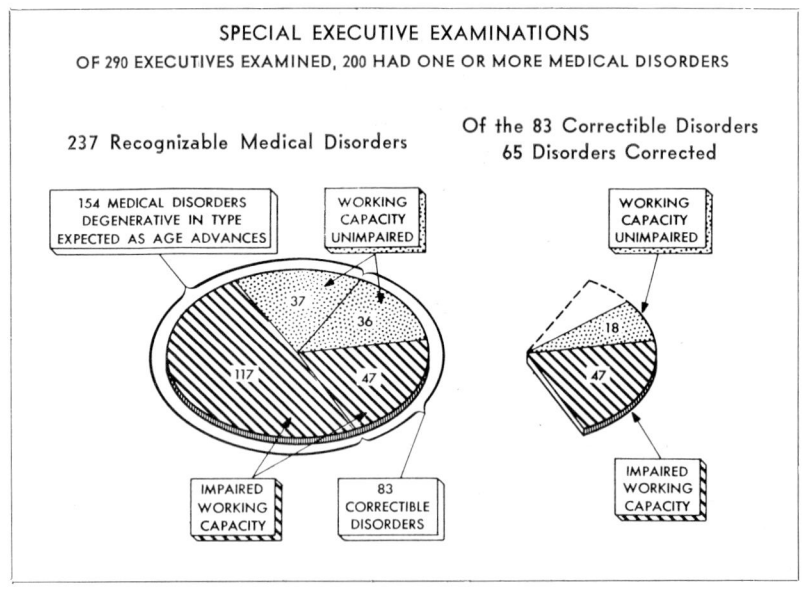

SPECIAL EXECUTIVE EXAMINATIONS
OF 290 EXECUTIVES EXAMINED, 200 HAD ONE OR MORE MEDICAL DISORDERS

237 Recognizable Medical Disorders

Of the 83 Correctible Disorders
65 Disorders Corrected

154 MEDICAL DISORDERS DEGENERATIVE IN TYPE EXPECTED AS AGE ADVANCES

WORKING CAPACITY UNIMPAIRED

WORKING CAPACITY UNIMPAIRED

37
36
117
47

18
47

IMPAIRED WORKING CAPACITY

83 CORRECTIBLE DISORDERS

IMPAIRED WORKING CAPACITY

Graph XIV. — Disability of executives. Of 290 executives examined, 200 had one or more medical disorders — 237 recognizable defects in all. Of these defects, 154 were degenerative in type; of the remaining 83 correctible disorders, 65 were corrected. The scheme on the right shows the proportion of those whose disorders were not correctible (154) and those whose disorders were correctible (83). Also shown in each scheme are the numbers of persons of each group whose working capacity was impaired, and the numbers whose working capacity was not impaired.

What a Thorough Health Audit Includes

A thorough health audit will include the sociological history of the individual. Best results will be obtained if the audit is first made at the time of selection and kept carefully up to date throughout the crucial first 10 years of employment. Nonorganic factors brought to light and recorded at this stage will often prove to have considerable bearing on incipient organic problems detected after the so-called onset of aging. The audit will round out its sociological picture by an exploration of emotional factors, the individual's reaction to stress and strain, the nature of his home life, and the range of his nonoccupational activities.

With thorough periodic inventories there is no need for the individual to ask for a yearly check-up. Furthermore, there is a positive advantage in his not doing so, particularly in his younger period, since undue preoccupation with nonexistent disorders may have the effect of inducing real problems.

The spacing of inventories will depend upon the individual. At the time of examination, it can be determined when the individual, for his own good, should present himself again. In the interim, the employee is encouraged to visit the medical department only when he is ill or in trouble. During the first 10 years of employment, if the above procedures are carried out, it is possible to make a total medical profile for each employee. It is possible, too, in the light of individual findings, to determine what deficiencies in the individual's health education stand in need of correction. The anxious individual may learn the truth about the physiological changes compatible with the aging process. He may discover, perhaps for the first time, that minor conditions such as alopecia (baldness), presbyopia (far-sightedness), erratic elevations of blood pressure, and the like, are to be considered as slight deviations from the normal instead of early symptoms of impending serious disease.

The Supervisor's Responsibility

The supervisor's responsibility with regard to the periodic audit is as large as his concern in the matter, except in cases where the company offers a cradle-to-grave medical package deal, such as has been seen in tropical areas. Here, of course, the supervisor has little to say in the matter. The periodic health audit either is or is not a matter of high-level policy, depending upon the size of the facilities, the time at the disposal of the medical people, and the interest or lack of interest on the part of top management.

In the up-to-date, humanologically sound company, where the practice of medicine remains outside the company, and where the prime factor in the auditing process will be the individual's private family physician, or other physician of choice, the company may undertake to foot the entire bill for the periodic evaluation, but the examinee himself will have to go to it under his own steam, so to speak, and an able, respected supervisor can provide that steam if all else fails.

The average individual, as noted previously, has a certain element of maverick in him. If his wife tells him to go quickly and take advantage of the company's offer to pay for a health audit, he will balk. He doesn't want it to be thought that he is henpecked. Why should he do something just because his wife tells him, he asks himself. If top management sends down a brochure or other propaganda device telling him how lucky he is to be working for the benevolent Blank Company, and why he should run immediately to the nearest family doctor to be audited at company expense, he will write the whole thing off as just another public relations snow job. Only a very gullible individual, he will say to himself, could be taken in by such obvious hanky-panky.

On the other hand, if his immediate supervisor, presumably an able, understanding person with an ability to inspire team spirit, says to him: "Get smart, friend. In the first place, this is your chance to get a couple of hundred dollars worth of medical service for nothing, and in the second place, if we're going to have the best team in the plant here, we can't afford to put up with anyone who isn't going to keep himself at the peak of health." The maverick will listen, and respond.

From a day-to-day point of view, the supervisor is a front-line captain in the battle against illness, psychological as well as physiological. Health education, at a basic level, is a continuing responsibility. It is worthwhile for him to develop a keen eye for deviations, seemingly minor ones, from routine health habits. Carelessness with regard to oral health is often an indicator of significant underlying problems. The individual with unclean, decayed teeth, who cannot be persuaded to visit a dentist, is quite likely to be disorganized in a number of other respects. Such a person is probably more in need of understanding and a friendly helping hand than of contempt or overt distaste. The alert supervisor will enlist the aid of the plant nurse in an attempt at straightening out this individual. The same is true of the individual who is careless with regard to nutrition. Sometimes this latter is in need of simple education as to what constitutes a healthy diet. Quite possibly, on the other hand, he knows that a gulped doughnut washed down with scalding coffee is not a healthy lunch; he simply cannot bring himself to care. He is not oriented.

At the root of this lack of personal organization may be some form of fatigue. The supervisor will look carefully to see if he can discern a correlation between carelessness in health habits and decreased quantity or, decreased quality, or increase of accidents in work performance. In many cases, it will be found that the job itself is at the root of fatigue. In most cases, some factor in the individual's attitude toward his job will be responsible. If the supervisor can discover what the real problem is and how to correct it in the course of a friendly man-to-man talk, he should by all means do so. If he finds that the problem is beyond the reach of lay common-sense he will seek help at once.

The Supervisor's Role in Preventive Psychiatry

The supervisor's role with regard to preventive psychiatry is a limited but highly important one. The limitations must not be minimized. Under no circumstances is a lay supervisor equipped to diagnose a psychiatric problem. From ethical considerations, a good supervisor will bend over backward to avoid even giving the appearance of making an informal diagnosis. The statement, or even implication, by a supervisor that a given subordinate is not normal, can be highly damaging to the subordinate, and unjustly so, whether or not the implication is based upon something that is later determined to be truth.

Normality Is a Delicate Concept

Normality is a delicate concept. It is not a word to be tossed about lightly, or even seriously, by the nonexpert. An indication of the care required in using the word "normal" is seen in Kubie's definition:

A contrast between normal and neurotic can have nothing to do with the statistical frequency of any act. The fact that 99 per cent of the population has dental caries does not make cavities in the teeth normal. Nor has it to do with the legality of an act or its conformity to social mores or its divergence from them, since one can be good or bad, conformist or rebel, for healthy or neurotic reasons. Even the apparent sensibleness or foolishness, the usefulness or uselessness, of an act is not the mark which distinguishes health from neurosis; since one may do foolish things for sensible reasons (for instance as an initiation stunt), and one may do sensible things for very foolish reasons indeed, as for instance out of phobic anxiety. All of this will seem strange only to those who think only of neurotic as synonymous with queer or eccentric or foolish or weak or immoral or rare or useless. We must learn instead that there is literally no single thing that a human being can think or feel or do which may not be either normal or neurotic, or, and more often, a mixture of the two; and the degree to which it is the one or the other will depend not only upon the nature of the act, but upon the nature of the psychological forces that produce it. This is true of work and play, or selfishness or generosity, of cleanliness or dirtiness, of courage or fear, of a sense of guilt or a sense of virtue, of activity or indolence, of extravagance or penuriousness, of ambition or indifference, of ruthlessness or gentleness, of conformity or rebellion, of playing poker or writing poetry, and even of fidelity or infidelity. Determining all of these there is a continuous, unstable, dynamic equilibrium of psychological forces; and in this flux it is the balance of power between conscious or unconscious forces which determines the degree of normality or the degree of neuroticism of the act or feeling or trait.

The Interpretation of Atypical Behavior

It is clear from the above that the supervisor and even the plant nurse must exercise rigorous caution in interpreting what may seem to be atypical behavior on the part of an individual. On the other hand, such behavior, if the cause of preventive psychiatry is to be served and the preventive approach is the only one that industry or the average affected individual can afford, must never be overlooked or simply allowed to run its course or blow over. At the risk of being called a busybody, the assiduous supervisor will inquire into every possible indication of a disorientation of the individual subordinate from his job, his teammates, or life in general. If tactful questioning doesn't reveal an answer and a solution within the range of operation of the supervisor, he will bring the plant nurse into the picture.

Emotional disorder which can lead to a psychiatric state is by no means rare in industry. An employed individual lives a considerable part of his life within the bounds of his job. His goals; his picture of himself; his justification for demanding love or approbation all are bound up in his job. It is not surprising that industry is considered the front line in preventive psychiatry. Refer to pages 169, 486 and 539.

The importance of the supervisor is underlined by the statement of a practicing psychiatrist, L. E. Himler, in the November, 1944, *Bulletin of the Industrial Hygiene Foundation,* that 74 per cent out of 90 cases referred to him from industry were general problems of personality and adjustment *which could be handled by adequate lay supervision.* The remaining 26 per cent were problems with medical and neuropsychiatric aspects. Himler presented the following breakdown:

General Problems of Personality and Adjustment..............74 Per Cent
    Relatively simple job adjustment problems............24 per cent
    Difficulties in interpersonal relationships................13 per cent
    Mild personality disorders..21 per cent
    Family, home and marital problems........................16 per cent
Problems with Medical and Neuropsychiatric Aspects........26 Per Cent
    Functional nervous conditions associated with
        physical disorders...11 per cent
    Psychoneuroses... 2 per cent
    Severe personality disorders...................................... 9 per cent
    Epilepsy and hypnolepsy.. 3 per cent
    Psychosis... 1 per cent

It should be pointed out that the problems which are listed as being susceptible to supervisory action, problem drinking might be added to this list, can, if unchecked, become full-blown psychiatric problems in the latter category. When this happens, the individual often passes beyond all human help, and the final solution, for the sake of his unhappy family, is to institutionalize him. Even when help is not entirely out of the question, it is likely to be extremely costly, long-drawn-out, and perhaps little more than palliative. The full-blown psychiatric case who is capable of being rehabilitated 100 per cent is the exception to the rule. Such exceptions do exist however, and the supervisor is once again charged with the important responsibility of being the chief "rehabilitationist." Rehabilitation means reacceptance on the job. If the expatient is evaluated and back at work under an understanding team captain he feels wanted, worthwhile, and confident. His acceptance by the team does a great deal to remove any lingering cobwebs of the recent past, and it is the supervisor who can make this acceptance a reality.

The Supervisor's Function

A great deal has been said above, about the importance of morale and teamwork. To sum up: it should be underlined that it is the supervisor's *primary function* to inspire high morale and enthusiastic team action. If he does this, his other responsibilities will fall easily into line as a matter of course.

The Quality of Supervision

The quality of supervision should always be judged not so much by the performance of the supervisor himself but by the performance and spirit of those under him. The supervisor who is thrown off by difficult people, the boss who simply cannot get cooperation — say — out of the women workers in his department, or who brings out the maverick in his male subordinates, needs considerable reorientation in human motivation.

A Constructive Approach

The New York State Banker's Association has developed a scientific approach to this problem in its "Employee Morale Survey" involving an exhaustive procedure of attitude analysis. Trained researchers question employees as well as supervisors regarding their opinions and feelings as to certain aspects of the occupational setup, and the resultant cards and scored sheets are processed electronically to give answers to the following questions:
1. Are supervisors tuned in to the thinking of those they supervise? How well do they know the attitudes of their subordinates? Do they know the reasons why employees become discontented and look for other jobs?

2. Do employees have confidence in their supervisors? Do they respect their judgment? Fairness? Methods of criticism and promotion?
3. Are two-way communication channels from management (a) to supervisors (b) to employees — direct and clear? Do employees believe they are told what they should know quickly and fully? Is morale suffering from a feeling of "I don't know what's going on?"
4. Are employees satisfied with their pay?
5. Do employees fully understand and appreciate fringe benefits given to them in addition to their salary? What is their attitude toward these benefits?
6. Do employees feel the work load is evenly and fairly distributed? What is their attitude toward work pressure? Overtime? Working conditions?
7. Do employees believe there is opportunity for advancement in position and salary? Do they have a feeling that their jobs are secure?
8. Do employees have confidence in management? Do they believe management has an interest in their personal welfare and progress? If they stick to banking as a career, will it pay off?
9. Is morale good in some departments and/or branches, poor in others? Is the company getting a good day's work for a good day's pay?
10. Which employee policies are strong and need little attention? Which are weak and require immediate attention?

Another scientific approach, which casts light on some classic misconceptions on the part of supervisors, was made by the Labor Relations Institute of Newark, New Jersey. Foremen in 24 plants were asked to rank 10 key elements in order of their importance to the employees under them. The workers in the same companies were asked to rate the same items. The results were as follows:

Job element
Rating

	workers	/	foremen
Full appreciation of work done	1st	/	8th
Feeling "in" on things	2nd	/	10th
Sympathetic help on personal problems	3rd	/	9th
Job security	4th	/	2nd

Good wages	5th /	1st
Work that keeps you interested	6th /	5th
Personality and growth in company promotion	7th /	3rd
Personal loyalty to workers	8th /	6th
Good working conditions	9th /	4th
Tactful disciplining	10th /	7th

The Key to Supervision

A restudy of the above figure should be rewarding to the humanologist who seeks to find the key to supervision — the key that the supervisors questioned in this survey apparently lacked. The job elements given top rating by the foremen were those that appear most often upon the collective bargaining table. These foremen, then, were thinking of their subordinates as holders of union cards — as breadwinners with primarily economic motives.

The employees on the other hand were thinking of themselves all along in the way the really shrewd supervisor or humanologist would think of them — as people!

REFERENCES

AMERICAN MANAGEMENT ASSOCIATION. *The Man in Management, A Personal View:*, General Management Series No. 189, New York: American Management Association, 1957.

COMBS, ARTHUR W., M.D. "Why Executives Become Frustrated." *Office Executive,* (April, 1955) *in Management Guide.*

CRUICKSHANK, W. H., M.D. "Mental Hygiene in Industry." *Industrial Medicine and Surgery,* XXVI (No. 10, October, 1957), pp. 477-481.

DILL, D. B. "Nature of Fatigue." *Geriatrics,* X (October, 1955), pp. 474-478.

EDWARDS, JOSEPH DEAN. *Executives: Making Them Click.* New York: University Books, 1956.

FRANCO, S. CHARLES, M.D., F.A.C.P. "The Early Detection of Disease by Periodic Examination." *Industrial Medicine and Surgery,* XXV (No. 6, June, 1956), pp. 251-257.

GAFAFER, WILLIAM M., EDITOR. *Manual of Industrial Hygiene.* Philadelphia: W. B. Saunders Company, 1943, p. 156.

GREENLY, R. J., AND MAPEL, E. B. "The Trained Executive: A Profile." *The Development of Executive Talent.* New York: American Management Association, 1952.

HOLMAN, D. V., M.D. "Psychiatric Problems in Industry." *Medical Bulletin,* X (April, 1950), pp. 149-158.

HUNTINGTON, T. W. "Freedom from Fatigue." *California Monthly,* (January, 1950), Alumni Publication, University of California.

JACOBSON, EDMUND, M.D. "Tension Control for Today's Executive." *The Man in Management — A Personal View.* New York: American Management Association, 1957.

KUBIE, L. S., M.D. "The Neurotic Potential and Human Adaption." *Adaption.* Ithaca: Cornell University Press, 1949, edited by John Romano, M.D.

MCGEE, LEMUEL, M.D. "Report of the Medical Department." *Manual of Industrial Medicine.* Philadelphia: University of Pennsylvania Press, 1956.

MCGINNIS, PATRICK B. "The Top Management Man: His Selection and Development." *American Management Association Journal,* CXLIV.

NATIONAL INDUSTRIAL CONFERENCE BOARD. *Company Health Programs for Executives.* Studies in Personnel Policy, No. 147. New York: National Industrial Conference Board, Incorporated, 1955.

NORO, L., M.D. "Some Social Aspects of Sickness Absenteeism." *Medical Bulletin,* X (No. 3, July, 1950), pp. 282-286.

PAGE, ROBERT COLLIER, M.D. "Health Maintenance of Key Personnel." *Industrial Medicine and Surgery,* XX (July, 1951), pp. 325-330.

_____. "A Message for the Wife of the Ambitious Business Man." *It Pays to Be Healthy.* New York: Prentice-Hall, 1957, p. 124.

PAGE, ROBERT COLLIER, M.D.; THORPE, JOHN J., M.D.; CALDWELL, D. W., M.D. "The Problem Drinker in Industry." *Quarterly Journal of Studies on Alcohol,* XIII (No. 3, September, 1952), pp. 370-396.

PORTIS, SIDNEY A., M.D.; ZITMAN, IRVING H., M.D.; LAWRENCE, CHARLES H., M.D. "Exhaustion in the Young Business Executive." The *Journal of the American Medical Association,* CXLIV (No. 14, December 2, 1950), pp. 1162-1166.

SCIMECA, ALFRED, M.D. "Efforts to Reduce Sickness Absenteeism Costs in the N. Y. Marketing Division." *Medical Bulletin,* XIV (March, 1954), pp. 35-39.

STRYKER, PERRIN. "Who are the Executives." *The Executive Life.* Chicago: Time Incorporated, 1956.

——————. "How Executives Delegate." *The Executive Life.* Chicago: Time Incorporated, 1956.

——————. "How to Treat V.P.'s." *The Executive Life.* Chicago: Time Incorporated, 1956.

THORPE, JOHN H., M.D. "Attendance Motivations." *Industrial Medicine and Surgery,* XXIV (October, 1955), pp. 450-452.

URWICK, LYNDALL F. *How the Organization Affects the Man.* General Management Association Series 189. New York: American Management Association, 1956.

WHYTE, WILLIAM H., JR. "How Hard Do Executives Work?" *The Executive Life.* Chicago: Time Incorporated, 1956.

16

Retirement

"... *There is an order of mortals on the earth, who do become Old in their youth, and die ere middle age, Without the violence of warlike death; Some perishing of pleasure, some of study, Some work with toil, some of mere weariness, Some of disease, and some insanity, And some of wither'd or of broken hearts. For this last is a malady which slays More than are numbered in the lists of Fate, Taking all shapes, and bearing many names.*"

BYRON — "MANFRED"

Retirement a Form of Severance

RETIREMENT IS LOOKED UPON from an expediency point of view as a form of "severance." This puts it in a category with "discharge" as well as resignation, where it quite patently does not belong. Both discharge and resignation have to do with dissatisfaction on the part of either the company or the individual concerned, resulting in a severance of relations. Retirement, on the other hand, has to do with the inexorable march of time. Far from being severance from an individual's life work, it is actually, if properly planned, just one more step in his career.

This is a difficult concept to communicate, and as a result, retirement planning with rare exceptions tends to be essentially negative. After a certain amount of financial planning and perhaps a few sessions devoted to advice and guidance, the retirable individual is given a gala send off, complete with gold watch; then, as far as the company is concerned, he ceases to exist.

To Replace Inadequate Concepts with Dynamic Ones

The first function of the humanologist in any industrial setting is to see to it that the practices that exist are carried out as intelligently and as constructively as possible. In the long run, he will endeavor to replace inadequate concepts with dynamic ones, even where this involves basic changes in organizational patterns, policies, and procedures, but in the meanwhile he must take business as he finds it and take whatever corrective steps are needed to prevent human wreckage owing to unsatisfactory execution of existing standard operating procedures.

Established Usages and Precedents

In general, management, in dealing with the retirable person, has a tendency to fall back on established usages and precedents.

Individuals invariably are retired under one of the four following categories:
1. The good of the service
2. Total and permanent disability
3. Total and permanent incapacity
4. Normal retirement (at a predetermined time, fixed by company policy; retirement at a set chronological age, such as 60 or 65, is common policy; though there is a trend toward establishment of flexible policies which equate "retirability" with certain measurable

standards of performance modified by individual wants and needs)

For the Good of the Service

Retirement *for the good of the service* means that the individual concerned, though not measurably disabled or incompetent, cannot be fitted into the company tables of organization. Perhaps there has been a merger, which has reduced the number of job categories available, and it has been necessary to fit a younger man into this individual's position. Whatever the reason, the company has come to the conclusion that this individual's usefulness to the organization is at an end, but that his record of loyal service and his age entitle him to retirement on pension rather than cold-blooded discharge.

Retirement under this category is undoubtedly better than discharge when complete severance is clearly unavoidable, but even so, it should be considered a last resort, subject to minute humanological scrutiny. There is always the chance that severance is not as unavoidable as it may have appeared at first to the hasty organizational planner. To this individual, retirement for the good of the service may be the easy way to solve a problem of expediency. However, both for the stockholders and the individual himself, the extra effort involved in finding some other productive full-time or part-time position within the company may prove more satisfactory than condemning a well and capable person to idleness on the pension roll. If the extra effort proves fruitless, and no alternative can be found to severance, humanological scrutiny is still to the point; a nonhumanological approach to this type of severance can result in a public relations fiasco for the company involved.

Arnold B., retired for the good of the service at age 55, is a case in point. For 20 years, Arnold had been somewhat disgruntled. Organizational changes which occurred 10 years after the date of his original employment, had pigeonholed him to a relatively insignificant working assignment, which he performed with relative efficiency. His spare time was taken up with political activities in the neighborhood town in which he lived. As the years went by, more and more of his time was spent in this work. No one from the company restricted his activities in this field, and he thought he had company-approval.

Suddenly one day without any kind of warning or *build-up*, after 30 years service he was informed that in accordance with accepted policy his services would be dispensed with as of such and such a date. A routine severance medical examination showed no organic health problem. Physiologically, he appeared to be no older than his chronological age of 55.

Arnold had lived somewhat beyond his monthly income; so his predetermined retirement pay was not sufficient for his acquired mode of living. He and his family experienced adjustment difficulties immediately. They made it known widely that he had been badly treated by the company. His friends, who had no reason not to agree with him, saw that the word was disseminated. It would be difficult to assess, in dollars and cents, just what damage was done to the company by this. A budget item of several thousands of dollars was earmarked annually for public relations, and a considerable portion of this effort was directed — for recruitment purposes — towards giving the company a public reputation as a place where an employee could count on a fair shake. How much of this effort was cancelled out by this one hasty severance action can only be surmised.

Total and Permanent Disability

Retirement under the second category, *total and permanent disability,* may or may not be a necessary evil. If the individual in question is truly totally and permanently disabled, the humanologist can only do what has to be done, and see that the severance procedure is as equitable and considerate as possible. The disability may have been preventable once upon a time, but now it is too late.

Nonetheless, the humanological team has a clear obligation to bend over backward to avoid retiring anyone under this category if he is still able to perform his job or if he can be rehabilitated. Carelessness in this respect can be considered wanton cruelty.

Only an understanding and qualified medical authority should be called upon to determine whether an individual is totally and permanently disabled. Based upon his complete knowledge of all factors in each individual case and his understanding of concrete medical findings, a qualified physician is able to render a considered diagnosis. He will of course, as a matter of routine, offer a prognosis too. His innate knowledge, judgment and past experience should enable him to prognosticate the individual's life expectancy in terms of weeks, months, and, in some instances, years. There are no slide rule calculations for this distressing work; only an experienced, able professional can do it.

The cases which are considered under this category are, commonly:

1. Cancer which has been detected too late, so that when surgery is attempted, metastasis is in evidence at a distant site or involving an adjacent organ.
2. Decompensation of the cardiovascular system which no longer responds to standard methods of treatment.
3. Complete dysfunction of the renal system, often accompanied by hypertension of alarming proportions.
4. Metabolic diseases such as diabetes which no longer respond to prescribed therapy. At this stage, related disturbances of the cardiovascular (circulatory) system are invariably in evidence.
5. Insanity, meaning a psychiatric condition in which the individual is no longer in the realm of reality. Chronic alcoholism is often logically disposed of under this category, though it does not strictly belong here.

Total and Permanent Incapacity

The third category of retirement, *total and permanent incapacity*, poses a number of extremely delicate problems. It is difficult to imagine how a physician can ethically or morally put his name to any document stating that an individual is totally and permanently *incapable,* not disabled, with regard to his job performance.

No physician can — as a physician — have all the facts relating to an employed person's incapacity; this is management's business. The physician can recommend to management that from a *strictly medical* point of view an employee ought to be retired for his own good. If the physician or medical consultant goes beyond this he is trespassing in management's bailiwick. His position should only be that of a counselor to both employee and management. His prime consideration must be what is best for the employee *in terms of health!*

In the past, professional people connected with business have been known to find themselves in awkward positions after signing documents urged upon them by management, without asking themselves certain fundamental questions. When a doctor is asked to approve the severance of any employed individual he must obtain satisfactory answers to the following questions:

"If I sign this document and get management off the hook, what will happen to the employee? Will he wither on the vine? Am I creating iatrogenic disease? Will he become a public charge? What could be done, in terms of habilitation, that has not been done? Whose fault is it that the individual is no longer able to carry out his work assignment to the satisfaction of his superiors?"

These so-called incapable individuals are not necessarily physically disabled. They may live on and on after severance, burdens to themselves, to their families, and to the community at large.

No one physician or one member of management can be expert in dealing with the problem of incapacity — genuine or apparent. Teamwork on the part of management and all members of the humanological team is what is needed to consider each individual problem from a *total situational* point of view.

A Team Approach Will Bring Realistic Answers

Out of this team approach will come realistic answers. These answers, and whatever course of action is based on these answers, can be explained rationally to the individual in question in a language he can understand. If severance is the only logical course of action, based on study of all pertinent facts, it can be presented to the individual as something sincerely decided upon for his own good, and it can be offered hand in hand with truly helpful advice and concrete assistance in the planning of a satisfactory adjustment to retirement.

The Four Subcategories

There are four subcategories of persons retirable under the heading of *total and permanent incapacity:*
1. Those who *cannot* function because of (a) ill-health, (b) situational maladjustments that have become acute after years of neglect, or (c) changes in the nature of the work that cannot be coped with by the individual.
2. Those who are gradually becoming incapable because of the advance of premature senility.
3. Those who are still capable but who (a) cannot be placed in a suitable job for lack of a vacancy, or (b) cannot carry the full load because of uncertain health.
4. Those who are chronically ill and necessarily absent a good deal of the time.

The Twilight Zone

An important step for dealing with employees in this twilight zone is to separate the sheep from the goats early in the game. Frequently, loyal workers, who come to the point where they can no longer carry a full work load, find themselves judged by standards formulated to deal with malingerers and clock-watchers. In any employed group, these latter absence-prone individuals (some true malingerers, others merely carelessly organized personalities) make up one-third of the total of individual absentees, though their absences make up 75 per cent of the total absences.

The Absent-Prone Cause Precedents to Be Set

These absence-prone persons usually make themselves known to their superiors before they have been on the job very long. They may not be as obvious as the classic baseball fan who went to the funerals of three grandmothers, but an experienced supervisor or foreman can pick them all out in time. They should be singled out from their conscientious fellows and studied in meticulous detail. Otherwise precedents will be set which will be unfair to the nonabsence-prone individual who may be forced to absent himself from work periodically owing to circumstances beyond his control.

A Decision Requires Specific Answers

Undue loss of time from work or failure to perform at a satisfactory level brings the older employee to the attention of those who have to decide on the proper disciplinary or remedial action to be taken. Before a decision is taken, answers to the following questions ought to be obtained:
1. Is the individual ill?
2. If so, is the illness an adequate explanation of his absences?
3. Is he under a doctor's care, following the doctor's advice?
4. Is his condition likely to improve or get worse with time?
5. Is his health or life expectancy likely to be adversely affected by continued performance of his job assignment?
6. Is he an effective and desirable employee despite his present illness?

7. Is his absenteeism or poor job performance related in any way to an industrial illness or injury?
8. Is his absenteeism or poor job performance related in any way to the fact that he has been misassigned over the years as a result of an inadequate company placement policy?
9. Is he in a job where his under-par performance, owing to poor health, may contribute to or cause substantial deviation from top productivity of a production unit or department?
10. Can he be used effectively in some other assignment more nearly compatible with his abilities to perform?
11. Is his absenteeism or poor work performance typical of his department? If so, what environmental factors (physical, social, psychic) are influencing his and his associates' lack of enthusiasm for the job? Can these be corrected?
12. Will reprimanding, on one or more occasions, change his attitude?
13. Should he be discharged or retired for the good of the service?

Standards of Efficiency Can Be Graded Realistically

Enlightened management strikes a balance between maximum efficiency and the fact that some people are better workers than others. Standards of proficiency can be realistically graded, taking into consideration not only that proficiency differs from person to person but also that it differs from age-period to age-period in the same person.

The Desired Goals

All twilight zone problems should be approached with the following goals in mind:
1. To determine individual work proficiency

2. To keep all proficient individuals on the job
3. To reclassify others to tasks to which they are suited

In any case, the individual worker is kept posted. Even if he is not of A-1 caliber, he wants to know periodically where he stands. Frequently it will be found that the worker can correct his own shortcomings or suggest constructive changes in his work pattern. Retirement, under such circumstances, unless voluntary, is unthinkable.

Expediency-Based Thinking

Sometimes it is tempting, in an examination of one of these twilight zone lame ducks to assume the presence of a psychogenic problem, simply because an organic or clear-cut psychiatric one cannot be found. This is expediency-based thinking. The interplay between psychogenic and other factors must be determined, since the causes of many incapacitating problems are multiple. When conflict or stress is found to exist and to have an overt effect on the health of the worker, open discussion must come into play.

As more and more employers come to feel a moral obligation to their long-service lame ducks, they also come to realize that the *worth of people at all levels of employment must be fully appreciated*. Appreciation must be based on comprehensive knowledge and understanding of all pertinent facts, which can be gained from a running appraisal of each individual's work from the start of his employment.

The following cases of retirement owing to total and permanent incapacity are drawn from the files of leading United States' industrial organizations. They are illustrative of how nonhumanological this type of retirement can be unless kept in vigilant check.

CASE I: A 44-year-old laborer with 25-years service was retired because of "syphilis with generalized neuromuscular degeneration."

One year later he was still unable to work because of "leprosy and questionable diagnosis of generalized syphilis."

The true facts of the case are:

He *had* had syphilis; leprosy was ruled out. In appearance, he was 10 years older than his stated age. His degree of intelligence was low. The patient's eye ground examination revealed fairly marked arteriosclerosis with no definite vascular spasm. All of his teeth were present. He had a very narrow pulse with a blood pressure of 118/100 mm. hg. and his heart was normal. The remainder of his physical examination revealed normal health for age and circumstances. This meant that the individual was not suffering from any severe organic disease except for a rather premature arteriosclerosis. He showed a normal degree of alertness, and his memory and intellectual capacities were compatible with his cultural and educational background and opportunities. There were no medical grounds for retirement. He had been with the same company 13 years. Surely there was some means of gainful occupation for this man within the organizational structure. Allowing him to remain on the sickness roll with limited pay hardly seems to have been a soundly conceived solution.

CASE II: A 50-year-old man with 33-years service was retired prematurely because of psychoneurosis, anxiety, and bronchial asthma. He remained under the care of his family physician who presented noncommittal reports as evidence of total and permanent incapacity. Here are the facts:

The individual had been mildly neurotic most of his life in accordance with his personality pattern. In his 33-years service, he had developed considerable expertness and skill in his line. Difficulty with the supervisor was the cause of his trouble. The supervisor fostered premature retirement, and the physician cooperated. Iatrogenic disease was the end result.

CASE III: A 38-year-old female with 18-years service was certified as totally and permanently disabled because of "neuritis." The real problem had to do with unsatisfactory placement. Here again an unsuspecting doctor cooperated by performing unnecessary medical tests and not evaluating the total person.

CASE IV: A 43-year-old female, after 19-years service, was retired on grounds of undulant fever and was continued on the disability payment rolls for eight years. When her case came up for a thorough review she refused to cooperate and accepted a termination settlement without dissent. Such a tragedy as this is the result of lack of humanological understanding.

Points of Similarity

The implications in the foregoing cases are obvious. The points of similarity are striking:
1. All of the individuals concerned had relatively long service. Their exact status of work proficiency must have been long known by management.
2. All were retired on semimedical grounds without an expert interpretation of the exact nature of the complaint.
3. All were relatively young (50 or under). The life expectancy of the individual could not be considered lessened by the nature of his problem.
4. A doctor's signature was required in all cases before retirement benefits could be given.
5. No thought of a constructive approach with respect to habilitation was instituted at the time of the original certification for retirement.
6. The *passive role* of the physician was a notable feature of each case. He was not appraised of all the facts, nor was he allowed to consider the implications in terms of iatrogenic disease or, prolonged costs in sickness benefits and retirement pay.
7. The effect on morale of other employees who were aware of the treatment received by these employees could hardly be placed under the heading of satisfactory human relations.

Contributing Factors

Thoughtlessness, ignorance, haste, lack of understanding, lack of tact, lack of communication, or cold-blooded expediency can combine to make retirement a "kick out the door" instead of the culmination to a fruitful career. The humanologist is duty-bound to make a strong stand when the question of total and permanent incapacity comes up.

Normal Retirement

The fourth category of retirement, so-called *normal retirement,* is the goal of the true humanologist. Standards of normal are, however, not uniformly high throughout industry. Unfortunately, it is oftener than not, considered normal retirement when an individual reaches the predetermined retirement age still in harness, outwardly in middling good health. He is given a last-minute medical examination, a handful of brochures telling him how to adjust to retirement and what hobbies to pursue, some advice as to where and how to collect his pension and other old-age benefits (i.e.: social security), a testimonial banquet, and perhaps taxi or bus fare to eternal boredom and desolation.

This kind of *normal* retirement is actually subnormal. It is not based on constructive planning. It is cut-and-dried severance based on a mixture of expediency and a last-minute qualm of conscience.

Retirement Planning Can Be Constructive

It is not widely realized that retirement planning *can* be constructive, both in the interests of the retired individual and the stockholders of the company.

The representative of management who has charge of over-all retirement plans and policies will make this statement in all sincerity:

"Of course we owe our people a decent retirement. We want to see to it that they're all set up, with a home and enough to live on after they leave us. That's not only good human relations; it's good community relations. But that's as far as retirement planning can go. One can keep on improving pension plans and retirement insurance programs, but what can one do that's *really* constructive, outside of building up the people who are to replace the ones that are going into retirement?" Let us examine the problem carefully.

What Retirement Means

To answer this it is necessary to analyze what retirement really means (not what it seems to mean) to an employed individual in terms not only of his wants and needs but of his productivity.

Today there are more than 15,000,000 Americans over 65, and their numbers are steadily growing as life expectancy goes up. This increase of the ratio of oldsters to the citizens of other age groups is of tremendous significance to society at large.[1] One may even find evidences of widespread concern with the problem in such youth-directed publications as *The Junior League Magazine.*

Yet our society is still geared to youth, as it was necessarily in pioneer times, when there was a premium on sheer muscular stamina. The aging individual may be one of many, but he is still somewhat lonely in a world that does not quite understand him. This is partly the fault of a social organization which has not kept pace with scientific progress. Medical science has made it possible for the average 65-year-old today to be about as young, physiologically, as was the average 45-year-old at the time of the Spanish-American War. Yet he is fortunate if he is not obliged to play the role of a *decrepit old man.*

However, part of the blame must be laid on the fact that many present-day senior citizens simply have not learned how to adjust to the facts of aging. They seem to wake up and find themselves old without having given the matter much prior thought. They find suddenly that there is more time to worry about trifles that used to be forgotten in the haste and confusion of pursuing that success-dream while keeping a family on the right track. The children are gone now, and the evenings are long.

---

[1] HOBBS, G. WARFIELD. "Economic and Political Implications of Aging." An address given at the 1958 Meeting of the National Committee on Aging, Washington, D.C.

Then, at this age, there is the decline of the *success-dream* itself. Before 50, everything was in the future. Now the aging individual says: "I've been as high as I will go, and I'm on the way down. I suppose I never will accomplish the things I thought I was going to." To complicate his feeling of despair is the fact that he cannot communicate his real problems to those who are young and full of hope. They neither understand nor tolerate the whinings of an old *has-been*.

There has been an inevitable emotional reaction to unpleasant physical changes such as wrinkles, baldness or grey hair, declining energy, lapses of memory, and the like. Friends and loved ones begin to drop out of his life, leaving him with fewer and fewer emotional supports. He finds that he is expected to take a back seat to make room for the young person on the way up. He begins to feel unwanted, shoved aside. He is apprehensive about the future.

Other symptoms or reactions are a feeling of inadequacy and a sense of insecurity, emotional as well as economic. The only thing that can prevent this developing into a serious health problem is some kind of gratification of his deep-seated and now exacerbated need to be useful, to feel productive and important, to respect himself.

Seven Sins Against Older People

The Age Center of New England, which has done considerable research into the problems of aging, cites "Seven Sins Against Older People" which might well be incorporated into humanological business policy regarding the aging worker:
1. Giving helpful advice to older people who are wrestling with problems of unprepared-for retirement, loss of a loved-one or unexpected isolation
2. Failing to respect older person's sincere wish to live alone
3. Treating older people as older people; not as human beings

4. Regimenting older people
5. Institutionalizing older persons for wrong reasons i.e., just to get them out of the way
6. Physically or emotionally pampering older people
7. Accepting old-age as an ending of something

The Humanologist, as Well as the Potential Retiree, Must Understand

In other words, a normal approach to retirement entails a certain amount of education of the humanologist as well as the retirable individual himself.

It is unfortunate that the average business man seems to have to reach the age of 60 — after he has personally weathered the storm of competition, had his stomach resected, survived his first coronary attack, or narrowly escaped death from cancer — before he begins to give serious thought to the vital importance of human relations. Up until this time, he has been happy to delegate the responsibility for human relations to younger subordinates, many of whom have a tendency to *follow the book* of policy prepared entirely from the standpoint of monetary operating expediency. It requires no judgment of people to go by the *book*. The book allows no leeway for practical understanding. It seems to fit well into a mechanical age, but acts as a disturbing agent among working people who now have leisure time to think and understand, and who are on the threshold of anxiety and apprehension.

From the point of view of the individual himself, it is generally conceded that *he* can be constructive or otherwise in his *own* retirement planning. He can arrange his finances wisely or foolishly; he can conserve his health or burn himself out ahead of schedule; he can develop a philosophy of life that will keep him in harmony with the facts of aging, or he can fight life. In the end, however, all elements of society are in accord that what he does about such matters is up to him. His company, his doctor, and his family can counsel him and that is all.

Therefore, the man of management is wholly sincere when he asks how retirement planning can be constructive and effective, and redound to the benefit of the individual, management, and the stockholders.

The Least Important Part of Retirement

The answer: first, the severance aspect of retirement should be trimmed down to size. It is the least important part of the retirement story. To make it the be-all and the end-all of retirement is like taking a novel and tearing up all but the last few pages on the grounds that all the rest is nothing but an elaborate build-up to the ending.

Constructive retirement begins many years before the actual act of severance, and ends many years after it. The severance itself is, or should be, merely an incident along the way.

To bring the significant aspects of retirement planning into proper focus, it might be well to forget severance altogether, and think in terms of *Aging in Harness.*

Aging in Harness

Retirement procedures should be *operating* procedures, having to do with people who are producing, but whose productivity is undergoing, or is about to undergo, changes in keeping with the aging process. Retirement planning should deal with each individual from the onset of aging (when he has not yet reached his peak of productivity) through the twilight years when duties must be modified periodically in keeping with changing ability to perform.

It is impossible to set an arbitrary figure for this onset of aging that will apply equally to each employed individual.

It varies from person to person. The irreversible physiological changes that mark the beginning of the aging process may appear in some individuals in their late 30's, and may in others be postponed until the middle 40's. For all practical purposes, it is usually most expedient to assume a round figure of 40 or 45 as the date of onset of aging, and institute a program of retirement planning for each employed person as he reaches that age.

Arbitrary Figures for Bookkeeping Only

Assumption of an arbitrary figure of this sort is expedient for bookkeeping purposes only, and should not enter into any assessment of individual proficiency. This figure is merely a record of *chronological* age; not an adequate indicator of true *physiological* age.

Chronological Age

Chronological age is a simple matter that may be estimated from one's birth certificate. It is age by the clock. Physiological age is far more complex. It involves mind, hand and heart, arteries and the thickness of their walls, the pituitary gland, back muscles, and eyesight. It is not an even process and all our organs do not age in step with each other. The same person may have a young heart, an old stomach, and a middle-aged colon. Physiological age is the sum total of all of these factors. It cannot be measured in an isolated physician's office or laboratory; it *can* be measured in the occupational environment over a period of days, weeks, or months.

Physiological Age

Measurement of true physiological age will surely be a major function of the humanologist of tomorrow. The best measurement of age will always be, not what a statistical chart or a pathologist's report says is true of individuals of a given age, but what the individual himself can do, without unusual strain. Frequently, a dominating factor in *physiological* proficiency in later life, is psychological health. An older person who has psychological and intellectual resources to draw upon may often find ways to compensate for specific physiological deficiencies in order to perform in accordance with standards ordinarily set for younger workers.

The Harvard School of Public Health has done considerable research into the effects of aging on persons employed in occupations classed as youthful; notably aviation and truck-driving. A study of airline pilots revealed that while older persons tended to take more time to assimilate new material and to react in a situation calling for snap decisions, many of them "are particularly well suited for operations which demand a high degree of accuracy. . .Their deficiency in organizing new material is often more than offset by gains in quality and accuracy of performance once the material is assimilated." Moreover, evidence has tended to support the belief that the ability of the psychologically healthy older pilot to foresee emergencies and to plan a course of action in advance is more valuable than the younger pilot's swift reaction when the emergency occurs. Similar findings have been made with regard to truckdrivers.

Humanology will, in the near future, develop different standards of age-measurement, correlated with proficiency standards, for different types of occupation. Anyone who studies the obituary columns of the newspapers will notice the striking connection between early death from serious gastrointestinal or circulatory emergencies and the demands, self-imposed or otherwise, upon the deceased individual. The hard-driving public figure, the fiercely competitive entrepreneur or organization man who dies in harness at the age of 45 or 50, turns out to have had an ulcer, hypertension, atherosclerosis, or other health problems not shared by thousands of his less-tense contemporaries who expect justly to live and enjoy life for at least another baker's dozen of years. Therefore, the nature of the individual's job must enter into determination of his true age. A 50-year-old farmer with angina pectoris is considerably younger than a 50-year-old president of a company that is just emerging from bankruptcy into a highly competitive field, if the latter has the same health problem.

An authority on physical education at a midwestern university has even gone so far as to fix age 26 as the onset of aging in some individuals, citing that age as the period at which many persons cease to exercise or develop physical or mental facilities, becoming old in attitude as well as muscular proficiency.

The Development of Reliable Standards — The Gateway to Flexible Retirement

An important result of fundamental research into this matter of true age with relation to occupation will be the development of reliable standards for determining what should be the proper retirement age in each individual case. There is currently a tendency to make a fetish of the chronological figure, 65. As indicated above, some persons at age 65 are physiologically in their 50's or are able to compensate for certain degenerative deficiencies to the extent of performing in accordance with standards set for younger workers. On the other hand, an alarming number of individuals in this hard-driving civilization are ready to be turned out to pasture before age 65.

In the foreseeable and fairly near future, environmental research will arm the humanologist with the facts for making an accurate interpretation in each individual case. It will be possible to say of one given individual:

X is 40. He has not yet reached the peak of performance. He will reach it by age 45, and become department head before he is 55, barring unforeseen emergency. His heart is strong, and his family has a history of longevity. X has a tendency to overeat and imbibe somewhat immoderately upon occasion. After 55 he may have a minor alcohol problem affecting home life, but not his job. His strength of character will keep the problem within bounds. Slight diminution of job capacity may reflect home tensions. When he reaches 60, some minor signs of psychological senility may be expected — partly result of alcohol. Memory lapses will occur, plus some hypersensitivity to imagined criticism. Otherwise his judgment will be good. However, it may be necessary to decelerate him at age 60, and replace him as department head. Mr. X will be in fair health at 65, but ready for retirement. He is intelligent and has insight in most matters.

Or:

Y is 38. He is a star salesman and will head the sales department before 45 if organizational turnover permits. He is capable of rising to policy-making level. Y is a go-getter, but he is disciplined. He delegates intelligently and inspires subordinates to perform well. He relaxes easily. He has an innate sense of leadership. His arteries are somewhat older than his individual chronological age, and he may have a heart attack before 60. Odds: he will survive same and remain an asset to the company. At 70 he will still be competent to perform in directoral capacity, and he will adjust easily to retirement. He has a number of outside interests.

The Theorists' Point of View

Management theorists, particularly those under 50 or 55, seem to be agreed that the virtue of having an arbitrarily fixed retirement age is that it shoves the *old goats* out periodically, creating a personnel turnover and starting a chain of morale-building promotions all the way down the line.

Often when these theorists approach the age of 65 they begin to see matters in a different light. Many find themselves in good health; they love their jobs and are almost convinced they will live forever. At this time, they may wish to persuade their companies to raise retirement age to 68 or 70, and find themselves pitted against younger organization men who are prototypes of themselves some 15 years back.

It is difficult for a humanologist without all the facts to diagnose the atrophy of hope and understand what it means to the individual concerned to be let out, or pushed out, after 30 or more years of helping to build a company, department, or work-team. Unless a person has begun to prepare early for retirement he will find it almost impossible to gear himself for the change of pace. When he tries to embark upon new activities, he finds that he is in a rut; he simply doesn't know what to do or how to do it. Time weighs upon him.

This seems to be no secret, and yet it is surprising how many members of the policy-making team still view retirement merely as a termination and are quite literally shocked at the thought of applying retirement procedures to anyone under 55, at the youngest.

The enlightened, and truly well-meaning man of management will say: "We bring them into the pension plan as soon as they join the company so that they can build up an equity for themselves. Otherwise, it doesn't seem right to worry about their retirement problems until they are 55 or 60 or thereabouts and ready to slow down a bit. After all, an employee in his 40's hasn't even started to produce at top capacity yet, unless he is a ditch-digger, hod-carrier, or messenger boy. Our best, most reliable officers and supervisors are in their 50's or better. Most of our top salesmen are over 40. Our most valuable skilled workers in the shop are in their late 40's and our most dependable clerical people go as high as 65. Why should we start thinking about retiring people who haven't even hit their stride yet?"

This is how the idea of severance comes to dominate retirement planning.

The 40-Year Old Is a Potential Candidate for a Health Setback

It should be borne in mind that any individual who has passed 40 is a potential candidate for a health setback. Illness at this age is not likely to come (and go) with dramatic suddenness, as is usually the case in youth; it tends to creep inexorably, making its presence known gradually, by degrees. The most-feared in this category is cancer, a regenerative disease. This chapter, however, is not the appropriate one for a detailed discussion of cancer problems as they relate to humanological practices. The subject has been explored in Chapter 8 in connection with the periodic health audit. Humanologically speaking, the basic problem of cancer is not one of adjusting to the aging process. Many individuals, drawing on personal reserves, have adjusted heroically to irreversible cancer problems, but this is outside the scope of humanology; it has more to do with background and breeding. The humanological goal is to detect cancer early, when it can be rooted out altogether.

Circulatory Problems and Adjustment

Circulatory problems, on the other hand, though they are kept at bay only through programs of diagnosis and early detection, are best considered in relation to *aging in harness* and adjustment to later life. Few degenerative circulatory problems can be rooted out altogether, even if detected in their earliest stages. At best they can be slowed down, and lived with. Eventually, such problems will occur in any individual who does not die first from some other cause. The circulatory system was not designed to last forever. Thus, it is humanologically proper to deal with circulatory problems — present, anticipated, or purely hypothetical — in the normal course of preparing the 40-year-old employed individual for the facts of aging in harness.

The Circulatory System No Great Mystery

To the medically-trained humanologist, the circulatory system is no great mystery, and least mysterious of all is that powerful, and essentially uncomplicated muscle, the heart. For this reason, there is often a lack of communication between the humanologist and the layman with a real, or feared, circulatory problem. To be truly effective, therefore, the humanologist who has to do with the employed individual should know something of the fears and misunderstandings that race through the mind of the layman when circulatory problems, particularly those affecting the heart, occur. Moreover, he should know how to interpret these problems in language that the layman can understand. Mere understanding is not the final goal. The interpretation should be put in terms that do not frighten the individual who has not had a medical education; it should not go into unnecessary details which may be of interest to the physician but only serve to clutter the untrained mind, and it should offer a basis for constructive, palliative action on the part of the individual concerned.

The Humanologist — A Counselor

The employed individual with disturbing symptoms is human in that he tends to be at least somewhat apprehensive of the unknown. Perhaps he goes to a diagnostician who hasn't the time or the knack for putting his diagnosis in terms readily comprehensible by the layman. This physician manages to convey the idea that the patient has something wrong with his heart, and says in all sincerity that the patient need not worry so long as he lives within certain specified limits.

The physician may not be aware of what goes on in the patient's mind in subsequent days. This patient goes home to a series of sleepless nights and confused days. All of his life he has known that his heart was an intricate, unknowable thing that thumped rhythmically and endlessly in his left side above his stomach and somehow held within its surface the secret of life. Heart disease, he has read, is the nation's number one killer. He knows too that the way to kill an antagonist is to shoot or stab him in the heart. Finally, he knows that his doctor in spite of some impressive-sounding vocabulary has not offered any cut-and-dried cure for this mysterious *heart* condition.

He feels that he is fighting in the dark with a magical opponent who can see him clearly, and who can finish him off easily any time the notion happens to strike him.

If things continue this way for a long period of time, the patient's tortured imagination begins detecting symptoms that do not exist at all. These symptoms may make themselves known only at crucial moments of stress. The patient becomes preoccupied with these symptoms, and the quality of his work deteriorates.

If he has let it be known that he is suffering from a heart condition, his decline in work performance is attributed to this, by other laymen who are as mystified by the circulatory system as he is. Eager for a plausible alibi for his letdown, the patient is happy to corroborate this opinion. He has sold *himself* on it.

How Heart Myths Get Into Currency

This is the way heart myths get into currency. In actuality, it is not the heart condition that has caused the decline in work performance; it is the patient's lack of understanding of his heart, and his resultant terror of the unknown. He is not only incapacitated for his job, but for life itself.

The employed individual nearing the threshold of aging needs to be made to understand, first, that the heart is in actuality only a part, by no means the most intricate part, of his total circulatory system. Its processes are so simple that almost any blacksmith could, with instructions, manufacture a crude but workable heart. Science has not yet been able to construct a fully practical human brain, liver, reproductive system, or stomach. To date, man has not even con-

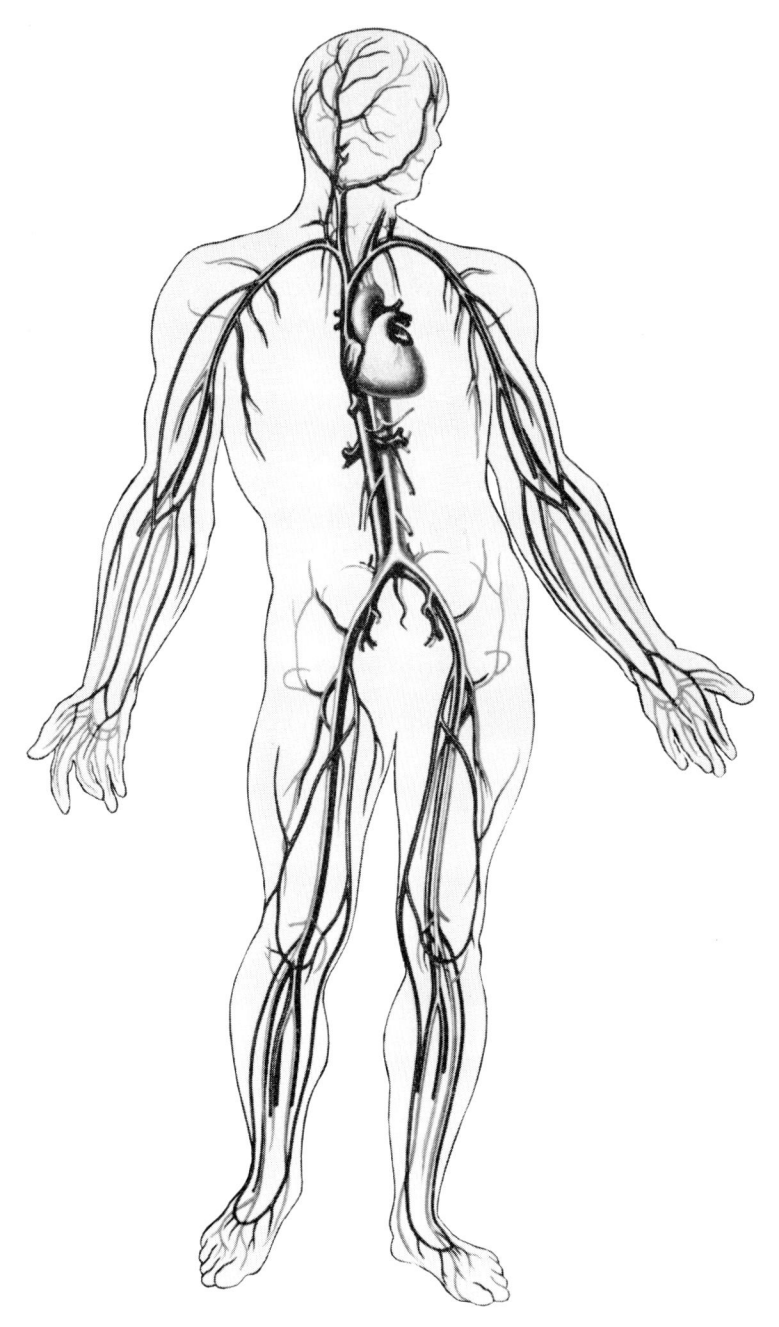

Fig. 16 — THE CIRCULATORY SYSTEM: HEART AND BLOOD VESSELS FROM HEAD TO TOE BELONG TO ONE INTRICATE SYSTEM OF THE HUMAN BODY

structed, artificially, a single living cell (though developments in the synthesis of proteins indicate that the possibility of this accomplishment is no mere science fiction fantasy). Nevertheless, man *has* built an artificial heart and *made it work in living human beings!*

To understand the nature of anything is a step toward learning to take it in stride. The farmer who knows the nature of his hand-operated water pump will know how to cope with it (by priming, for instance) when it becomes old and eroded and fails to draw up water by normal manual operation. The individual who knows his heart and understands the significance of its aging problems will learn to adapt himself to its limitations just as the farmer adapts himself to the limitations of his pump.

Facts for the Layman to Understand

A tried and true method of familiarizing the layman with the workings of his circulatory system in comprehensible language is as follows:

The layman does not need to have it explained to him that every living thing needs nourishment to stay alive. He knows this almost instinctively. He knows too that his body is made up of thousands of separate cells, each of which is a living entity, demanding nourishment for its own life. Nourishment is in food and air, which is taken in through the mouth or nasal passages, and must be distributed by some agency to every cell in the body. The distributing agency is the blood. Freshly aerated and bright red blood from the lungs is pumped outward from the *left* side of the heart, acquiring digested nutrients on its way, via the arteries to the outer extremities. Then it picks up various waste materials and returns, dark in color, via the veins, to the *right* side of the heart, whence it is pumped through the veins to be repurified, aerated, and pumped outward again. On its return, it disposes of considerable waste material in the kidneys, which are essential parts of the circulatory system.

The heart does nothing but keep this distributing agent circulating: out and back, out and back. It is merely a great muscle; not a chemical laboratory. The faster nourishment and energy is burned up in various parts of the body, the faster the heart must work to supply the demand for more nourishment in those parts. This extra work does not injure nor shorten the life of a sound heart (Cross-country-runners in their 60's have been found to have circulatory systems that would be envied by many in their 40's), but it may cause a weak heart to protest violently.

Like any pump with a two-way action, the heart has valves. When the heart contracts to expel blood from both sides (the dark blood in the right side going to the lungs for oxygenation, and the bright blood on the left going out to nourish the body) the exit valves must be open, and the entrance valves must be closed; otherwise the blood will regurgitate into the vessels by which it is supposed to enter the heart. By the same token, when the heart is expanding (the right side filling up with dark blood from the veins; the left filling up with purified blood from the lungs), the entrance valves must be open and the exit valves closed.

The Mechanical Types of Heart Disorder

The three principal mechanical types of heart disorder may be explained to the layman as follows:
1. Failure of one or more of the valves to open or close properly at the right time.
2. Weakness of the heart muscle, or lack of coordination, resulting in inadequate circulation to the extremities.
3. Blockage of one of the vessels through which the blood flows, cutting off nourishment from a part of the body (a heart-attack is the result of blockage of one of

the "coronary" vessels sending nourishment to the heart muscles itself).

To clarify nomenclature, it may be advisable to give the layman a simplified picture of the two principal types of valve dysfunction, as follows:

When the valve won't open wide enough to let a normal head of blood pressure through, it is called *stenosis*. A common disorder resulting from rheumatic fever in early life is stenosis of the *mitral* valve — the exit valve in the heart's left chamber. When the valve will not close all the way, and allows regurgitation of blood, it is known as an *insufficiency*. Many insufficiencies today can be corrected by surgery.

Mechanical weakness of the heart muscle may be owing to infection, or cumulative exhaustion as a result of overwork caused by valvular disorder or impediments, or irregularities elsewhere in the circulatory system. A common, and, to the uninformed person, rather frightening form of weakness has to do with *fibrillation*. This is, put simply, a lack of rhythmic coordination between the two chambers of the heart. One chamber, it may be explained in lay terms, gets out of step with the other; hence power is lost. Modern medicine has many ways of coping with this situation. One of the oldest is the use of digitalis, which blocks the nervous impulse that controls the beat of the heart and brings about some degree of coordination and rhythm at least temporarily. The layman who has at least a simplified picture of his mechanical problem plus knowledge of how to cope with it, will neither be frightened by it nor abnormally obsessed by it. It is not painful, and except in extreme cases should not interfere with high quality job performance.

Medical Terminology Complicates the Picture

When blockage of an artery occurs, the tissue to which that artery is designed to carry nourishment is starved and put out of commission quite rapidly. If a vital area is not affected and the patient survives, the task of the humanologist is to reassure the sufferer, explain to him what has happened, and what it means to him in terms of his future way of life. The best-known result of blockage is the classic heart attack (myocardial infarction, coronary thrombosis, occlusion, and the like). Patients often do not realize that the above terms are simply names for different aspects of the same things. Harold Mac, a contractor age 57, who suffered a mild attack involving one of the anterior coronary arteries, overheard his doctor refer to both "coronary thrombosis" and "myocardial infarction" and was somehow convinced that he had been stricken simultaneously by *two entirely unrelated* heart disorders. Both he and his wife were so frightened by the seeming dual nature of his problem, that he felt obliged to ask for premature retirement, wholly unnecessary in his case, in order to spend his "last days" with his children in a distant city. Another doctor who examined him in this city introduced the words "occlusion," referring to the past accident, and "atherosclerosis," referring to his present condition. Harold began to think of himself in terms of Oliver Wendell Holmes' "one hoss shay" which fell apart all at once, despite the fact that his problem was by no means a major one and was easily manageable. He went into a complete decline, and although he survived to the age of 70, his life was that of a wretched, neurotic invalid.

It should be explained to such a patient that a certain part of the heart muscle, owing to lack of nourishment, has died and has been or is about to be replaced by scar tissue, after which the heart will work quite adequately if not pushed beyond certain limits. The problem of adjustment, after recovery from the actual attack, has nothing to do with the scarred but comparatively healthy heart; it has to do with the condition which brought about the blockage in the first place: atherosclerosis (involving the gradual accumulation of a "mushy" substance inside hardened artery walls). This condition can be kept under control by an individually prescribed regimen (not too rigorous) of diet and general behavior.

When atherosclerosis is discovered in an individual *before* blockage has occurred in any part of his circulatory system, he may consider himself fortunate. It should be explained to him that this discovery actually gives him a new lease on life; he may now take steps to keep at bay the occlusion that otherwise might have occurred at almost any time. Painful as it may be, angina pectoris is a result of atherosclerosis of the coronary arteries. This pain occurs when the heart muscle requires extra oxygen owing to exercise or excitement, and cannot get it — because of the narrowness of the vessel — and expresses its protest in the form of a painful cramp. This is a useful warning signal that a heart attack may be in the offing. The forewarned individual must now take steps to avert or postpone the attack. In the meanwhile, the pains are easily controllable by prescribed medication.

The anxious layman with a real or imagined heart problem may have learned somewhere that a heart attack is more often fatal than not. It should be communicated to him that even if he does suffer an actual heart attack his chances of surviving are better than 7 in 10, and the odds are improving all the time.

No Single Symptom Adequately States the Case

No single one of the symptoms of circulatory dysfunction is *necessarily* a symptom of such dysfunction. The nurse or other front-line counsellors, who must deal with the frightened possessor of one of the following symptoms, will know how to offer reassurance that will allay fright without creating a lackadaisical attitude:

1. Irregularity of pulse
2. Edema (evidenced by swelling of the lower extremities)
3. Dyspnea (shortness of breath)

4. Increased wetness of the lungs
5. Pain in the chest
6. Cyanosis (blueness of complexion)

The first symptom, irregularity of pulse, may be purely psychosomatic; it may be the result of too many cups of black coffee. In any event, if noncirculatory causes are not wholly obvious, the front-line humanologist will arrange or advise a medical examination for the individual with the symptom, explaining at the same time that the symptom is not a frightening one, and may well be meaningless.

In the case of edema, it should be explained that this can be seen in a mild form in healthy individuals, particularly in hot weather. A checkup is of course in order, but even if a circulatory problem is found to be the cause, it is a manageable one, and need not interfere with job or leisure fun.

Dyspnea always accompanies extra effort. If it accompanies minor effort, it should be viewed with mild suspicion, and medical advice should be sought; though the root problem is not necessarily circulatory. A respiratory problem may be involved.

Wetness of the lungs, marked by rales, may be a sign of bronchitis; it may be simply the aftermath of a minor head cold. If it is serious enough to induce a patient to stay home from work or seek the advice of the nurse or other front-line humanologist, the possibility of a circulatory basis should not be ruled out until a medical examination has been undertaken. This condition can be symptomatic of inadequate blood pressure through the lungs owing to heart irregularity, weakness, or valvular disorder.

Pain in the chest may possibly indicate *angina* — or even a mild heart attack. It may indicate inflammation of some part of the heart muscle. More probably, however, it is a symptom of pleurisy, pneumonia, or some digestive disturbance. All of these problems call for some degree of medical attention; none need cause undue alarm.

Cyanosis is caused by inadequate oxygenation of the blood. Whether or not this is caused by a circulatory disorder is something to be determined by a physician. Pneumonia, poor adjustment to rarefied air at high altitudes, or exhaustion following strenuous exercise can cause cyanosis.

All of the above discussion of circulatory problems is undoubtedly old stuff to the medically trained humanologist as well as to the average front-line paramedical member of the team. It has not been presented for the purpose of teaching a presumably well-grounded student anything about the circulatory system; it is offered as a way of presenting the problem to the apprehensive layman in terms that he can grasp and put to constructive use. It is important to diagnose and prescribe suitable medication for the circulatory problems of the aging individual. All this is worth precisely nothing, if there is no real communication between humanologist and layman. The former must adequately interpret the significant aspects of the problem in such a way that the latter is neither confused nor frightened when he has to cope with this problem.

Recommendations Must Be Stated Clearly

If the individual with the problem is employed and expects to return to work after sufficient recovery, the humanologist representing the company should see to it that the attending physician puts into writing precisely what the individual can and cannot do. A recommendation such as "light work for (x) days (or weeks)" is not very useful. No one knows exactly what is meant by "light" work. The employer, if he feels obligated to keep the individual on the payroll at all, may decide to dump him into some sort of corporate wastebasket. That is, he may assign him to some almost useless, tedious task, far below his mental level, and proceed to forget about him. In addition to a circulatory problem, the individual will now find himself suffering from a kind of occupational dry rot. His colleagues will pity him and treat him as an invalid.

In some cases, by no means the majority, the individual returning to work after a serious circulatory accident, or the individual currently facing some acutely progressive degenerative circulatory problem, may be considered, after meticulous examination, to be permanently incapable of carrying out his former work assignment. When this occurs, the ideal solution is to fit him into a modified version of the same kind of work or find a lighter assignment somewhere in his old department, where he has friends and a feeling of belonging, and where his *experience and judgment* may be of value.

Where a union contract is involved, a transfer may conflict with written provisions regarding seniority. Ordinarily, an informal conference between representatives of the union and humanology team, with the employee present, will solve this problem; the function of a union is not to penalize long-service members who happen to have circulatory problems. Frequently a provision can be inserted into the contract agreement making it possible for certain, long-service employees to carry their accumulated seniority with them in the case of transfers that are necessary for medical reasons.

If the individual cannot be fitted into his former department or into a job assignment somewhat similar to his former one, he should be painstakingly reassessed; his opinion as to his wants, needs, and capabilities should be elicited and given consideration. An assignment that seems to him to be degrading, or that is so foreign to his past experience as to be difficult to assimilate, may complicate his health problem beyond the limits of manageability.

Constructive Retirement Planning

The above discussion has dealt with after-the-fact handling of specific after-40 health problems. Constructive retirement planning deals with the effective control of these problems *before* they happen. A routine program of diagnosis and early detection beginning when the employee reaches the age of 40, will tend to keep problems like the above from occurring in their acute form at least during the years of employment, and, more likely than not, for several years after retirement.

The company that fails to establish such a program is flirting daily with disaster in some form or other. The promising, and almost indispensable junior executive, may call up the day he is to step into his new job as sales manager and announce that his doctor has told him he can never work again. The irreplaceable technician may have a crippling stroke in his laboratory at a time when the shortage of men with his kind of training was never so acute. The expensively trained, newly-appointed shop foreman may have a cerebral hemorrhage before any of the green men under him have learned to do their *own* jobs adequately, let alone step into *his*.

The Little Insidious Mishaps

These are the big mishaps. Then there are the little, insidious ones. The loyal and efficient file clerk who has been with the company for 20 years may become snappish and slovenly in her work. The indispensable secretary to a key executive may develop an astonishing record of absenteeism, with a wide range of implausible excuses. The dynamic department head may show a progressive tendency to make ill-advised snap decisions and wear a chip on the shoulder when reprimanded. The eager beaver vice-president who has always put the company ahead of wife, children, and friends may suddenly begin to appear in various downtown saloons when he is supposed to be off on certain vague, self-assigned public relations missions.

Assuming that all of these people were once highly competent, and assuming that there has been no change in the nature of their jobs, the changes must be taking place within the individuals themselves. If, upon analysis, it is determined that these changes are progressive and have to do with the aging process, then it can be seen that these people are moving toward retirement at varying rates of speed, and should therefore be dealt with as potential retirable persons — not as of age 65 — but as of any time, perhaps, as of tomorrow.

The aging process brings about, in everyone, physiological and psychological changes which can, if ignored or uncompensated for, result in total disability with shocking rapidity. Hence it follows that everyone past 40 ought to be looked upon as someone who might have to be replaced without warning at any minute. From this, it follows further that everyone past 40 (being still *on his way* to his peak of performance) should be maintained as scrupulously as the most valuable piece of machinery, and assessed periodically with an eye to postponing retirement — or deceleration — as long as is practically possible. Many companies set up constructive depreciation programs with respect to machinery, under which the emphasis is on the maintenance and retardation of depreciation of the equipment; *not* on mere cold-blooded preparation for depreciation itself. Such constructive thinking must be brought to bear on the problem of retirement planning.

At this point, the old-fashioned retirement planner may feel impelled to protest: "True enough. People *do* have sudden illnesses, and these can cause considerable extra expense and trouble. That still need not come under the heading of retirement. We put it under the 'sickness and accident' category. All of our people are covered by insurance. Some have their own policies in addition to company group insurance."

The trouble with this argument is, for one thing, that it does not distinguish between younger and older employees. In the majority of cases, when a younger person takes advantage of medical benefits, it is because of some infectious disease which is completely curable and which puts him out of action only temporarily. There are, of course, exceptions to this, but they are not common.

On the other hand, when a person in his 40's or 50's has to take advantage of health insurance, it is more likely than not that his illness has to do with aging, and that it is not wholly curable. Every illness is likely to bring him an irretraceable step closer to retirement.

From the point of view of everyone concerned i.e., the employer, the stockholders, and the employee himself, the ideal solution is to prevent such illness and to arrest the aging process insofar as this is humanly possible. If the illness is unpreventable and the aging process uncontrollable in other words: if premature retirement is inevitable, it is at least desirable for all concerned to know in advance *what* physiological or psychological changes are going to take place and *when* these changes are likely to occur.

Obviously no mere financial coverage or indemnification plan can accomplish either of these ends. Such coverage serves a useful purpose in that it removes some of the financial complications of illness and tends to eliminate self-deception i.e., "I can't afford to be ill; therefore I am not ill." All this, however, is secondary, and does not get to the heart of the matter. What is important in the case of those who are aging in harness is not to treat them hopefully, though often futilely, *after* illness strikes, but to detect illness in its incipiency while it still may be eliminated or kept at bay.

Health Planning and Retirement Planning Go Hand in Hand

In other words, health planning and retirement planning go hand in hand for the individual who has crossed the threshold of aging. Health and retirement cannot be put into two separate pigeonholes i.e., one crammed with insurance benefit plans, the other with pension arrangements. Insurance plans and pension arrangements are important, but they should be a routine matter supplementary to the primary goal of planning, which is to keep the individual employee eager and fully productive within his limitations. The questions to ask about any person past 40 are not "how is he provided for in case of sickness?" nor "when will he retire, and on how big a pension?" but: "What are his peak capabilities, how can he be kept at his peak for the longest possible period of time, and how can he be most gradually and painlessly decelerated after he has passed his peak?"

Incidentally, the time will come, in the near future when this is the principal question that is asked of the over-40 job-hunter. Unfortunately, there is little the humanologist can do to speed up this inevitable revolution. The bookkeeping problems, plus some real monetary problems, of incorporating the older job-hunter into pension, insurance, and other fringe-benefit programs, is responsible for the fact that thousands of eminently capable individuals going slightly bald or grey at the temples are pounding the pavements hopelessly, drawing unemployment compensation. The excuse often given by representatives of management for their part in perpetrating this social and economic waste is that older people are harder to absorb into modernized programs, or that they are set in their ways, or incapable of performing the work a younger person can perform. The humanologist may point out that none of these objections are valid, but he cannot eliminate the bookkeeping or cost barrier. As the proportion of older workers to the total population increases, enlightened management will either find ways and means to lower this barrier on its own initiative, or the combined pressure of organized labor and government will force the barrier down in a manner rather unpleasant for all concerned.

An example of constructive retirement planning is seen in the case of Herman S., composing room foreman in a printing shop since age

34. Herm's employer knew the value of health maintenance and construction; seasoned experts like Herm, he knew, were not easily replaced. In the front office, there was a complete file on Herm, covering his aptitudes, his strong and weak points, his known likes and dislikes, his mental attitudes, and his health. After Herm's 40th birthday, the boss saw to it that this file was taken out and brought up to date at fairly regular intervals. This involved medical evaluation by Herm's own doctor from time to time, at a prearranged cost which was not out of line. Since Herm's doctor knew him and saw him at sensible intervals, he did not need to give him the works in the course of each visit. The doctor knew what facets of Herm's health needed careful watching, and what facets could stand a once-over-lightly or could safely be taken for granted.

In the course of one of these routine visits, the doctor noticed the early signs of incipient diabetes. Herm was alerted, and modified his life accordingly. In a subsequent visit, it was noticed that suppressed resentment over a misunderstood job assignment was threatening to undermine Herm's value as an employee. Since this was discovered in its early stages, the problem was easily ironed out.

Herm is now 62 and is still an efficient employee. He knows the limitations imposed by his age and condition, and is gradually relinquishing his authority to a younger man who will be fully able to take over when Herm comes up for retirement. Herm's mental attitude toward retirement is good. He knows what he can and cannot do, and he knows what he wants out of life. He acquired this knowledge as a result of a constructive retirement program that was initiated before he began aging in harness. His value to his boss and to the stockholders of his company has far exceeded the modest sums expended in *keeping* him vertical and productive. Not only has he been able to maintain high standards of job performance, but he has been absent only four days out of eight years.

If Herm's employer had been content to handle Herm's potential health problem merely by setting up some sort of group medical costs indemnification plan, Herm would in all probability have acquired a full-blown case of diabetes. This would have come as a nasty surprise, perhaps with serious complications, and Herm would have found himself, despite his insurance coverage, deeply in debt to his physician and hospital. Moreover, he would have lost many days, maybe months, on the job, creating innumerable snarls in the shop schedule. Had he survived all this, it is difficult to estimate the magnitude of the problems that would inevitably have been triggered by the later emotional problem. Since it was in his nature to suppress such matters, the situation would not have come to light in its early stages. By degrees Herm would have become an inefficient, senile, tendentious, and thoroughly sick person. Just how much this would have meant in dollars and cents to his company cannot be determined with precision; there is no doubt that Herm's rating as a liability would have exceeded his rating as an asset.

Accent on Mantalent Development

Every company has more than its share of Herm's who are either burning out ahead of schedule or are carrying within them the unseen seeds of irreversible illness. For the sake of the stockholders, for the sake of management, for the sake of the employed person himself, humanology must bend every effort to bring about the inauguration of retirement programs which will put the accent on *mantalent development,* on human maintenance, not on after-the-fact patch jobs. This philosophy belongs at the core of all humanological planning, whether at the stage of recruitment or that of retirement. It needs *special* emphasis, however, with regard to aging in harness, because it applies then to the most important human resources of any given company, and because it strikes at the heart of the growing social problem that is threatening to become the most perplexing one of modern times — that of the social waste of America's expanding population of senior citizens. This is a problem which will inevitably be placed on the shoulders of business, either directly as an organizational challenge or indirectly as a tax burden. Industry will gain by taking the initiative and choosing the former alternative.

REFERENCES

FRANCO, S. CHARLES, M.D. "Clinical Ballisto-Cardiography." *Industrial Medicine and Surgery,* XXI (May, 1952), pp. 197-205.

GERTHER, M. M., M.D.; WOODBURY, M. A., M.D.; GOTTSCH, L. G., M.D.; WHITE, PAUL D., M.D.; RUSK, HOWARD A., M.D. "The Candidate For

Coronary Heart Disease." The *Journal of the American Medical Association*, CLXX (No. 2, May 9, 1959), pp. 87-149.

GOFMAN, JOHN W., M.D. "Some Concepts of the Problem of Coronary Heart Disease in Industry." *Industrial Medicine and Surgery*, XXIV (No. 4, April, 1955), pp. 151-156.

GOLDWATER, LEONARD J., ET AL. "The Cardiac in Industry." *Industrial Medicine and Surgery*, XXI (February, 1952), p. 75.

HANSEN, HOWARD, M.D., AND WEAVER, N. K., M.D. "Arteriosclerotic Hearts at Work." *Medical Bulletin*, XXI (No. 2, July, 1955), pp. 84-94.

HYMAN, ALBERT SALISBURY, M.D. "Heart Disease and Industrial Medicine." *New York State Journal of Medicine*, LI (No. 13, July, 1950), pp. 1603-1606.

ISRAEL, MURRAY, M.D. "An Effective Therapeutic Approach to the Control of Atherosclerosis." *American Journal of Digestive Diseases*, XXII (No. 6, June, 1955), pp. 161-168.

KAMMER, A. G., M.D. "Optional Retirement Plans — Their Implications for the Industrial Physician." *Industrial Medicine and Surgery*, XXI (July, 1952), pp. 343-344.

KIENZLE, T. C., M.D. "Retention-Retirement Program of the Standard Oil Company (New Jersey); The Practical Application." *Medical Bulletin*, XII (No. 1, February, 1952), pp. 168-172.

——————————. "The Medical Interpretation of Retirement Cases from the Company Viewpoint." *Medical Bulletin*, X (No. 1, January, 1950).

MCFARLAND, ROSS A., PH.D. "Psycho-Physiological Problems of Aging in Air Transport Pilots." Harvard School of Public Health, *The Journal of Aviation Medicine*, XXV (June, 1954), pp. 210-220.

MCFARLAND, ROSS A., PH.D.; MOSELEY, ALFRED L., M.A.; FISHER, M. BRUCE, PH.D. "Age and the Problems of Professional Truck Drivers in Highway Transportation." Harvard School of Public Health, *Journal of Gerontology*, IX (July, 1954), pp. 338-348.

MORGAN, JOE ELMER. "So You're Over Forty." *Senior Citizen Leaflet*, No. 1, 1701 Sixteenth St., N.W., Washington 9, D.C.

NEUGARTEN, BERNICE L. "Social Change and Our Aging Population." *A.A.U.W. Journal*, (January, 1957): *Junior League* Magazine, (March-April, 1958), p. 1.

OLSON, CARL T., M.D. "Maintenance of Health in the Elderly Work Force." *Industrial Medicine and Surgery*, XXI (No. 12, December, 1952).

Page, Robert Collier, m.d. "Without Health — Why Live to Retire?" *Industrial Medicine and Surgery,* XXV (April, 1956), pp. 189-190.

――――――. "Aging in Harness." *Surgical and Medical News,* (Goregaon, Bombay, India), (October, 1958).

――――――. *Good Tips to Good Health.* Retirement Planning Guide Book. Stanford: Retirement Council Publication, 1957.

――――――. "The Aging Executive." *Occupational Health Review,* XII (No. 1, 1960), p. 7.

Poutasse, Eugene F., m.d., and Dustan, Harriet P., m.d. "Arteriosclerosis and Renal Hypertension." *The Journal of the American Medical Association,* CLXV (No. 12, November 23, 1957), pp. 1521-1525.

Pratt, Laurence O. "Seven Sins Against Older People." *This Week Magazine,* (December 8, 1957), pp. 7-10, 36.

Still, Joseph W. "Boredom — The Psychological Disease of Aging." *Geriatrics,* XII (September, 1957), pp. 557-560.

Thorpe, John J., m.d. "The Optimum Time for Return to Work Following Various Major Cardiovascular Disabilities." *Industrial Medicine and Surgery,* XV (July, 1956), pp. 329-336.

Ungerleider, Harry E., m.d. "The Borderlands of Heart Disease." *Industrial Medicine and Surgery,* XXIII (December, 1954), pp. 525-527.

United States Department of Labor. *Older Workers Under Collective Bargaining.* Part 1: Hiring, Retention, Job Termination — Part 2: Health and Insurance Plans, Pension Plans. (Bulletins 1199-1 and 1199-2), Washington, D.C.: U.S. Government Printing Office, 1956.

――――――. *Pension Costs in Relation to Hiring of Older Workers.* (BES No. E-150), Washington, D.C.: U.S. Government Printing Office, 1957.

――――――. *Job Performance and Age — A Study in Measurement.* (Bulletin No. 1203), Washington, D.C.: U.S. Government Printing Office, 1956.

Wade, Leo, m.d. "Medical Preparation for Retirement." *Medical Bulletin,* XII (No. 3, August, 1952), pp. 385-391.

Walmer, C. Richard, et al. "A Symposium of Heart Disease in Industry." *The American Journal of Cardiology,* (March, 1958).

White, Paul Dudley, m.d. "The Epidemiology of Heart Disease." *New York Academy of Medicine Bulletin,* XXXIII (No. 12, December, 1957), pp. 819-837.

Wilson, F. W., m.d. "Problems in Industrial Cardiology." *Journal of Occupational Medicine,* II (No. 12, December, 1960), p. 585.

17

A Constructive Mantalent Development Program

THE FOLLOWING factors in summary highlight the need for a well-conceived humanological program in any given business.

1. The mantalent pinch is particularly acute in the age group from 25 to 45, from which the managers, supervisors, technicians, and specialists for tomorrow must be recruited and trained.... This shortage will increase — never decrease.

2. The costs of benefits *not included* in take-home pay have *trebled* during the last decade. The size of an employee's pay check is far from being a reliable measure of what his employer is investing in him. The average employee does not *see* these benefits in his pay envelope. Being unaware of them, he does not appreciate them.

3. Operating costs shoot up whenever new benefits are introduced. If mantalent development techniques are not applied to turn these increased costs into improved productivity, to promote efficiency, and to eliminate wasted man-hours, then the new costs must be passed on to the consumer. If the product is an essential item, a boost in price may result in public pressure for some kind of inflationary relief measure on the part of the government.

4. Nonscientifically planned "wage-inducement battles" tend to be self-perpetuating and self-accelerating. In the long run every business is hurt by "wage-inducement battles."

Specific Concerns of Humanology

An absent employee adds to production costs. So does an employee who is present but inefficient or half-hearted in the performance of his job.

Each human is an individual. When an employer establishes or follows precedents in which the individual is dealt with as a piece of machinery or a pin on the map and not as a *whole* person, a slackening of efficiency is inevitable.

In order to meet this challenge, to keep the individual employee *vertical*, and *wholeheartedly* on the job, the employer must endeavor to:

1. Understand all the causes of absenteeism; know to what extent it can be prevented and how, and what each individual case means in terms of cost accounting.

2. Estimate in advance the probable wants, needs, and demands of those employed. This will enable the employer to anticipate and preplan management-employee negotiations, just as potential markets for a product are preplanned. If this is done thoughtfully, these negotiations will result in policies in keeping with the philosophy of mantalent development and will, incidentally, leave control of operations where it belongs.

3. Determine what per cent of the payroll is represented by money paid out for nonproductive man-hours in the case of employees who are horizontal, absent, retired, or otherwise noneffective. Such a determination will make it possible to establish ways and means to allocate the lion's share of this money toward a *workable* program for keeping all employees vertical and in jobs suited to them.

4. Fix constructive policies for dealing with each employed person at each of the different stages in his working career. This entails the ability to differentiate clearly between each stage, beginning with recruitment and ending with severance or retirement; no one policy will fill the bill for all.

Before setting forth the particulars of a humanological program in industry, it seems advisable to review the seven stages of progression of the employed individual, with a list of the questions a humanologist in any given business should ask himself regarding each stage.

1. Recruitment

Recruitment is a big business in itself. There are two principal kinds of recruitment programs. One is continuous and probably comes under the heading of public relations. It consists of creating a sort of *esprit de corps* among the young men who are on their way up in the organization, so that these young men will tell the world that their *Company* is the best in the country to grow up in. It is similar to the spirit engendered in certain military outfits, schools, and colleges and is one of the best ways of attracting outsiders.

The other form of recruitment is the direct form. Company representatives haunt college campuses trying to encourage young hopefuls to come their way. A considerable amount of recruitment is also carried on through private and government employment services.

Management's answers to the following questions are requisite to a well conceived recruitment program:

a) How do you attract the best possible labor force?
b) How do you attract the best possible sales force?
c) How do you attract the best possible administrative and clerical staff?
d) How do you attract the best possible technical and professional men?
e) How do you attract the best possible executives?
f) How much time and money is spent weeding out undesirables, bluffers, potential dead wood, etc., because of poorly planned recruiting efforts?

2. *Selection and Placement*

It is essential for the company that is selecting and placing a young person to become fully acquainted with that person — not only with his or her abilities and limitations on the job, but also with his other wants, hopes, staying power, avocational interests and talents, family and social adjustments, and background. In other words, what is sought is a record of how he is adjusting himself to his total environment. Psychological tests when coordinated help in this endeavor.

History taking becomes of prime importance in the placement examination. Common sense is brought into the picture too. The employee is evaluated in a number of informal-seeming talks which cast light upon his true self.

The following questions should be asked of himself by the businessman who is assessing his company's program:

a) How do you put each applicant in the place where he will work most happily and efficiently to produce the most?

b) Do you know in advance which individuals will move up and, for your benefit, ought to be moved up? Do you know in advance which individuals will crack and under what kind of strains?

c) Do you know which individuals will leave you before they have learned to produce as much as you are paying them?

d) How many thousands of dollars would such advance knowledge be worth to you?

3. *Training*

The purpose of mantalent development, specifically training, is to *construct* an employee from the materials already in his makeup — not to *reconstruct* him according to a preconceived plan that has no relation to the individual equation. This involves fitting the job to the man as well as fitting the man to the job. Since no two individuals are alike, it stands to reason that no one job assignment can be absolutely the same for two different people. The stubborn employer who insists that Jones perform his job exactly as his predecessor, Smith, did, in every detail, is being unrealistic, and may be driving a good man to the arms of a rival company. Jones is not Smith. Many of the things Smith did well and with enthusiasm may be drudgery to Jones; on the other hand Jones may have talents Smith never dreamed of. The thing to decide is whether Jones' talents on the job are worth as much money to the company as Smith's were. If so, the job should be tailored to Jones, not to the man who has gone before. If not, Jones oughtn't to have been picked for the assignment in the first place. Answers to the following questions will be of value:

a) How many sudden vacancies on your staff occur because of premature resignation or retirement?

b) How much does it cost you to train one person to fill any given vacancy which might occur at any echelon?

c) What does it cost to fulfill pension obligations to someone prematurely retired?

d) Do you train men for specific jobs, or do you also teach them to live and develop proper attitudes, so that they will not lose their grip or drop out ahead of schedule?

4. Motivation and Promotion

What Makes an Executive? "The courage to dream, the ability to organize, the strength to execute" — these are the qualities that distinguish a true executive from a mere titled supervisor hanging on to a responsible job by luck or by the sufferance of the bottle, the ulcer, or the migraine headache.

This is the most crucial aspect of mantalent development. Errors in human evaluation can be costly at any level of employment. At the executive level they can be ruinous.

The above quotation is lacking an important word, "health." It would be more to the point if revamped as follows: "The courage *through health* to dream, the ability *through health* to organize, the strength *through health* to execute. . . ."

Often it is luck that singles a man out for an executive job. Sometimes politics help. But in the long run integrity, hard work, personal enthusiasm, and innate ability are the things that count. These — and health.

"To dream . . . to organize . . . to execute . . ."

To dream requires facility of relaxation, with time to read and think. With a sick or poorly functioning body, normal judgment becomes blurred, and ideas are tempered by personal whims. Dreams become, to some degree, nightmares.

To organize calls for innate appreciation, knowledge, and understanding of people and goals, plus an eye to the future. A human sick in mind or body can no longer judge people, goals, or trends with proper facility or effectiveness.

To execute is out of the question when fear or personal security as a result of an unhealthy mind or body becomes a prime, though possibly hidden, force in day-to-day behavior.

Executive motivation involves goals far more complex than mere money, fear of higher authority, or badges of power. The soldier who dies in battle is not motivated by his monthly pittance nor by the frown of his top sergeant, nor by any chevrons or shoulder patches he might happen to be wearing. He is impelled by subtler motives: love of country, pride in his outfit, belief in his mission, or identification of himself with the goals of his regiment, division, army, or nation. The same things hold true in the case of the real executive. He works extra hours, carries extra burdens, faces greater risks in terms of health as well as job security than does the nonexecutive. For this he is paid more, is driven more, and is given more impressive titles, but above and beyond the salary, the demands of the board of directors, and the name in gold leaf on the frosted-glass door, there is a more impelling motive. As a real executive, he has become identified with his company; he is imbued with its goals. When his company pursues a certain line, he says: "We are doing such and such. . . ." NOT: "The Company is doing such and such."

This is the kind of motivation that counts. The company that gears its mantalent-development program to this kind of motivation, the company that understands the meaning of *esprit de corps* and how to instill it in its executive team, will not have to prod its leaders into accepting more leadership than they want to take. Responsibilities will be accepted as a matter of course by the men who can bear them. The men who are due for promotion and who are fully fit to be promoted will move ahead almost without being told what is wanted. The following questions are to the point:

a) Do your unpromotable employees accept their role in the scheme of things?

b) Do they have the kind of morale that keeps them producing to the best of their ability?

c) Do your potential supervisors and executives *want* to move ahead to the more demanding jobs for which you are grooming them?

d) How do you do this?

e) Can you measure your losses due to low morale on the part of disgruntled unpromotables?

f) Can you measure your losses in terms of waste when your valuable men refuse to advance to the high posts you have planned for them?

g) How much time and money is spent in organization turnover when a promoted employee fails to deliver the goods?

h) How many otherwise competent people ought not to be promoted for health reasons?

i) How many good, able men have folded after promotion?

j) What is your system for preventing the resignation of

MANTALENT DEVELOPMENT PROGRAM 599

valuable men who *must* be passed over for promotion?
k) How many such resignations have you had in the last five years and how much did they cost you in money and morale?

5. *Supervision*

It falls to the supervisor to deal *directly* with people, as individuals, in the accomplishment of some tangible company goal. On a small scale, he is now actually exercising executive control, and is a part of management.

If the job of selection and placement was thoroughly done, it is no difficult matter to find out what is the proper supervisory slot for him — the one in which he will best develop along the lines predicted for him. However, if flaws in his character were overlooked in the preliminary evaluation, this stage in his development will surely bring them out. His employers need to exercise caution now. Misassignment can produce a chain reaction at this stage. Anything the *supervisor* does on the job will affect those under him, for better or worse. If he is not big enough for the job, he will not only damage himself, but will do injury to those who are obliged to take orders from him. The end may be chaos, costly to the company and extremely disagreeable to the individuals directly involved.

The following questions cover major supervisory areas:
a) Do you know exactly how many employees will be full-blown problem drinkers within five years, if the right steps aren't taken? Do you know the right steps?
b) How much do problem drinkers cost you in terms of inefficiency, turnover, or rehabilitation?
c) How many lazy employees have you and how do you deal with them?
d) How much do they cost you?
e) Is it cheaper to replace them or reactivate them?

f) How do you deal with maladjusted and frustrated employees?
g) Can you pick out the ambitious bluffer who is getting beyond his depth?
h) Can you straighten him out before his driving ambition gains him a promotion ruinous to him and costly to you?
i) How much damage has been caused by such men in your company?

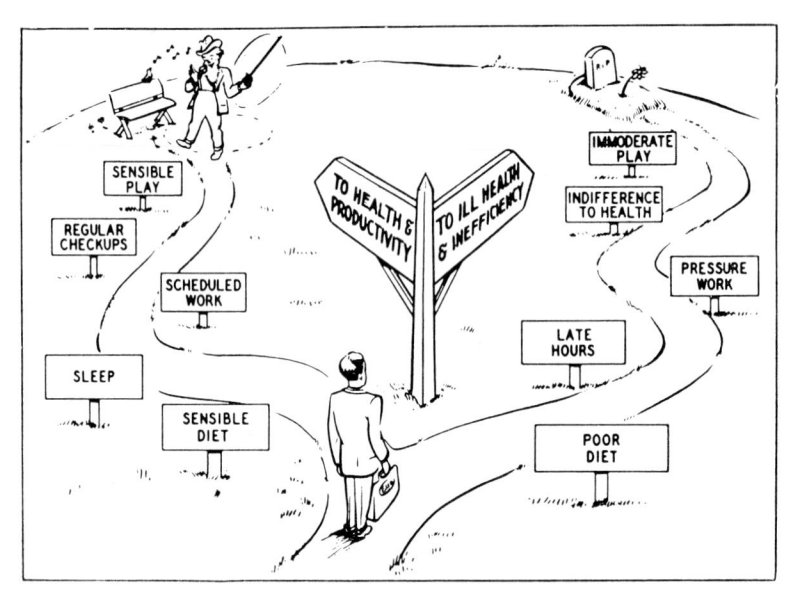

Fig. 17 — It's Your Choice

6. Direction

The period of top-line officership is when the aging process begins to bring about inexorable changes. It is the period when the leader in business needs to know exactly where he stands healthwise. The judgment and experience of a wise physician is most important here to alert him to his limitations, healthwise and to help him focus his thinking on what he can accomplish efficiently with the health budget nature has given him. It is in his own best interest that his knowledge be determined accurately. It should be emphasized that maintenance of his health is his own personal responsibility.

From a corporate point of view, every effort should be made to assist the leader in the determination of this health factor, as the desired goal of the employer is to have each employee age *normally* in harness. The goal of the individual in this connection is to reach the inevitable age of *substitution and compromise* in a satisfactory state of health, and with the knowledge and understanding necessary for the constructive planning of his future years.

Lack of self-knowledge can lead to tragic errors. One frequently hears of men in their mid-forties suffering heart attacks while shovelling snow. The tendency is to blame the snow-shovelling, but this is an error. These men had ailing circulatory systems before they began to shovel snow. A health examination would have revealed this fact. It was lack of self-knowledge that caused the damage; not the exercise. Exercise is good medicine for healthy people. Lack of self-knowledge is also a major factor in the injurious (to self, company, and colleagues) course of the individual who finds himself in a job too big to handle.

At this stage, the company should see to it that the executive is fully *evaluated* from time to time. A rigid program of annual examinations is rarely necessary. Individual factors, competently interpreted, will determine what the frequency rate should be. The following questions will aid management in developing its leaders:

   a) What yardstick do you use for measuring potential executives?

   b) Can you tell what each of your executives will be worth to you five years from now in terms of ability and willingness to produce?

   c) Can you distinguish between:

      (1) the natural executive who will, in the end, leave

his department in better shape than he found it, and who actually enjoys responsibility?
(2) the loyal, intelligent manager, who will produce for a while but who will eventually wither under responsibility?
(3) the convincing bluffer who will lean entirely on favorites until the lid blows off, and he and his hotbed of office politics are exposed?

d) How much have (2) and (3) cost you in the past?

7. *Relinquishment of Power*

After an employee has passed the age of 55 or 60, the chances are that even the best worker has lost a certain amount of staying power. Even if he hasn't, or thinks he hasn't, there are younger men behind him ripe for a turn at the wheel. What is more, he has earned a little leisure. Personally he has reached the sixth stage of life — that of *substitution and compromise*. He can't do all the things he could 10 or 20 years ago, and if he is intelligent, he has learned to take pleasure in the things he *can* do. Both at home and on the job he is *decelerating*.

The rate of this deceleration, as far as the company is concerned, is not difficult if the mantalent audits have been kept up to date. An employee's limitations are known. His physiological age has been measured, and his load of responsibility is lightened accordingly. It would not be fair to this loyal individual to take too much away too quickly, if he is able and willing to bear a fair load. It is just as injurious to such a person to ask him to do *less* than he can do comfortably as it would be to demand that he do *more* than he can do.

Retirement planning. — For each employee's own good, he should be encouraged to think of what he is going to do after retirement. He should be given vacations from time to time, for two reasons: (1) to enable him to accustom himself to the idea that the company can function without him; and (2) to let him wet his feet in the world outside the company which has been his life. Some firms have experience with deceleration plans under which aging individuals cut down on their attendance at work by degrees: four days a week for a few months, then 3½ days, then 3 days, and so on.

If all of these stages have been planned wisely, retirement, often considered a major business problem, becomes no problem at all. The employee goes through it happily, as through an Arch of Triumph, to golden years of earned leisure, satisfying reminiscences, and opportunities to pursue the many goals in life he has been too busy to pursue while still in harness.

He is a happy man and an excellent testimonial to his company, which will always be *his* company, in harness or out of it.

The following questions point up some basic goals of a constructive retirement program:

a) What is the physiological, not chronological, age of all your employees past 40?

b) What is your company policy regarding normal retirement age?

c) Can you predict roughly how many employees will retire at this normal age — how many will not? — If you cannot make these predictions, why not?

d) How many past 40 have made concrete plans for after retirement?

e) What do you consider your function with regard to helping individuals plan for retirement, and for the years leading to retirement?

f) What is your deceleration system for individuals approaching retirement?

g) What do premature retirements cost in money and in public relations?

Requirements for Inaugurating a Constructive Humanological Program

The keystone of success in the establishing of a workable program is *sincere interest and active participation on the part of top management.* Humanology is founded on management's understanding and appreciation of each individual human being from the time of his recruitment as raw material to the time of his severance as a member of the company *family.*

A coordinated team approach is necessary. Principal members of the team should be:

1. An officer from the top echelon of management
2. A *professional* advisor, recognized in the mantalent development field
3. An economic realist with experience in large-scale cost-accounting

General Functions and Qualifications of Principal Team Members

1. *Management representative.* — This individual should be expected to captain the team and devote considerable time and effort to it. He should be a *top* management man, of about 55 years of age, who has reached his pinnacle of success in an orderly fashion. He should now be ready and willing to relinquish graciously the power he has gained during his acquisitive years and consolidated since arriving at the top.

The ideal person would be one who has planned constructively for his later years; an individual in a position to be above the struggle for material wealth and power, and who is prepared to delegate, substitute, and compromise. In other words, he is one on the threshhold of his golden years who, by virtue of his place in the scheme of things, can see beyond the expediency of the moment, and who has no compulsion to follow the line of least resistance.

He should be an individual respected in his organization, who, though delegating authority to the younger men who will fill his shoes, can still wield it when the need arises. A leader who can make recommendations from strength and judgment.

2. *Professional advisor.* — This individual should be recognized as having a thorough knowledge of the physical and psychological make-up of people. His background of experience and education should be such as to command the respect of his teammates as well as his professional colleagues. In addition to technical knowledge, this person should have a well-founded familiarity with the working environment and its effect on human beings.

Such a person would be expected to be cognizant of his own limitations and willing to work with departments within a given organization. He should know the channels for obtaining information from outside agencies, such as universities, when needed, and should be willing to do so.

He should be qualified to interpret data that are obtained in an agreed-upon pilot undertaking, and recommend effective policies which will remedy existing problems.

3. *Economic realist.* — This individual should be experienced in dealing with large-scale cost-accounting problems of large multifaceted organizations. He should be able to interpret human factors in terms of dollars and cents. Having determined, analyzed, and classified management's total investment in mantalent he recommends what portion should be allocated from year to year for a continuing mantalent development program.

The Team Approach

This program can be carried on effectively from year to year by subcommittees made up of individuals who will devote their efforts to and make annual assessments of separate aspects of the total mantalent investment picture.

Seminars for discussion at annual or semiannual intervals will keep the mantalent development effort up to date. Pertinent facts would be understood and more generally appreciated before applied to policy. This would prevent the "underground" growth of new precedents which would, in the long run, result in needless expenditure.

For the sake of the stockholders, the managers and supervisors, the employed human beings, and the economy at large, constructive mantalent development philosophy needs to permeate American business before the chain reaction of ill-advised precedents reaches an irreversible stage. The future depends upon the leadership exercised by major corporations.

What a Humanology Team Can Accomplish

The foundation of any undertaking in the field of mantalent development must be an exhaustive audit of the hidden extra operating costs caused by absenteeism, low morale, bad placement, ineffective organization, and other deviations from normal, efficient functioning.

This audit should not only determine what these extra costs amount to now but what they *will* amount to in the predictable future, or what they can amount to if certain given courses of action are followed.

Such an audit will, upon analysis, reveal means of predicting future trends. Precedents formerly established can be altered before they cause further trouble, and future demands or needs can be studied in pilot undertakings before they become manifest or before some sort of compulsory action is brought to bear from the outside.

A first-rate humanology team can determine the following from this audit:
1. At the stage of recruitment:
    a) How to attract *desirable* and *durable* personnel.
    b) How to cut to a minimum the time-consuming process of weeding out undesirables before they have

an equity in an established benefit program.
2. At the several stages of mantalent *development* beginning with *placement:*
    a) How to prevent the resignation of potentially valuable employees before they have begun to bring back a return on the time and money invested in them.
    b) How to tell in advance which promising employees will live up to expectations after promotion, and establish policies for selecting and promoting durable and competent men (thus eliminating the short and long range confusion caused by ill-advised promotion).
    c) How to assess the nature of jobs currently considered man-killers, and what to do about them.
    d) How to keep the unpromotables content with their place in the scheme of things, and how to motivate these easily embittered persons to do the best work of which they are capable.
    e) How to keep key men from burning out ahead of schedule.
3. At the stage of aging in harness, or preretirement:
    a) How to compute the actual physiological, not chronological age of all older employees; one man at 55 may be fully capable of doing the work of a man of 40, while another at 45 may be ripe for retirement. Research into this vitally important question can only be carried on within a business, since an individual's physiological age can only be revealed when he is vertical and on the job.
    b) How to assess the true potential productivity of each older individual. It is neither good business nor good relations to put any individual in a position where he is expected to do less than his best, or where he is forced to do more than his best.
    c) How to establish constructive retirement policies

which will be appreciated by those employed as well as by the communities in which they live.
- *d*) How to decelerate each aging individual at his own particular proper rate.
- *e*) How to make it possible for the high pressure key men, as well as the members of the rank and file, to live long enough to reminisce upon past accomplishment, and go willingly and happily to a fruitful golden age of retirement.
- *f*) How to outline a practicable program for keeping each employee vertical and on the job. Such a program will cost only a fraction of the amount currently spent on the horizontal employee in sick pay and other benefits. Many current policies for dealing with the horizontal employee have evolved out of a chain of precedents, not out of constructive planning. There is no need to follow the leader in compounding bad precedents already established.
- *g*) How to educate the recipients of various benefits; so they will know what they are getting and appreciate it. Also, how to accomplish a pruning job in the way of eliminating costly benefits which are neither wanted by those employed nor are in the interests of management.

4. From the standpoint of intercompany relations:
   - *a*) How to predict which new benefit program will start a chain reaction harmful to smaller or more heavily manned companies.
   - *b*) How to lead the field in the establishment of constructive mantalent-development programs, in such a way that other companies will happily follow suit so that American business *as a whole* will benefit.

5. From the standpoint of the stockholders:
   - *a*) How to make all money invested in mantalent pay *tangible* dividends.

b) How to make these dividends discernible to stockholders in the form of annual reports, profit-and-loss statements or other means of communication.

Methods

Each individual will be assigned to a job in keeping with his wishes and abilities to perform. From the point of view of total evaluation, this means that he is able to perform from a mental, emotional, and physical point of view, his work assignment in an efficient manner.

This will be accomplished roughly as follows:

Periodic appraisal will begin at recruitment and will be followed through at required intervals until time of severance.

At time of recruitment, there will be a total evaluation of education, previous employment, habits, desires, and potentials considered in terms of physical status, past and future environment.

At time of selection and placement, specific aptitudes and training will be clearly interpreted in terms of the job to be done and recorded for periodic study and review.

At the time of training, audits will be made and considered in light of desired potential, with a constructive effort to open the right doors for subsequent advancement.

During period of supervision, effects of stress and strain will be interpreted by means of periodic inventory.

Management will be cognizant at all times of the true meaning of motivation. It will think in terms of who have *not* been promoted, as well as those who are going forward. Management will further be careful not to direct individuals into positions beyond their capacities to perform.

Management will prepare employees for severance in keeping with a constructive, well thought out plan. Such a plan adopted in terms of what is best for the employee will in turn be best for the employer.

Evaluation Center

For management or labor to perform this function, there is the need for an evaluation center located in any given central area whereby those employed can receive efficient service during working hours.

Purpose
1. To keep the vertical individual vertical and on the job.
2. To detect insipient maladjustment and disease states early — at a time when something constructive can be done at a price the employed person can afford to pay.
3. Prevent unnecessary and costly long-term absence from work.

Requirements
1. Performance of competent placement and periodic evaluation in keeping with the specific requirements of the individual and the job.
2. Render competent emergency needs.

Staff needed to perform such work (a well rounded professional team):

1. Board of consultants
2. Competent diagnosticians
3. Psychologist
4. Trained counsellors
5. Registered nurse
6. Qualified technicians

Note: Therapy other than emergency care rendered by physician of choice at his office or hospital concurred in by patient.

Comment

To assemble such a staff is more than any one company, regardless of size, can substantiate from the standpoint of their individual needs, or amount of cost involved. Furthermore any company that should embark upon such a constructive program would be practicing corporate medicine.

On the other hand, it is feasible, practical, and efficient for a number of companies both large and small to underwrite such an undertaking and obtain specific services in keeping with their individual needs at a predetermined and controlled cost.

Advantages

Services available at all times. Employees would be attracted by virtue of the spirit of personal, confidential service being promulgated throughout. Caliber of professional services controlled by professional people. Findings made known to all concerned in accordance with established policy.

Space and Equipment

A relative matter which can be predetermined and expressed in terms of year to year operating and maintenance costs.

REFERENCES

Bower, M. "Shortage of Key Men: What Can We Do About It?" *Sales Management*, (May 20, 1949) p. 37: (June 1, 1949), p. 89.

Canfield, G. W., and Soash, D. G. "Presenteeism — A Constructive View." The *Industrial Medicine and Surgery*, XXIV (September, 1955), pp. 417-418.

Edwards, Joseph Dean. "Psychological Testing." *Executives — Making Them Click*. New York: University Books, 1956, Chapter IX, pp. 169-173.

Maisel, Albert D. "The Health of People Who Work." The Report to the National Health Forum, 1959, The National Health Council, 1790 Broadway, New York 19, New York, 1960.

Page, Robert Collier, M.D. "Constructive Medicine and Industry." The *Journal of the American Medical Association*, (November 4, 1944).

—————. "Industry Calls in the Doctor." *Harvard Business Review*, (September, 1953), pp. 109-117.

—————. "Executive Health for Company Wealth." *Dun's Review*, (April, 1952), p. 16.

—————. "Mantalent Development and Occupational Health." *Industrial Medicine and Surgery*, (December, 1960), pp. 582-587.

Suman, John R. Speech. Jersey Standard, 30 Rockefeller Plaza, New York, New York, 1953.

Vosburg, G. L., M.D. "Blueprint for Health." *The Management Review*, (July, 1955).

18
Reports

SPACE TRAVEL points up the true and ultimate potential of mantalent development and epitomizes the significance of occupational health.

To efficiently record a given individual's reaction to his total environment, is a task in which each member of the humanological team must work in complete harmony and understanding with one another.

In Chapter 17, fifty individual points are mentioned and numerous questions asked. Each of these has a particular organizational, academic, economic, and personal interest to the policy maker.

To determine truths, to answer a pertinent question in a given instance, to compare like facts with similar and dissimilar occupational endeavours, to profit by past efforts and experiences, to formulate an efficient pattern for the future, a standard record and recording system is paramount.

Since many of the thoughts and ideas expressed throughout this work are new and provocative, the following suggestions and recommendations are made for those members of the humanological team whose interests have been commanded by the philosophy expressed in terms of Occupational Health and Mantalent Development.

Guiding Principles

(1) The fundamental purpose of any report is to record factually all pertinent details *to* the interest and *in* the interest of the recipient of the given report.

(2) Simplicity is of paramount importance in all reporting having to do with the various personal and environmental aspects of an occupation as related to the total needs of the employed person.

(3) *Efficient* recording is essential, with recognition that factual data must interpret and answer the multiple questions and problems that arise with respect to these subjects, individually and collectively.

Such data will permit judgment to operate on a *long-range* rather than an *expediency* basis in the formulation and guidance of policy in relation to the numerous environmental needs of those employed (including health needs).

(4) Practical explanations which will afford means for solutions are the essence of an efficient reporting system.

(5) *Nomenclature* — Until a standard nomenclature for all problems inherent to a constructive occupational health and mantalent development program is universally accepted, the glossary in this chapter is recommended for use by each member of the humanological team.

This approach promises a standard beginning, allows for practical additions, and facilitates mutual communication, coordination, understanding, and productive collaboration.

Glossary

To facilitate the understanding of certain terms used, an explanation of their meaning follows.

Lost-time cases: Any person whose absence, disability, sickness, or injury results in absence from work of one day or more is considered a lost-time case.

Frequency rate: This rate indicates the average number of absences per 1,000 persons per annum. The rate is determined by dividing the number of cases reported by the average number of employees, and is expressed in terms of every 1,000 employees. A frequency rate of 28.86 would mean, therefore, that of every 1,000 employees, approximately 29 were absent, sick or disabled. This would not necessarily indicate that 29 individual employees were absent at some time during the year, for several of them may have been ill or absent several times during that year, and they are, of course, included in the totals each separate time.

Noneffective rate: This figure reflects both the frequency and the length of a particular absence, disease or disability, and serves as a reliable index of the effect of disability or absenteeism from an economic point of view. It indicates the number of employees who will, on an average, be absent on any given day of the year. Thus, if 500 employees are required to perform a given task within a stated period of time, and the noneffective rate for this group of employees has been determined as 5.24, it would indicate that 505 men should be employed if the task is to be completed by the predetermined date. This rate is determined by dividing the average days of disability per employee by the number of days in the work year.

Days of disability: These represent the actual, or absolute, number of work days from the first day of absence to the end of a disability by return to duty, retirement, or death.

Days of disability per case: This is the severity rate, or average number of days absence per case, and is determined by dividing the total work days of disability by the total number of cases for the same period of time.

Days of disability per person: This is the disability rate, or average number of work days of disability per person, and is determined by dividing the total work days of disability per period by the average number of persons for the same period.

Files

The importance of a practical method for the filing of correspondence and reports in a manner which will permit the rapid and accurate location of such material when required cannot be overemphasized. Much time and effort will be saved when serious thought is given to the type of system to be used. This is a problem that must be determined at the time an occupational health and mantalent development program is begun.

Obviously the type of system adopted will depend largely on the size of the establishment, extent of operations, and the volume of correspondence and reports involved. One system which might prove to be perfect for one organization could well prove to be too complicated and unwieldy for another organization.

Where the organization is largely centralized, the number of employees small, and the amount of correspondence is not excessive, a chronological filing system, based on main topic headings, will serve the purpose.

In the preparation of monthly and annual reports, the basic filing system can be established with the following six main divisions:

      00. Administrative Medicine
      20. Preventive Medicine
      40. Constructive Medicine
      60. Educative Medicine
      80. Curative Medicine
    100. Mantalent Development

Under each of these divisions the material pertinent to each can be filed. If the volume of material to be filed is not large, an alphabetical system by subtopics under each heading would doubtless suffice; where the volume of material is considerable, the assignment of additional subtopic code numbers could be considered. Whatever the system adopted as required by local activities, it should be simple and easily understood by all concerned.

The Filing Guide which follows illustrates how this system lends itself to practical application with respect to main divisions and subtopics as needed.

Filing Guide

00. ADMINISTRATIVE MEDICINE
   .01 Policy
   .02 Procedures, Methods
      .021 Procedure Manuals
      .022 Medical Service in Industry Guide
      .023 Policy Manual and Guide for Medical Reports
   .03 Regulations and Laws
   .04 Plans and Projects
   .05 Organization
   .06 Personnel and Salaries
   .08 Requisitions and Purchases
   .09 Medical Activities Reports
      .091 Monthly Reports
      .092 Annual Reviews

20. PREVENTIVE MEDICINE
   .01 Preventive Measures
      .011 Vaccination and Inoculation
      .012 Industrial Hygiene Measures
      .013 Sanitation
   .02 Epidemic Control Measures
   .03 Selection of New Field Areas

40. CONSTRUCTIVE MEDICINE
   .01 Placement and Periodic Health Inventories (Patient's Files)
   Alphabetical — by age — by department or any other suitable combination dependent upon the specific interest of those in charge.

60. EDUCATIVE MEDICINE
   .01 Medical Department Courses for Employee
   .02 Postgraduate Training Courses
   .03 Medicine in Industry Courses
   .04 Reprints of Educational Literature
   .05 Medical Training Films
   .06 Professional Meetings
   .07 Public Relations

80. CURATIVE MEDICINE
   .01 Patient's Clinic Files
   .02 Methods of Treatment
   .03 Hospitals — Dispensaries

100. MANTALENT DEVELOPMENT
   .01 Efficiency
   .02 Promotion
   .03 Career Incentives
   .04 Methods of Evaluation
   .05 What others are Doing
   .06 References

Diagnoses and Codes

The following is a list of the more common causes of absences from work, together with their individual workable code number.

Abortion	7x2–009
Abscess (1)	010–1002
Agranulocytosis (2)	502–790
Allergic rhinitis (hay fever)	310–392
Allergy, other types	014–393
Amebiasis	010–151
Amputation, whole or part of extremity	010–405
Anemia, pernicious	501–702
Anemia, other types	501–736
Appendicitis	661–100
Arthritis, degenerative	240–912
Arthritis, infectious	240–100
Arthritis, rheumatoid	240–1x0
Arthritis, traumatic	240–400
Asthma	350–390

Benign Tumors

Nevi	010–8885
Fibroids of uterus	782–866
Polyps	010–8024
Others	010–8A
Birth, premature	790–004
Birth, term	790–000
Bronchiectasis	350–1006
Bronchitis	350–100
Brucellosis	010–117
Burns	010–4411
Bursitis	250–190

Cancer (for non-malignant tumors, *see* Benign Tumors)
    Carcinoma of brain   901–899
    Carcinoma of breast   190–899
    Carcinoma of G. I. system   600–899
    Carcinoma of respiratory system   300–899
    Carcinoma of skin   130–899
    Carcinoma of urogenital system   700–899
    Hodgkin's disease   500–832
    Leukemia   500–832
    Lymphosarcoma   550–830
    Polycythemia   501–8271
    Sarcoma   010–879
Carbuncle   180–1003
Cardiovascular Renal Disease
    Angina pectoris (due to coronary arteriosclerosis)   41x–942
    Atherosclerosis, generalized   460–942
    Atherosclerosis, peripheral   460–952
Cardiac Rhythm Abnormalities
    Auricular fibrillation   431–x26
    Auricular flutter   431–x24
    A-V block   444–x36
    BB block   4442–x37
    Other types   443–x32
    Cerebral hemorrhage   94x–5337
    Cerebral thrombosis   94x–9427
    Congenital heart disease   400–010
    Essential vascular hypertension   47x–x30
    Hypertensive C-V disease   400–533
    Hypertensive vascular disease   460–533
(1) Give location in writing
(2) Give cause, if known, in writing
    Myocardial infarction   430–9x7
    Pericarditis   420–100
    Raynaud's disease   47x–582
    Rheumatic heart disease   410–932
    Syphilitic heart disease   410–147
    Thrombo-angiitic obliterans   402–930
    Thrombophlebitis   480–1007
    Varicose veins   480–9x6

Cataract	x20–797
Cellulitis	180–100
Chickenpox	010–161
Cholecystitis	687–100
Cholelithiasis	687–615
Cirrhosis	680–956
Colitis, ulcerative	660–951
Common cold	300–100
Conjunctivitis	x56–100
Cyst	010–8034
Cystitis	730–100
Cystocele	730–4x9
Dacryocystitis	x66–100
Deafness (3)	x70–042
Delirium tremens	000–3312
Dengue fever	010–162
Dermatitis	111–190
Diabetes mellitus	871–785
Diarrhoea (see Gastroenteritis)	
Diphtheria	010–125
Dislocation	240–406
Diverticulitis	660–642
Drug addiction	000–x642
Duodenal ulcer	651–951
Dupuytren's Contracture	296–9x6
Dysentery, amebic	660–151
Dysentery, bacillary	660–116
Dysmenorrhea	780–x32
Eclampsia	015–388
Eczema	111–390
Emphysema	368–1006
Empyema	370–100
Encephalitis	930–100
Epididymitis, gonorrheal	756–103
Epididymitis, non-specific	756–100
Epilepsy	930–x08
Erysipelas	130–1028
Filariasis	010–257
Fistula	010–1003
Foreign body in eye	x11–496

620   OCCUPATIONAL HEALTH AND MANTALENT DEVELOPMENT

Fracture, compound	200–418
Fracture, simple	200–416
Functional Disease	
Cardiospasm	641–580
Migraine	930–x40
Mucous colitis	660–557
Neurocirculatory asthenia	009–580
Pylorospasm	645–590
Spastic colon	660–580

(3) Give type or cause in writing

Ganglion	280–600
Gangrene	010–1001
Gastroenteritis	601–617
German measles	010–165
Glaucoma	x18–616
Glomerulonephritis, acute	712–100
Glomerulonephritis, chronic	712–190
Goiter, non-toxic diffuse	810–943
Goiter, non-toxic nodular	810–952
Goiter, toxic diffuse	810–9436
Goiter, toxic nodular	810–9526
Heat prostration	010–445
Hematoma	180–4x7
Hemorrhoids, internal and/or external	60x–641
Hepatitis	680–100
Hernia, femoral	0602–639
Hernia, inguinal	0601–639
Hernia, ventral	042–4x9
Herniation of intervertebral cartilage	2511–9x9
Herpes zoster	130–167
Hydrocele	757–9008
Hydrocephalus	920–063
Hydronephrosis	719–435
Ileitis	654–952
Impetigo	111–105
Infarction of lung	360–510
Infection of wound	010–100
Influenza	010–168
Injuries (4)	010–400
Intestinal adhesions	604–025

REPORTS 621

Intestinal obstruction	604–4x4
Intestinal parasites	604–200
Intestinal rupture	604–4x5
Intestinal strangulation	604–514
Intestinal tuberculosis	604–123
Intussusception	666–630
Iritis	x15–100
Jaundice, catarrhal (see Hepatitis)	
Keloids	130–8701
Keratitis	x12–100
Kidney, calculus	719–615
Kidney, polycystic	710–010
Kidney, tuberculosis	710–123
Labyrinthitis	x85–100
Laryngitis	330–100
Leprosy	010–124
Lichen planus	110–965
Ligament, tear of	260–412
Liver abscess	680–1002
Lumbosacral joint instability	2414–022
Lymphogranuloma venereal	550–198
Malaria	010–157
Malnutrition	014–711
Mastitis	190–190
Mastoiditis	x84–100
Measles	010–169
Meniere's disease	x85–9x8
Meningitis, meningococcic	912–104
Meningitis, tuberculous	910–123

(4) Give location in writing

Mononucleosis, infectious	101–1301
Mumps	621–170
Nephrosis	713–1009
Nerve, injury of (5)	980–402
Neuritis	980–100
Obesity	010–70x
Oophoritis	788–100
Orchitis	755–190
Osteo-myelitis	200–100

Otitis external	x75–100
Otitis media	x80–190
Otosclerosis	x86–7x9
Ovarian cyst	788–795
Paget's disease	200–9401
Pancreatitis	690–190
Paralysis agitans	9464–953
Paratyphoid fever	010–114
Paresis (dementia paralyticus)	930–1479
Paronychia	176–100
Pellagra	010–7623
Pernicious vomiting of pregnancy	013–388
Phimosis	753–017
Pilonidal cyst	058–029
Placenta previa	794–631
Pleurisy	370–190
Pneumonia, atypical	360–160
Pneumonia, broncho-	361–190
Pneumonia, lobar	360–100
Pneumothorax	3075–900
Poisoning, barbiturate	010–3371
Poisoning, kerosene	101–3342
Poisoning, lead	010–3112
Poisoning, mercury	010–3111
Poliomyelitis	972–171
Pregnancy, delivered	7x2–000
Pregnancy, tubal	7x7–789
Pregnancy, undelivered	7x2–789
Premature placenta separation	794–9005
Problem Drinking	
Dypsomania	000–331
Environmental intoxication	0091–331
Neuropsychiatric intoxication	0092–331
Pathological intoxication	0093–331
Prostatic hypertrophy	764–799
Prostatitis, gonorrheal	764–103
Prostatitis, non-specific	764–100
Psoriasis	111–961

REPORTS 623

  Psychiatric Disease
   Anxiety state   000–x01
   Depressive reaction   000–x06
   Manic-depressive reaction, depressive state   000–x12
   Manic-depressive reaction, manic state   000–x11
   Mental deficiency   000–x90
   Phobia reaction   000–x04
   Psychopathic personality   000–x60
   Schizophrenia   000–x25
  Pterygium   x58–956
  Pyelitis   722–100
  Pyelonephritis   719–100
(5) Give location in writing
  Rectal prolapse   668–631
  Rectal stricture   668–1004
  Rectocele   668–4x9
  Retinal detachment   x23–9x5
  Rheumatic fever   010–932
  Sacroiliac strain   2468–43x
  Salpingo-oophoritis   787–100
  Scabies   110–281
  Scarlet fever   010–102
  Scrub typhus   010–183
  Scurvy   010–763
  Sinusitis   320–100
  Situational state   001–x82
  Smallpox   010–176
  Spina bifida   2202–037
  Spondylolisthesis   2203–037
  Stomach ulcer   640–951
  Subarachnoid hemorrhage   91x–yx5
  Sunstroke   101–453
  Syphilis   012–147
  Tabes dorsalis   906–147
  Tenosynovitis   280–100
  Testes, undescended   755–0281
  Tetanus   010–119
  Tonsillitis   634–100
  Torticollis   272–017
  Tuberculosis of lung   360–123

Tuberculosis other types	y00–123
Typhoid fever	010–115
Typhus fever	010–184
Ureter, calculus	723–615
Urethra, calculus	740–615
Urethra, stricture	740–1004
Urethritis, gonorrheal	740–103
Urethritis, non-specific	740–100
Uterus, prolapse	782–631
Uterus, retroversion	782–63x
Urticaria	11x–390
Vaccinia	010–177
Varicocele	76x–6x6
Vincent's angina	610–141
Volvulus	604–637
Whooping cough	350–108
Wounds (see injuries)	
Yellow fever	010–178

Standardization for Statistical Purposes

The preceding list of the more common causes of absences from work permits practical standardization for statistical purposes. For a more detailed accounting, see the codes used in the *Standard Nomenclature of Diseases and Operations*.

For monthly and annual reports the various diagnoses can be readily regrouped in terms of the total incidence of:

(1) Addictive disease (problem drinking),
(2) Regenerative disease (cancer),
(3) Circulatory (cardio-vascular renal disorders),
(4) Mental disease,
(5) Metabolic disorders (obesity),
(6) Respiratory and other infectious diseases,
(7) Accidents,
(8) Elective and emergency surgical intervention,

for a given number of employed persons who can, in turn, be thought of in terms of their respective life cycle. For example, Years of Growing Independence, etc., see Chapter 9.

Periodic discussion by the various members of the humanological team of all factors inherent to lost time from work will shed a truer light upon the actual causes of absenteeism. Conclusions arrived at will prove to be of inestimable value to the policy making body of any organization.

EMPLOYEE REGISTER CARD						
Family Name	First Name		Middle Name		Employee No.	
Register No.	Occupation					Social Security No.
Code No.	Location Code	Age	Sex	Service No. Years	Status	Began Disability 19
Primary Diagnosis						Code No.
Secondary Diagnosis						Code No.
Operation						Code No.
Disposition				Days Disabled	Hospital Code	No. of Hospital Days

Fig. 18 — EMPLOYEE REGISTER CARD

Statistical Analyses

(A) A Register Card should be made out for each *employee* disability case completed during the month (whether hospitalized or not). These cards should be submitted with the monthly report.

(B) Entries on the Register Card must be legible — typewritten if possible.

> *Register Number* — Cards are numbered serially in the order in which cases are completed, commencing with Number 1 for the first case completed on October 1st, or following, and continuing serially throughout the year. When a company maintains more than one medical of-

fice, a separate series of initials can be assigned to each, and these must directly *follow* the Register Number.

Name — Enter the family name followed by the first name.

Company and Location — Each health unit area may be assigned a numerical code number, and each separate location in a given area may also have an identifying number. The appropriate code number for "Company" and "Location" should appear on each card.

Age — Show age in years on last birthday.

Sex — This space is used to indicate sex only, using the following codes:

Male employee – 11   Female employee – 12

Service — Report service in completed full years only. Leave blank if patient is not an employee or has less than 1 year service.

Status — Under this heading indicate by appropriate code number the status and type of disability incurred. The following code numbers apply:

 Employee Sickness   Code–1
 Employee nonindustrial injury Code–2
 Employee industrial injury* Code–3

*An *industrial injury* or *illness* is one in which the cause arises out of and in the course of employment, and is so accepted by the local laws in effect.

An *occupational* or *environmental disease* occurs so infrequently that, before a diagnosis of this type is made, the complete history should be sent to the next highest echelon for its review and recommendations.

Disability Began — Indicate in numerals in order of month, day and year, e.g.: 4/9/51 (April 9, 1951).

Primary Diagnosis — Indicate the title and code number of the disease or injury which was the primary

---

NOTE: A register card must be prepared for all deaths among employees, and a brief clinical report should be made.

cause of disability or death. The titles and code numbers will conform to the *Standard Nomenclature of Disease* (See page 617 for a list of the more common causes and code numbers). If during a patient's stay in hospital or quarters, the disability from which he is disabled terminates, but he continues to be disabled from some new condition, a Register Card will be prepared for the first condition on the day of its completion. An additional card will be prepared for the second condition upon its completion.

Operations — Report *all* surgical procedures performed.

Disposition — Code numbers to be used under this heading are:

Recovered	Code–1	Died	Code–4
Improved	Code–2	Terminated or Transferred	Code–5
Unimproved	Code–3		

When any case reaches the 365th day of disability, a register card will be made out. A second card will be prepared for the remainder of the time lost at the time the disability is terminated.

Determination of days of Absence or Disability — Medical departments per se should not be concerned that their reported disability days do not approximate those reported by personnel or annuity and benefits departments. Nonetheless, mutual collaboration is necessary to determine the true incidence and significance of total absenteeism. Medical department figures should reflect as closely as possible the number of full calendar days of sickness absence; they should represent actual days of morbidity, not actual working days of disability. In order to have any reliable means of comparing the medical statistics of one company with another, a uniform method must be concurred in for the reporting of disability days. The following formula for the determination of days of disability is recommended to give the uniform statistics needed for this purpose.

Only Full Calendar Days Of Absence Will Be Counted (Including Holidays, Days Off, etc.)

(A) If a person's first FULL day of absence is February 4th and his date of return to work is February 9th, his absence consists of 5 days (first date deducted from last date).

(B) If a person reports for work on June 4th, becomes ill and is sent home for the balance of that day, and returns to work the following day, there is no *FULL* day of absence and no report should be made.

(C) If a person reports for work on April 4th, becomes ill and is sent home for the balance of that day, and returns to work on April 17th, his absence consists of 12 days (the 4th is not a FULL day of absence, and is not counted).

It has been noted during many years experience with Register Cards of this type that, in practically all instances, the greater number of lost time disabilities consists of short term absences or illnesses such as minor respiratory and gastrointestinal complaints. In many instances such disabilities cannot be medically verified. When used to explain 1, 2, 3 day absences it gives rise to faulty statistics and no doubt contributes to hypochondriasis and a completely erroneous interpretation of absenteeism.

For true medical statistics no Register Card need be prepared for any losing time disability of *less than eight days* duration unless it has been medically verified by a competent doctor. The large number of short term disabilities, where verification of an actual diagnosis is impossible to obtain, although valueless for medical statistical purposes serves a valuable purpose in determining absenteeism. Sickness-benefit privileges are efficacious when legitimate, when misused they represent a costly fringe benefit, a problem more managerial in nature than medical.

In Hospital — Give the number of calendar days the case was actually hospitalized.

Monthly Report of Medical Activities

A *Monthly Report of Medical Activities* is a concise résumé of all pertinent activities for a given month. Its purpose is to provide current interest to the managing director, the medical consultant and to record pertinent facts for incorporation in the annual report.

Suggested Topics To Be Discussed

ADMINISTRATIVE MEDICINE

Organization and Administration: All matters pertaining to the organization and administration of the Health Unit including matters of Company and administrative policy, procedures, regulations, laws, etc.

Employee Benefits: All matters pertaining to policy, plans, projects, etc., affecting employee health or benefits.

Reports and Surveys: All reports including statistical reports, special studies, and medical surveys. All matters pertaining to policy or procedures regarding same. Reports of trips made, and consultations requested.

Requisitions and Purchases: All information regarding purchase requisitions for medical supplies and equipment. Requests for information regarding availability of items for purchase.

Personnel: A review of personnel additions, promotions, transfers, terminations, etc., with a listing of the effective dates. A qualification record for each newly employed person should accompany this page, with a photograph attached. Remarks concerning personnel problems, changes in personnel policy etc. Requisitions for the replacement of medical and paramedical personnel.

Construction and Fixed Equipment: Details of all plans for the alteration or construction of new medical facilities. Details regarding installation of, or change in, fixed medical equipment.

PREVENTIVE MEDICINE

All matters pertaining to industrial hygiene and sanitation.

CONSTRUCTIVE MEDICINE

Diseases: Reports of interesting or unusual case histories, epidemics, methods of handling, etc.

EDUCATIVE MEDICINE

All matters pertaining to training, postgraduate studies for personnel, medical department classes, special training projects, medical staff meetings, attendance at conferences, etc. Information concerning or requests for articles for company publication, etc.

CURATIVE MEDICINE

Reports concerning treatment of industrial accidents, occupational and environmental diseases.

MANTALENT DEVELOPMENT

When reports are made in terms of case studies or profiles they serve as excellent educational material for all members of the humanological team.

When management's interpretations of the individual's aptitudes, work performance, etc., is added to the findings recorded by a periodic health audit the humanologist has the required facts to formulate judgment and give constructive advice to all parties concerned.

When this is done in the spirit of honesty, true and efficient "human relations" is being practiced to the mutual benefit of both employee and employer.

Preparation and Distribution

In large companies, where there are several medical units, the person in charge of each medical unit should prepare monthly reports, and send the original together with all other suggested medical statistical reports to local management and copies to the medical consultant who coordinates all the activities of the medical department. This doctor will, in turn, use this information to compile the monthly report of medical activities of the department, directing the original to the members of management to whom he reports.

Confidential Medical Information

All reports of confidential-professional information are kept within professional channels.

Summary

Monthly Statistical Summary form to be used:

A. *Population:* It is important that the average population for the month be accurate if statistics are to be reliable. This information must be obtained from the most reliable source available, i.e., the personnel or payroll departments.

    1. *Average population by location:* Information under this heading is required only from companies having two or more groups of employees working and living under such different conditions that separate statistical studies are justified. The designations for locations *must not* be changed during the year. They may be revised as of a given date in any year with a report of such changes being sent forward in accordance with established policy.

B. *New lost time cases:* Report in the proper column all new sickness and disability cases losing one or more full calendar days during the month.

C. *New nonlost time cases:* Nonlost time cases are those which receive ambulatory treatment in the clinic or dispen-

MONTHLY STATISTICAL SUMMARY

Company __________ Month __________ 19__

Address __________

City State

EMPLOYEE POPULATION				MEDICAL DEPARTMENT PERSONNEL			
LOCATION OR DEPARTMENT	TOTAL	MALE	FEMALE	LOCATION OR DEPARTMENT	TOTAL	MALE	FEMALE
				Doctors - Full Time			
				Part Time*			
				Graduate Nurses			
				Technicians			
				First Aid Attendants			
				Sanitary Engineers			
				All Other			
TOTAL				TOTAL			

DISABILITY CASES	TOTAL	SICK	NONINDUSTRIAL	INDUSTRIAL
†New Losing Time Cases During Month				
New Nonlosing Time Cases During Month				

DISPENSARY AND CLINIC REPORT

	DISPENSARY ACTIVITIES					OUTSIDE VISITS	
	VISITS	X-RAY	PHYSIOTHERAPY	LABORATORY	OPERATIONS	CO. DOCTORS	CO. NURSES
Total Cases							
Sickness							
Nonindustrial							
Industrial							

ASSIGNED TO LIMITED DUTY

	BALANCE FROM PREVIOUS MONTH	ASSIGNED DURING CURRENT MONTH	REMOVED DURING CURRENT MONTH	BALANCE FORWARD
Employees of Under 1 Year Service				
Employees of Over 1 Year Service				
Total on Limited Duty				

PHYSICAL EXAMINATIONS	TOTAL	PREPLACEMENT					TROPICAL SERVICE	RETIRE-MENT	PERIODIC INV.	BENZOL LEAD	ALL OTHERS
		A	B	C	D	E					

REGISTER CARDS FOR MONTH

FROM NO. _____ TO _____ INCLUSIVE TOTAL REGISTER CARDS _____

*Total Hours Devoted to Department During Month

†Frequency Rate

Noneffective Rate

Signed: __________

Date: __________

Fig. 19 — MONTHLY STATISTICAL SUMMARY

sary, or are disabled for less than one full day. The figures appearing in these columns should represent the *first visit* only for the treatment of any one disease only. The number of cases, naturally, will never exceed the total number of "Dispensary Visits" (E).

D. *Medical service:* Under "Dispensary Visits" include all visits *after* the first visit for treatment of any one disease only. "Outside visits" represent all visits made by a doctor or a nurse to a patient at work or at home.

E. *Medical examinations:* Information given in these columns pertains to examinations performed for the purpose indicated and a full report should be given whether performed by company personnel or outside private doctors.

F. *X-ray, laboratory and operations:* Show the total number of such examinations or operations performed by Company personnel in Company facilities *only*.

G. *Frequency rates**

H. *Noneffective rates**

The Placement Health Inventory

Constructive medicine has to do with vertical health as opposed to horizontal sickness. It is the art of elevating the supposedly well individual to *optimum* health by searching for signs and symptoms of insipient maladjustment before it becomes firmly established.

Constructive medicine controls and minimizes certain forms of maladjustment by understanding their mode of onset and forestalling or delaying their development. A wide variety of health problems including the degenerative diseases, occupational neuroses, neuropsychiatric conditions, nutritional and endocrine disorders, are favorably influenced by periodic health audits.

Vigorous attack upon factors causing or influencing the progress of the total man is of vital importance. In order to carry out the program properly, the medical department per-

---

* Per 1,000 employees per month.

forms the following examinations and functions:

A. *Placement examinations:* Placement examinations are

PLACEMENT EXAMINATION

ISSUING OFFICE TO COMPLETE THIS SECTION

Family Name | First Name | Middle Name | Birth Date Mo. Day Year | Sex

Home Address | City | State | Home Phone

Previous Employment

Applicant for Company | Location | Position

REMAINDER OF FORM TO BE FILLED OUT BY PHYSICIAN OR APPLICANT

ENVIRONMENTAL HISTORY

PRE-EMPLOYMENT HISTORY (BIRTHPLACE, HOME LIFE, EDUCATION, ETC.)

EMPLOYMENT HISTORY

SOCIAL HISTORY AND SERVICE IN ARMED FORCES

HABITS — Tobacco | Drugs | Alcohol | Sleep | Diet

AVOCATIONS

HEALTH HISTORY TO DATE

FAMILY HISTORY

	Age	PRESENT HEALTH OR CAUSE OF DEATH
FATHER		
MOTHER		
SIBLINGS		
WIFE		
CHILDREN		

DETAILS OF ANY FAMILY HISTORY (EPILEPSY, TUBERCULOSIS, DIABETES, MENTAL OR NERVOUS DISORDERS, ETC.)

Family Physician | Name | Address | Telephone

(SEE REVERSE SIDE)

Fig. 20 — PLACEMENT EXAMINATION *(Front)*

given to all individuals before permanent assignment. Every worker must be physically and emotionally able to perform his duties in such a manner that he is not and does not become a hazard to himself or to his fellow workers whatever his job assignment.

PHYSICAL EXAMINATION

Physical Appearance							Temp.	Height	Weight
Deformities Noted									
SKIN									
EYES	VISION	WITHOUT GLASSES Right	WITHOUT GLASSES Left	WITH GLASSES Right	WITH GLASSES Left	Color Blindness	Disease		
	Near	J #	J #	J #	J #	Yes ☐ No ☐			
	Far	20/	20/	20/	20/				
EARS	Hearing Whispered Voice at	Right Ft.	Left Ft.	Canals		Ear Drums		Disease	
MOUTH AND UPPER RESPIRATORY	Tongue and Tonsils						Sinuses		
	TEETH Missing	Carious		Six Opposing Pairs Including Dentures ☐ Yes ☐ No			Disease		
LYMPHATIC AND ENDOCRINE	Lymph Nodes						Thyroid		
RESPIRATORY	Thorax and Breasts						Lungs		
CARDIO-VASCULAR	Pulse	Heart Size	Blood Pressure	Murmurs Organic	Murmurs Functional		Arteries / Veins		
ABDOMEN	Organs		Masses				Hernia		
GENITOURINARY	Venereal Disease ☐ Yes ☐ No	External and Internal Genitalia					Other		
RECTAL	Sphincter	Hemorrhoids		Masses			Bleeding	Prostate	
MUSCULO-SKELETAL	Muscle Tone	Posture		Atrophy			Extremities		
	Edema	Peripheral Pulses	Spine	Abnormalities					
NERVOUS SYSTEM AND REFLEXES	Abnormalities						Reflexes	Right	Left
	Romberg		Temperament				Remarks		
LABORATORY AND X-RAY	URINALYSIS Spec. Gr.	Albumin	Sugar	Microscopic			Chest X-ray		
	Hgb.	RBC	WBC	Polys	Lymphos	Monos	Eosins	Serology	
	Other Procedures								
EXPLANATION OF ALL ABNORMAL FINDINGS									
VACCINATIONS AND INOCULATIONS		DATE VACCINATED	TYPE OF REACTION Immune	Vaccinoid	Vaccinia		DATE REACTION READ		
	Small Pox								
	Typhoid Paratyphoid	1	2	3	Booster				
	Date and Kind Other Inoculations								
CONCLUSIONS AND RECOMMENDATIONS IN EXPLANATION OF CLASSIFICATION									
CLASSIFICATION	A B C D E		Qualified for: ☐ Domestic Service		☐ Foreign Service				
TICKLER FOR	Periodic (Date)	Other (Date)	Reason						
MEDICAL EXAMINER	Date	Address				Signature			

Fig. 21 — PLACEMENT EXAMINATION *(Back)*

The examining physician being cognizant of any existing disability gives recommendations for work limitations where indicated. These duties delegated to an individual must be within his physical and temperamental capabilities. An individual having a communicable disease must receive proper treatment before being introduced into a group of healthy employees.[1]

The Periodic Health Inventory

Examinations of this type are done during the course of employment. These examinations are thorough and exhaustive but also flexible and individualized. The medical history includes not only the usual inquiries concerning personal and familial disease but also the occupational, social, and neuropsychiatric information necessary for the practice of constructive medicine. In addition to a thorough clinical examination, the health inventory includes:

(1) Complete blood count
(2) Urinalysis
(3) X-ray examination of the chest
(4) Serological tests
(5) Any other laboratory procedure indicated as a result of the history or physical examination.

The frequency of the periodic health audit is individualized, depending upon the age, occupation and health status of the employee. Many individuals, especially in the older age groups, will have progressive disorders that require frequent consultations. Others, especially before the age of 40, may safely be seen at intervals of two to three years. Specialty consultations may be indicated at the time of the periodic inventories. Such consultations are left to the discretion of the examining physician and are given at company expense.

---

[1]NOTE: (1) That ample space must be allowed for a detailed environmental history. This is important. When recorded in a logical and precise fashion, it will preclude the necessity of repetition upon subsequent audits.

(2) The findings upon physical examination are recorded on the reverse side.

(3) There is space for:
Conclusions
Classification and recommendations
And the recommended date for a subsequent follow-up (tickle date)

REPORTS 637

In order that these periodic health audits may be performed competently, the company provides the necessary personnel,

PERIODIC HEALTH INVENTORY				
Family Name	First Name	Middle Name	Birth Date	Mo. Day Year
Company	Location		Years of Service	
Department	Position			
Family Physician — Name	Address			

ENVIRONMENTAL HISTORY (1)

INTERVAL MEDICAL HISTORY (2)

VACCINATIONS AND INOCULATIONS		Date Vaccinated	REACTION			Date Reaction Read
	Small Pox		Immune	Vaccinoid	Vaccinia	
	Typhoid Paratyphoid	1	2	3	Booster	
	Date and Kind Other Inoculations					

(1) The environmental history must be up to date and include a statement regarding employee's attitude toward work, co-workers, supervisors and the employee's progress. Changes in ambitions, social adjustment, self-estimate, home life and relations with family must also be noted. If no environmental history has been previously recorded, a complete history should now be recorded.

(2) The Interval Medical History must include: Number of dispensary visits and reasons, details of interval medical problems, details of time lost due to certified disability, and a résumé of outside consultations.

(SEE REVERSE SIDE)

Fig. 22 — PERIODIC HEALTH INVENTORY *(Front)*

space, and equipment. These facilities are located strategically within the working area.

Periodic health audits are only a part of an active program for protecting the health of the individual. It is important that the physician establish a friendly relationship with the employee so that he will be willing, and will understand the

PHYSICAL EXAMINATION						
Physical Appearance					Temp.	Height	Weight
Deformities Noted							
SKIN							
EYES	VISION	WITHOUT GLASSES		WITH GLASSES		Color Blindness	Disease
^	^	Right	Left	Right	Left	^	^
^	Near	J#	J#	J#	J#	Yes No	^
^	Far	20/	20/	20/	20/	^	^
EARS	Hearing Whispered Voice at	Right	Left	Canals		Ear Drums	Disease
^	^	Ft.	Ft.	^^	^	^
MOUTH AND UPPER RESPIRATORY	Tongue and Tonsils					Sinuses	
^	TEETH				Six Opposing Pairs Including Dentures	Disease	
^	Missing		Carious		Yes No	^	
LYMPHATIC AND ENDOCRINE	Lymph Nodes						
RESPIRATORY	Thorax and Breasts					Lungs	
CARDIO-VASCULAR	Pulse	Heart Size	Blood Pressure	Murmurs		Arteries
^	^	^	^	Organic	Functional	Veins
ABDOMEN	Organs			Masses			Hernia
GENITOURINARY	Venereal Disease	External and Internal Genitalia				Other
^	Yes No	^^^^^
RECTAL	Sphincter	Hemorrhoids		Masses		Bleeding	Prostate
MUSCULO-SKELETAL	Muscle Tone	Posture		Atrophy		Extremities	
^	Edema	Peripheral Pulses	Spine	Abnormalities			
NERVOUS SYSTEM AND REFLEXES	Abnormalities						Reflexes	Right	Left
^	Romberg	Temperament			Remarks	
LABORATORY AND X-RAY	URINALYSIS					Chest X-ray
^	Spec. Gr.	Albumin	Sugar	Microscopic			
^	Hgb.	RBC	WBC	Polys	Lymphos	Monos	Eosins	Serology
^	Other Procedures						
EXPLANATION OF ALL ABNORMAL FINDINGS						
CONSULTATIONS						
CONCLUSIONS AND RECOMMENDATIONS						
TICKLER FOR	Periodic (Date)	Other (Date)		Reason		
MEDICAL EXAMINER	Date	Address			Signature	

Fig. 23 — PERIODIC HEALTH INVENTORY *(Back)*

importance of reporting early complaints or symptoms.

Special Examinations for Employees

It may be necessary on occasion because of some special occupational hazards to arrange special types of examinations. Insofar as is possible, such examinations will be coordinated with the periodic health inventories. The physician who performs these examinations must be cognizant of the type of hazard involved. The following are examples of workers requiring such special consideration:

Painters	Tetraethyl lead handlers	Litharge handlers
Divers		Sheet lead handlers
Welders	Residual products handlers	Sand blasters
Lead burners		Pilots and co-pilots
Benzol handlers	Paraffin plant workers	
Aniline handlers		
Cable splicers	Alcohol plant workers	
Metallizing or grip blasting workers	Guniting workers	

Maintenance of Adequate Nutrition

When periodic health inventories reveal evidence of malnutrition, advice is given the employee for correction of this disorder. When periodic health inventories reveal evidences of malnutrition in large numbers of employees, a public health problem may exist. If this is found to be the case, management is advised concerning the best way to correct the existing deficiencies.

Reassignment of Duties for Medical Reasons

Because of individual handicaps, susceptibilities, or diseases found at the time of the preplacement or periodic health examinations, it may be necessary to limit the type of work assigned an individual. Recommendations of a general nature are made on the basis of the findings, and reassignments are arranged when possible.

Premature Medical Retirement (Because of Illness)

When it becomes apparent that ill health precludes an employee from being productive in any job available within the

640 OCCUPATIONAL HEALTH AND MANTALENT DEVELOPMENT

company, or when any such employment will be likely to aggravate the existing defect of illness, the Medical Department recommends premature retirement.

Fig. 24 — CONFIDENTIAL REPORT

Periodic Health Inventory Summary

When a total audit of a given department is required or, to avoid repetition, a summary of all pertinent data can be transferred to the form shown in Figure 24 at stated intervals (5, 10, 15, 20 years).

This form should have spaces to record these facts concerning the employee:
(1) His present exact total health status.
(2) To what extent his state of health influences his work efficiency.
(3) The effect of work environment on disease process.
(4) The effect of disease process on private life.
(5) The influence of private life on the disease process.

These facts with respect to a given employee at any particular time during his occupational career will point the way to optimum health.

With respect to employer needs, it will focus the required attention upon the status of health construction that is present among any group of employees at stated intervals.

Estimating Expenses and Initial Capital Requirements

Suggested Work Sheet for estimating expenses and initial capital requirements is shown in Figure 25 on page 642.

Annual Report

This report should cover all pertinent activities related in any way to the overall program of Occupational Health and Mantalent Development.
  (a) Recommendations — recorded numerically in order of their importance and should appear first.
  (b) The factual basis for each recommendation follows.
    (1) If the number of placement examinations is out of keeping with the number of employees still on the payroll — some adjustment or change in either the recruitment or placement policy may well be indicated. Statistical facts will substantiate such a recommendation.

Suggested Work Sheet for Estimating Expenses and Initial Capital Requirements

Annual operating and maintenance will equal approximately fifty per cent of the original capital expenditure

Monthly operating expenses	Monthly estimate, amount	Instruction	Amount
	Dollars		Dollars

Physician's fee or salary
Nurses' salaries
Administrative salaries
Occupancy (including rent, light, heat, water, building service)
Depreciation of equipment
Medical supplies
Posters, pamphlets, films, etc., for health education
All other expenses:
   Travel expenses for nurses' car
   Telephone
   Miscellaneous supplies, stationery, record cards, etc.
   Laundry
   Unemployment insurance
   Old Age and Survivors' Insurance
Other

     Average monthly operating expenses

     Nonrecurring initial capital requirements
Purchase of real estate
Fixtures and equipment
Installation of fixtures and equipment
Remodeling
Other

     Total estimated initial capital requirements

Fig. 25 — Suggested Work Sheet

(2) The "repeaters" for dispensary care may indicate poor supervision in a given department. Correlation of such facts with actual medical findings at the time of periodic health audit will add additional factual data, and pave the way for constructive recommendations.

(3) Absenteeism — broken down in terms of actual health causes, i.e., peptic ulcer, cardiovascular disease, etc., will afford the necessary data to periodically constructively evaluate the quality of a given health construction program and in addition allow management to assess this effort in terms of their particular overall fringe benefit commitments.

(c) The "meat" of the report to be arranged in sections in the same manner as the monthly report.

Suggested Topics To Be Discussed

ADMINISTRATIVE MEDICINE

Organization and Administration: All matters pertaining to the organization and administration of the Health Unit including matters of company and administrative policy, procedures, regulations, laws, etc.

Employee Benefits: All matters pertaining to policy, plans, projects, etc., affecting employee health or benefits.

Reports and Surveys: All reports including statistical reports, special studies, and medical surveys. All matters pertaining to policy or procedures regarding same. Reports of trips made, and consultations requested.

Requisitions and Purchases: All information regarding purchase requisitions for medical supplies and equipment. Requests for information regarding availability of items for purchase.

Personnel: A review of personnel additions, promotions, transfers, terminations, etc., with a listing of the effective dates. A qualification record for each newly employed person should accompany this page, with a photograph attached. Remarks concerning personnel problems, changes in personnel policy, etc. Requisitions for the replacement of medical or paramedical personnel.

Construction and Fixed Equipment: Details of all plans for the alteration or construction of new medical facilities. Details regarding installation of, or change in, fixed medical equipment such as X-ray machines, etc.

PREVENTIVE MEDICINE

All matters pertaining to industrial hygiene and sanitation.

CONSTRUCTIVE MEDICINE

Diseases: Reports of interesting or unusual case histories, epidemics, methods of handling, etc.

EDUCATIVE MEDICINE

All matters pertaining to training, postgraduate studies for personnel, medical department classes, special training projects, medical staff meetings, attendance at conferences, etc. Information concerning or requests for articles for company publication, etc.

CURATIVE MEDICINE

Reports concerning treatment of industrial accidents, occupational and environmental diseases.

MANTALENT DEVELOPMENT

Reports to be prepared in such a way that they can be coordinated and used in conjunction with the overall effort in the realm of Mantalent Development. The two forms, Figures 26 through 29, found on the following pages are recommended to help give uniform information on each employee in the mantalent development program.

Figure 26, Guide for Evaluating Mantalent Potential, should provide sufficient space for: (A) Appraisal of

GUIDE FOR EVALUATING MANTALENT POTENTIAL

Family Name	First Name	Middle Name	Age	Years of Service	Company
Position			Location		
Evaluation by					Date

The purpose of this guide is to facilitate discussion of the qualifications, potential for advancement, and developmental needs of key and executive personnel. A separate list of "Qualities Associated with Actual Performance" can be used as a work sheet to aid in an objective evaluation.

A. APPRAISAL OF PERFORMANCE

On the scale below, indicate your evaluation of the employee's over-all work performance on his most recent assignments:

Outstanding ☐ Excellent ☐ Good ☐ Fair ☐ Poor ☐

B. ANALYSIS OF ACTUAL QUALIFICATIONS

Based on consideration of the extent to which the employee exhibits the "Qualities Associated with Mantalent Development," describe briefly in your own words:

1. The strongest aspects of his performance.

2. What should receive particular attention in his further development.

3. Any weaknesses he may have which have persisted in spite of efforts to correct them, and which in your opinion will probably be apparent in future jobs, but which are overshadowed by outstanding strengths in other areas and therefore would not block his advancement.

4. To the best of your knowledge, does he follow a constructive health maintenance program? (Yes ☐) (No ☐) Comment further if necessary.

C. DEVELOPMENT. What specifics are planned to provide additional development?

Fig. 26 — GUIDE FOR EVALUATING MANTALENT POTENTIAL *(Front)*

Performance, (B) Analysis of Actual Qualifications, and (C) Development. It is important that ample space be provided under each heading so that the employee can be truly evaluated.

Figure 27 illustrates the back of the form in Figure 26. Spaces are provided for: (D) Estimate of Potential, (E)

D. ESTIMATE OF POTENTIAL

Based on your knowledge of this individual's ability, which of the following best describes your estimate of his potential for advancement?

1. ☐ Appears to have qualifications for advancing to significantly higher level positions.
2. ☐ Has the potential to advance one level and possibly higher.
3. ☐ Handles responsibilities on present position adequately or better, but is not likely to advance to a higher level.
4. ☐ Is not handling responsibilities of present position adequately, must improve or be removed.
5. ☐ Do not have sufficient knowledge of this employee's abilities to justify making an estimate of potential.

E. PRINCIPAL REASON FOR THE ABOVE ESTIMATE (Note: This may be some quality in the employee, perhaps different from anything previously mentioned in this appraisal, which strongly influenced your estimate.)

F. EMPLOYEE'S NEXT STEP

1. What position or alternate positions do you believe should be the next step for the employee? ______________________

2. Is he ready now? ______________ If not, when will he be ready? ______________

G. REPLACEMENT. Regardless of whether or not this man is promotable, furnish the following information on his probable replacement, if you have one:

(NAME) (POSITION) (AGE) (YRS. SER.)

H. COMMENT BY APPRAISER'S SUPERIOR (Note: This item is applicable in all situations where the original appraiser's superior is not the Recommending Executive.)

SIGNATURE POSITION

I. COMMENT BY RECOMMENDING EXECUTIVE (Particularly in areas of question or difference of opinion.)

SIGNATURE

Fig. 27 — Guide for Evaluating Mantalent Potential *(Back)*

Principal Reason for the Above Estimate, (F) Employee's Next Step, (G) Replacement, (H) Comment by Appraiser's Superior, and (I) Comment by Recommending Executive.

It is recommended that the form illustrated in Figures 28 and 29 be used in conjunction with the form illustrated

QUALITIES ASSOCIATED WITH MANTALENT PERFORMANCE

Family Name	First Name	Middle Name	Age	Years of Service	Company
Position			Location		
Evaluation By				Date	

This list is intended for use primarily as a guide or work sheet in evaluating key potential. Wherever any of these terms are used in such evaluations they will be presumed to have the meanings given in these definitions. The list is not meant to be restrictive. If terms other than these are used, their meaning should be made quite clear.

Guiding Philosophy: Embracing the basic idea that good managing is establishing and conducting an efficient and profitable operation aimed at survival and growth of the enterprise. Alertness to economic, political, and social forces which affect the Company; understanding the Company's aims, goals, organizational relationships, and the necessity of teamwork in consolidated company operations.

Organizing: Visualizing and effectuating an efficiently structured formal organization with appropriate accountability and clear delineation of responsibilities and authorities; staffing the key positions under his direction; constantly re-evaluating and modifying the existing structure in the light of changing conditions.

Delegating: Willingly and skillfully delegating to subordinates the optimum degree of responsibility and authority for making decisions and taking action; accepting and retaining ultimate accountability; determining and using appropriate controls to assess progress; concerning himself principally with major problems and decisions and limiting to a minimum his attention to detail.

Mantalent Development: Motivating people to develop to their full capacity through objective appraisal of their work performance, counseling, coaching, and providing appropriate opportunities conducive to individual growth, thus constantly developing qualified manpower for key positions within his own sphere of operations and for other executive positions in the Company.

Planning: Planning both current and long-range activities on a continuing basis; anticipating conditions indicated by accurate analysis of trends, forecasts, etc.; setting realistic objectives and devising adequate projects or programs to meet them.

Coordinating: Developing balanced operations within his own organization and between his function and other functions of the Company; recognizing the value of staff departments to successful operation and utilizing their services to maximum advantage.

Communicating: Sensing the need, timing, and content of communications to others; listening and comprehending what people are saying which may have an impact on any phase of corporate activity; clearly communicating his ideas to others both orally and in writing.

Leadership: Gaining and holding the respect and confidence of others, inspiring in them a desire to do their best.

Cooperativeness: Working harmoniously with and for others toward the best interests of all concerned; acceptance and active support of final management decisions.

Fig. 28 — QUALITIES ASSOCIATED WITH MANTALENT PERFORMANCE *(Front)*

in Figures 26 and 27.

Again, the form should provide adequate space for a *total* evaluation of the employee's performance.

All mantalent forms should be filled out in as objective a manner as possible.

QUALITIES ASSOCIATED WITH MANTALENT PERFORMANCE	
Drive: A real desire for and willingness to assume executive responsibilities which results in constant self-motivation and industry to achieve outstandingly successful performance in his present position and the highest attainable position in the consolidated company, through efforts which are at all times in harmony with the best interests of the Company.	
Initiative: Recognizing what needs to be done, independently taking timely and appropriate action, and following through to a successful conclusion.	
Mental Alertness: The ability to learn rapidly, to grasp new and complex ideas with relative ease, to analyze quickly and accurately.	
Originality: The ability to think from an independent viewpoint, to produce and develop ideas, either new or derived from past or familiar practice, which contribute substantially to achieving Company objectives.	
Judgment: Ability to render, with reasonable promptness, sound decisions based on recognition and accurate evaluation of all aspects of a matter, including the opinions of others.	
Intellectual Honesty: The courage to express his ideas freely, the willingness to examine and evaluate his decisions and actions, and when in error, to admit his mistakes.	
Integrity: Having and observing the highest standards of honesty, sincerity, loyalty to the company, and personal conduct.	
Maturity: Behaving and expressing himself in an adult manner; exhibiting skillful control of his emotions in his business activities and relationships.	
Stamina: Good general health, including the medical capacity to provide sufficient energy to meet the requirements of his job.	

Fig. 29 — QUALITIES ASSOCIATED WITH MANTALENT PERFORMANCE *(Back)*

Preparation and Distribution

In large companies where there are several medical units, the person in charge of each medical unit should prepare annual reports, and send the original together with all other suggested medical statistical reports to local management and copies to the medical consultant who coordinates all the activities of the medical department. This doctor will, in turn, use this information to compile the annual report of medical activities of the department, directing the original to the members of Management to whom he reports.

REFERENCES

MacDiarmid, Richard. "Medical Department Records." The *Medical Bulletin*, XVI (No. 1, March, 1956), p. 31.

Page, Robert Collier, m.d. "Vital Statistics." The *Medical Bulletin*, V (No. 4, September, 1942).

The Standard Nomenclature of Diseases and Operations. New York: 4th ed., edited by Dr. R. J. Plunkett and A. C. Hayden, Blakiston Company, 1952.

The Standard Oil Company (New Jersey) and Affiliated Companies. "Ten Year Biometic Review: 1937-1946." The *Medical Bulletin*, VII (No. 1, April, 1947).

AUTHORS' INDEX

Italicized numbers appearing after the author's name indicate a direct quotation or reference in text.

A

Abrams, Frank W., *123*
Adams, James H., 77
Albert, Roy E., 77, 266
Albornoz, F. F., 350
Allen, Raymond B., 6
Allman, David B., 77
Anderson, H. W., *123*
Andrews, John B., 77
Arcaro, Eddie, 506
Arthurs, R. G. S., 346
Ashe, William F., 34

B

Babb, E. A., 519
Bacon, S. D., 266
Baehr, George, 206
Baker, James P., 349
Ballou, Charles, 349
Bateson, William, 34
Bauer, Louis H., 104
Baysen, J. E., 346
Beatty, C. Francis, 104
Beck, Dorothy F., 347
Bennett, Richard J., 79
Benson, Otis, Jr., 350
Berlin, Louis, 265
Berry, Clyde M., 77, 265
Birminham, Donald J., 350
Blough, Roger M., 133
Bonkalo, K., 346
Bower, M., 611

Bradley, William R., 77, 265
Bridges, Clark D., 265
Brindle, James, 78, 206
Britton, James L., XVI
Brown, Mary L., 34
Brunn, John M., 206
Buchan, Donald F., 78
Bulger, T. J., 34
Bullis, Harry A., 156
Burk, Samuel L. H., 1
Burney, Leroy, 347
Burwell, C. Sidney, 265
Byron, Lord, *550*

C

Caesar, Julius, 319
Calderone, Frank A., 265
Caldwell, D. W., 548
Campbell, Paul A., 347
Canfield, G. W., 611
Chapman, Loring F., 265
Clark, Robert J., 78, 347
Cluver, E. H., *4*, 6, *26*, 34
Combs, Arthur W., 547
Conant, James B., *30*, 34
Conant, Roger G., 347
Cook, Warren A., 78, 265
Crowder, Thomas Reid, *40*, 78
Cruickshank, W. H., 265, 547
Crumpacker, Edgar L., 349
Cuber, John F., 265

D

D'Alonzo, C. A., *263*, 265
Dastur, H. P., 34
Davies, J. A. Lloyd, *264*
Davis, Stanley W., 436
Day, Emerson, 347
Deichmann, William B., 266
Descartes, Rene, *218*, *219*, 266
Diamond, E. Gray, 6
Dickinson, Frank G., 347
Dietz, David, 130
Dietz, Nicholas Jr., 492
Dill, D. B., 547
Draper, Warren F., 206, 207
Drinker, Philip, 34
Dubos, Rene J., 266
Dustan, Harriet P., 591
Duvalier, Francois, 296

E

Eckardt, Robert, 104, 266
Edwards, Joseph Dean, 547, 611
Eells, Richard S. F., 172
Efferson, Carlos A., 459
Eller, Joseph Jordan, 266
Eller, William Douglas, 266
Elsom, Kendal A., 347
Emerson, Ralph Waldo, 500
Ewalt, Jack R., 436
Ewing, David W., 78, 347

F

Farnsworth, Dana L., 266
Farnsworth, Paul R., 401
Feinberg, Mortimer, 459, *461*
Felton, Jean S., 34
Ferlaino, Frank R., 347
Fidler, Anna, 207
Fisher, M. Bruce, 590
Fleming, A. J., 34, *263*, 265
Flynn, Robert H., 437
Folger, John H., 78
Follmann, Joseph F. Jr., 78
Forcher, K. R., 347
Ford, Benson, 1
Ford, Henry, *8*
Ford, Henry II, *505*
Foulger, John H., 266
Fox, Jack V., 78
Franco, S. Charles, 547, 589
Francone, Mario P., 347
Frank, T. M., *25*
Freeman, G. L., 401
Fryer, Douglas, 459, *461*
Funston, Keith G., 167

G

Gafafer, William M., 547
Gage, Alfred, 347
Gerther, M. M., 589
Gofman, John W., 590
Goldberg, J. A., *263*
Golden, Clinton S., 177
Goldwater, Leonard J., 590
Gompers, Samuel, *179*, *180*, *205*
Goodman, Kennard E., 207, 266
Goodman, J. I., 492
Gordon, Gerald, 348
Gorgas, William G., *292*
Gottsch, L. G., 589
Grace, William J., 268, 350
Gray, Robert D., 78
Greenhill, Stanley, 378
Greenly, R. J., 548
Gregory, S., 348
Guidotti, F. P., 78, 348
Gundersen, Gunnar, 172
Guthrie, Thomas C., 265

H

Halsey, George D., 459
Hamilton, Alice, *40*, 78
Hansen, Howard, 590
Hardy, Harriet Louise, 78, 266
Harriman, Averill, *505*
Hatch, Theodore F., 78
Hawkins, Edward R., 173
Hayhurst, Emery Roe, *40*, 78
Helen, S., 437
Henderson, R. M., 266
Hendricks, N. V., *250*, *255*, 348
Hess, Elmer, 206
Hewitt, J. G., 79
Hillman, Sidney, *181*
Himler, M. E., *544*
Hinkle, Lawrence E. Jr., 266
Hoag, A. E., 437
Hobbs, G. Warfield, 57, 563
Holman, D. V., 348, 548
Holmes, Oliver Wendell, *579*
Holt, J. P., 267
Horan, Joseph C., 79
Hough, Cass, *124*
Hunter, Donald, 6, *263*
Huntington, T. W., 548
Hurtado, Alberto, 348
Hyman, Albert Salisbury, 590

I

Irvin, E. A., *98*
Irwin, P. C., 79
Israel, Murray, 590
Ives, Irving M., *229*

J

Jackson, H., 131
Jacobson, Edmund, 548
Jenkins, I. Dent, *123*
Johnstone, Rutherford T., *4*, 6, *263*
Jordan, Edwin P., 266

K

Kafka, M. M., 348
Kammer, A. G., 34, *97*, 590
Kaplan, Harold I., 437
Kehoe, Robert A., 79, *248*, *249*, *251*, 266
Kelly, A. D., 132
Kemp, Hardy A., 266
Kestenbaum, Meyer, 6, *122*, 130
Kienzle, T. C., 590
Killian, James R., 267
Kirkpatrick, A. L., 130
Knudsen, William S., 97
Kober, George Martin, *40*, 79
Krynicki, F. X., 348
Kubie, L. S., 548
Kuhn, Hedwig S., 348

L

Lanza, A. J., *263*
Lapiere, Richard T., 401
Larabee, Byron H., 173
Laveran, Alfonse, *292*
Lawrence, Charles H., 548
Legee, Robert T., 79
Leven, Aaron S., 267
Levinson, Harry, 267
Lewis, John L., *181*
Luango, E. P., 348

M

McAnally, William F., 79
McCahan, J. F., 79
McCahill, William P., 348
McCord, Carey P., 79
McCord, W. S., 459
McCormick, John F., 267
MacDonald, George, 349
McFarland, Ross A., 590
McGee, Lemuel C., 34, 267, 548

McGinnis, Patrick B., 548
McGuinness, Aims C., 206
McLean, Allan O., 349
McMurray, Robert N., 349
McNamara, William J., 79
Maddock, Charles S., 79, 207
Makover, Henry B., 207
Mallery, O. T. Jr., 34
Mapel, E. B., 548
Maslow, Abraham Harold, *500*, 519
Mathieson, Olin, 517
Menninger, William C., 267
Metcalf, Wendell O., 349
Meyer, Norman C., 79
Misel, Albert D., 611
Mock, Harry Edger, *38*, 80
Monk, R. M., 34
Moore, William L., 207, 266
Morgan, Joe Elmer, 590
Morhous, Eugene J., 349
Moseley, Alfred L., 590
Moskin, J. Robert, 130
Mott, Frederick D., 177
Muller, John N., 349
Murray, D. N., 6, 401
Murtagh, John M., 159
Myers, Charles A., 130, 437, 492, 519

N

Neugarten, Bernice L., 590
Newquist, M. N., 349
Noro, Leo, 81, 104, 548
Norris, Edgar H., *10*, 35

O

Oates, James F., *530*
Oesterle, John A., 267
Olson, Carl T., 590
Osler, Sir William, *40*, 80

P

Page, Robert Collier, 1, 6, 33, 34, 35, 46, 80, 104, 130, 144, 165, 173, 207, 225, 267, 272, 349, 378, 401, 437, 492, 519, 548, 591, 611
Penalver, Rafel, 349
Pigors, Paul John W., 130, 437, 492, 519
Planty, Earl G., 459
Pollack, Jerome, 207
Portis, Sydney A., 548
Potter, H. Phelps, 347
Poutasse, Eugene F., 591
Pratt, Laurence O., 591
Price, Leo, 207

R

Rankin, L. M., 492
Rappleye, Willard C., *230*, 267
Rathbone, M. J., *121*, 130
Regan, J. M., *264*
Reuther, Walter, *200*, 201
Rhoads, C. P., *254*
Rigney, Thomas G., 104, 347
Roberts, H. S., 350
Robertson, Reuben B. Sr., 121
Robertson, Logan T., 80
Rockefeller, John D., *505*
Rogan, J. M., 267
Rosen, George, 350
Ross, Donald, *292*
Rubinstein, Hyman S., *228*, 268
Rusk, Howard A., 589
Rustein, David D., 268

S

Sackman, Morris, 207
Samuelson, Paul A., *183*, 185, 207, 268

ns' INDEX

Sataloff, Joseph, 350
Saunders, Philip K., *491*
Sawyer, William A., 80, 207
Scheele, Leonard Andrew, 350
Schepers, G. W. H., 80
Schereschewsky, J. W., *40*, 80
Schoenleber, Alvin W., 35
Schottstaedt, William W., 268
Schrider, B. D. L., 350
Schwartz, Louis, 350
Scimeca, Alfred, 548
Selleck, H. B., 104
Selye, Hans, 378
Seymour, William M., 38
Shaw, George Bernard, 504
Shilling, Jerome W., 80
Singer, Henry A., 437
Slavik, Fred, 208
Sloan, Alfred Pritchard Jr., 173
Smillie, Wilson G., *276*, 277, 350
Smyth, Henry F. Jr., 80, 268
Snyder, John C., 268
Soash, D. G., 611
Spoont, Stanley, 347
Sterner, James H., 80, 268, 350
Stessin, Lawrence, 459
Stevenson, Robert L., 35
Stewart, Matthew J., XIX
Stieglitz, Edward J., *4*, *6*, *226*, 268
Still, Joseph W., *27*, 35, 591
Stovall, W. R., 350
Stryker, Perrin, 549
Sullivan, A. M., 519
Sullivan, J. D., 350
Suman, John R., 611

T

Tabershaw, Irving R., 80, 350
Tahka, Aleksis, 81
Taylor, Clifford F., 350

Taylor, E. K., 401
Taylor, Graham C., 349
Tead, Ordway, 519
Terhune, William B., 268
Thayer, W. R., 350
Thetford, William M., 265
Thompson, Doris M., 81
Thorpe, John J., 548, 549, 591
Tommey, William A. Jr., 208
Tulipan, Louis, 350
Turell, Robert, 350
Turner, Homer W., 135

U

Ungerleider, Harry E., 591
Urwick, Lyndall F., 549

V

Vorwald, A. J., *264*
Vosburg, G. L., 611

W

Wade, Leo, 229, 268, 591
Wagner, O. J. M., 401
Wagner, Sara P., 268
Walmer, C. Richard, 35, 591
Wampler, Frederick J., 81
Warner, Nathaniel, 437
Warshaw, Leon J., 350
Watts, Isaac, *215*, 268
Weaver, N. K., 590
Webster, H. T., *211*
Weider, Arthur, 268, *319*
Weinerman, E. Richard, 208
White, C. S., 350
White, Paul Dudley, 589, 591
Whittaker, A. H., 104
Whittaker, Paul J., 268
Whyte, William H. Jr., 401, 549
Wilkins, George G., 81, 350, 437

Williams, C., 35
Wilson, A. M., 138
Wilson, F. W., 591
Wilson, Robert E., *122*
Wilson, R. N., 350
Wolf, Stewart, 350, 492
Wolff, Harold G., *223*, 265, 266, 268, 269, 350, 492
Woodbury, M. A., 589
Woodcock, Leonard, *200*, *201*, 202
Woody, McIver, 81
Worthy, James C., *473*, 492

Y

Yingling, Doris B., 81
Yoder, Dale, 130, 401

Z

Zaleknik, A., 459
Zalkind, Sheldons, 459, *461*
Zapp, J. A., 81, *263*, 266
Zimmerman, Charles J., 268
Zitman, Irvin H., 548

SUBJECT INDEX

A

A Future for Preventive Medicine, 4
Abortion, code number of, 617
Abscess, code number of, 617
  of liver, code number of, 621
Absence-prone persons, 557
Absentee, short term, 116-117
Absenteeism
  cost to industry, 298
  reasons for, 521-522
  short term, 116-117
  study by New York Telephone Company, 299
  true sickness, 520
Accent on mantalent development, 589
Acceptance of others, 501-502
Accidents at work, World Health Technical Report, 214
Achievement, desire for, 463
Acquisition, period of, 353, 362-363
Activity, desire for, 463-464
Actual training areas, 456-457
Acute glomerulonephritis, code number of, 620
Addict, chronic alcohol, 529-530
Addiction, drug, code number of, 619
Adequate nutrition, maintenance of, 639
Adhesions, intestinal, code number of, 620
Adjustment, social, 299
Adult experimentation, period of, 353, 359-362
Advertising to find source of manpower, 399-400
  in newspapers, 399
  on radio, 400
Advisor, professional, 605
After-the-fact medical care, 176

Age and habits, 457-458
Age of automation, 440-441
Age Center of New England, 564
Aged, chronically ill, 156
Ages of man
  physical growth, 28
  psychic growth, 28
Aging in harness, 566-567
Agitans, paralysis, code number of, 622
Agranulocytosis, code number of, 617
Air Hygiene Foundation of America, 92
Airs, Waters and Places, 274
Akron Plan, 53-54
Alcohol, chronic addiction to, 529-530
Allergic rhinitis, code number of, 617
Allergy, code number of, 617
  as cause of deafness, 330
Amalgamated Clothing Workers Union, 181
Amebiasis, as environmental disease, 44
  code number of, 617
Amebic dysentery, code number of, 619
American Academy of Occupational Medicine, 90-91
American Association of Industrial Dentists, 94
American Association of Industrial Nurses, 94-97
American Association of Industrial Physicians and Surgeons, 37, 97
American Association of Railway Surgeons, 89
American Board of Preventive Medicine, 77
American College of Surgeons, 319

[657]

American Conference of Governmental Industrial Hygienists, 94
American Federation of Labor, 179, 180, 181
American Industrial Hygiene Association, 264
American Journal of Nursing, 236
American Management Association, 402
American Medical Association, 38, 85, 217, 229, 288
American Medical Association's Council on Hospitals, 38
American Medical Association's Council on Industrial Health, 38, 85
American Medical Association's Council on Medical Education, 38
American medicine, future of, 169-172
American Public Health Association, 37, 93
American Red Cross, 158
Amputation, code number of, 617
Anemia, as cause of deafness, 330
code number of, 617
other types, 617
pernicious, 617
Angina, 581
Angina pectoris, in case study, 13
code number of, 618
Angina Vincent's, code number of, 624
Annual report, 641-649
Anopheles, mosquitoes, 292, 293, 295
Anthro Silicosis Among Hard Coal Miners, 263-264
Anxiety state, code number of, 623
Appendicitis, code number of, 617
Applicant, coordinating data on, 244
Applied psychology and personality, lectures on, 242-243
Appreciation
of simple pleasures, 503-504
of work, 505-507

Approach to therapy, 432-434
Aptitudes, significance of, 412-414
Art of training, 441-442
Arthritis
code number of, 617
degenerative, 617
infections, 617
most common types, 329
osteoarthritis, 329
rheumatoid, 617
traumatic, 617
Articles in employees' magazines, 245
Asian influenza, 287-289
Assembly line doctors, 177
Assignments, reading, 244
Associated health-conservation units, 69
Asthenia
in case history, 429
neurocirculatory, code number of, 620
Asthma, code number of, 617
Atherosclerosis
code number of, 618
generalized, 618
in case study, 478
peripheral, 618
Attendance
competing interests in, 525
degenerative diseases and, 535, 537-538
five basic motives for, 521
gastrointestinal disturbances and, 536-537
job factor in, 524
nuisance diseases and, 534-535
problem drinking and, 526-533
respiratory diseases and, 536
Atypical behavior, 543-545
Atypical pneumonia, code number of, 618
Auricular fibrillation, code number of, 618
Auricular flutter, code number of, 618
Automation, age of, 440-441
Automotive industry, 52-53
Average population by location, 631

SUBJECT INDEX 659

Aviation medicine, 305
A-V block, code number of, 618
Award for Health Achievement in Industry, 99

B

Bacillary dysentery, code number of, 619
Back pain
 allergy as cause, 331
 as common symptom, 330
 malingering and, 331
Barbiturate poisoning, code number of, 622
Basic principles of maturity
 acceptance of others in, 501-502
 appreciation of simple pleasures in, 503-504
 appreciation of work in, 505-507
 enjoyment of present in, 504-505
 self-acceptance in, 500-501
 sense of humor in, 502-503
BB block, code number of, 618
Behavior
 atypical, 543-545
 interpretation of, 543-545
 psychologist as student of, 238
Benign tumors
 code number of, 617
 fibroids of uterus, 617
 nevi, 617
 others, 617
 polyps, 617
Bennett Mechanical Comprehension Test, 240
Bibliotherapy, 245
Birth
 code number of, 617
 premature, 617
 term, 617
Birth of National Safety Council, 39
Blue Cross, 119
Blue Cross and Blue Shield, 119
 in case study, 17
Blue Shield, 119
Board of consultants in evaluation center, 610
Board of Industrial Advisors, 100
Brain carcinoma, code number of, 618

Bronchiectasis
 causes of, 328
 code number of, 617
Bronchitis
 causes of, 328
 code number of, 617
Broncho-pneumonia, code number of, 622
Brucellosis, code number of, 617
Bulletin of the Industrial Hygiene Foundation, 544
Burns, code number of, 617
Bursitis, code number of, 617
Business and health causes
 committee wall in, 133
 guiding principles of action in, 133-134
 how to consider, 135-136
 the policy makers in, 138-139

C

Calculus
 kidney, code number of, 621
 ureter, code number of, 624
 urethra, code number of, 624
Cancer, code numbers of
 carcinoma of brain, 618
 carcinoma of breast, 618
 carcinoma of G.I. system, 618
 carcinoma of respiratory system, 618
 carcinoma of skin, 618
 carcinoma of urogenital system, 618
 Hodgkin's disease, 618
 leukemia, 618
 lymphosarcoma, 618
 polycythemia, 618
 sarcoma, 618
Cancer research
 classical example in industry, 251-259
 shows significance of early detection, 323-325
  Papanicolaou smear, 323
Captain Brassbound's Conversion, 504
Carbuncle, code number of, 618
Cardiac rhythm abnormalities, code numbers of
 auricular fibrillation, 618

auricular flutter, 618
A-V block, 618
BB block, 618
cerebral hemorrhage, 618
cerebral thrombosis, 618
congenital heart disease, 618
essential vascular hypertension, 618
hypertensive C-V disease, 618
hypertensive vascular disease, 618
myocarditis, 618
other types, 618
pericarditis, 618
Raynaud's disease, 618
rheumatic heart disease, 618
syphilitic heart disease, 618
thrombo-angiitic obliterans, 618
thrombophlebitis, 618
varicose veins, 618
Cardiospasm, code number of, 620
Cardiovascular renal disease, code number of
angina pectoris, 618
atherosclerosis, generalized, 618
atherosclerosis, peripheral, 618
Case histories, situational state, 429-432
Case studies
analyzed, 24-25, 335
in occupational health program, 333-335
case 1, 333
case 2, 333-334
case 3, 334-335
in selection and placement, 412-413, 415-416, 419-420, 429-432
man of management in, 12-15
obesity in, 337
scientist in, 21-24
secretary in, 17-19
summarized, 24-25
union man in, 19-21
with constructive medicine, 337-338
worker in, 15-17
Cataract, code number of, 619
Catastrophic illness, 58
Caterpillar psycho-graph profile, 240-241
Caterpillar Tractor Company Plan, 239-246

Cause of fatigue, 422-425
Cellulitis, code number of, 619
Cerebral hemorrhage, code number of, 618
Cerebral thrombosis, code number of, 618
Certificate of health maintenance, 100
Character, significance of, 412
Check list of emotional growth, 507-508
Chickenpox, code number of, 619
Cholecystitis, code number of, 619
Cholelithiasis, code number of, 619
Cholera, vaccination against, 283
Chronic alcohol addict, 529-530
Chronic excessive drinker, 529
Chronic glomerulonephritis, code number of, 620
Chronically ill and aged, 156
Chronological age, 567
Churches as sources of manpower, 339
Circulatory problems and adjustment, 572-583
Circulatory system, 572-583
Cirrhosis, code number of, 619
Clubs as sources of manpower, 339
Coal-mining industry, 51
Code for professional nurses, 236-237
Colitis, mucous
as functional disorder, 475
code number of, 620
relationship with diverticulitis, 326
ulcerative, code number of, 619
Collective bargaining and health, 49-51
Colon, spastic, code number of, 620
Comments on evaluation center, 610-611
Common cold
code number of, 619
discussion of, 285-287
Communications, 472
Company safety program, 270
Competent diagnosticians, 610

SUBJECT INDEX 661

Competent employees, 374-375
Competing interests and attendance, 525
Complex goals, 469
Compound fracture, code number of, 620
Compulsive drinker, 529
Concept of disease, 331-332
Conditioned reflexes, 355
Conference Board of Physicians in Industrial Practice, 90
Confidential report, form for, 640
Conflict between staff and line officer, 106-110
Congenital heart disease, code number of, 618
Congress of Industrial Organizations, 181
Congress on Industrial Diseases, 37
Conjunctivitis, code number of, 619
Considerations in deciding penalty, 518
Considering health causes
 community facilities, 136, 137
 professional and technical education in, 135
 public education in, 135-136
 research in, 135
Constructive approach to morale, 545-546
Constructive health maintenance, 33
Constructive medicine, 31, 127-128, 226-227, 308-309
Constructive retirement planning, 562, 583-584
Continuous recruitment, 378-380
Contracture, Dupuytren's, code number of, 619
Control of malaria, 292-295
Coordinating data on applicant, 244
Cornell Index, 240
Cornell word form, 240
Corporate giving for health causes, 131-172
Cough, whooping, code number of, 624

Council on Industrial Health of the American Medical Association, 85-86, 273
Counseling, 239, 241-242, 307-308
Cover-up, results of, 532
Cracking the whip, 516-517
Culex mosquitoes, 292, 293, 295
Cult of the positive, 509
Curative medicine, 128-129
Curative symptomatic medicine, 126
Cyanosis, 581, 582
Cyst, code number of, 619
 ovarian, code number of, 622
 pilonidal, code number of, 622
Cystitis, code number of, 619
Cystocele, code number of, 619

D

Dacryocystitis, code number of, 619
Dangerous employee to promote, 374, 375-376
Days of disability, 614
Days of disability per case, 614
Days of disability per person, 614
Deafness
 classifications of, 329-330
 code number of, 619
Dealing with training resistance, 453-454
Decelerating on the job, 602-603
Deceleration and retirement, 354
Deciding penalty, considerations in, 518
Definition of a Supervisor, 493
Degenerative arthritis, code number of, 617
Degenerative chronic illness, 407
Degenerative disease, 535, 537-538
Delirium tremens, code number of, 619
Delivered pregnancy, code number of, 622
Dengue fever, code number of, 619
Depressive reaction, code number of, 623

Dermatitis
  as occupational disease, 313-315
  code number of, 619
  ivy, 314
Dermatoses, 313-314
Desire for achievement, 463
  for activity, 463
  novelty, 464
  variety, 464
Detachment, retinal, code number of, 623
Detection unit for problem drinking, 532-533
Determination of days of absence or disability, 627-629
Developing People in Industry, 461
Diabetes mellitus, code number of, 619
Diagnosis and codes, 617-624
Diagnostic level, medical services, 314
Diagnosticians, competence of, 610
Diarrhoea (*see* gastroenteritis)
Dictionary of Occupational Titles, 381, 382, 383
Diphtheria
  code number of, 619
  in quadruple vaccine, 285
  in triple vaccine, 285
Dipsomania, 528-529
  code number of, 622
Direction, 354
  humanology role in, 600-602
Disability
  and retirement, 553-554
  compensation, 186-188
  days of, 614
  nonoccupational, 434-435
Discipline
  that controls, 517-518
  that prevents, 517
  that punishes, 518
Disciplines of humanology
  economics, 221, 222
  epistemology, 218-219
  ethics, 216-217
  human ecology, 222-223
  logic, 215-216
  sociology, 220
Disease
  concept of, 331-332
  degenerative, 535, 537-538
Diseases of Occupation, 4, 263
Dislocation, code number of, 619
Disorganization, pathological, 25
Dispensary level
  educative medicine at, 309-310
  services at, 65
Distress, cause of, 478-480
Distribution of annual report, 648-649
Divergent views of occupational health, 1
Diverticulitis
  code number of, 619
  explanation of, 326
  relationship with mucous colitis, 326
Division of Foreign Quarantine, 288
Doctor in industry, 230-231
Dollars for health, 132
Donald E. Cummings Memorial Award, 94
Dorsalis, tabes, code number of, 623
D.O.T., 381, 382, 383
Drinker
  chronic alcohol addict, 529
  chronic excessive, 529
  compulsive, 529
Drinking, problem, categories of
  dipsomania, 528
  environmental habituation, 528
  neuropsychiatric, 529
  pathological, 528
Drug addiction, code number of, 619
Duality, 355-357
Duodenal ulcer
  code number of, 619
  in case study, 488
  relation to situational state, 430
Dupuytren's contracture, code number of, 619
Dysentery
  amebic, code number of, 619
  bacillary, code number of, 619
Dysmenorrhea, code number of, 619
Dyspnea, 580

SUBJECT INDEX 663

E

E. I. du Pont de Nemours and Company, 263
Eclampsia, code number of, 619
Economic realist, 605
Economics, 221-222
Economics: An Introductory Analysis, 183
Eczema, code number of, 619
Edema, 580
Education
  as a health cause, 156-157
  mental hygiene, 239, 245-246
  of humanologist, 215-223
Educative medicine
  at dispensary level, 309-310
  versus constructive medicine, 308-309
Emotional end point, 419-420
Emotional maturity, 499-508
Emotional stress, need for release from, 464
Emphysema, code number of, 619
Employee
  competence of, 374-375
  goals of, 469-470
  magazine, 245
  morale survey, 545-546
  register card, 625-626
Employment
  of college graduates, 396-397
  of older workers, 395-396
Empyema, code number of, 619
Encephalitis, code number of, 619
Endicott Johnson Corporation, 75
  medical plan of, 75-76
End point, emotional, 419-420
Enjoyment of the present, 504-505
Environmental
  habituation, 528
  history, 315-317
  intoxication, code number of, 622
Envy as a motive, 470-471
Epidemiology, 277-278
Epididymitis
  gonorrheal, code number of, 619
  non-specific, code number of, 619
Epilepsy, code number of, 619

Episodic medicine, 31
  versus constructive medicine, 31
Epistemology, 218-219
Equipment in evaluation center, 611
Equitable Life Assurance Society, 530
Erysipelas, code number of, 619
Esprit de corps, 474
Essential vascular hypertension, code number of, 618
Essentials of medical policy, 127-130
Esso Laboratories, 254
Established Usages and precedents of retirement, 551-552
Estimating expenses, 641
Ethics, 216-217
Evaluation
  of man, 353
  of the selectee, 404-405
Evaluation center
  comments on, 610-611
  purpose of, 610
  requirements of, 610
  space and equipment in, 611
Evolution of occupational medicine, 36-77
Examination
  pre-employment, 434
  preplacement, 434
Example of fundamental research, 251-259
Executive responsibility, 5
Expediency-based thinking in retirement, 559-560
Expenses, estimating of, 641
Explanation a counseling function, 490-491
External hemorrhoids, code number of, 620
Eye, foreign body in, code number of, 619

F

False security, 472
Family doctor and major illness, 524

Family Service Association, 165-166
Father of occupational medicine, 37
Fatigue, 420-422
Fear as a motive, 467-468
Federal funds, hospital construction, 140
Federation of Unions, 183
Femoral hernia, code number of, 620
Fever
　Dengue, code number of, 619
　paratyphoid, code number of, 622
　rheumatic, code number of, 623
　scarlet, code number of, 623
　typhoid, code number of, 624
　typhus, code number of, 624
　yellow, code number of, 624
Fibrillation, 578
Fibroids of uterus, code number of, 617
Filariasis, code number of, 619
Filing guide, 616-617
First
　Industrial Insurance Act, 37
　industrial nurse, 37
　plant surgeon, 37
　ten years of employment, 362
Fistula, code number of, 619
Five
　basic motives for attendance, 521
　categories of health plans, 190
　motives for nonattendance, 521-522
　potential patients, 225-226
Flexible retirement, 569-570
Food and physical welfare needs, 462-463
Ford Facts, 53
Foreign body in eye, code number of, 619
Form
　confidential report, 640
　employee register card, 625
　guide for evaluating mantalent potential, 645-646
　monthly statistical summary, 632
　periodic health inventory, 637-638
　placement examination, 634-635
　qualities associated with mantalent performance, 647-648
　suggested work sheet, 642
Formulating occupational health policy, 111
Four sides of man's nature, 11
　body, 11
　mind, 11-12
　socid, 11-12
　spirit, 11-12
Fourth dimension, 352-353
Fracture
　compound, code number of, 620
　simple, code number of, 620
Frequency rate, 613-614
Fringe benefits, 186
　goals, 189
Functional disease
　cardiospasm, code number of, 620
　migraine as a, 480-481
　migraine, code number of, 620
　mucous colitis, code number of, 620
　neurocirculatory asthenia, code number of, 620
　pylorospasm, code number of, 620
　spastic colon, code number of, 620
Functional disorders
　early detection of, 476-478
　migraine headache as a, 480-481
　obesity as a, 483-485
Functions
　of line officer, 106
　of operating officer, 106
　of staff officer, 106
　of supervisor, 493-547
Fundamental research, 167-168
　genetics and, 168-169
Fundamentals of motivation, 461
Future of American medicine, 169-172

G

G.I. system carcinoma, code number of, 618
Ganglion, code number of, 620
Gangrene, code number of, 620

SUBJECT INDEX

Gastroenteritis, code number of, 620
Gastrointestinal disturbances, 536-537
General Motors Corporation, 97
Generalized atherosclerosis, code number of, 618
Genetics and fundamental research, 168-169
Gentle reins, use of, 516
German measles, code number of, 620
Glaucoma
  as cause of headache, 482
  code number of, 620
  early detection of, 325-326
Glomerulonephritis
  acute, code number of, 620
  chronic, code number of, 620
Goals of mental hygiene, 246-247
Goiter
  non-toxic diffuse, code number of, 620
  non-toxic nodular, code number of, 620
  toxic diffuse, code number of, 620
  toxic nodular, code number of, 620
Gonorrheal
  epididymitis, code number of, 619
  prostatitis, code number of, 622
  urethritis, code number of, 624
Good discipline, basic principles of, 517-519
Good of the service, retirement for, 552-553
Governmental Employment Services, 397-399
Gratitude as motive, 468-469
Gregarious habits of man, 26-27
Group clinic idea, 61-62
Growing independence, 353, 357-359

H

Habilitative medicine, 129-130
Habits and age, 457-458
Half man, 299-300
Handicapped, 317-319
Handling the unsuitable, 408-409
Happiness as a motive, 468-469
Harness, aging in, 566-567
Harvard School of Public Health, 568
Harvard University's School of Public Health, 92
Headache
  clinics, 243-244
  functional disorder, 475
  migraine, 480-481
  nonfunctional, 481-482
  tension, 483
  with physical cause, 482
Health
  and collective bargaining, 49-51
  benefits, defined, 189-190
  conservation center, 69
  control, 63-64
  education calendar, 311
  facilities for small business, 69
  inventory, periodic, 319-320
  maintenance, constructive, 33
  planning and retirement, 586-589
  problems, off-the-job, 42
  program essentials, 192-194
  requirements of supervisor, 496-499
  safeguards, 446
  setback, 571-572
  team, 2, 3, 4, 5
  total, XIII
Health and Welfare Fund, United Mine Workers' Union, 51
Health insurance, 56
  cooperatives, 119
  private insurance companies, 118
  service organizations, 119
Heart disease, congenital, code number of, 618
Heart myths, 574-576
Heat prostration, code number of, 620
Hematoma, code number of, 620
Hemorrhage
  cerebral, code number of, 618
  subarachnoid, code number of, 623

Hemorrhoids
  external, code number of, 620
  internal, code number of, 620
Hepatitis, code number of, 620
Hernia
  femoral, code number of, 620
  inguinal, code number of, 620
  ventral, code number of, 620
Herniation of intervertebral cartilage, code number of, 620
Herpes zoster, code number of, 620
Hill-Burton Construction Act, 140-141
Hippocratic work, 274
History
  environmental, 315-317
  of the individual, 353-354
Hodgkin's disease, code number of, 618
Honest self-analysis, 489-490
Horizontal care, 152
  insurance plans, 33
  system, 32
  the patient and, 141-142
Hospital
  construction, federal funds, 140
  design, 143-157
    architect's role in, 144-146
    hospital team's role in, 146
    management's role in, 146-147, 150-151
    technological devices used in, 147-150
  for tomorrow, 152-156
  horizontal-care system, 32
  management methods, 150-151
  trustees, 142-143
    function of, 142
Hughes-Esch Act, 40
Human
  being, importance of, 224-225
  ecology, 222-224
  engineering program, 239-246
  maintenance, 589
  needs, nine basic, 462-466
  relations, 112, 517
    stimulation of, 112
Humanologists, 8
Humanology
  in business, 8-9

  team, 2, 3, 209-264
    in detecting potential illnesses, 29
    in preventing human problems, 29
    professional members of, 209-264
    the doctor's role in, 214-231
Humor, sense of, 502-503
Hydrocele, code number of, 620
Hydrocephalus, code number of, 620
Hydronephrosis, code number of, 620
Hygienic standards, 264
Hypertension, code number of, 618
Hypertensive
  C-V disease, code number of, 618
  vascular disease, code number of, 618
Hypertrophy, prostatic, code number of, 622

I

Iatrogenic diseases, 320-321
Ideal
  condition of man, 25
  medical facility, 67-68
Ileitis, code number of, 620
Illness
  catastrophic, 58
  statistics, 137-138
Immigrant workers, 387-388
Immunization
  against cholera, 283
  against influenza, 287-289
  against plague, 284
  against poliomyelitis, 284-285
  against smallpox, 280, 282
  against tetanus, 285
  against typhoid fever, 282
  against typhus fever, 283
  against yellow fever, 284
Impetigo, code number of, 620
Importance of morale, 27
Incapacity and retirement, 554-555
Index, Cornell, 240
Individual
  conference with personnel consultant, 244

SUBJECT INDEX 667

health conservation contracts, 69-73
Induction, 239, 241
Industrial Health Conference, 94
Industrial hygiene, 42, 43, 91-92
  engineer, 247-251
  in action, 251
  in petroleum industry, 255-263
  other industries, 263-264
  recommendations of, 255-263
Industrial Hygiene, 263
Industrial Hygiene Association, 93-94
Industrial Hygiene Association of the American Public Health Association, 93
Industrial Hygiene Engineer, 247-251
  co-workers, 247-251
Industrial Hygiene Foundation, 9, 92
Industrial Hygiene World Health Technical Report, 213
Industrial Medical Association, 94, 97-99
Industrial Medicine, International Commission, 88-89
Industrial Physiology, World Health Technical Report, 213
Industrial Practice, Conference Board of Physicians, 90
Industrial Technology, World Health Technical Report, 214
Ineffectual health planning, 139-140
Infarction
  myocardial, code number of, 618
  of lung, code number of, 620
Infection of wound, code number of, 620
Infectious
  arthritis, code number of, 617
  mononucleosis, code number of, 621
Influence of environment on morale, 29
Influencing factors and basic motives, 466-467
Influenza
  code number of, 620
  discussion of, 287-289
  vaccination against, 288-289
Information Bulletin, 290
Inguinal hernia, code number of, 620
Injuries, code number of, 620
Injury of nerve, code number of, 621
Insipient maladjustment, 610
Institute of Industrial Health of the University of Cincinnati, 93
Institute of Industrial Health of the University of Michigan, 93
Institute of Industrial Medicine of New York University, 93
Institute of Occupational Health, 92
Insufficiency, circulatory, 578
Insurance, health, 56
  cooperatives, 119
  private insurance companies, 118
  service organizations, 119
Intercompany relations, 608
Internal hemorrhoids, code number of, 620
International Certificate of Vaccination, 280, 281
International Commission on Industrial Medicine, 88-89
International Committee on Occupational Health Services of the World Medical Association, 83-85
International Labor Office, 88
International Labor Organization, 87, 88, 213
International Labor Organization, Petroleum Committee, 88
Interpretation of atypical behavior, 543-545
Intervertebral herniation, code number of, 620
Interviewers, training of, 239, 242-244
  and counseling, 241-242
Intestinal
  adhesions, code number of, 620
  obstruction, code number of, 621

parasites, code number of, 621
rupture, code number of, 621
strangulation, 621
tuberculosis, 621
Intoxication
chronic alcohol, 529-530
environmental, code number of, 622
neuropsychiatric, code number of, 622
pathological, code number of, 622
Intussusception, code number of, 621
Iritis, code number of, 621

J

Jaundice, catarrhal (see Hepatitis)
Job
classification and definition, 381-382
decelerating on the, 602-603
discipline on the, 517-518
factor in appreciation of work, 505-507
factor in attendance, 524
fatigue, 420-422
Joint instability, lumbosacral, code number of, 621
Journal of Industrial Medicine and Surgery, 97
Journal of Occupational Medicine, 97
Journal of the American Medical Association, 217, 288
Judicial Committee of the American Medical Association, 217
Junior League Magazine, 563

K

Keloids, code number of, 621
Keratitis, code number of, 621
Kerosene poisoning, code number of, 622
Kidney
calculus, code number of, 621
polycystic, code number of, 621
tuberculosis, code number of, 621
Knights of Labor, 179
Knudsen Award, 97

Korean War, studies in, 418, 419

L

Labor
and medical policy, 174-206
organization, sources of manpower, 397
Labor Relations Institute of Newark, 546
Labyrinthitis, code number of, 621
Laryngitis, code number of, 621
Lead poisoning, code number of, 622
Lectures on personality and applied psychology, 242-243
Legal advice in health programs, 195
Leprosy, code number of, 621
Leukemia, code number of, 618
Levels of occupational health program, 272
Lichen planus, code number of, 621
Ligament, tear of, code number of, 621
Limitations
of job classifications, 382
significance of, 412-414
Line officer, functions of, 106
Liver abscess, code number of, 621
Lobar pneumonia, code number of, 622
Lodges as sources of manpower, 399
Logic, 215-216
Long service employee, 451-452
Lost-time cases, 613
Lucas County Academy of Medicine, 75
Lumbosacral joint instability, code number of, 621
Lung
infarction of, code number of, 620
tuberculosis of, code number of, 623
Lymphosarcoma, code number of, 618
Lymphogranuloma venereal, code number of, 621

SUBJECT INDEX 669

M

Maintenance
  constructive health, 33
  of adequate nutrition, 639
Major
  illness and family doctor, 524
  medical catastrophe, 176-177
Maladjustment, insipient, 610
Malaria
  as environmental disease, 44
  code number of, 621
  control of, 292-295
  effect of D.D.T. program on, 294
Malnutrition, code number of, 621
Man
  as tetrahedron, 10-11
  as total being, 10-12
  gregarious habits of, 26-27
  ideal conditions of, 25-26
  morale and, 27, 29
  performance of, 27-29
  potential of, 27-29
  seven stages of, 5
  social side of, 10
Management
  health commitments of, 120-121
  policy, 113-115
    history of, 113
  representative, 604-605
  role in hospital design, 146-147, 150-151
  sets occupational health policy, 105-130
Manager, case study of, 12-15
Manic-depressive reaction
  depressive state, code number of, 623
  manic state, code number of, 623
Manpower
  and World War I, 41
  sources of, 399
  versus mantalent, 443
Mantalent
  potential, form for, 645, 646, 647, 648
  versus manpower, 443
Marine Medical Service, 68
Mastitis, code number of, 621
Mastoiditis, code number of, 621

Meaning
  of administration, 194-195
  of supervisor, 493-495
Measles
  as cause of bronchiectasis, 328
  as cause of deafness, 330
  code number of, 621
  German, code number of, 620
    as cause of deafness, 330
Mechanical
  efficiency, 473
  types of heart disorders, 577-578
Medical activities
  annual report of, 641-649
  monthly report of, 629
Medical affairs
  private enterprise in, 54
  stockholder interest in, 55
Medical consultant, role of, 60-61
Medical dependents in Latin America
  in Aruba, 46
  in Columbia, 46
  in Peru, 46
  in Venezuela, 46
Medical education, 166-167
Medical program of the Endicott Johnson Corporation, 75-76
Medical reasons for reassignment of duties, 639
Medical retirement, premature, 639-640
Medical services at diagnostic level, 314
Medically handicapped, 317-319
Medicolegal Problems, Social Security World Health Technical Report, 214
Mellitus, diabetes, code number of, 619
Mellon Institute, 9
Members of humanology team, 3-4, 8, 209-264
  captain of, 4-5
  doctor as, 214-231
  industrial hygiene engineer, 212, 245-251
  nurse in, 236-237
  psychologist, 212

Memorial Hospital for the Treatment of Cancer and Allied Diseases, 254
Meniere's disease, code number of, 621
Meningitis
  meningococcic, code number of, 621
  tuberculosis, code number of, 621
Meningococcus meningitis, code number of, 621
Mendel's Law, 26
Mental deficiency, code number of, 623
Mental health, 297-298
Mental hygiene
  education, 239, 245-246
  goals of, 246-247
Mercury poisoning, code number of, 622
Migraine
  causes of, 480-481
  code number of, 620
Minimum standards of employability, 443-444
Misassignment, 417-418
Miscellaneous sources of manpower, 399
Modern Occupational Medicine, 263
Money as a motive, 467
Mononucleosis, infectious, code number of, 621
Monthly report of medical activities, 629
Monthly statistical summary, 631-633
Moonlighting, 310, 313
Morale
  factors in, 458
  importance of, 27
  survey of, 545-546
Motivation
  aspects of, 460-461
  envy or pride as a, 470-471
  fear and, 467-468
  fundamentals of, 461
  gratitude and happiness as, 468
  money as a, 467
  nine basic needs influencing, 462-466

desire for achievement, 463
desire for activity, variety, and novelty, 463-464
food and physical welfare needs, 462-463
personal development, 463
release from emotional stress, 464
security of status, 464-465
sense of participation, 466
sense of personal worth, 465
worthy group membership, 465
physiological and social, 461
team spirit and, 473-475
title as a, 467
Mucous colitis
  code number of, 620
  relationship with diverticulitis, 326
Mumps
  cause of deafness, 330
  code number of, 621
Myocardial infarction, code number of, 618
Mystery of problem drinking, 526-529

N

National Conference on Labor Health Services, 177
National Foundation for Cerebral Palsy, 165
National Foundation for Infantile Paralysis, 165
National Foundation for Multiple Sclerosis, 165
National Fund for Medical Education, 37
National Health Council, 165
National Health Service, 178
National Industrial Conference Board, 90
National Rehabilitation Association, 91
National Safety Council, 39, 93
National Society for Crippled Children and Adults, 91
National Society for the Advancement of Management and Industry's Advisory Board for Hospitals, 150

SUBJECT INDEX

National Society for the Prevention of Blindness, 91
Needs, nine basic, 462-466
  desire for achievement, 463
  desire for activity, variety, and novelty, 463-464
  food and physical welfare, 462-463
  personal development, 463
  release from emotional stress, 464
  security of status, 464-465
  sense of participation, 465
  sense of personal worth, 465
  worthy group membership, 465
Nephrosis, code number of, 621
Nerve, injury of, code number of, 621
Neuritis, code number of, 621
Neurocirculatory asthenia, code number of, 620
Neuropsychiatric
  intoxication, code number of, 622
  problem drinking, 529
Nevi, code number of, 617
New
  lost time cases, 631
  non-lost time cases, 632-633
New York Hotel Trades Council and Hotel Association Insurance Fund, 74-75
New York State Banker's Association, 545
New York University-Bellevue Medical Center, 254
Nine human needs, 462-466
Nonabsence-prone persons, 557
Nonattendance, motives for, 521-522
Noneffective rate, 614
Nonfunctional headache, 481-482
Nonoccupational disability, 434-435
Nonoccupational illness, 116, 534
Non-specific
  epididymitis, code number of, 619
  prostatitis, code number of, 622
  urethritis, code number of, 624
Non-toxic
  diffuse goiter, code number of, 620
  nodular goiter, code number of, 620
Normal retirement, 562
Novelty, desire for, 463-464
Nuisance diseases, 534-535
Nurse
  code for, 236-237
  role in counseling, 302-308
Nutrition, maintenance of, 639

O

Obesity
  as a functional disorder, 483-485
  code number of, 621
  personal hygiene and, 309
Objectives
  and policies, 510-514
  of recruitment, 380
Obstruction, intestinal, code number of, 621
Occupational environments, 12
Occupational health
  abroad, 43-51
  as everybody's business, 2
  as part of the whole, 2
  captain of team, 4-5
  defined, 1
  divergent views of, 1
  introduction to, 1-5
  management sets policy for, 105-130
  policy, formulation of, 111-112
    sound basis for, 125-126
  professional consultant in, 2-3
  program, 112, 270-346
  reference materials for, 4
  seven stages of progression in, 5
Occupational Health Institute, 99-102
Occupational Medicine, American Academy of, 90-91
Occupational Medicine and Industrial Hygiene, 4, 263
Occupational medicine comes of age, 76
Occupational Pathology and Toxicology, World Health Technical Report, 213
Occupational Psychologist, 237-239
Occupational Psychology, World Health Technical Report, 214

Off-the-job health problems, 42
Old-age and survivors insurance, 188
Older worker, 57
Olin Mathieson Chemical Corporation, 517
On-the-job training, 457
Oophoritis, code number of, 621
Open health letter, 312
Operating officer, 106
Orchitis, code number of, 621
Organization and administration, 629
Organization and Administration, World Health Technical Report, 214
Osteoarthritis, 329
Osteomyelitis, code number of, 621
Otitis
  external, code number of, 622
  media, code number of, 622
Otosclerosis, code number of, 622
Ovarian cyst, code number of, 622
Overplacement, 414-416

P

Paget's disease, code number of, 622
Pancreatitis, code number of, 622
Paralysis agitans, code number of, 622
Parasites
  intestinal, code number of, 621
  malarial, 292-295
Paratyphoid fever
  code number of, 622
  vaccination against, 282
Parental dominance, 353, 354-357
  conditioned reflexes in, 355
  duality and, 355
Paresis, code number of, 622
Paronychia, code number of, 622
Pathological disorganization
  environmental, 26
  genetic, 25-26
Pathological intoxication, 528
  code number of, 622
Patients, five potential types of, 225-226

Peak attainment, 353, 364-366
Pellagra, code number of, 622
Penalty, considerations in deciding, 518
Pennsylvania Department of Industry and Labor, 40
Peptic ulcer, 486-487
Period of
  acquisition, 353, 362-363
  adult experimentation, 353, 359-360
  growing independence, 353, 357-359
  parental dominance, 353, 354-357
  peak attainment, 353, 364-366
  reminiscence and leisure, 353, 371-374
  substitution and compromise, 353, 367-371
Periodic health audit, 534
  and degenerative disease, 537-538
Periodic health inventory, 319-320, 534
Peripheral atherosclerosis, code number of, 618
Permanent disability, 553-554
Permanent incapacity, 554-555
Permissible margin of error, 442
Pernicious vomiting of pregnancy, code number of, 622
Personal
  development needs, 463
  hygiene, 309
  worth, sense of, 465
Personality
  and applied psychology, lectures on, 242-243
  psychopathic, code number of, 623
  significance of, 412
Personnel
  consultant, individual conference with, 244
  recruitment, 378-380
  reports, 629-630
Petroleum Committee, International Labor Organization, 88
Phimosis, code number of, 622
Phobia reaction, code number of, 623

SUBJECT INDEX 673

Phosphorus necrosis, 40
Phossy jaw, 40
Physical
  cause of headaches, 482
  growth and ages of man, 28
Physiological
  age, 567-569
  and social motives, 461-462
  limit, 419-420
Pilonidal cyst, code number of, 622
Place of performance, 317
Placement
  examination, 314, 435
  interview, 406
Placenta
  privia, code number of, 622
  separation, premature, code number of, 622
Plague, vaccination against, 284
Planning in retirement, 562
Planus, lichen, code number of, 621
Pleurisy, code number of, 622
Pneumoconiosis, Beryllium, Bauxite Fumes Compensation, 264
Pneumonia
  as complication of bronchitis, 328
  atypical, code number of, 622
  broncho-, code number of, 622
  cause of lobar, 287
  influenza baccillus as cause of, 287
  lobar, code number of, 622
Pneumothorax, code number of, 622
Poisoning
  barbiturate, code number of, 622
  kerosene, code number of, 622
  lead, 40
    code number of, 622
  mercury, code number of, 622
Policy
  and objectives, 510-514
  consistency of, 110-111
  defined, 110
  flexibility of, 110-111
  makers of, 138-139
Poliomyelitis
  code number of, 622
  foundation for, 165
  in quadruple vaccine, 285
  vaccination against, 284-285

Polycystic kidney, code number of, 621
Polycythemia, code number of, 618
Polyps, code number of, 617
Poor placement, 422-425
Population, average by location, 631
Positive program of occupational health, 270-346
Potential retiree, 565-566
Power, relinquishment of, 602-603
Practice of Industrial Medicine, 264
Precedents of retirement, 551-552
Pre-employment examination, 434
Pregnancy
  delivered, code number of, 622
  tubal, code number of, 622
  undelivered, code number of, 622
Premature
  birth, code number of, 617
  medical retirement, 639-640
  placenta separation, code number of, 622
Preparation of annual reports, 648-649
Preplacement examination, 434
Present-day service units, 65-67
Preventive
  constructive medicine, 126
  medicine, 127, 276-277
  medicine and public health, 276-277
  or diagnostic services, 58
  psychiatry, 542-545
Pride as a motivating force, 449-450, 470-471
Primary diagnosis, 626-627
Prime motivating force, 449-450
Principles
  of maturity, 500-507
  of occupational nursing, 234-236
Principles and Practice of Medicine, 40
Principles of Medical Ethics, 217
Principles of Occupational Medicine, World Health Technical Report, 213

Prison associations as sources of manpower, 399
Private
  employment services, 397, 398
  enterprise in medical affairs, 54
Problem drinker, 159-165, 529-530
  chronic alcohol addict, 529-530
  chronic excessive, 529
  compulsive, 529
Problem drinking
  and personality disorders, 486
  as a health cause, 159
  basic medicolegal problem, 159
  categories of, 528-529
  definition of, 526
  detection unit and, 532-533
  dypsomania, code number of, 622
  environmental intoxication, code number of, 622
  human relations in, 164-165
  law enforcement role in, 164-165
  locating problem drinker, 159
  mystery of, 526-527
  neuropsychiatric intoxication, code number of, 622
  pathological intoxication, code number of, 622
  stake of community in, 160-161
  statistics in, 527-528
  top management and, 530-533
Proctologic disease, 331
Professional advisor, 605
Professional consultant, 2-3
Program of occupational health, 270-346
Prolapse, rectal, code number of, 623
  uterus, code number of, 624
Promotion and the unpromotable, 471-472
Properties of work environment, 271
Prostatic hypertrophy, code number of, 622
Prostatitis
  gonorrheal, code number of, 622
  non-specific, code number of, 622
Psoriasis, code number of, 622
Psychiatric disease
  anxiety state, code number of, 623
  depressive reaction, code number of, 623
  manic-depressive reaction, depressive state, code number of, 623
  M.D.R., manic state, code number of, 623
  mental deficiency, code number of, 623
  phobia reaction, code number of, 623
  psychopathic personality, code number of, 623
  schizophrenia, code number of, 623
Psychiatry, preventive, 542-545
Psychic growth and ages of man, 28
Psychological tests, 410-412
Psychologist, occupational, 237-239
Psychopathic personality, code number of, 623
Pterygium, code number of, 623
Public health, 275-277
  and preventive medicine, 276-277
  service, 264
Pulmonary tuberculosis, 327-328
Purchases and requisitions, 629
Purpose
  of evaluation center, 610
  of training, 458-459
Pyelitis, code number of, 623
Pyelonephritis, code number of, 623
Pylorospasm, code number of, 620

Q

Quadruple vaccine, 285
Qualities associated with mantalent performance, 647, 648
Qualifications
  and background of registered nurse, 232-233
  of industrial hygiene engineer, 250-251
  of technicians, 610

R

Railway Surgeons, American Association, 89

SUBJECT INDEX 675

Raynaud's disease, code number of, 618
Reading assignments, 244
Realist, economic, 605
Reassignment of duties, 639
Rectal
 prolapse, code number of, 623
 stricture, code number of, 623
Recruitment, 5, 379-400, 594
Rectocele, code number of, 623
Reflexes, conditioned, 355
Register card, 625-629
Register number, 625-626
Registered nurse
 as key person, 272-273
 her code, 236-237
 her place in industry, 233-234
 qualifications and background, 232-233
 standard orders for, 273
Release from emotional stress, 464
Relinquishment
 of power, 602-603
 of responsibility, 5
Reminiscence and leisure, 353, 371-374
Reports
 annual, 641-649
 diagnosis and codes used in, 617-624
 files used in, 614-615
 filing guides used in, 616-617
 glossary used in, 613-614
 guiding principles used in, 612-613
 monthly, 629, 631-641
 preparation and distribution of, 631-649
 register card used in, 625-629
 statistical analysis of, 625-629
Required health program, 176
Requirements of evaluation center, 610
Requisition and purchases, 629
Research, 239, 246
Research Institute of America, 507
Resentment, 452-453
Respiratory disease, 536
Respiratory system carcinoma, code number of, 618

Results of cover-up, 532
Retinal detachment, code number of, 623
Retirement
 as form of severance, 550-551
 established usages and precedents of, 551-552
 expediency based thinking in, 559-560
 for good of the service, 552-553
 for permanent disability, 553-554
 for permanent incapacity, 554-555
 health planning and, 586-589
 meaning of, 563-564
 planning and, 562, 583-584
Retroversion uterus, code number of, 624
Rheumatic
 fever, code number of, 623
 heart disease, code number of, 618
Rheumatoid arthritis, code number of, 617
Rising costs of sick time, 118
Rockefeller Foundation, 293
Role of medical consultant, 60-61
Routine pre-employment test battery, 240
Rubber industry, 53-54
Rupture, intestinal, code number of, 621
Rural-urban migration, 388-391

S

Sacroiliac strain, code number of, 623
Salpingo-oophoritis, code number of, 623
Sarcoma, code number of, 618
Scabies, code number of, 623
Scarlet fever
 as cause of deafness, 330
 as influenza complication, 287
 code number of, 623
Schizophrenia
 code number of, 623
 relation to situational state, 430
Scientist, case study, 21-24
Scopes of service units, 65-67

Scrub typhus, code number of, 623
Scurvy, code number of, 623
Secretary, case study, 17-19
Security of status, need for, 464-465
Selectee's wants, 404
Selection and placement, 5, 239, 240-241, 402-434, 595
Self-acceptance, 500-501
Self-training, 447
Sense of humor, 502-503
Sense of participation, need for, 466
Sense of personal worth, need for, 465
SERPIAN, 296
Service, social, 239, 246
Service units, present day, 65-67
Services
  at dispensary level, 65
  preventive or diagnostic, 58
Seven cycles of life, 352-378
Seven sins against older people, 564-565
Seven stages of progression
  executive responsibility, 5
  motivation and promotion, 5, 460-492
  recruitment, 5, 379-400
  retirement, 5, 550-589
  selection and placement, 5, 239, 240-241, 402-434
  supervisory responsibility, 5
Severance, retirement form of, 550-551
Short-term absentee, 116-117
Sickness absenteeism, true, 520
Significance of stress, 418-419
Simple fracture, code number of, 620
Simple pleasures, appreciation of, 503-504
Sins against older people, 564-565
Sinusitis, code number of, 623
Situational state
  case histories of, 429-432
  code number of, 623
Six basic principles of maturity, 500-507

Skin carcinoma, code number of, 618
Sloan-Kettering Institute for Cancer Research, 254
Small businesses, health facilities, 69
Smallpox
  code number of, 623
  vaccination against, 280, 282
Sobering effect of statistics, 137-138
Social
  adjustment, 299
  motives, 461-462
  service, 239, 246
  side of man, 10
Social Medicine, 4
Socid, 10
Sociology, 220
Socony-Vacuum Oil Co., Inc., 434
Sources of manpower
  advertising and, 399-400
  employees within organization, 383
  employment services
    governmental, 397
    private, 397, 398
  immigrant workers, 387-388
  labor organizations, 397
  miscellaneous, 399
  older workers, 395-396
  persons filing employee applications, 385-386
  persons nominated by employees, 383-385
  rural-urban migration, 388-391
  schools and colleges, 396-397
  women, 391-395
Space and equipment in evaluation center, 611
Space medicine, 305-307
Spanish-American War, 563
Spastic colon, code number of, 620
Special examinations, 638-639
Special Medical Problems, World Health Technical Report, 213
Special test batteries, 241
Specialty Board of Occupational Medicine of the American Medical Association, 229

SUBJECT INDEX

Specific concerns of humanology
  direction, 600-602
  motivation and promotion, 596-599
  recruitment, 5, 379-400, 594
  relinquishment of power, 602-603
  selection and placement, 5, 239, 240-241, 402-434, 595
  supervision, 599-600
  training, 595-596
Spina bifida, code number of, 623
Spondylolisthesis, code number of, 623
Staff officer, 106
Stanacola Corporation, 73
Stanacola Plan, 73
Standard Nomenclature of Disease, 627
Standard Oil Company (New Jersey), 46, 253
Standardization for statistical purposes, 624-625
Standards of medical service, 190-191
Standing Orders for Nurses in Industry, 273
State
  of becoming, 509-510
  situational, code number of, 623
"Statement of Principles to Govern Management's Relationship with Industrial Nurse," 95-96
Static impression of the individual, 515
Statistical analysis, 625
Statistical Methods, World Health Technical Report, 214
Statistics in problem drinking, 527-528
Stenosis, 578
Stockholder interest in medical affairs, 55
Stockholders, 608-609
Stomach ulcer, code number of, 623
Strain, sacroiliac, code number of, 623
Strangulation, intestinal, code number of, 621

Stricture
  rectal, code number of, 623
  urethra, code number of, 624
Subarachnoid hemorrhage, code number of, 623
Substitution and compromise, 353, 367-371
Sunstroke, code number of, 623
Supervised interviewing, 244
Supervision, 493-579
Supervisor, definition of, 493
  nurse and, 302-305
  responsibilities to top management, 514-515
  role of, 301
  training of, 239, 244-245
Symptoms of situational state, 426-429
Syphilis, code number of, 623
Syphilitic heart disease, code number of, 618

T

Tabes dorsalis, code number of, 623
Team approach in industrial hygiene, 253-255
  spirit, 472, 473-475
Tear of ligament, code number of, 621
Technical Report No. 135 of World Health Organization
  accidents at work, 214
  industrial hygiene, 213
  industrial physiology, 213
  industrial technology, 214
  medicolegal problems, social security, 214
  occupational pathology and toxicology, 213
  occupational psychology, 214
  organization and administration, 214
  preventive medicine, 214
  principles of occupational medicine, 213
  special medical problems, 213
  statistical methods, 214
Technique for living, 445-446
Tenosynovitis, code number of, 623

Tension headache, 483
Testes, undescended, code number of, 623
Tetanus
 as cause of deafness, 330
 code number of, 623
 in quadruple vaccine, 285
 in triple vaccine, 285
 vaccination against, 285
The Diseases of Occupation, 4, 263
The Ship's Medicine Chest and First Aid at Sea, 69
Thrombo-angiitic obliterans, code number of, 618
Thrombophlebitis, code number of, 618
Thrombosis, cerebral, code number of, 618
Time, 131
Title, a motive, 467
Today's labor leaders, 181
Toledo and Lucas County Academy of Medicine, 75
Toledo Willys Unit plan, 75
Tonsillitis, code number of, 623
Top management
 and problem drinking, 530-533
 supervisor responsibility to, 514
Torticollis, code number of, 623
Total health, 204, 205
Total man, 10-11
Toxic diffuse goiter, code number of, 620
Toxic nodular goiter, code number of, 620
Training
 age of automation and, 440-441
 areas of, 456-457
 art of, 441-442
 employee need for, 442
 for long service employee, 451-452
 habits and age in, 457-459
 methods, 447-449
 morale factors in, 458
 motivating forces in, 449-450
 needs, 454, 455
 of interviewers, 239, 242-244
 of supervisors, 239, 244-245
 on-the-job, 457
 program, 454
 purpose of, 458-459
 research, 445
 resistance, 453-454
Traumatic arthritis, code number of, 617
Tremens, delirium, code number of, 619
Triple vaccine, 285
True sickness absenteeism, 520
Tubal pregnancy, code number of, 622
Tubercle bacilli, 327
Tuberculosis
 cause of deafness, 330
 intestinal, code number of, 621
 meningitis, code number of, 621
 number of deaths from, 327
 of lung, code number of, 623
 other types, code number of, 624
 pulmonary, 327-328
Twilight zone, 556-557, 558-559
Types of employees,
 competent and promotable, 374-375
 dangerous to promote, 374, 375-376
 misfit, 374, 376-378
Types of patients, 225-226
Types of service units, 65-67
Typhoid fever
 as environmental disease, 44
 code number of, 624
 vaccination against, 282
Typhus fever
 as cause of deafness, 330
 code number of, 624
 vaccination against, 283

U

Ulcer
 duodenal, code number of, 619
 statistics on, 223
 stomach, code number of, 623
Ulcerative colitis, code number of, 619
Undelivered pregnancy, code number of, 622

SUBJECT INDEX

Underplacement, 414-416
Undescended testes, code number of, 623
Unemployment compensation, 188
Unfit employees, 374, 376-378
Unhappiness, 166
UNICEF, 296
Union man, case study, 19-21
Union organization, 182-183
United Auto Workers, C.I.O., 52, 53, 59, 60, 75, 186, 200, 201, 202
United Auto Workers' Plan, 59
United Mine Workers, 181
United Mine Workers' Union, 51
United Nations, 87
United States Department of Interior, Bureau of Mines, 254
United States Public Health Service, 37, 40, 68, 69, 93, 280, 288
United States Sanitary Commission, 292
Ureter, calculus, code number of, 624
Urethra
  calculus, code number of, 624
  stricture, code number of, 624
Urethritis
  gonorrheal, code number of, 624
  non-specific, code number of, 624
Urogenital system carcinoma, code number of, 618
Urticaria, code number of, 624
Use of gentle reins, 516
Uterus, prolapse, code number of, 624
  retroversion, code number of, 624

V

Vaccina, code number of, 624
Vaccination
  cholera, 283
  influenza, 287-289
  plague, 284
  poliomyelitis, 284-285
  smallpox, 280, 282
  tetanus, 285
  typhoid fever, 282
  typhus fever, 283
  yellow fever, 284
Value of maturity tests, 508
Vanguard, supervisor as a, 495-496
Varicocele, code number of, 624
Varicose veins, code number of, 618
Variety, desire for, 463-464
Vascular hypertension, essential, code number of, 618
Veins, varicose, code number of, 618
Venereal, lymphogranuloma, code number of, 621
Ventral hernia, code number of, 620
Vertical individual, 3
Vincent's angina, code number of, 624
Visuotherapy, 245-246

W

What retirement means, 563-564
Whole man, 7-34
Whooping cough, code number of, 624
Women employees, 391-395
Wonderlic personnel test, 240
Work environment, properties of, 271
Work fatigue, 420-422
Worker, case study of, 15-17
Workmans' compensation, 188
World Health Assembly, 87
World Health Organization, 86, 87, 213, 280, 296
World Medical Association International Committee, 83-85
World War I, 445
World War II, 444
Worthy group membership, need for, 465
Wounds (see injuries)
  infection of, code number of, 620

X

X-ray technicians
  as team members, 75
  part of Stanacola plan, 73
  part of Toledo Willys Unit Plan, 75

Y

Yaws
  control of, 296-297
  penicillin used in, 296
  regions of, 296
Yellow fever
  belt of, 283
  code number of, 624
  foci of, 283
  vaccination against, 284

Z

Zone, twilight, 556-557, 558
Zoster, herpes, code number of, 620

RC963
P27

NO LONGER THE PROPERTY
OF THE
UNIVERSITY OF R. I. LIBRARY